ALLE ZEIT WACH 1842

PALAEONTOLOGIA CATHAYANA

Volume 5

Edited by

Nanjing Institute of Geology and Palaeontology, Academia Sinica

Science Press, Beijing, China

Springer-Verlag
Berlin · Heidelberg · New York · London · Paris
Tokyo · Hong Kong
1990

Responsible Editor *Li Zengquan*

Published by Science Press, Beijing

Distribution outside the People's Republic of China exclusively by Springer–Verlag, Berlin · Heidelberg · New York · London · Paris · Tokyo · Hong Kong.

ISBN 3-540-52145-3 Springer-Verlag, Berlin · Heidelberg · New York
ISBN 0-387-52145-3 Springer-Verlag, New York · Berlin · Heidelberg

Printed in Hong Kong

Printing and binding: C & C Joint Printing Co.,(H.K.) Ltd.

Typesetting: Shanghai No.3 Printing Works, Shanghai, China

ISBN 7-03-001630-0 / P · 312

2132 / 3140-543210

PALAEONTOLOGIA CATHAYANA

Volume 5

CONTENTS

PALAEONTOLOGIA CATHAYANA Volume 5, 1990, p.1 – 84

TRILOBITES ACROSS THE CAMBRIAN-ORDOVICIAN BOUNDARY OF THE TRANSITIONAL REGION OF SANDU, SOUTHEASTERN GUIZHOU

Lu Yanhao

(*Nanjing Institute of Geology and Palaeontology, Academia Sinica*)

Zhou Tianrong

(*The 8th Geologic Research and Prospecting Team, Ministry of Geology and Mineral Resources, Guiyang*)

Abstract

The strata and fossils between the Cambrian and Ordovician boundany outcropping in the northwest region of the Sandu district are transitional both in biofacies and lithofacies. Two sections, about 8 km apart, were measured. One at Pu'an, here the late Upper Cambrian Maotian Formation contains dolomites, limestones, argillaceous dolostones, dolomitic and sandy mudstones with about 75% non-agnostid trilobites of the Yangtze Subprovince of the North China Faunal Province, 16.7% of the Jiangnan Subprovince of the Southeast China Faunal Province and 8.3% of endemic forms; while the early Tremadocian Guotang Formation is chiefly composed of striped dolostones and argillaceous dolostones bearing 22.2% non-agnostid trilobites of the Yangtze Subprovince, 72.2% of the Jiangnan Subprovince and 5.6% of the endemic elements. In the other section at Shizilu, the late Upper Cambrian Sandu Formation consists of thin-bedded striped argillaceous dolostones, calcareous mudstones, arenaceous dolostones and limestones with about 20% nonagnostid trilobites of the Yangtze Subprovince, 70% of the Jiangnan Subprovince and 10% of endemic forms; whereas the early Tremadocian Guotang Formation is almost entirely composed of thin-bedded striped argillaceous dolomites with very few mudstones containing 12.5% non-agnostid trilobites of the Yangtze Subprovince and 87.5% of the Jiangnan Subprovince. 57 species and 33 genera and subgenera of trilobites are described, among them, 18 species, 3 genera and 2 subgenera are new.

INTRODUCTION

In the vicinity of the Sandu city at a locality between the Sandu Middle School and the Electric Power Station, the Upper Cambrian and Lower Ordovician strata are well developed, and the name the Sandu Formation of the Upper Cambrain and the name the Guotang Formation of the Lower Ordovician were proposed by Lu in 1954 and by Chien, Lee and Lee in 1959 respectively. The Cambrian and Ordovician boundary is tentatively drawn on the base of the *Hysterolenus* Zone and above the *Hedinaspis* Zone by Lu and Qian (in Lu and Mu, 1980). But the contact line between the Lower Ordovician and Upper Cambrian was later found to be a fault. While in the geological mapping work from 1977 to 1980 in the Pu'an area, about 18 km northwest of the Sandu city, Messrs. Wang Pu, Zhang Shu-Yuan, Lin Cao-an, Peng Yen-wei and the junior writer discovered two sections in which the Cambrian-Ordovician boundary is conformable in contact and they are monocline in stratigraphic succession, These sections are restudied in detail and the fossils are collected bed by bed in the boundary interval by the junior writer and his colleagues in the end of 1982.

As it has long been known in China, the Cambrian and Ordovician faunas are classified into two major provinces: The North China Faunal Province and the Southeast China Faunal Province. Region between them is the transitional area which possesses a mixture of the characteristic bio-facies and litho-facies of both provinces. According to the sedimentation of different rock-types, the North China Province is again subdivided into the North China-Southern NE China Subprovince and the Yangtze Subprovince, whereas the Southeast China Fannal Province can be subdivided into the Jiangnan-NW China Subprovince and the Zhujiang (the Pearl River) Subprovince (Lu et al., 1974; 1976; Lu, 1984). The Pu'an area occupies a small portion of an elongate, bow-shaped transitional belt, about 1600 km long and 50-300 km wide, running from southwestern Guangxi to central Zhejiang passing through southeastern Guizhou, northern Guangxi, south-eastern and central Hunan, northern Jiangxi and southern Anhui. This transitional belt is bounded in north, northwest and northeast by the Yangtze Subprovince, and in southwest, south and southeast by the Jiang-nan Subprovince (Lu, 1984, Map 1).

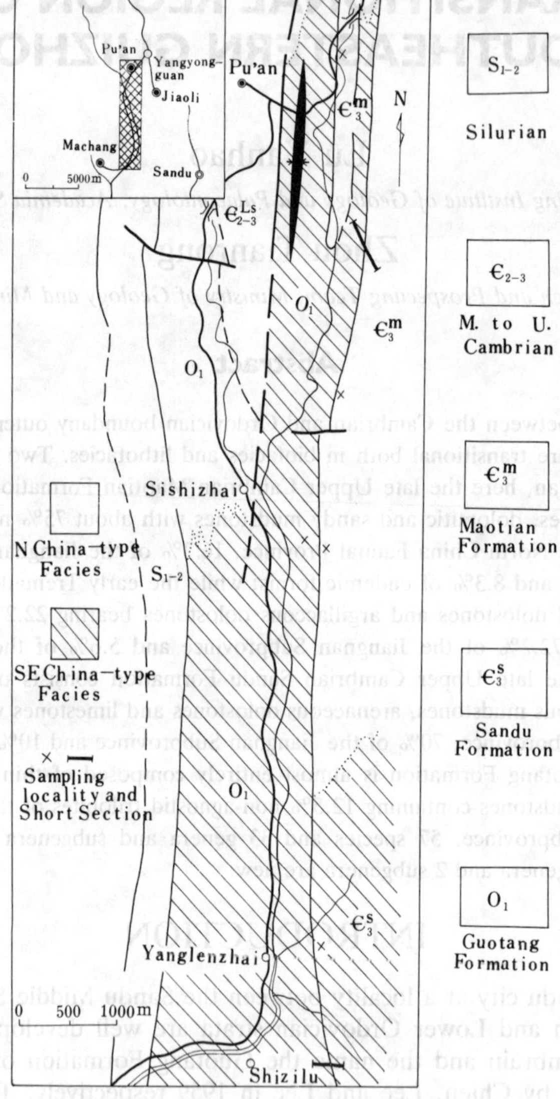

Text-fig. 1. Sketch map showing the location of the Pu'an and Shizilu sections.

Two sections, the Pu'an section and the Shizilu section, were measured at the Pu'an area, where the facies changes in the faunas and lithology of both provinces are rapid along a N-S line directed from east of Pu'an southwards to Shizilu, a small village about 8 km from Pu'an (Text-fig. 1).

57 species and 33 genera of trilobites are described in this paper, including 18 new species, 3 new genera (*Puanocephalus, Shuizuia* and *Hunjiangites*) and 2 new subgenera (*Lorrettina* (*Puanella*), *Mictosaukia* (*Mictosaukioidia*)).

The writers express many thanks to Messrs. Wang Pu, Chen Yonkuang, Yu Kaifa, Shang Kedong, Jian Shaoti and Yang Xiulan for the cooperation in the field.

STRATIGRAPHY

The Pu'an Section (Text-fig. 2)

This section is situated at a locality about 1.6 km southeast of Pu'an (Text-fig. 1). Here the interval between the Cambrian-Ordovician boundary crops out excellently, the succession in descending order is given as follows.

Lower Ordovician Guotang Formation (lower part):

17. Grey and deep gray thin-bedded, striped argillaceous dolomites. 5m.
16. Grey thin-bedded striped argillaceous dolomites, becoming very loose, yellow-earth coloured thin-bedded mudstone after weathering, middle part yielding: *Proteuloma subquadratum* (Lu et Chien), *Troedssonia wimani* (Troedsson), *Macropyge* (*Promacropyge*) *carinata* Lu, *M.* (*P.*) *ciliensis* Peng, *M.* sp., *Golasaphus Carinatus* Lu et T. R. Zhou (sp. nov.), *Pharostomina sanduensis* Zhou, *Diceratopyge sanduensis* Lu et T.R. Zhou (sp. nov.), *Protarchaegonus* sp., *Shumardia puanensis* Lu et T. R. Zhou (sp. nov.), *Neoagnostus bilobus* (Shaw), *Geragnostus* (*Micragnostus*) *elongatus* Chien, *G.* (*M.*) sp., *Acrocephalina sanduensis* Lu et T. R. Zhou (sp. nov), *Diceratopyge agenor* Lu et Chien, *Puanocephalus guizhouensis* Lu et T. R. Zhou (gen. et sp. nov.), *P. rectangula* Lu et T. R. Zhou (gen . et sp. nov.), *Leiostegium* (*Leiostegium*) *punctatum* Zhou et Zhang, *L.*(*L.*) sp., *Taoyuania xenisma* Liu, *T.* sp., *Hospes guizhouensis* Zhou, *H. humilis Lu et T. R. Zhou (sp. nov.), H. carinata Lu et T. R. Zhou (sp. nov.), Palaeoharpes spectabilis* Lu et T.R. Zhou(sp. nov.) (82Fg-24), (78Fg-12), *Pseudokoldinioidia encrinuroides* Lu et T.R. Zhou (sp. nov.), *Callograptus* sp. (82Fg-23). 23.3m.
15. Grey thin-bedded striped argillaceous dolomites, becoming yellow mudstone after weathering, the middle-upper part containing abundant fossils: *Hospes guizhouensis* Zhou, *H. carinata* Lu et T. R. Zhou (sp. nov.), *Shumardia puanensis* Lu et T. R. Zhou (sp. nov.), *Proteuloma subquadratum* (Lu et Chien), *Leiostegium* (*Leiostegium*) sp., *Callograptus dendriticus* Lin, *C.* sp. (82FG-22). 5.Om.
14. Grey thin-bedded argillaceons dolomites, the middle-upper part with nodular limestones, containing the following fossils on the top: *Neoagnostus sanduensis* Lu et T. R. Zhou (sp. nov.), *Shumardia puanensis* Lu et T.R. Zhou (sp. nov.), *Hospes sanduensis* Zhou, *H. guizhouensis* Zhou, *Proteuloma subquadratum* (Lu et Chien), *Diceratopyge sanduensis* Lu et Zhou (sp. nov.), *Pseudokoldinioidia encrinuroides* Lu et Zhou (sp. nov.) *Troedssonia wimani* (Troedsson), *Leiostegium* (*Leiostegium*) sp., *Callograplus cognotus* Zhao (82Fg-21). 8.5m.
13. Grey and pale grey papery to thin-bedded striped argillaceous lenticular limestones. 2.1m
12. Deep grey thin-bedded argillaceous limestones with striped argillaceons limestones, middle part yielding fragments of brachiopods and following trilobites: *Troedsonia elongata* Lu, *Harpides* sp. (82Fg-17). 3.1m
11. Pale grey thin-bedded arenaceous and dolomitic mudstones, middle part with siliceous mudstone of 10 cm thick, containing rich trilobites and graptolites: *Proteuloma subquadratum* (Lu et Chien), *Macropyge sanduensis* Lu et Chien, *Diceratopyge* sp., *Niobella* sp., *Troedssonia wimani* (Troedsson), *Leiostegium* (*Leiostegium*) sp., *Shumardia puanensis* Lu et T.R. Zhou (sp. nov.), *Hospes guizhouensis* Zhou, *Pseudokoldinioidia encrinuroides* Lu et T. R. Zhou (sp. nov.), *Dictyonema flabelliforme sociale* Salter, *D. flabelliforme qiannanense* P. Wang, *D. flabelliforme yichangense* Wang, *D. flabelliforme regulare* Lee et Chen, *D. flabelliforme contractum* P. Wang, *D.* sp. (82Fg-16). 0.5m

———— Conformity ————

Upper Cambrian Maotian Formation.

10. Yellowish grey and grey thin-bedded arenaceous mudstones with thin-bedded dolomitic mudstone, bearing abundant trilobites: *Geragnostus* (*Micragnostus*) sp., *Neoagnostus shihziluensis* Lu et T.R. Zhou (sp. nov.), *Akoldinioidia hewenensis* (Zhou), *Mictosaukia* (*Mictosaukioidia*) *transita* Lu et T. R. Zhou (sp. nov.), *Linguisaukia spinosa* Peng, *Loreettina* sp.,(?) *Lorrettina* sp., *Puanocephalus orientalis* Lu et T. R. Zhou (gen. et sp. nov.), *Eosaukia rectangula* Lu et T. R. Zhou (sp. nov.), *Shuizuia laevigata* Lu et T.R. Zhou (gen. et sp. nov.), *Shuizuia incerta* Lu et T. R. Zhou (gen. et sp. nov.), *Pseudokoldinioidia expansa* Lu et T. R. Zhou (sp. nov.), *P. encrinuroides* Lu et Zhou (sp. nov.), *Hospes transversalis* Lu et T. R. Zhou (sp. nov.) (82Fg-10). 1.0 m
9. Grey and yellowish grey thin-bedded dolomitic mudstones with layers of mudstones, yielding: *Geragnostus sp.*, *Neoagnostus* sp., *Akoldinioidia sanduensis* Lu et T. R. Zhou (sp. nov.), *Mictosaukia* (*Micto-*

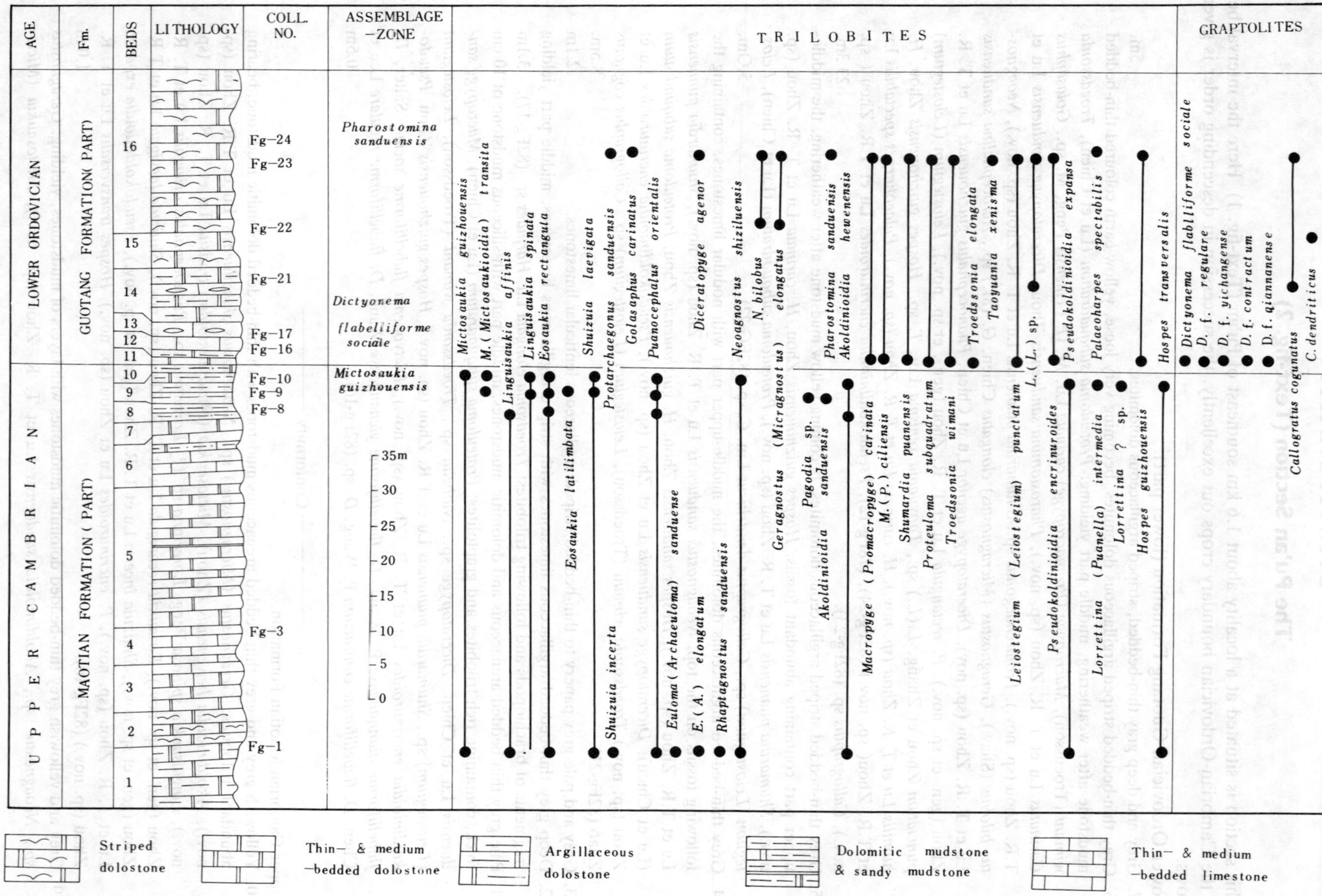

Text-fig. 2. Columnar section showing the distribution of trilobites and graptolites of the Pu'an section.

saukioidia) transita Lu et T. R. Zhou (gen. et sp. nov.), *Linguisaukia affinis* Peng, *Eosaukia rectangula* Lu et T. R. Zhou (sp. nov.), *E. latilimbata* Lu, *Puanocephalus orientalis* Lu ey T. R. Zhou (gen. et sp. nov.), *Pagodia* sp. (82Fg-9). 6.0 m

8. Grey thin-bedded arenaceous dolomitic mudstons, becoming yellow arenaceous mudstone after weathering, bearing rich trilobites: *Geragnostus* (*Micragnostus*) sp., *Shumardia, Akoldinioidia hewenensis* (Zhou), *Linguisaukia affinis* Peng, *Eosaukia rectangula* Lu et T.R. Zhou (sp. nov.), *Puanocephalus orientalis* Lu et T. R. Zhou (gen. et sp. nov.), *Diceratopyge* sp., *Shuizuia* sp., *Pseudokoldinioidia* sp. (82FG-8), (F1-79-88). 1.5m
7. Grey and deep grey medium- to thick-bedded dolomites. 2.1m
6. Deep grey thick-bedded dolomites interbedded with thin-bedded sandy, fine crystalline limestones, partially dolomitization laterally. 8.6m
5. Dark grey thin-to medium-bedded dolomites, lower part interbedded with striped limestones, middle part containing a bed of limestone about 1.5-2m in thickness. 20.3m
4. Grey and pale grey medium-bedded limestone interbedded with thin-bedded fine crystalline limestones, upper part with fragments of fossils, top part bearing *Coelocerodontus burkei* Druce et Jones (82Fg-3). 5.0m
3. Dark grey thick-bedded microcrystalline to fine crystalline dolomites containing pyrite grains. 7.1m
2. Dark grey thin-bedded, lenticular striped argillaceous dolomites. 4.5m
1. Grey and deep grey thin-bedded argillaceous dolomites interbedded with dolomitic mudstones, yielding rich trilobites: *Geragnostus* (*Micragnostus*) sp., *Rhaptagnostus sanduensis* Lu et T .R. Zhou (sp. nov.), *Neoagnostus shihziluensis* Zhou, *Akoldinioidia hewenensis* (Zhou), *A. sanduensis* Lu et T. R. Zhou (sp. nov.), *Mictosaukia guizhouensis* Lu et T. R. Zhou (sp. nov.): *M.* (*Mictosaukioidia*) *transita* Lu et T. R. Zhou (subgen. et sp. nov.), *Linguisaukia spinata* Peng, *L. affinis* Peng, *Eosaukia rectangula* Lu et T.R. Zhou (sp. nov.), *Shuizuia laevigata* Lu et Zhou (gen. et sp. nov.), *Hospes transversalis* Lu et T. R. Zhou (sp. nov.), *Euloma* (*Archaeuloma*) *sanduensis* Lu et T. R. Zhou (sp. nov.), *E.* (*A.*), *elongatum* Lu et T. R. Zhou (sp. nov.), *Pseudokoldinioidia expansa* Lu et T. R. Zhou (sp. nov.) *and Puanocephalus orientalis* Lu et T. R. Zhou (gen. et sp. nov.) (82Fg-1). 9.7m

It is worth to note in this section that the uppermost part of the Cambrian is mainly composed of a series of thin-bedded to medium-bedded dolomites, limestones and arenaceous-argillaceous dolomites; the lithology is obviously different from the typical shale series of the Sandu Formation distributed in the suburb of the Sandu city (Lu, 1954; Chien, Lee and Lee, 1959) and the Shizilu section described in the forthcoming paragraphs, but rather close to the Maotian Formation of NE Guizhou. The Maotian Formation is much alike to the Cambro-Ordovician Loushankuan Dolomites in N. Guizhou of the Yangtze Subprovince. Therefore, the uppermost Cambrian at Pu'an can be used as the name of the Maotian Formation. The lower part of the Ordovician consisting mainly of striped dolomites can be called the Guotang Formation erected by Chien, Lee and Lee in 1959.

The distribution of the Cambrian and Ordovician faunas in the Pu'an section is briefly discussed as follows.

The Upper Cambrian: From the lower part passing upwards, five layers of fossils, Bed no. 1 (82Fg-1). Bed no 4 (82Fg-3), Bed no. 8 (82Fg-8), Bed no. 9 (82Fg-9) and Bed no. 10 (82Fg-10), have been collected from the lower part of the Maotian Formation, most of them are dominated in the uppermost part of this formation. Execpt a few conodonts, e. g. *Coelocerodontus burkei* Druce et Jones, obtained from the top of the Bed no. 4, the chief elements of the other four beds are trilobites, they are *Geragnostus* (*Micragnostus*), *Neoagnostus, Rhaptagnostus, Pseudagnostus, Leiagnostus, Mictosaukia, M.* (*Mictosaukioidia*), *Linguisaukia, Shuizuia, Euloma* (*Archaeuloma*), *Akoldinioidia, Pseudokoldinioidia, Hospes, Eosaukia, Pagodia, Puanocephalus* and *Lorrettina.*

The Lower Ordovician:It is just above the Bed no. 10(82Fg-10) that yields *Mictosaukia guizhouensis,* abundant graptolites were collected in association with the trilobites from the middle part of a dolomitic and siliceous mudstone (Bed no. 11) (82Fg-16) with a thickness of 0.3m. All the graptolites are *Dictyonema,* mainly belonging to the flabellate *Dictyonema,* altogether 5 subspecies are identified. Among the associated trilobites are *Leiostegium, Troedssonia, Proteuloma, Pseudokoldinioidia, Shumardia, Diceratopyge, Hospes, Macropyge* and *Niobella.* On the other hand, the elements of the Saukiidae, Kingstoniidae and Pagodiidae, characteristics of the underlying Maotian Formation, entirely disappeared in this bed.

Of all the graptolites, the most characteristic form is *D. f. sociale* which is regarded as one of the graptolites appearing in the beginning of the Tremadoc, therefore the Cambrian-Ordovician boundary is here placed at the base of the Guotang Formation which bears graptolites and trilobites.

Above the Bed no. 11 (82Fg-16), four layers of fossils, i.e. Bed no. 12 (82Fg-17), Bed no. 14 (82Fg-21), Bed no. 15 (82Fg-22), and Bed no. 16 (82Fg-23 and 82Fg-24) are similar to those genera and species yielded in the base of the Guotang Formation. They extend upwards into the overlying beds. Three species of *Callograptus* occur in the Bed no. 14 (82Fg-21), Bed no. 15 (82Fg-22) and Bed no. 16 (82Fg-23). The trilobites *Pharostomina sanduensis, Golasaphus, Taoyuania, Acrocephalina,* etc. are collected from the middle part of the Bed no. 16.

According to the faunal succession, the boundary interval of the Cambrian and Ordovician Systems in the Pu'an section can be divided into two fossil assemblage-zones: the upper is called the *Dictyonema flabelliforme sociale* Assemblage-Zone and the lower the *Mictosaukia guizhouensis* Assemblage-Zone.

The Cambrian-Ordovician boundary lies between these two assemblage-zones, that is drawn between the Bed no. 10 and Bed no. 11.

The Shizilu Section (Text-fig. 3)

This section is situated about 0.8 km east of Shizilu, a small village about 9.5 km south of Pu'an. The distance between this section and the Pu'an section is about 8 km. The succession of the Cambrian and Ordovician interval is given as follows:

Lower Ordovician Guotang Formation (lower part):

15. Grey and yellowish grey thin-bedded striped argillaceous dolomites, becoming thin-bedded mustone after weathering, with trilobites on the base: *Amzasskiella sanduensis* (Lu et Chien), *Troedssonia wimani* (Troedsson), *Proteuloma* sp. *Harpes* sp. (82Fg-66). 1.5m
14. Yellowish grey thin-bedded striped argillaceous dolomites, interbedded with thin-bedded mudstones, middle part containing *Neoagnostus* sp., *Leiagnostus* sp., *Proteuloma subquadratum* (Lu et Chien), *Hospes guizhouensis* Zhou, *Shumardia puanensis* Lu et T.R. Zhou (sp. nov.), *Diceratopyge sanduensis* Lu et T. R. Zhou (sp. nov.), *Leiostegium* (*Leiostegium*) *constrictum* Peng, *Pharostomina sanduensis* Zhou, *Macropyge* sp., *Pseudokoldinioidia expansa* Lu et T. R. Zhou (sp. nov.) (82Fg-65). 2.0m
13. Grey and yellowish grey thin-bedded striped argillaceous dolomites interbedded with thin-bedded mustones. 9.5m
12. Greyish yellow thin-bedded striped argillaceous dolomites, becoming yellow thin-bedded mustones after weathering, middle part yielding abundant trilobites: *Geragnostus sp.*, *Leiagnostus* sp., *Hospes sanduensis* Zhou, *H.* sp., *Parabolinella* aff. *latilimbata* Lu et Chien, *Troedssonia wimani* (Troedsson) (82Fg-64). 2.0m
11. Greyish yellow thin-bedded striped argillaceous dolomites. 37.2m

——— Conformity ———

Upper Cambrian Sandu Formation (upper part):

10. Grey and deep grey thick-bedded to massive limestones, thickness unstable, thinning out laterally to only 1.5m thick along strike at a distance within 20 m, containing conodonts in the middle to the upper parts: *Cordylodus proavus* Pander (82Fg-60). 11.3m
9. Grey and yellowish grey thin-bedded striped arenaceous and argillaceous limestones interbedded with thin-bedded calcareous mustones, lower part rather muddy, becoming yellow mudstones after weathering, middle and lower parts each yielding one layer of fossils: Middle part bearing *Paraeuloma* cf. *spinosa* Palmer, *Sanduspis sanduensis* Lu et T. R. Zhou (sp. nov.), *Macropyge* sp. (82Fg-63); Lower part containing *Euloma* (*Archaeuloma*) *sanduensis* Lu et T. R. Zhou (sp. nov.), *Macropyge* sp., *Niobella* sp., *Leiagnostus* sp. (82Fg-62). Total thickness 12.7m
8. Yellow and greyish yellow thin-bedded mudstones interbedded with thin-bedded striped argillaceous limestones. Top and base each with a layer of fossils:
   Top part: *Geragnostus* (*Micragnostus*) *elongatus* Chien, *Acrocephalina convexa* Lu et Chien, *Paraeuloma* cf. *spinosa* Palmer, *Sanduspis sanduensis* Lu et T. R. Zhou (sp. nov.), *Hedinaspis shengi* Lu, *Eosaukia rectangula* Lu et T. R. Zhou (sp. nov.), *Macropyge* (*Promacropyge*) *flexa* Peng, *M.* (*P.*) *carinata* Lu, *Linguisaukia affinis* Peng, *Niobella yangjiawanensis* Chien (82Fg-57).
   Basal part: *Geragnostus* (*Micragnostus*) sp., *Niobella* sp., *Promacropyge* sp. (82Fg-56). Total thickness 9.9m
7. Yellowish grey thin-bedded calcareous mudstones, with a layer of 2.5 m thick grey thin-bedded striped argillaceous dolomite. 7.4m
6. Yellowish grey thin-bedded striped arenaceous and argillaceous dolomites, rich in argillaecous material

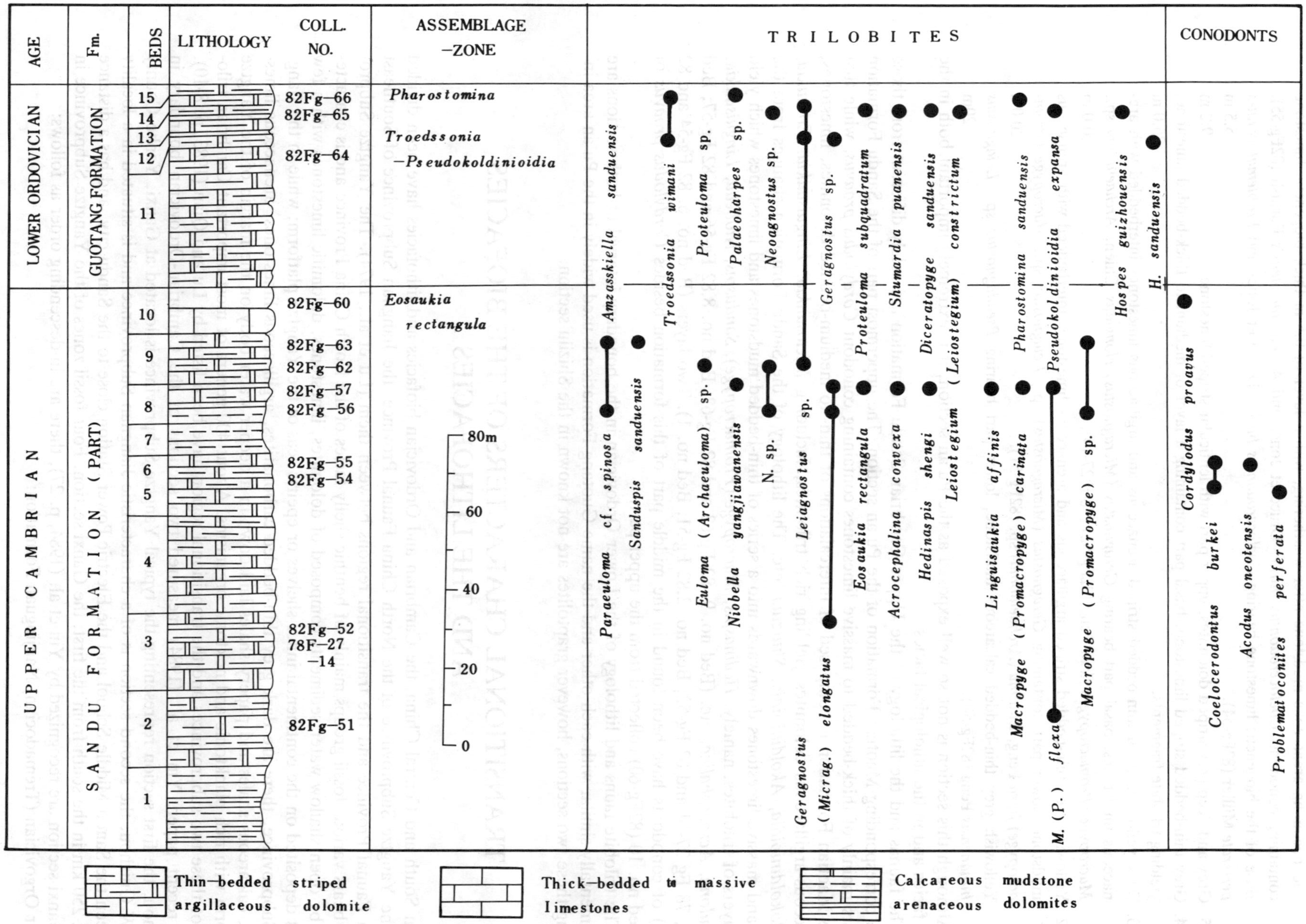

Text-fig. 3 Columnar section showing the distribution of trilobites and conodonts of the Shizilu section.

and becoming arenaceous mudstone after weathering, with a layer of 0.5 m thick biogenetic limestone containing conodonts *Coelocerodontus burkei* Druce et Jones and *Acodus oneotensis* Furnish (82Fg-55). Base of the biogenetic limestone bearing *Coelocerodontus burkei* Druce et Jones and *Problematoconites perferate* Müller (82Fg-54). 5.5 m

5. Grey and deep grey striped dolomites, upper part with lenticular striped limestone. 9.2 m
4. Grey thin-bedded striped limestone, basal part containing a layer of medium to thick-bedded limestone, yielding trilobite fragments. 25.0 m
3. Dark yellowish green thin-bedded striped arenaceous and argillaceous limestones interbedded with arenaceous mudstones, basal part bearing *Geragnostus* (*Micragnostus*) *elongatus* Chien, *Archaeuloma* sp., *Macropyge* (*Promacropyge*) *flexa* Peng (82Fg-52) (78-F 27-14). 20.0 m
2. Yellowish grey thin-bedded striped argillaceous and arenaceous limestones interbedded with arenaceous mudstone, lower part containing *Geragnostus* (*Micragnostus*) sp., *Leiagnostus* sp., *Macropyge* (*Promacropyge*) *flexa* Peng, *Euloma* (*Archaeuloma*) sp. (82Fg-51). 20.0m
1. Yellowish grey thin-bedded calcareous mudstone, lower part bearing *Pseudagnostus* sp., *Leiagnostus hunanensis* Peng (82Fg-50). 20m

Although this section is not so well exposed as the Pu'an section, it is extremely important both in the biotic facies and in the lithological facies.

The faunas and the lithology of the Upper Cambrian Sandu Formation are solely different from those of the corresponding Maotian Formation of the Pu'an section. The uppermost part of the Sandu Formation consists mainly of thick-bedded to massive limestones containing conodont *Cordylodus proavus,* while that of the Maotian Formation is composed of intercalations of thin- to medium-bedded dolomites, limestones, arenaceous-argillaceous dolomites yielding rich trilobites, including *Mictosaukia, Linguisaukia, Eosaukia, Pseudokoldinioidia, Akoldinioidia, Shuizuia,* etc. The lithology of the Sandu Formation changes from the thick and massive limestones downwards into a series of thin-bedded mudstones and limestones which yield five layers of trilobites, namely: *Hedinaspis, Macropyge* (*Promacropyge*), *Sanduspis, Niobella, Linguisaukia, Paraeulona, Acrocephalina,* etc. (Bed no. 9, 82 Fg-63 and 82 Fg-62; Bed no. 8,82 Fg-56 and 82 Fg-57; Bed no. 3, 78 Fg 27-14 and 82 Fg-52; Bed no. 2,82 Fg-51; Bed no. 1). Two layers (Bed no. 6, 82 Fg-54 and 82 Fg-55) of conodonts have been found in the middle part of this formation besides *Cordylodus proavus* of the Bed no. 10 (82Fg-60) collected from the upper part.

The trilobite fauns and lithology of the Lower Ordovician in the boundary interval of both sections are fundamentally similar with each other and the name Guotang Formation is used widely in the Pu'an area including these two sections, however graptolites are not known in the Shizilu section.

TRANSITIONAL CHARACTERS OF THE BIOFACIES AND THE LITHOFACIES

In South and Central China, the Cambrian and Ordovician biofacies and lithofacies have been divided into the Yangtze Subprovince of the North China Faunal Province, the Jiangnan Subprovince of Southeast China Faunal Province and the transitional regions between them (Lu et al., 1974). The Yangtze Subprovince bears various fossil groups mainly of benthic shelly facies of the North China Province and is characterized by open shallow water sediments composed of dolomites, limestones, dolomitic limestones with a few shales deposited on the continental inner shelves or epeiric seas of the Yangtze platform; while in the Jiangnan Subprovince, there are dark grey thin-bedded carbonatites, argillaceous limestones, arenaceous limestones, calcareous shales and papery shales laid down in a deeper water body on the margin of the Yangtze platform with rich planktonic and nektonic organisms. About 400 genera of most important Cambrian trilobites of these two subprovinces and the transitional region have been listed by Lu and Qian (1983, p. 8-10). In a recent paper, Yin et al. (1984) measured two sections of the Cambrian-Ordovician boundary in Guizhou, the first section representing the typical Yangtze Subprovince is located at Ganxi, Yanhe County of NW Guizhou, the second section is of a characteristic Jiangnan Subprovince and is situated in a locality between the Sandu Middle School and the Electric Power Station close to the Sandu city, and has a distance about 250 km in the south from the first, the Ganxi section. Four fossil zones of the Yangtze Subprovince in the Ganxi section are recognized by Yin et al. (1984, p. 27), there are in descending order as follows:
Lower Orgovician (Tremadocian) Nanjinguan Formation:

4. *Asaphellus inflatus-Dactylocephalus dactyloides* Zone: *Asaphellus inflatus* Lu, *A. yanheensis* Yin, *A.* sp., *Dactylocephalus dactyloides* Hsu, *D.* sp., *Shumardia pentagonalis* Lu, *Szechuanella* sp., *Tritoechia* sp.

3. *Wanliangtingia latilimbata-Finkelnburgia meitanensis* Zone: *Wanliangtingia latilimbata* Yin, *W. yanheensis* Yin, *W.* sp., *Finkelnburgia meitanensis* Wang, *F.* sp., *Nanorthis* sp., *Imbracatia* sp., *Ellesmeroceras* sp.

Upper Cambrian Maotian Formation:

2. *Yanheceras endogastrum* Zone: *Yanheceras endogastrum* Chen, *Y.* sp.

1. *Metacalvinella typica* Zone: *Metacalvinella typica* Lee et Yin, *M. elongata* Lee et Yin, *M. robusta* Lee et Yin, *M.? yanheensis* Lee et Yin, *M.? ganqiensis* Lee et Yin, *M.* sp.

In the second section between the Sandu Middle School and the Electric Power Station nearby the Sandu city, Yin et al. (1984, p. 28) separate the late Upper Cambrian and the early Lower Ordovician faunas of the Jiangnan Subprovince into the following three zones:

Lower Ordovician (Tremadocian) Guotang Formation:

3. *Callograptus? taitzehoensis* Zone: *Callograptus? taitzehoensis* Mu, *Herrmannograptus uniforme* Mu, *Amzasskiella elongata* (Lu et Chien), *Proteuloma subconicum* (Lu et Chien), *P. subquadratum* (Lu et Chien), *Troedssonia lubrica* Lu et Chien, *Pilekia transversalis* Zhou, *Leiagnostus* sp., *Szechuanella* sp.

2. *Hysterolenus-Dictyonema flabelliforme sociale* Zone: *Hysterolenus asiaticus* Lu, *Troedssonia lubrica* Lu et Chien, *Szechuanella* sp., *Proteuloma subconicum* (Lu et Chien), *P. subquadratum* (Lu et Chine), *Harpides* sp., *Hospes guizhouensis* Zhou, *Shumardia* sp., *Dictyonema flabelliforme sociale* Salter, *D. flabelliforme regulare* Lee et Chien, *D. jugatum* Lin, *Callograptus dendriticus* Lin, *C. salteri* Hall.

Upper Cambrian Sandu Formation:

1. *Hedinaspis kueichouensis* Zone: *Hedinaspis kueichouensis* Lu, *H.* cf. *shengi* Lu, *H.* sp., *Jegorovaia expansa* Lu, *J.* sp., *Acrocephalina reticulata* Chien, *Diceratopyge* sp., *Pseudoyuepingia modesta* Chien, *Niobella yangjiawanensis* Chien, *Oñchonotina vigilans* Lu, *Guizhoucephalus* sp., *Charchaqia* sp., *Yuepingia* sp., *Proteuloma subconicum* (Lu et Chien), *P. subquadratum* (Lu et Chien), *Hospes guizhouensis* Zhou, *Geragnostus* (*Micragnostus*) sp.

The succession of the Cambrian and Ordovician strata of the Ganxi section seems to be normal, but the boundary between the Upper Cambrian Sandu Formation and the Lower Ordovician Guotang Formation of the second section in the suburb of the Sandu city is not comformable, there exists a fault as it has been pointed out in the foregoing paragraph in this paper. The *Eosaukia-Cordylodus proavus* Assemblage-Zone of the uppermost part of the Sandu Formation in the Shizilu section is missing in the second section, the lowermost zone, the *Hysterolenus-Dictyonema flabelliforme sociale* Zone, of the Lower Ordovician lies directly upon the medial Upper Cambrian *Hedinaspis kueichouensis* Zone.

As it has been pointed out that in the palaeozoogeographical division of the Cambrian-Ordovician period (Lu, 1984, Map. 1), the Pu'an area is situated in the alternating region between the Yangtze Subprovince of the North China Faunal Province and the Jiangnan Subprovince of the Southeast China Faunal Province, or as it is called the transitional region between the two major provinces. This character can be revealed regardless of the distribution of the lithofacies or that of the biofacies (Text-fig. 1). In the vicinity of Pu'an chaizi (meaning the "stockaded village") and from here westward is the dolomitic sedimentary region of the Middle-Upper Cambrian Loushankuan Group, the Lower Ordovician Tungtzu Formation and the Hunghuayuan Formation representing the characteritics of the Yangtze Subprovince. The Tungtzu and the Hunghuayuan Formations and large part of the Dawan Formation extend southward to the vicinity of Shishizhai, and then change in facies to thin-bedded striped limestone of the Guotang Formation and shales of the Tonggao Formation, representing the characteristics of the Jiangnan Subprovince, one of the Southeast China Faunal Province. In the vicinity of Sishizhai, this line of facies change is undulated. Same lithofacies of the lower part of the Guotang Formation extends further northward. The line of facies change of the top of the Cambrian lies roughly at Yanglenzhai, running in a direction to north.

It should be noted that the transitional characters of this region is clearly shown. If it is dominated by the North China lithofacies, the faunas are mixed with only a few elements of the Southeast China faunas and vice versa. The depositional environments in the boundary interval are changeable due to the alternating marine transgression and regression or, in the other words, to the oscillation of the sea level in different time and different places in the inner carbonate belt (the Yangtze-type deposits on the Yangtze platformal area), and the outer carbonate and detrital belts (the Jiangnan-type deposits on the marginal belt of the Yangtze platformal area).

The fossils collected from the Pu'an and the Shizilu sections are separately listed in the Table 1. This table reveals that in these two sections there are typical elements of both the North China type (which are

Table 1. Showing the trilobites of the transitional Cambrian-Ordovician boundary interval of the Pu'an section.

		NORTH CHINA PROVINCE	ENDEMIC FAUNA	SOUTHEAST CHINA PROVINCE
LOWER ORDOVICIAN	GUOTANG FORMATION (PART)	*Troedssonia* *Pseudokoldinioidia* *Leiostegium (Leiostegium)*	*Taoyuania*	*Proteuloma* *Niobella* *Parabolinella* *Diceratopyge* *Troedssonia* *Macropyge (Promacropyge)* *Hospes* *Pharostomina* *Protarchaegonus* *Golasaphus* *Shumardia* *Geragnostus (Micragn.)* *Neoagnostus* *Pseudokoldinioidia* *Acrocephalina* *Palaeoharpes*
UPPER CAMBRIAN	MAOTIAN FORMATION (PART)	*Mictosaukia* *M. (Mictosaukioidia)* *Eosaukia* *Linguisaukia* *Pseudokoldinioidia* *Pagodia* *Puanocephalus* *Shuizuia* *Akoldiniodia* *Neoagnostus* *Rhaptagnostus* *Leiagnostus* *Geragnostus (Micragnostus)*	*Lorrettina (Puanella)*	*Euloma (Archaeuloma)* *Hospes* *Neoagnostus* *Rhaptagnostus* *Leiagnostus*

Table 2. Showing the trilobites of the transitional Cambrian-Ordovician boundary interval of the Shizilu section.

		NORTH CHINA PROVINCE	ENDEMIC FAUNA	SOUTHEAST CHINA PROVINCE
LOWER ORDOVICIAN	GUOTANG FORMATION (PART)	*Troedssonia* *Pseudokoldinioidia* *Leiostegium (Leiostegium)*		*Proteuloma* *Niobella* *Diceratopyge* *Troedssonia* *Macropyge (Promacropyge)* *Hospes* *Pharostomina* *Parabolinella* *Arocephalina* *Pseudokoldinioidia* *Shumardia* *Hysterolenus* *Amzasskiella* *Geragnostus (Micragnostus)* *Neoagnostus* *Leiagnostus*
UPPER CAMBRIAN	SANDU FORMA (PART)	*Linguisaukia* *Eosaukia* *Geragnostus (Micragnostus)* *Neoagnostus*	*Sanduspis*	*Hedinaspis* *Euloma (Archaeuloma)* *Paraeuloma* *Sanduspis* *Acrocephalina* *Niobella* *Macropyge (Promacropyge)* *Geragnostus (Micragnostus)* *Neoagnostus* *Leiagnostus*

included in the left column) and of the Southeast Chine type (which are included in right column) with some endemic elements (which are included in the middle column) between them. In the Pu'an section, except agnostids, the Maotian Formation of the late Upper Cambrian contains about 75% trilobites of the North Chine elements, 16.7% of the Southeast China elements and 8.3% of the endemic elements; while the Guotang Formation of the early Lower Ordovician contains about 22.2% trilobites of the North China elements, 72.2% of the Southeast China elements and 5.6% of the endemic elements. The percentages of the late Upper Cambrian faunas of the Sandu Formation change rapidly in the Shizilu section, where the trilobites of the North China type occupy only 20%, the Southeast China elements increase up to 70%, and the endemic forms are 10%. Whereas the percentages of the early Lower Ordovician trilobites are rather close to those of Upper Cambrian of the same section and also to those of the Lower Ordovician of the Pu'an section, here the North China elements occupy about 12.5%, and the Southeast China elements reach up to 87.5%. (Text-fig.4). The variation of percentages in these two sections is certainly due to the fact that in the late Upper Cambrian time, Pu'an is located closer to the region of the North China Province and here is occupied by the marine water carrying more North China faunas than the Southeast China ones, thus the faunal percentage of the former is higher than that of the latter. In the Lower Ordovician time, the condition changed, the marine water bringing more Southeast China elements and less North China elements transgressed into the Pu'an area, therefore the percentage of the former is higher than that of the latter. At Shizilu, no great change of the faunal constituents takes place from the late Upper Cambrian to the early Lower Ordovician time, this area is surrounded by the Southeast China Fannal Province and is continuously occupied by the marine water containing more Southeast China elements and less North China ones. As a result, the percentage of the Southeast China faunas is high and there keep rather stable through the interval between the Cambrian and the Ordovician boundary. In the other words, the transitional character is variable from place to place and also from time to time. The early Ordovician faunas dominated by the Southeast China type overlap the late Cambrian faunas of both the North China type and the Southeast China type although they are different in their percentages. Besides, there are also European and North American elements, e.g. *Hospes* and *Pharostomina* occurring in the basal part of the Ordovician, and *Puanocephalus*, a genus very close to *Pseudosalteria,* and *Eosaukia,* a genus similar to the North American *Stigmaspis* on the top of the Cambrian. The mixture of several kinds of faunas provides evidences not only for the correlation of the Cambrian-Ordovician boundaries of the two majar types in China, but also for a world-wide correlation.

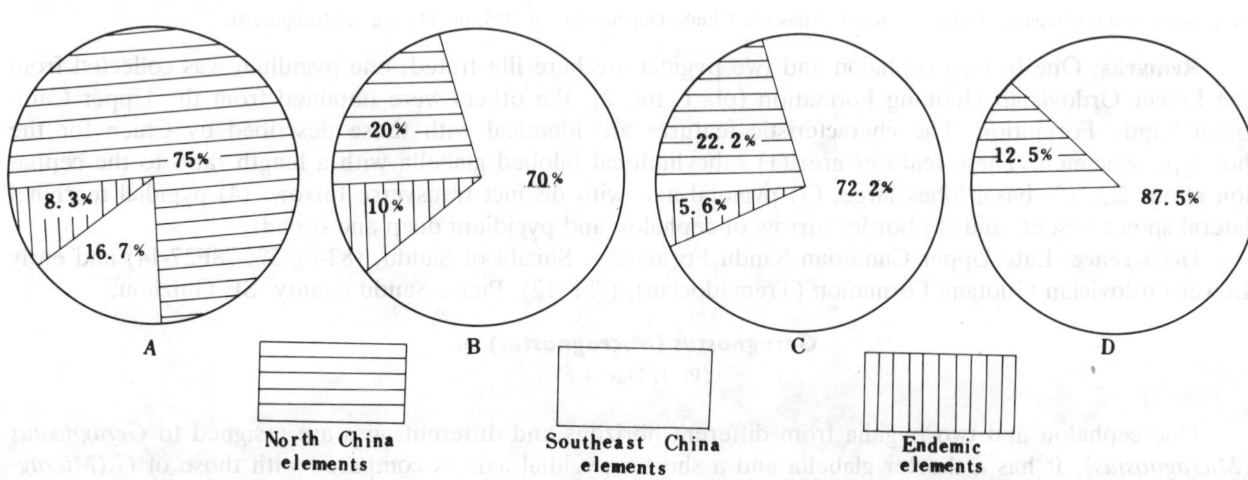

Text-fig. 4. Variation of the percentages of the North China elements, Southeast China elements and the endemic elements of the non-agnostid trilobites in the Pu'an and Shizilu sections. A. Maotian Formation (ϵ_3^m) of the Pu'an section; B. Sandu Formation (ϵ_3^s) of the Shizilu section; C. Guotang Formation (O_1) of the Pu'an section; D. Guotang Formation (O_1) of the Shizilu section.

CORRELATION OF THE CAMBRIAN-ORDOVICIAN BOUNDARIES (THE BASAL BOUNDARY OF THE ORDOVICIAN)

In the Pu'an section, the *Dictyonema flabelliforme sociale* Assemblage-Zone has been taken as the basal boundary of the Ordovician, which is approximately coincident with the bottom of the European Tremadocian. Since *Hysterolenus* has been found in association with *Dictyonema flabellaforme sociale* in the lowermost part of the Ordovician in the suburb of the Sandu city, this basal boundary can be correlated with that of the *Hysterolenus* Zone in the Lower Ordovician Yinchupu Formation of western Zhejiang (Lu et Lin, 1984) and that of the *Onychopyge-Hysterolenus* Assemblage-Zone in the Lower Ordovician Pengjiazui Formation of the Taoyuan-Cili area of northwestern Hunan (Peng, 1984). *Troedssonia* has a long geological range from the base of the Guotang Formation to the *Pharostomina* Assemblage-Zone. This genus is widely distributed in the Early Ordovician of Eastern Tianshan (Troedsson, 1937), North China and Northeast China (Kuo et al., 1982; Zhou et al, 1985; Chen et al., 1986), W. Zhejiang (Lu et Lin, 1984), NW Hunan (Peng, 1984), and Kazakhstan (Apollonov et Chugaeva, 1984). Like that of *Troedssonia,* the genus *Leiostegium* also ranges from the lowermost part of Ordovician upwards to the younger *Pharostomina* Assemblage-Zone, but it is only restricted to the North China Faunal Province with a few entering the transitional region in China, and no representatives are known in the Southeast China Faunal Province. On the other hand, this genus is widely distributed in the Lower Ordovician of North America, showing the close relationship between the North China Faunal Province and the North American Faunal Province.

SYSTEMATIC DESCRIPTIONS

Order Miomera Jaekel, 1909
Suborder Agnostina Salter, 1864
Family Agnostidae M'Coy, 1849
Subfamily Agnostinae M'Coy, 1849
Genus *Geragnostus* Howell, 1935
Subgenus *Micragnostus* Howell, 1935

Geragnostus (Micragnostus) elongatus Chien

(Pl. 1, Figs. 1–3)

1961 *Geragnostus elongatus* Chien, Acta Palaeont. Sinica, vol. 9, no. 2, p, 93, pl. 1, fig. 1.
1964 *Geragnostus elongatus,* Lu, Index Fossils of South China, p. 32, pl. 4, fig. 9 (refigured).
1978 *Geragnostus elongatus,* Chien, Palaeont. Atlas SW China, Guizhou (1), p. 385, pl. 144, fig. 24 (refigured).

Remarks: One broken cephalon and two pygidia are here illustrated, one pygidium was collected from the Lower Ordovician Guotang Formation (pl. 1, fig. 2), the others were obtained from the Upper Cambrian Sandu Formation. The characteristic features are identical with those described by Chien for the holotype specimen. These features are: (1) subcylindrical bilobed glabella with a length ratio to the cephalon about 2:3, (2) basal lobes large, (3) pygidial axis with distinct transverse furrow, (4) pygidial marginal lateral spines absent, and (5) border furrow of cephalon and pygidium deep and broad.

Occurrence: Late Upper Cambrian Sandu Formation, Shizilu of Sandu, (82-Fg-57; 78F27-14) and early Lower Ordovician Guotang Formation (Tremadocian), (78F-12), Puan, Sandu county, SE Guizhou.

Geragnostus (Micragnostus) sp.

(Pl. 1, Figs. 4–6)

One cephalon and two pygidia from different horizons and different ages are assigned to *Geragnostus* (*Micragnostus*). It has a shorter glabella and a shorter pygidial axis as compared with those of *G.*(*Micragnostus*) *elongatus* Chien illustrated on pl. 1, figs. 1–3 in this paper. Because of the bad state of preservation, other details are not clearly shown.

Occurrence*:* *Mictosaukia guizhouensis* Assemblage-Zone of the Upper Cambrian Maotian Formation (82 Fg-1) and the early Lower Ordovician Guotang Formation (82 Fg-24), Pu'an of Sandu County, SE Guizhou.

Table 3. Correlation chart of the Cambrian-Ordovician boundary of Pu'an, Shizilu, Sandu, NE. Sichuan, W. Zhejiang, North and Northeast China.

	Jiangnan Subprovince				Transitional Regions				Yangtze Subprovince		North and Northeast China Subprovince			
	Western Zhejiang		Sandu, SE Guizhou		Pu'an		Shizilu		Ganxi, NE Sichuan		North China		Northeast China	
	Lu and Lin, 1984		Lu, 1983, Yin et al., 1984		This paper		This paper		Yin et al., 1984		Zhou et Zhang, 1985		Kuo et al., 1982	
LOWER ORDOVICIAN	YINCHUPU FORMATION	*Dichelepyge sinensis* Z. *Staurograptus dichotomus* Z. *Hysterolenus*–*Dictyonema flabelliforme* Z.	GUOTANG FORMATION	*Callograptus* ? *taitzehoensis* Z. *Hysterolenus*–*Dictyonema flabelliforme sociale* Z.	GUOTANG FORMATION	*Pharostomina sanduensis* *Dictyonema flabelliforme sociale* A.–Z.	GUOTANG FORMATION	*Troedssonia wimani*–*Pseudokoldinioidia expansa* A.–Z.	NANJINGUAN FORMATION	*Asaphellus inflatus*–*Dactylocephalus dactyloides* Zone *Wanliangtingia latilimbata*–*Finkelburgia meitanensis* Z.	YEHLI FORMATION	*Wanliangtingia* Zone *Yosimuraspis* Zone *Pseudokoldinioidia perpetis* Zone	YEHLI FORMATION	*Aristokainella*–*Onychopyge* A.–Z *Pseudokainella* (*Parakainella*) A.–Z. *Richardsonella*–*Troedssonia* A.–Z.
UPPER CAMBRIAN	SIYANGSHAN FORMATION	*Lotagnostus hedini* Z. *Acaroceras*–*Antacaroceras* Z. *Lotagnostus punctatus* Z.: *Lophosaukia* Subzone, *Hedinaspis* Subzone	SANDU FORMATION	*Hedinaspis guizhouensis* Z.	MAOTIAN FORMATION	*Mictosaukia guizhouensis* A.–Z.	SANDU FORMATION	*Eosaukia rectangula*–*Cordylodus proavus* A.–Z.	MAOTIAN FORMATION	*Yanheceras* Zone *Metacalvinella typica* Zone	FENGSHAN FORMATION	*Mictosaukia orientalis* Z. *Sinoeremoceras* Zone *Quadraticephalus* Zone	FENGSHAN FORMATION	*Mictosaukia*–*Fatocephalus* A.–Z.

Family Diplagnostidae Whitehouse, 1936
Subfamily Pseudagnostinae Whitehouse, 1936
Genus *Neoagnostus* Kobayashi, 1955

Neoagnostus shiziluensis Lu et T.R. Zhou (sp. nov.)

(Pl. 1, Figs. 7–11)

Description: Cephalon subrectangular, slightly longer than wide. Glabella convex, parallel-sided posteriorly, rounded angulate anteriorly. Frontal lobe rhomboid in outline, defined posteriorly by V-shaped anterior furrow. Median lateral furrows running inwards and forwards, meeting anterior furrow axially to isolate a pair of triangular antero-lateral lobes. Glabellar node situated behind median lateral furrows. Median preglabellar furrow present, deep. Cephalic acrolobe convex, unconstricted laterally, gently sloping outward. Marginal furrow deep; bordor narrow and slightly convex. A pair of cephalic spines present at genal angles. Pygidium subquadrate, slightly wider than long. Pygidial acrolobe unconstricted. Axis gently convergent rearward, occupying about 1/2 total pygidial length (sag.), with 2 pairs of transverse furrows. Anterior transverse furrows running transversely from axial furrow to pygidial node; posterior ones running inwards, then around posterior margin of pygidial node and meeting sagittally. Pygidial node pointed anteriorly, expanded rearward from back of first segment, rounded posteriorly. Terminal pygidial node prominent. Marginal furrow deep; border narrow, bearing a pair of postero-lateral spines.

Remarks: The new species is most closely similar to *N. aspidoides* Kobayashi (1955, p. 473, pl. 7, fig. 5, non. fig. 4; Shergold, 1977, p. 91, pl. 16, fig. 16). However, the former can be distinguished by having a longer cephalic proportion, larger anterolateral lobes, a more pointed frontal lobe and a more slender glabella. The new species differs from *N. hangulensis* Lu et Z. Q. Zhou (1981, p. 4, pl. 2, fig. 1) in having a longer, more pointed glabella, a wider (tr.) pygidium and a longer pygidial axis. *N. longicollis* (Kobayashi, 1966, p. 283, fig. 7; Shergold, 1977, pp. 79, 99; Z. Y. Zhou et J. L. Zhang, 1984, p. 68, pl. 5, fig. 7; pl. 27, figs. 4–7) is distinguished from our new species by its rounded-subquadrate instead of rectangular cephalon, by the longer and narrower glabella which is acute anteriorly and by the straight posterior margin of the pygidium. *Neoagnostus kentauensis* Ergaliev (Ergaliev, 1983, p. 43, pl. 1, fig. 24; pl. 3, figs. 1–4; Apollonov, Chugaeva and Dubinina, 1984, p. 10, pl. 15, figs. 1–4) differs from *N. shiziluensis* (sp. nov.) in the subcircular instead of subrectangular cephalon, the wider (sag.) preglabellar area, shorter glabella and pygidial axis, and in having radiating surface sculptures both on cephalon and pygidium.

Occurrence: Upper Cambrian Maotian Formation of Pu'an and Sandu Formation of Shizilu, Sandu, Guizhou (78F2-6, 82Fg-1, 82Fg-5).

Neoagnostus bilobus (Shaw)

(Pl. 1, Figs. 12–14; Pl. 2, Figs. 1–2)

1951 *Pseudagnostus bilolus* Shaw, p. 112–113, pl. 24, figs. 17–22.

1977 *Neoagnostus bilobus*, Shergold, pp. 79, 93, pl. 6. figs. 7–8.

1984 *Neoagnostus bilobus*, Peng, p. 318, pl. 6, fig. 11.

Remarks: Three cephala and two pygidia are figured here. They are badly pressed by vertical compression, which caused some irregular grooves on the surface of the cheek, the posterior portion of glabella and the pygidium (pl. 1, figs. 12, 13, 14; pl. 2, fig. 1). Like the paratype cephalon of *Neoagnostus bilobus* (Shergold, 1977, pl. 16, fig. 7), the cephalon is subquadrate in outline, a little wider than long, the glabella tapers slightly forward, rounded in front and surrounded by deeply incised circumglabellar furrow. The large and elongate axial glabellar node lies behind the antero-lateral glabellar lobes (spectaculate) which are well defined by deep, oblique antero-lateral glabellar furrows. The anterior lobe of glabella is separated from the posterior portion by a deep, backwardly arched furrow. The preglabellar median furrow is shallow but distinct. The border is broad, clearly defined by deep border furrow. The pygidium is characterized by a pair of very pronounced terminal lateral axial lobes and three pairs of lateral axial lobes in front of the terminal lobes. The border and border furrow are especially broad in the postero-lateral corners which produce a pair of small pygidial spines. All the above features are fitted well to those of the holotype pygidium of *bilobus* (Shaw, 1951, p.112. pl. 24, fig. 20; Shergold, 1977, pl. 16, fig. 8), thus the present specimens from Guizhou are regarded to be conspecific with *bilobus*.

Occurrence: Uppermost part of the Upper Cambrian Maotian Formation and Lower Ordovician (Tremadocian) Guotang Formation of Pu'an, Sandu, SE Guizhou (78F26-12, 82Fg-24).

Genus *Rhaptagnostus* Whitehouse, 1936

Rhaptagnostus changshanensis Lu et Lin

(Pl. 2, Figs. 3, 4, 5)

1980 *Rhaptagnostus changshanensis* Lu et Lin, p. 123, pl. 1, figs. 3–4.

1984 *Rhaptagnostus changshanensis* Lu et Lin, p. 55-57, pl. 2, figs. 11–13; pl. 3, figs. 1–3.

Remarks: One dorsal shield, one cephalon and one pygidium are illustrated here. The dorsal shield is ovate in outline. Cephalon is larger than the pygidium, wider than long, moderately convex, short, with a length about 3/4 its width, broadly rounded in front, tapering from the widest portion at the mid-length posteriorly. The glabella is moderately convex, broad and short, less than 2/3 the total cephalic length, slightly expanded at the anterior third of its length, obtusely rounded in front which is slightly pointed in middle. The antero-lateral glabellar lobes and furrows are indistinctly defined. The median node lies between the antero-lateral lobes (papilionate), long, prominent, with a length about 3/5 of the glabellar length in the middle portion. Preglabellar median furrow and dorsal furrows are narrow and very shallow, almost effaced. The basal lobes are large and equilateral triangular-shaped. The cheeks are uniform in width, a little narrower than the glabellar width in mid-length, moderately convex, gently sloping downwards towards the narrow but pronounced marginal furrow. The marginal border is narrow and uniform in width throughout.

Thoracic axis broad, first segment wider than the basal width of glabella, devoid of longitudinal or oblique lateral furrows; first pleural segment short, equal to or slightly less than 1/4 the axial width; second axial segment morphologically similar to the first, but narrower, and wider than the axis of pygidium; second pleural segment about 1/2 the axial width. Pleural furrows shallow and narrow.

Pygidium is subsemi-elliptical in outline, slightly tapers posteriorly, moderately convex. Axis short, pentagonal-shaped, the length is a little greater than 1/3 the pygidial length, well-defined laterally by deep dorsal furrows and a pair of inwardly oblique lateral furrows. The median node is long, extending from the anterior extremity of the axis and protruding at a short distance to the frontal portion of the pseudolobe. The pseudolobe is highly vaulted above the pleural lobes. The auxiliary furrows run directly backwards and slightly outwards, disappear before reaching the marginal furrow. The pleural lobe is narrow, about 1/2 the axial width, strongly sloping down laterally. The marginal border is narrow, widest at the mid-length of pygidium where it produces a pair of tiny postero-lateral spines. The marginal furrow is wide, broadest near the spines, sloping down inwards laterally and posteriorly.

The above description calls the similarity of the Guizhou specimens to those of *Rhaptagnostus changshanensis* Lu et Lin from the late Upper Cambrian Siyangshan Formation of western Zhejiang. A comparison of this species with the type-species of *Rhaptagnostus, Pseudagnostus cyclopygeformis* Sun (Sun, 1924, p. 26, pl. 2, figs. la-h; Lu et al., 1965, p. 42, pl. 4, figs. 9–12) and with the American species *Rhaptagnostus convergens* (Palmer, 1955, pp. 96-97, pl. 19, figs. 14–15; Shergold, 1977, p. 87, pl. 16, figs. 1–2) from the Upper Cambrian Trempleauan has been made by Lu et Lin (1984, p. 56). The Australian form, *Rhaptagnostus clarki patulus* (Shergold, 1975, pp. 62–64, pl. 1, figs. 1-6; pl. 2, figs. 1–2) of the later Upper Cambrian differs from *changshanensis* in the narrower and longer glabella which has a length greater than 2/3 of the total length of cephalon, and in the shorter pygidial medial node which springs rearwards from the middle of the axis.

Occurrence: Late Upper Cambrian Maotian Formation, at Pu'an of Sandu county, SE Guizhou (78F2-7, 78F2-14, 82Fg-1, F5-53).

Family Phalacromidae Hawle et Corda, 1847

Genus *Leiagnostus* Jaekel, 1909

Leiagnostus (?) *insolitus* Lu et T.R. Zhou (sp. nov.)

(Pl. 2, Figs. 6, 7, 8)

Description: Cephalon ovate in shape, as long as wide, widest at the posterior third of the sagittal length, moderately convex. In the exfoliated portion, the glabella is very inconspicuously defined by feebly dorsal furrows which are visible at the posterior part of the cephalon and may be traced anteriorly as a pair of slightly convergent furrows to the sagittal midlenth, then tapering forwards to meet with each other at 1/4 the axial line. A slightly backwardly arched transverse furrow divided the glabella into an equilateral triangular anterior lobe and a longer trapezoidal posterior lobe, On both sides of the latter is a pair of small, faintly defined triangular basal lobes. Marginal border very narrow, wire-like.

Pygidium moderately convex, as long as wide. Surface badly weathered. A short axial furrow can be recognized on the right side of the specimen. Border broad. nearly uniform in width. Other details not known.

Remarks: The exfoliated specimens show the glabellar structures not known in other smooth agnostids except *Trilobagnostus* Harrington, 1938, a subgenus of *Lotagnostus* Whitehouse, 1936, which bears a similar glabella with very shallow, faintly marked dorsal furrows and transversely curvelinear furrow. The shape of cephalon and pygidium of this species is more like that of *Leiagnostus* than that of *Trilobagnostus*. The cephalic border of *Trilobagnostus* is much wider than that of *Leiagnostus*, but the latter provides a very broad pygidial border like the present species and never seen in *Trilobagnostus*. Whether the combination of the pygidium and cephalon into a single species named as *Leiagnostus* (?) *insolitus* is correct remain to be clarified after further materials, especially a complete carapace, are obtained.

Occurrence: Upper Cambrian Sandu Formation., at Pu'an of Sandu (82Fg-50, 82Fg-57).

Leiagnostus sp.

(Pl. 3, Fig. 1)

Only one small carapace, measuring 4.6 mm long and 2.3 mm wide (widest portion of cephalon), is available for study. It differs from *Leiagnostus bexelli* Troedsson (1937, p. 32, pl. 5, figs. 13, 14) from the Lower Ordovician of Eastern Tianshan and *Leiagnostus* (?) *insolitus* Lu et T. R. Zhou (sp. nov.) from the Upper Cambrian Sandu Formation of Pu'an, SE Guizhou described in this paper (pl. 2, figs. 6, 7, 8) in the straighter instead of strongly curved lateral border of cephalon, the relatively broader glabella and in the longer pygidium with more rounded posterior marginal border. From *L. erraticus* Jaekel, the type species of the genus, it is distinguished by its shorter and rounded cephalon, by the presence of ill-defined axial furrows and by the remarkable but narrow cephalic and pygidial border furrows. Because the present specimen is too small for a detail comparison with the large adults of other species of *Leiagnostus*, it seems better not to give a specific name for this specimen until more material is available.

Occurrence: Early Lower Ordovician Guotang Formation of Shizilu, Sandu county, SE Guizhou (82Fg-64).

Order Ptychopariida Swinnerton, 1915
Suborder Ptychopariina Richter, 1933
Superfemily Ptychopariacea Matthew, 1887
Family Eulomidae Kobayashi, 1955
Subfamily Eulominae Kobayashi, 1955
Genus *Euloma* Angelin, 1854
Subgenus Archaeuloma Lee, 1978

The subgeneric diagnosis given by Lee in Chinese (Lee, in Yin and Li, 1978, p. 459) was translated into English by Peng (1984, p. 319). Both the type species, *E.* (*A.*) *guizhouensis* Lee (Lee in Yin and Li, 1978, p. 459, pl. 160, fig. 16) and *E.* (*A.*) *taoyuanensis* Peng (1984, pp. 319–320, pl. 2, fig. 3c; pl. 4, figs. 6, 7) are represented by cranidia only, therefore the finding of a dorsal shield (no librigenae attached) of *E.* (*A.*) *sanduensis* (sp. nov.) makes it possible to give an emended generic diagnosis of this subgenus as follows:

Heteropygous. Dorsal shield oval, with a length ratio of cranidium, thorax and pygidium about 3.2:4.2:1. Cranidium broad, with a length and width ratio measured between the extremities of the posterolateral limbs about 1:1.8. Glabella broad and short, tapering slightly forwards, truncated or broadly rounded anteriorly, with three pairs of shallow glabellar furrows, well-defined by deep axial furrows. Fixigena between the eyes approximately one-half the width (tr.) of glabella. Palpebral lobes medium-sized, slightly curved outwards, with posterior end situated at a distance from the axial furrows about one-third of the glabellar width. Ocular ridges slightly elevated, almost transversely extending from the axial furrows outwards. Posterior limb long (tr.) and narrow (exsag.); postero-lateral furrows deep and wide, groove-like. Preglabellar field wide (sag.), slightly wider than one half of the glabellar length. Anterior border furrow deep, with or without a row of faint pits. Anterior border narrow (sag.) and convex. Anterior facial sutures subparallel or slightly divergent forwards. Thorax composed of 13 segments. Axis as wide (tr.) as the pleural portion. Pleural segment with deep and broad pleural furrow, terminated by slightly oblique backward pleural spine. Pygidium fusiform, with a length less than 1/2 its width. Axis convex, with 2 to 3 rings and a terminal piece, about one-third the pygidial width in anterior portion. Pleural field gently convex, composed of three segments, pleural furrows very shallow.

Type species: *Euloma* (*Archaeuloma*) *guizhouensis* Lee

Geological and Geographical Distributions: Late Upper Cambrian, SE Guizhou and NW Hunan.

Discussion: This subgenus is closely allied to *E.*(*Euloma*) Angelin in the short and broad glabella, the broad (sag.) preglabellar field and narrow (sag.) anterior border, but the latter is distinguished by the larger and strongly curved palpebral lobes, the deeply incised glabellar furrows and well developed palpebral furrows, the forwardly oblique ocular ridges, and by the very prominent pits on the anterior border furrow. *Pareuloma* Rasetti, 1954, differs from *E.* (*Archaeuloma*) in the smaller glabella, wider fixigenae, wider (sag.) preglabellar field, smaller palpebral lobe and narrower (sag.) anterior border. *Proteuloma* Sdzuy, 1958, may be differentiated from *E.* (*Archaeuloma*) by the longer glabella which occupies 1/2 instead of less than 1/2 the total length of cranidium, the longer palpebral lobes, the deep glabellar furrows and by the remarkable and large pits of the anterior border furrow.

Euloma (Archaeuloma) elongatum Lu et T.R. Zhou (sp. nov.)

(Pl. 2, Figs. 9–10; Pl. 6, Figs. 6–7)

Diagnosis: An *Euloma* (*Archaeuloma*) with elongate glabella, medium-sized palpebral lobes and relatively narrower cranidium.

Description: Cranidium subrectangular. Glabella over half as long (sag.) as cranidium, slightly tapering forwards, rounded anteriorly. Three pairs of glabellar furrows present: anterior pair short: middle and preoccipital furrows relatively long and prominent, oblique backwards. Axial, preglabellar and occipital furrows deeply defined. Occipital ring widest mesially, narrowing laterally, with a tiny mesial tubercle. Preglabellar field wide (sag.), gently convex. Anterior border equally-widened, convex and gently upturned. Ocular ridge thick. Palpebral lobes moderate-sized. Fixigena slightly over half the width (tr.) of glabella between palpebral lobes. Anterior sections of facial sutures divergent forwards; posterior sections running outwards, then obliquely backwards to cut posterior borders, enclosing narrow (exsag.), posterior limbs. Posterior border furrow deep and broad (exsag.). Librigena of small specimens (Pl. 6, figs. 6,7) with broad and deep lateral and posterior border furrows and broad doublure, genal spine short.

Remarks: The new species differs from *E.* (*A.*) *guizhouensis* Lee, the type species (Lee in Yin and Li, 1978, p. 459, pl. 160, fig. 11), *E.* (*A.*) *taoyuanensis* Peng (1984, p. 319, pl. 2, fig. 3c; pl. 4, figs. 6,7) and *E.* (*A.*) *sanduense* described in this paper in having a longer (sag.) cranidium and glabella, deeper glabellar furrows and shorter ocular ridges. Furthermone, the size of the occipital node of the new species is similar to that of *E.* (*A.*) *taoyuanensis,* but it is situated more anterior than that of *E.* (*A.*) *guizhouensis.*

Occurrence: Sandu Formation and Maotian Formation, Upper Cambrian: Shizilu and Pu'an, Sandu, Guizhou (78F2-7=82Fg-1, 82Fg-51)

Euloma (Archaeuloma) sanduense Lu et T.R. Zhou (sp. nov.)

(Pl. 2, Figs. 11–12)

Diagnosis: An *Euloma* (*Archaeuloma*) with tapering glabella, broadly rounded anterior glabellar margin, wide fixigenae and comparatively small palpebral lobes.

Description: Exoskeleton oval in outline. Cranidium subrectangular. Glabella short, occupying a little less than one half of total cranidial length (sag.), convex transversely, rapidly tapering forwards, evenly rounded anteriorly, well-defined by deep axial furrows and relatively shallow preglabellar forrow. Three pairs of glabellar furrows present, very shallow and short. Preoccipital furrow slightly longer than anterior ones, running inwards and backwards with its inner end curved rearwards. Occipital furrow deep; occipital ring widest mesially and widening abaxially, provided with a small median node located close to its anterior margin (pl. 2, fig. 12). Ocular ridge conspicuous, running transversely from frontal end of palpebral lobe to antero-lateral corner of glabella. Palpebral lobe moderate-sized, crescentic in outline, gently convex. Fixigena with a width a little over half as wide (tr.) as the glabellar width between palpebral lobes, sloping down towards axial furrow. Preglabellar field wide, gently convex, separated from highly convex, equally widened anterior border by very deep anterior border furrow. Anterior sections of facial sutures divergent forwards;

posterior section straight, obliquely outwards and gently backwards. Posterior furrow narrow (exsag.) at inner end, expanding distally.

Thorax with 13 segments. Axis as wide (tr.) as pleural field. Pleural segments projecting spines of moderate length.

Pygidium relatively small, transversely spindle-shaped. Axis convex, consisting of 2-3 rings and a terminal piece. Pleural field gently convex, with 3 pairs of ill-defined segments. Pygidial border inconspicuous, undeveloped.

Remarks: The new species resembles the type species, *E.* (*A.*) *guizhouensis* Lee (in Yin et Li, 1978, p. 459, pl. 160, fig. 11) in having a short and wide (tr.) glabella and a wide (sag.) preglabellar field. However, in the new species, the glabella is more rapidly tapered forwards and more rounded anteriorly, the preglabellar field is more convex, and the fixigenae are wider than those of *E.* (*A.*) *guizhouensis* Lee. In addition, the palpebral lobes of the new species are relatively small. *E.* (*A.*) *taoyuanensis* Peng (1984, p. 319, pl. 2, fig. 3c; pl. 4, figs. 6, 7) from the *Euloma* (*Archaeuloma*) *taoyuanensis-Leiagnostus* cf. *bexelli* Assemblage-Zone of the late Upper Cambrian Shenjiawan Formation, NW Hunan is closly similar to *E.* (*A.*) *sanduensis* in the outline of cranidium, the shape of glabella and the ratio of width of glabella at the base and the postero-lateral border (tr.), but *E.* (*A.*) *taoyuanensis* has a longer glabella, a longer (sag.) preglabellar field and truncated anterior margin of glabella.

Occurrence: The lower part of the *Mictosaukis guizhouensis* Assemblage-Zone, late Upper Cambrian Sandu Formation, Pu'an of Sandu, Guizhou (78F2-6, 82Fg-1).

Euloma (Archaeuloma) sp.

(Pl. 3, Fig. 2)

A single ill-preserved cranidium is the only material for study. It is characterized by the truncato-conical glabella with very short and strongly oblique lateral glabellar furrows, by the very broad (sag.) and convex preglabellar field, by the narrow, thread-like anterior border which is strongly arched forwards in the middle and by the prominent, elevated palpebral lobes defined by deep palpebral furrows. A new species is clearly represented, but until further material is available, we leave it under open nomenclature.

Occurrence: Late Upper Cambrian Sandu Formation, at Shizilu of Sandu County (82Fg-51).

Genus *Proteuloma* Sdzuy, 1958
(=*Mioeuloma* Lu et Qian 1977)
Proteuloma subquadratum (Lu et Qian)

(Pl. 3, Figs. 3–4; Pl. 6, Fig. 5)

1977 *Euloma* (*Mioeuloma*) *subquadrata* Lu et Qian in Zhou, Liu, Meng and Sun, "Fossil Atlas of Central-South China", (1), p. 141, pl. 44, fig. 14.

1978 *Euloma* (*Mioeuloma*) *subquadrata*, Lu et Qian, in Yin and Li "Fossil Atlas of Southwest China", Fasc. 1, (Guizhou), p. 458, pl. 160, fig. 17 (refigured).

1983 *Euloma* (*Mioeuloma*) *subquadrata*, Lu et Qian, Palaeont. Cathayana, no. 1, p. 40, pl. 5, fig. 7; pl. 6, fig. 1.

1984 *Proteuloma* cf. *subquadratum*, Peng, Cambrian-Ordovician Boundary (1), p. 321, pl. 14, figs. 1, 2.

Remarks: Several nearly complete carapaces and detached cranidia and pygidia are in the present collection. Three of them are illustrated in this paper. They are identical in all features with those of the type specimens described fully by Lu a Qian in 1983 (p. 40, 41). The pits on the anterior border furrow are clearly shown in pl. 3, fig. 3, the number of them is about 18-22, similar to that of the type species of *Proteuloma*, *Euloma geinitzi* Barrande (Sdzuy, 1955, p. 12–14, pl. 3, figs. 77–89; pl. 4, figs. 90–92; 1958, p. 271, pl. 3, figs. 4,5, 7–11).

Occurrence: Lower Ordovician (Tremadocian) Guotang Formation, Pu'an and Shizilu, Sandu County (82Fg-21, 82Fg-65, 82F26-12, 82Fg-24, 82Fg-66).

Proteuloma sp.

(Pl. 3, Fig. 5)

Cranidium transversely broad, about twice as long as wide including postero-lateral limb. Glabella short, about 1/2 the length of cranidium and with a width more than 1.5 times that of the fixigenae between the palpebral lobes. Anterior border moderately convex, slightly arched forwards, as wide (sag.) as the brim. Anterior border furrow deep and wide, with a row of 30 pits. Occipital ring provided with a small median node near the posterior margin.

This form is distinguished from *Proteuloma subquadratum* (Lu et Qian) (Lu et Qian 1983, p. 40, pl. 5, fig. 7; pl. 6, fig. 1) by the shorter and broader glabella, the less forwardly arched anterior border and by the straighter anterior border with about 30 pits instead of 18-22 in *subquadratum*.

Occurrence: Early Lower Ordovician Guotang Formation, Pu'an and Shizilu, Sandu county (82Fg-24, 82Fg-66).

Genus *Pareuloma* Rasetti, 1954

Pareuloma cf. *spinosa* Palmer

(Pl. 3, Figs. 6–7)

cf. 1968 *Pareuloma spinosa* Palmer, p. 76, figs. 1–9.

Description: Cranidium trapezoidal in outline. Glabella short, about 1/2 the cephalic length, moderately convex transversely, gently tapering forwards, truncated anteriorly, defined by deep axial and shallow preglabellar furrows, bearing 3 pairs of glabellar furrows. Occipital furrow deep; occipital ring partially destroyed. Fixigena wide, slightly less than the width of the preoccipital lobes between the palpebral lobes. Ocular ridge well defined, running transversely inwards from the palpebral lobes to meet the glabella at its antero-lateral corner. Palpebral lobe medium-sized, situated at the mid-length of cranidium. Preglabellar field wide (sag.), swollen in central portion, separated from the relative narrow, slightly convex anterior border by deep, straight, transverse border furrow. Anterior sections of the facial sutures slightly diverging forwards; posterior sections obliquely outwards, enclosing wide (tr.) triangular posterior limb. Posterior border furrow deep and broad, defined by equally-wide, narrow posterior border.

Remarks: The present cranidia are closely similar to those of *Pareuloma spinosa* Palmer (1976, p. 76, figs. 1–9) from Alaska in the shape of cranidium and glabella, the nature of the ocular ridges, the shape and size of the palpebral lobes and the nature of the anterior border. However, they differ from *P. spinosa* in the less forwardly tapering glabella, narrower (sag.) fixigena, wider anterior border and in lack of surface ornamentation.

Occurrence: The uppermost part of the Upper Cambrian Sandu Formation, Shizilu, Sandu, Guizhou (82Fg-57, 82Fg-63).

Superfamily Conocoryphacea Angelin, 1854

Family Shumardiidae Lake, 1907

Genus *Shumardia* Billings, 1862

Shumardia puanensis Lu et T. R. Zhou (sp. nov.)

(Pl. 3, Figs. 8–10)

Diagnosis: A *Shumardia* with faintly defined antero-lateral lobes on the antero-lateral corners of glabella, obsolete lateral glabellar furrows and rounded preglabellar furrow.

Description: Cephalon semi-circular in outline, with sharply pointed genal spines. Glabella strongly convex, subparallel or gradually widening forwards, slightly expanding antero-laterally. Lateral glabellar furrows obsolete, only a pair of very shallow and broad antero-lateral glabellar furrows (the accessory furrows) to separate the antero-lateral corners into a pair of ill-defined eye-like antero-lateral lobes. Dorsal furrows very deep and broad, narrowing anteriorly into the narrow but prominent preglabellar furrow. Occipital ring nearly uniform in width (sag.) medially, narrowing laterally; occipital furrow narrow and straight. Cheek strongly convex, steeply sloping down laterally, continuously tapering forwards and inwards to unite in front of glabella as a narrow and convex preglabellar field.

Thorax composed of six segments. Axis narrower than the pleural lobe, gradually tapering posteriorly. Pleural segments grooved by broad and deep pleural furrows, fourth pleural spines large and long, extending backwards beyond the pygidium.

Pygidium semi-elliptical in outline, twice as long as wide. Axis conical, narrow, about 1/2 the width of pleuron, not reaching the posterior margin, divided by transverse furrows into 3–4 rings and a very small terminal piece. Pleural lobe flat, with 3 pleural segments, pleural furrows broad, interpleural furrows distinct. Marginal border narrow, convex, thread-like, posterior margin straight or broadly sinuated.

Remarks: This species is similar to the genus *Akoldinioidia* Zhou and Zhang, 1985, in the faintly defined eye-like antero-lateral glabellar lobes, the presence of distinct preglabellar furrow and in the pointed genal spines. It differs from the latter in the obsolete lateral glabellar furrows, in having 6 instead of 7 thoracic segments and in the sinuated instead of broadly curved posterior margin of pygidium. *Koldinioidia* as typified by *K. typicalis* Kobayashi and regarded as a subgenus of *Shumardia* by Zhou and Zhang (1985, p. 71) is identical with the present species in having a pair of antero-lateral glabellar lobes and in the obsolete lateral glabellar furrows on the flanks of glabella, but the former is distinguished by the stout genal spines and by more or less fading preglabellar furrow. In other species of *Shumardia,* the eye-like antero-lateral glabellar lobes are fairly well defined by deep accessory glabellar furrows, and the lateral glabellar furrows on the flanks of glabella are deeply incised.

Occurrence: Lower Ordovician (Tremadocian) Guotang Formation, at Pu'an (82Fg-21, 82Fg-22, 82Fg-24), and Limuzhai (82Fg-65, F5-102), Sandu County.

Genus *Akoldiniodia* Zhou et Zhang, 1985

Comments: Because no complete carapace among the species of *Akoldinioidia* has been known since the genus was erected in 1985, the discovery of several entire dorsal shields of two species, *Akoldinioidia hewenensis* (T. M. Zhou) and *Akoldinioidia sanduensis* (sp. nov.) obtained from SW Guangxi and SE Guizhou respectively is desirable. Based on the specimens of the two mentioned species, a new generic diagnosis of *Akoldinioidia* is subsequently given here.

Diagnosis: Shumardiid trilobites with oval outlined dorsal shield, tapering to subparallel-sided glabella defined by deep and wide dorsal furrows and narrow but distinct preglabellar furrow. Two or three pairs of very short lateral glabellar furrows present. Preglabellar area slightly convex, flat to slightly concave. Thorax with seven segments. Axis wide, fourth pleurae with long macropleural spines. Pygidium semi-circular to broad semi-elliptical in outline. Axis with 4 to 5 axial rings. Pleural lobes narrow, separated by 3 to 4 pleural and interpleural furrows. Marginal border absent or ill-defined.

Type Species: *Akoldinioidia pustulosa* Zhou et Zhang, 1985.

Remarks: In addition to the type species and an unnamed species, Zhou et Zhang include in the genus *Akoldinioidia* a number of species that Shergold (1975, p. 99) originally assigned to *Koldinioidia,* such as *Koldinioidia aspinosa* Kobayashi (1933, p. 100, pl. 10, figs. 5 & ? 6), *Koldinioidia yenchouensis* Resser and Endo (in Endo and Resser, 1937, p. 229, pl. 56, fig. 10), *Koldinioidia paiensis* Endo (in Endo et Resser, 1937, p. 299, pl. 71, fig. 5), *Koldinioidia sulcata* Robison et Pantoja-Alor (1968, p. 796, pl. 104, figs. 20–23), *Eoshumardia cylindrica* Shergold (1971, p. 65, pl. 18, figs. 7–8), *Koldinioidia* cf. *cylindrica* (Shergold, 1975, p. 100, pl. 58, figs. 1–4) and *Koldinioidia payntonensis* Shergold (1975, p. 101, pl. 58; figs. 6–10). Thus, *Koldinioidia*(*Koldinioidia*) cf. *sulcata* Robison et Pantoja-Alor (Shergold et Sdzuy, 1984, p. 98, pl. 6, fig. 72) is also a member of *Akoldinioidia.*

Shumardia is closely allied to *Akoldinioidia*, the former differs in having more anteriorly expanded glabella with a well developed anterior pair of lateral glabellar furrows (the accessory furrows) and lateral lobes, and in having six instead of seven thoracic segments, The genus *Koldinioidia* Kobayashi which is recently suppressed to a subgeneric level of *Shumardia* by Zhou et Zhang (1985, p. 71) can be easily distinguished from *Akoldinioidia* by almost fading preglabellar furrow and stout genal spines.

Geological and Geographical Distributions: Late Upper Cambrian and lowermost Lower Ordovician of China, Australia and Mexico.

Akoldinioidia hewenensis (Zhou)

(Pl. 3, Figs. 11, 12)

1977 *Shumardia? hewenensis* Zhou, T. M., Palaeontologic Atlas of Central and South China (1), p. 147, pl. 46, figs. 1–3.

Diagnosis: An *Akoldinioidia* differs from the type species, *A. pustulosa* Zhou et Zhang (1985, p. 74, pl. 25, figs. 1–3, 12), in the less angulated glabellar front, in the absence of an occipital node, in 2 instead of 3 pairs of lateral glabellar furrows and in the non-granulated surface of cranidium.

Description: The original description in Chinese of the holotype collected from Hewen, Debao County of Guangxi given by Zhou Tian-mei (1977, p. 147) is translated as follows: "Glabella convex, nearly parallel-sided, anterior portion contracted forwards, roundedly angulate in front, posterior portion with two pairs of very short and weakly defined glabellar furrows on flank; eye-like antero-lateral lobes very obscure or absent. Occipital furrow narrow, deeply incised. Occipital ring uniform in width, slightly narrowing on lateral sides. Cheeks convex, without genal spines. Thorax composed of 7 or 8 (?) segments. Fourth pleurae bearing a pair of backwardly directed, long pleural spines. Pygidium semi-circular. Axis broad, conical, divided into 4-5 axial rings; pleural lobes narrow. Marginal border narrow but distinct".

Remarks: According to the above description, there has no genal spines in the Guangxi specimens, but the genal spines are present and clearly shown in the Guizhou specimen (pl. 3, fig. 11, left side.). This species is readily differentiated from *Shumardia* (*Shumardia*) by the undeveloped eye-like antero-lateral lobes where they are not expanded laterally, and by the presence of 7 instead of 6 thoracic segments. It differs from the type species of *Koldinioidia, K. typicalis* Kobayashi (Kobayashi, 1931, p. 187, pl. 22, figs. 8, 9b; Resser and Endo in Endo, 1931, p. 86, fig. 14; Resser and Endo in Endo and Resser, 1937, p. 229, pl. 57, fig. 14; Zhou and Zhang, 1985, pp. 72–73, pl. 1, figs. 15–17; pl. 3, figs. 6,7; pl. 18, figs. 6, 7; pl. 20, figs. 9–11; pl. 22, fig. 20; pl. 24, fig. 14; pl. 26, figs. 15–20; pl. 28, fig. 8) in the more or less angulated instead of rounded glabellar front, in the shorter (sag.) occipital ring and in the better defined preglabellar furrow.

Occurrence: The upper part of the Maotian Formation, late Upper Cambrian, at Pu'an of Sandu county, SE Guizhou (78F2-7, 78F2-6, 82Fg-10).

Akoldinioidia sanduensis Lu et T. R. Zhou (sp. nov.)

(Pl. 4, Figs. 1,2)

Diagnosis: An *Akoldinioidia* characterized by the short anterior glabellar lobe which is broadly rounded in front, well-developed glabellar furrows, narrow preglabellar field, and shorter (sag.) occipital ring.

Description: Dorsal shield (pl. 4, fig. 1) ovate, measuring 4.3 mm long and 3.8 mm wide between genal angles. Cephalon semi-elliptical, with a length and width ratio about 6:15. Glabella strongly elevated above the cheeks, convex from side to side, gently sloping down anteriorly. Anterior glabellar lobe short, with a length a little more than 1/2 its width, the length and width ratio about 4:7, separated from the posterior glabellar lobe by short, deeply incised lateral furrows on flank, a pair of short, deep, curved preoccipital lateral furrows separated a pair of small basal glabellar lobes. Dorsal furrows subparallel, broad and deeply incised, slightly contracted at middle of glabella, shallowing and narrowing forwards into the weakly defined preglabellar furrow. Cheeks smooth, as wide (tr.) as the glabellar base, rapidly narrowing anteriorly, very narrow in front of glabella. Posterior border furrows of moderate width, deep. Postero-lateral border strongly convex, very narrow in inner portion, widening rapidly outwards and merged into the genal spines. Occipital furrow nearly straight, faintly bending forwards on sides, deep and broad. Occipital ring short (sag.), with a length less than 1/3 its width (tr.), narrowing very gradually laterally and slightly arching backwards, central part destroyed, seemingly without an occipital node or spine.

Thorax composed of 7 segments. Axis wide, tapering rapidly and regularly from 4th segment rearwards, the width (tr.) ratio of axis and pleurae is about 3:2, the 4th is the macropleural segment, carrying long pleural spines extending directly backwards far behind the posterior end of pygidium, pleural furrows deep and wide, geniculating at the half way of the pleural length.

Pygidium broad, semi-circular in outline. Axis convex, composed of 4 axial rings and a terminal piece which extends close the posterior margin. 4 pairs of pleural segments well separated by distinct pleural and interpleural furrows. Marginal border absent.

Remarks: This species is closest to *A. hewenensis* (Zhou) described in foregoing paragraphs, but it differs in the short and broad anterior glabellar lobe, the narrower preglabellar area, the deeper glabellar furrows, and in the absence of a well defined and broad marginal border of pygidium. *Akoldinioidia aspinosa* (Kobayashi, 1933, p. 100, pl. 10, fig. 5), *A. cylindrica* (Shergold, 1971, p. 65, pl. 18, figs, 7-8) and *A.* cf. *cylindrica* (Shergold, 1975, p. 100, pl. 58, figs. 1-4) have subparallel-sided glabellae and broad cranidia similar to the present species, but the lateral glabellar furrows of them are indistinctly defined. *A. yentaiensis* Resser et Endo (in Endo et Resser, 1937, p. 229, pl. 56, fig. 10), *A. paiensis* Endo (in Endo et Resser, p. 329, pl. 71, fig. 5) and *A. payntonensis* (Shergold, 1975, p. 101, pl. 58, figs. 6-10) form a group with conical or tapering glabellae and may be fundamentally differentiated from *A. sanduensis.*

Occurrence: The upper part of the Maotian Formation, late Upper Cambrian, at Pu'an of Sandu County, SE Guizhou (78F2-6).

Genus *Hospes* Stubblefield in Stubblefield et Bulman, 1927

Hospes transversalis Lu et T. R. Zhou (sp. nov.)

(Pl. 4, Figs. 7-8; Pl. 5, Fig. 3)

Diagnosis: A *Hospes* characterized by broad cephalon and subcylindrical glabella with 4 pairs of lateral furrows.

Description: Cranidium semi-circular or subtriangular in outline, slightly curved forwards, with a width about 1.8 times its length, convex from side to side, sloping down laterally. Glabella subcylindrical, twice as long as wide, tapering very slowly forwards, strongly convex transversely, broadly rounded anteriorly, with 4 pairs of very shallow and weakly defined lateral glabellar furrows, among them the 1p and 2p appear to unite across the glabella, 3p and 4p very short, only discernible on both sides of glabella. Axial furrows deep and broad. Preglabellar furrow shallow. Anterior area gently convex, about 1/6 the length of glabella. Occipital ring wider (tr.) than the basal width of glabella, uniform in width in middle portion, narrowing laterally, separated from the glabella by a deep, broad and straight occipital furrow. Palpebral lobe small, situated opposite the antero-lateral corners of the glabella. Ocular ridge wire-like, running transversely, about 1/3 the breadth (tr.) of postero-lateral limb. Fixigenae and postero-lateral limb broad, about 1.7 times the basal glabellar width. Anterior section of facial sutures short, convergent forward; posterior section long, oblique backwards to enclose a wide (tr.), large posterior limb. Posterior border furrow deep, obliterating distally. Genal angle rounded.

Remarks: The new species closely resembles *Hospes guizhouensis* Zhou (Zhou, 1981, p. 242, pl. 1, figs. 1–4, 6, 7a and Lu et T. R. Zhou in this paper pl. 4, figs. 3-6; pl. 5, figs. 1, 2) in the presence of 4 pairs of glabellar furrows, in the relative length and width of glabella and in the rounded genal angles, but it differs in having a larger and wider (tr.) postero-lateral limb, in the deeper lateral glabellar furrows, in the presence of ocular ridges and in lack of a pair of tiny tubercles at the antero-lateral corners of the glabella. *Hospes ciliensis* Peng (1984, p. 325, pl. 15, fig. 11) also possesses distinct lateral glabellar furrows, broader cranidial anterior area and similar outline of glabella like the new species, but the former has narrower (tr.) fixigenae and postero-lateral limb.

Occurrence*:* *Mictosaukia guizhouensis* Assemblage-Zone, late Upper Cambrian Maotian Formation at Pu'an, Sandu of SE Guizhou (82Fg-1, 82Fg-10).

Hospes guizhouensis Zhou

(Pl.4, Figs. 3–6; Pl. 5, Figs. 1–2)

1981 *Hospes guizhouensis* Zhou, p. 242, pl. 1, figs. 1–4, 6, 7a.
1984 *Hospes guizhouensis* , Peng, p. 323, pl. 15, fig. 4.

Remarks: The present specimens agree in all respects with the types of *H. guizhouensis* Zhou which are also collected from the Sandu area, Guizhou and identical with the cranidium from the Tremadoc *Onychopyge-Hysterolenus* Assemblage-Zone of NW Hunan (Peng, 1984).

Occurrence: Lower part of the Guotang Formation, Lower Ordovician, Pu'an and Shizilu, Sandu, Guizhou (82Fg-16, 82Fg-22, 82Fg-24, 82Fg-64, 82Fg-65).

Hospes sanduensis Zhou

(Pl. 5, Fig. 4)

1981 *Hospes sanduensis* Zhou, p. 243, pl. 1, fig. 5.

Remarks: A single cranidium from Pu'an with deep and wide axial furrows and relatively short glabella is the identical one collected from the same locality and originally described by the junior author. One dorsal shield and three cranidia from the *Onychopyge-Hysterolenus* Assemblage-Zone of NW Hunan are assigned to *Hospes* cf. *sanduensis* Zhou by Peng (1984, p. 324–325, pl. 15, figs. 7–10). The cranidia in most respects identify with that of *H. sanduensis* Zhou, but their glabellae seem to be more slender and the preglabellar furrows are relatively shallower than those of the specimens from Guizhou.

Occurrence: Lower Ordovician (Tremadocian) Guotang Formation, Pu'an, Sandu, Guizhou (82Fg-21).

Hospes sp.

(Pl. 5, Fig. 5)

A very poorly preserved cranidium differs chiefly from other species of *Hospes* in having a conical glabella and in a norrower preglabellar field. The specimen is recorded and illustrated here for further comparison of new material.

Occurrence: Lower Ordovician (Tremadocian) Guotang Formation, Shizilu of Sandu county, SE Guizhou (82Fg-64).

Genus *Puanocephalus* Lu et T. R. Zhou (gen. nov.)

Type species: *Puanocephalus orientalis* Lu et T. R. Zhou (gen. et sp. nov.) from the *Mictosaukia guizhouensis* Assemblage-Zone of the Upper Cambrian Maotian Formation, at Sandu, Guizhou.

Diagnosis: A small-sized, isopygous genus of the family Shumardiidae. Exoskeleton oval. Cephalon semi-circular. Glabella gently convex, subquadrate in outline, parallel-sided and rounded anteriorly, with flat or depressed frontal lobe and 3 pairs of lateral glabellar furrows: anterior pair (the accessory furrows) very shallow, extending inwards and forwards from the dorsal furrows to cut the antero-lateral corners of the frontal lobe into a pair of small eye-like antero-lateral lobe; second ones also very shallow, pit-like, not reaching the dorsal furrows; posterior pair similar to the second, but continuously extending rearwards and reaching the occipital furrow. Occipital ring wide sagittally, narrowing abaxially, with a thick occipital spine; occipital furrow shallow and relatively wide (sag.), straight mesially and oblique forwards laterlly. Preglabellar field jointed with fixigenae to from a convex semicircle surrounding glabella. Fixigena with a greatest width about 2/3 the width of glabella. Palpebral lobes and ocular ridges absent. Posterior border narrow at its inner end, widening abaxially rearwards, projecting a short genal spine. Thorax with 4 segments. Axis as wide (tr.) as pleural field, slightly tapering rearwards. Pleural segments narrow (sag.), bearing relatively wide (sag.) pleural furrows. Pygidium semi-circular, downsloping posteriorly. Axis conical, as wide as pleural field, containing 4 rings and a terminal piece, reaching close the pygidial margin. Pleural field divided into 4-5 pleurae, with poorly defined interpleural furrows. Border furrows absent.

Discussion: The new genus resembles *Shumardia* Billings, 1862, *Koldinioidia* Kobayashi, 1931, and *Akoldinioidia* Zhou and Zhang, 1985, in the presence of eye-like antero-lateral lobes on the antero-lateral corners of the frontal glabellar lobes and in the absence of palpebral lobes. Except it possesses only four thoracic segments (six in *Shumardia* and seven in *Akoldinioidia*) and its pygidium is as large as the cephalon, it differs from *Shumardia* in the unexpanded frontal glabellar lobe and in the isolated pit-like lateral glabellar furrows. The unexpanded frontal glabellar lobe is one of the characteristic features of *Koldinioidia* and *Akoldinioidia*, but the preglabellar furrow is faded out in front of glabella in the former and the laterl glabellar furrows are connected with the dorsal furrows in the latter. *Puanocephalus* may be compared with some non-shumardiid trilobites without eye-like antero-lateral lobes on the frontal glabellar lobe as discussed below. It resembles *Pseudosalteria* Raymond (1924, p. 398, pl. 12, fig. 11) from the Upper Cambrian of North America. However, in *Pseudosalteria*, the palpebral lobes are very conspicuous and the glabella is relatively narrow (tr.), the occipital and posterior furrows are much shallower and the occipital spine is poorly developed. The new genus is also similar to *Leiobeinvillia* Resetti, 1954, especially to the less effaced species, such as *L. terranovica* (Rasetti) (1954, p. 582, pl. 61, figs. 7–12), in having a semi-circular cephalon and nearly quadrate glabella, but the latter has well marked palpebral lobes and shallower glabellar furrows, all of which are jointed to axial furrows, and devoid of the occipital spine. *Leiobeinvillia leonesis* Winston et Nicholls (1967, p. 87, pl. 11, figs. 16, 20, 21), which is subsequently included into his new genus *Larifugula* by Ludvigsen (1979, p. 75), is also similar to the new genus, but it differs in having narrower fixigenae, prominent palpebral lobes, a more convex pygidial axis with a large number of segments, and a well developed pygidial border. The new genus can be distinguished from *Strotocephala* Raymond (1937, p. 1098, pl. 1, fig. 26) in lacking of palpebral lobes and in having broader (sag.) occipital furrow and posterior cephalic border furrows.

Geological and Geographical Distributions: Upper Cambrian, SE Guizhou.

Puanocephalus orientalis Lu et T.R. Zhou (gen. et sp. nov.)

(Pl. 5, Figs. 6–9; Pl. 6, Figs. 1–4)

Description: The specific description is the same as the generic diagnosis.
Occurrence: *Mictosaukia guizhouensis* Assemblage-Zone of Maotian Formation, Upper Cambrian, Pu'an, Sandu, Guizhou (78F2-6, 82Fg-1, 82Fg-8, 82Fg-9, 82Fg-10, F1-79–87).

Superfamily Olenacea Burmeister, 1843
Family Olenidae Burmeister, 1843
Subfamily Triarthrinae Ulrich, 1930
Genus *Parabolinella* Brögger, 1882

Parabolinella aff. *latilimbata* Lu et Chien

(Pl. 7, Fig. 9; Pl. 17, Fig. 12)

1978 *Parabolinella latilimbata* Lu et Chien, in "Fossil Atlas of Southwest China", Fasc. 1 (Guizhou), p. 478, pl. 164, fig. 7.
1983 *Parabolinella latilimbata* Lu et Chien, Palaeont. Cathayana, 1, p. 49, pl. 6, fig. 2.

Description: Cranidium trapezoidal in outline. Glabella subquadrate, obtusely rounded anteriorly, tapering gently forward, with 3 pairs of lateral glabellar furrows. Anterior glabellar furrows short, situated at antero-lateral corners of glabella; median lateral furrows long, running inwards from axial furrows; preoccipital furrows longest, running parallelly to the median one and meeting with each other sagittally. Axial furrows deeply incised and preglabellar furrow relatively shallow. Preglabellar field wide (sag.), crossed by radial lines; anterior border very narrow and slightly upturned. Ocular ridge conspicuous, running inwards and gently rearwards from palpebral lobe to meet axial furrow opposite the distal end of anterior glabellar furrow. Palpebral lobe small-sized, situated opposite anterolateral corner of glabella. Fixigena gently convex, posterior limb broad (tr., exsag.) bearing a deep, transverse posterior border furrow. Posterior border equally-widened.

Remarks: A single poorly preserved cranidium with its external mold on hand is characterized by a slightly and forwardly tapering glabella and wide (sag.) preglabellar field, and is extremely similar to *P. latilimbata* Lu et Chien (in Yin and Li, 1978, p. 478, pl. 164, fig. 7), however, the latter has a shorter, parallel-sided glabella. The present cranidium can be compared with *P. constracta* Lu et Z. Q. Zhou (in Lu, Zhou and Zhou, 1981, p. 7, pl. 2, figs. 2–3), but differs in having wider (sag.) preglabellar field and more evenly rounded glabella. The present specimen also resembles *P. variabilis* Robison et Pantoja-Alor (1968, p. 790, pl. 102, figs. 17–21), but differs in having more forwardly oblique ocular ridge and broader (sag.) preglabellar field.

Occurrence: Lower part of the Guotang Formation, Lower Ordovician; Shizilu of Sandu, SE Guizhou (82Fg-64).

Family Triplacephalidae Lu et Chien, 1977
Genus *Amzasskiella* Poletaeva, 1960
(=*Triplacephalus* Lu et Chien, 1977)

1960 *Amzasskiella* Poletaeva, in Petrunina et al., p. 431.
1977 *Triplacephalus* Lu et Qian, in Zhou, Liu, Meng et Sun in "Fossil Atlas of Central-South China". p. 154.
1978 *Triplacephalus* Lu et Chien, in Yin and Li, in "Fossil Atlas of Southwest China", Guizhou(1), p. 474.
1983 *Triplacephalus* Lu et Qian, Palaeont. Cathayana, no. 1, p. 53.

Type species: *Amzasskiella mirabilis* Poletaeva, 1960, from the Tremadocian of western Siberia, by original designation.

The diagnostic characters of the family Triplcephalidae and the genus *Amzasskiella* Poletaeva (= *Triplacephalus* Lu et Qian, 1977) are discussed by Lu and Qian in previous papers (1977, p. 154; 1983, p. 52, 53). In China, the following species may be assigned to the genus *Amzasskiella*:

1. *Triplacephalus sanduensis* Lu et Qian, in Zhou, Liu, Meng and Sun, 1977, p. 154, pl. 48, fig. 54; Lu et Chien in Yin and Li, 1978, p. 474, pl. 163, fig. 1 (refigured); Lu et Chien, 1983, p. 54, pl. 7, figs. 1-4; pl. 10, fig. 8.

Lower Ordovician Guoting Formation, Guotang of Sandu, SE Guizhou and Lower Ordovician Panjiazui Formation of Taoyuan, NW Hunan.

2. *Triplacephalus latilus* Liu, in Zhou, Liu, Meng and Sun, 1977, p. 155, pl. 48, fig. 5.
Lower Ordovician Panjiazui Formation of Taoyuan, NW Hunan.

3. *Triplacephalus elongatus* Lu et Chien, in Yin and Li, 1978, p. 475, pl. 163, fig. 16; Lu et Qian, 1983, p. 55, pl. 7, fig. 5.
Lower Ordovician Guotang Formation, Shuiniu of Sandu, SE Guizhou.

4. *Amzasskiella ciliensis* Peng, 1984, p. 330, pl. 7, figs. 8a, 8b.
Lower Ordovician *Onychopyge-Hysterolenus* Assemblage-Zone, Shenjiawan, Taoyuan, NW Hunan.

5. *Amzasskiella* sp., Peng, 1984, p. 331, pl. 7, fig. 8.
Lower Ordovician *Onychopyge-Hysterolenus* Assemblage-Zone, Shenjiawan of Taoyuan, NW Hunan.

Amzasskiella sanduensis (Lu et Qian)

(Pl. 8, Figs. 5, 6)

1977 *Triplacephalus sanduensis* Lu et Qian in Zhou, Liu, Meng and Sun in "Fossil Atlas of Central-South China". (1), Early Palaeozoic, Trilobites, p. 154, pl. 48, figs. 1–4

1978 *Triplacephalus sanduensis* Lu et Chien, in Yin and Li in "Fossil Atlas of Southwest China", Guizhou (1), p. 474, pl. 163, fig. 1 (refigured).

1983 *Triplacephalus sanduensis*, Lu et Qian, Palaeont. Cathayana, no 1, p. 54, pl. 7, figs. 1–4; pl. 10, fig. 8.

Only two specimens in the present collection are available for examination. They agree in all details with those described from the *Dictyonema flabelliforme* Zone of the Guotang Formation (Tremadocian) at Shuiniu, Wangdaozhai and Guotang of Sandu, SE Guizhou and from the Lower Ordovician Panjiazui Formation at Jiuxi of Taoyuan, NW Hunan. No additional comments are need.

Occurrence: Lower Ordovician Guotang Formation, at Shizilu of Sandu county (78F2-4, 82Fg-66).

Family Papyriaspididae Whitehouse, 1939

Genus *Hedinaspis* Troedsson, 1951

Hedinaspis shengi Lu

(Pl. 8, Fig. 7)

1964 *Hedinaspis shengi* Lu, p. 36, pl. 7, fig. 2.

1974 *Hedinaspis shengi* Lu et al., p. 85, pl. 2, fig. 3.

Diagnosis: An elongated *Hedinaspis* with forwardly arched anterior border, elongated glabella, relatively narrower fixgenae between eyes, narrow thoracic axis, broad thoracic pleural segments and small pygidium.

Comparison: This species is represented by a slightly distorted, nearly complete dorsal shield which is elongated ovate in outline, with a length of about 36 mm and a width of about 22 mm between genal angles of librigenae. The glabella is similar in shape to that of *Hedinaspis regalis* (Troedsson) (Troedsson, 1937, p. 56, pl. 3, fig. 5; pl. 7, figs. 1–5, 7–11; pl. 8, figs. 3–8; 1951, p. 695; Lu et al., 1965, p. 187, pl. 31, figs. 6–8), but the frontal margin of glabella is rounded in *H. regalis*, whereas a small median notch is in this species; the basal glabellar lobes are especially expanded laterally in *regalis*, while they are as wide as the occipital ring in this species. The width of the fixigenae between the eyes of the present species is relatively narrower than that of *H. regalis*. The anterior border is slightly arched forwards and the brim is wide in *H. shengi*, whereas the anterior border is sinuated backwards and the brim is narrow in *H. regalis*. *H. kueichouensis* Lu (1954, p. 147, pl. 1, fig. 9; pl. 2, figs. 1–8; pl. 3, figs. 1–4; Chien, 1961, p. 102, pl. 4, figs. 7–9) is comparable with *H. shengi* in the width ratio of the thoracic axis and pleurae, but the former has a shorter glabella, a backwardly concave anterior border and wider fixgenae between the eyes.

Occurrence: Late Upper Cambrian Sandu Formation, at Shizilu of Sandu county, SE Guizhou (78F-14).

Superfamily Solenopleuracea Angelin, 1854

Family Solenopleuridae Angelin, 1854

Subfamily Acrocephalitinae Hupé, 1953

Genus *Acrocephalina* Troedsson, 1937

Acrocephalina convexa Lu et Chien

(Pl. 8, Figs. 1–4)

1978 *Acrocephalina convexa* Lu et Chien, in Yin and Li, p. 487, pl.164, fig. 4.

Diagnosis: An *Acrocephalina* with trapezoidal cranidium, broad and short glabella, 2 pairs of glabellar furrows, narrow fixigenae, and a long and stout cranidial spine.

Description: Cranidium nearly trapezoidal in outline, with a thick and long anterior spine. Glabella convex, truncated anteriorly, with 2 pairs of deeply incised lateral glabellar furrows, well-defined by axial furrows and preglabellar furrow. Anterior lateral furrow short, oblique rearwards; preoccipital furrow long, running inwards and rearwards, bifurcated into a short anterior branch in its mid-point and a long posterior branch almost reaching the occipital furrow. Anterior border wide (sag.), gently convex. Anterior border furrow very faint. Occipital furrow deep and broad (sag.), slightly bended forwards sagittally. Occipital ring widest sagittally, narrowing abaxially, projecting backwards and upwards into a slender occipital spine. Ocular ridge raised, directed obliquely from the anterior end of palpebral lobe to the antero-lateral corner of glabella. Palpebral lobe reniform, with a length about 1/2 the sagittal glabellar length; separated from fixigena by a deep palpebral furrow. Fixigena narrow, about 1/4 the basal width of glabella. Anterior sections of facial sutures divergent forwards, curving inwards near antero-lateral corner of cranidium; posterior sections oblique outwards to enclose the triangular posterior limbs. Surface of glabella and fixigenae sparsely granulated.

Remarks: The observable characters of the present specimens are completely identical with the fragmental holotype cranidium of *A. convexa* Lu et Chien found in Sandu, and they are considered to be conspecific with it. In the trapezoidal cranidium, short glabella, stout cranidial spine and wide (sag.) preglabellar field, *A. convexa* agrees with *A. grandispinosa* Liu (Liu in Zhou, T. M. et al., 1977, p. 161, pl. 52, fig. 14) from the basal Ordovician of Cili, NW Hunan, but the fixigenae of the former are much narrower than those of the latter, there are 2 pairs of glabellar furrows in the former instead of 3 pairs in the latter. *A. reticulate* Chien (1961, p. 102, pl. 4, figs. 13–15; text-fig. 3; Lu et al., 1965, p. 218, pl. 38, figs. 5–6) from the late Upper Cambrian of Sandu, Guizhou, coincides with *A. convexa* in the short and stout glabella, in the straight cranidial frontal margin and in the presence of 2 pairs of glabellar furrows, the former differs in the broader fixigena which is about 1/2 the basal glabellar width instead of 1/4 in the latter, in the broader (sag.) preglabellar field, and in the thick and long cranidial spine. *A. armata* (Troedsson, 1937, p. 67, pl. 3, figs. 18-19; pl. 4, fig. 12), the type species of *Acrocephalina,* can be distinguished from *A. convexa* by the tapering preglabellar field, the broad fixigenae, the presence of 3 pairs of glabellar furrows and by the longer glabella. In the narrow fixigenae and 2 pairs of glabellar furrows, *A. convexa* may be compared with *A. acuta* (Troedsson, 1937, p. 69, pl. 3, fig. 20), but the preglabellar field of the latter is acutely pointed forward and the glabella is longer.

Occurrence: Late Upper Cambrian Sandu Formation (82Fg-57) to early Lower Ordovician Guotang Formation (78F26-12); Shizilu and Pu'an, Sandu, SE Guizhou.

Subfamily Hystricurinae Hupe, 1953
Genus *Taoyuania* Liu, 1977
(=*Batyraspis* Apollonov et Tschugaeva, 1983)

Taoyuania Liu, in Zhou, Liu, Meng and Sun, 1977, p. 161; Liu, 1982, p. 331.
Batyraspis Apollonov et Tschugaeva, 1983, Lower Palaeozoic Stratigr. and Paleont. of Kazakhstan, p. 85.
Taoyuania, Peng, 1984, p. 336.

Type species: *Taoyuania xenisma* Liu, Lower Ordovician of Taoyuan, Hunan, by original designation.

Remarks: Peng (1984, p. 336) gave a new diagnosis of this based on the reexamination of Liu's type specimens housed in the Geological Museum of Hunan in Changsha, Hunan Province and discussed the relationship of it with other hystricurid genera, in the same time he added a new species, *Taoyuania nobilis* Peng, from the Upper Cambrian and Lower Ordovician of the Taoyuan-Cili area, NW Hunan to it. The genus *Batyraspis* Apollonov and Tschugaeva, 1983, typified by *Batyraspis inceptoris* Apoll. et Tschug. from the Lower Ordovician of the Batyrbay section in Malyi Karatou Range of Kazakstan is junior synonymous of *Taoyuania*. Thus this genus includes four species: 1. the type species, *Taoyuania xenisma* Liu, 2. *Taoyuania nobilis* Peng, 3. *Batyraspis inceptoris Apoll.* et *Tschug.* and 4. *Taoyuania*? *affinis* Liu.

Geologic and Geographic Distributions: Upper Cambrian to Lower Ordovician, NW Hunan, SE Guizhou and Kazakstan.

***Taoyuania xenisma* Liu**

(Pl. 9, Figs. 1–3)

1977 *Taoyuania xenisma* Liu, in Zhou, Liu, Meng and Sun in "Fossil Atlas of Central-South China" (1), p. 161, pl. 49, figs. 1–3

Remarks: The cranidia illustrated here agree in all details with the type specimens described by Liu in 1977. The pygidium (pl. 9, fig. 3) in the present collection referred to this species is subelliptical in outline, occupying about 3 / 4 the total pygidial length, the axis is subconical, rapidly tapered rearwards, a little broader than the pleural lobe in frontal portion, divided by four transverse furrows into three rings besides the half-ring and the subtriangular terminal piece. On the lateral sides of the first and second axial rings are paired eye-like elevations, these peculiar structures are unusual in other groups of trilobites. Two anterior pleural segments separated by a narrow interpleural furrow are present, the area behind the second segment is flattened and subtriangular in outline, the pleural furrow of the first segment is deep and wide, where that of the second segment is very shallow and gradually merges into the posterior portion of pleural field. The border is wide and flat; the doublure is marked by terrace lines parallel to the lateral margin. Behind the axis the space between the terrace lines is narrower than that on the lateral portion, and the number of these lines decreases too. The posterior margin is nearly straight or very slightly sinuous inwards behind the axis. This pygidium is distinguished from that of *Taoyuania nobilis* Peng (1984, p. 337, pl. 9, fig. 5) from NW Hunan by the broad, subconical instead of subcylindrical axis and by the narrow pleural lobes. So far as the cranidium of *T. nobilis* Peng (1984, p. 337, pl. 9, figs. 1–4, 6) is concerned, the glabella is shorter and is outwardly expanded in the posterior portion, the palpebral area is relatively wider and the surface of cranidium bears larger and fewer pustules than those of *xenisma*. In *Taoyuania inceptoris* (Apollonov et Tschugaeva) (1984, pp. 10–11, pl. 17, fig. 4; pl. 21, figs. 10, 13) the shape and proportion of glabella and the wide palpebral area are more similar to those of *nobilis* than to *xenisma*. The Kazakstanian species is differentiated by the number of surface granules, they are about 45 to 50 in *inceptoris*, 95 to 100 in *nobilis* and 190 to 200 in *xenisma*. The brim in front of glabella is very narrow (sag.) in *xenisma* and broader in *nobilis* and *inceptoris*. The anterior border furrow is almost straight (tr.) (except a narrow triangular projection of the anterior border) in *inceptoris*, but it is strongly sinuous, curving rearwards sagittally and forwards exsagittally in *nobilis*. *Taoyuania* (?) *affinis* Liu (Liu in Zhou et al., 1977, p. 161, pl. 49, fig. 4) is separated from other species of *Taoyuania* by its cylindrical glabella, the lacking of brim and by the densely granulated anterior border.

Occurrence: Lower Ordovician Guotang Formation, at Pu'an of Sandu County (78Fg-12, 82Fg-24).

Family Dokimocephelidae Kobayashi, 1935
Subfamily Wuhuiinae Shergold, 1980
Genus *Lorrettina* Shergold, 1972

***Lorrettina* sp.**

(Pl. 6, Fig. 8)

Only one badly weathered cranidium on hand is not sufficient for precise identification. This specimen is similar to the genus *Lorrettina* Shergold, 1975, in some features such as the shape of glabella and anterior border, the position and size of the palpebral lobe, the narrow palpebral areas, and the converging and outwardly curved anterior sections of facial sutures. It differs from *L. macrops* Shergold (1972, p. 68, pl. 17, figs. 1–4; text-fig. 19; 1972, p. 104, pl. 53, figs. 1–8; text-fig. 43), the type species of this genus, in some subordinate characters, for example, the longer glabella, the flattened brim, and the backwardly arched instead of slightly forwardly arched occipital furrow. *Jingxiania* Chien, 1974, with *Jingxiania beigongliensis* as the type species (Chien 1974, in Lu et al., p. 102, pl. 4, fig. 3; Qian, 1985, p. 142, pl. 2, fig. 2; pl. 5, figs. 5a-c, 6), also resembles the present cranidium, but the former has large palpebral lobe, narrow palpebral areas and straight anterior border furrow. *Wuhuia* Kobayashi, 1933, is another genus comparable with the present specimen, but the palpebral areas of the former are narrow, the anterior facial sutures are divergent instead of convergent and the anterior border is wide (sag. et exsag.). From the subgenus *Lorrettina* (*Puanella*) described in this paper, the present cranidium differs in the longer and narrower glabella, in the outline of occipital ring and in the absence of granules on the surface.

Occurrence: *Mictosaukia guizhouensis* Assemblage-Zone, late Upper Cambrian Maotian Formation at

Pu'an, Sandu County, SE Guizhou (82F8-10).

Subgenus *Puanella* Lu et T.R. Zhou (subgen. nov.)

Diagnosis: A subgenus of *Lorrettina* charaterized by straight or slightly backwardly arched anterior border furrow, longer (tr.) postero-lateral border and strongly pustulosed surfæce.

Type species: *Puanella intermedia* Lu et T. R. Zhou (subgen. et sp. nov.)

Discussion: The new subgenus has some characters similar to *Lorrettina* Shergold, 1972, and *Jingxiania* Chien, 1974, for example, the shape of glabella, the glabellar furrows, the occipital ring and the occipital furrow. In addition, *Puanella* agrees with *Lorrettina* in the size of palpebral lobe, the narrow (tr.) anterior border and the width of fixigena between the palpebral lobes. The chief differences are in *Puanella* the anterior border furrow is slightly arched backwards instead of arcuted forwards in *Lorrettina*, the anterior border is subrhombic to subsemielliptical in outline instead of subcrescentic, the postero-lateral border is as wide as the glabellar basal width (tr.) instead of about 1 / 2 the basal width of glabella, and in the strongly pustulosed instead of smooth cranidial surface. From *Jingxiania*, the new subgenus differs in the shorter palpebral lobe, the shorter (exsag.) postero-lateral limb, the comparatively wider (tr.) anterior border, the subparallel instead of divergent anterior facial sutures and in having coarse pustulosed surface. The general shape of the anterior border and the width (tr.) of the postero-lateral border are alike in *Puanella* and *Jingxiania*. So *Puanella* possesses some features common to *Lorrettina* and *Jingxiania* on the one hand, and some features differing from the latter two genera on the other hand. Thus the specific name *intermedia* for the type species of *Puanella* is proposed.

Puanella is allied to *Wuhuia* Kobayashi, 1933, with *Solenopleura belus* Walcott as the type species from the Fengshanian of Shandong in the pustulosed cranidial surface, the size and location of the palpebral lobes, the shape of occipital ring, the course of occipital furrow, the width of the fixigenae between eyes and in the width and length ratios of the postero-lateral borders and posterior limbs. *Wuhuia* is distinguishable from *Puanella* by the relatively longer glabella with 2 instead of 3 pairs of glabellar furrows, the shorter (sag. and exsag.) stripe-like anterior border and by the divergent anterior facial sutures.

Chalfontia Shergold, 1982, of the subfamily Wuhuiinae shows some similarities to *Puanella*, but the former has a relatively longer glabella with four pairs of faintly impressed glabellar furrows and non-pustulosed cranidial surface sculpture.

The presence or absence of a median node on occipital ring is varied among genera and species of the subfamily Wuhuiinae and it is regarded as one of specific characters, e.g. *Jingxiania* Chien, 1974, as it is considered here as a member of the subfamily Wuhuiinae, has a distinct median node in the species *J. tangcunensis* Chien (in Lu et al., 1974, p. 103, pl. 4, fig. 15), but no median nodes are present in the species *J. beigongliensis* Chien (loc. cit. p. 102, pl. 4, fig. 13) and *J. transversa* Chien (loc. cit. p. 104, pl. 4, fig. 16).

Geological and Geographical Distributions: Late Upper Cambrian, SE Guizhou.

Lorrettine (*Puanella*) *intermedia* Lu et T.R. Zhou (subgen. et sp. nov.)

(Pl. 6, Figs. 9, 10)

Description: Glabella short, with a length ratio of glabella to the cranidium about 7:12, moderately convex from side to side and to front, tapering forwards, broadly rounded anteriorly, and surrounded by wide, deeply incised dorsal furrows laterally and anteriorly. Three pairs of glabellar furrows: anterior pair very faint; posterior pair deeply impressed, sigmoid in outline, curving strongly backwards-inwards, reaching and joining the dorsal furrows laterally; median pair also deeply incised, short, gently curving inwards and slightly backwards. Occipital furrow broad and deep, gently arched forwards and bending antero-laterally so as to form a wavy course connecting to dorsal furrows. Occipital ring wide (sag.) in center, narrowing laterally, provided with a remarkable median node. Brim almost flat, or slightly depressed in front of glabella, gently convex towards both sides, about 1 / 8 the width (sag.) of cranidial length. Anterior border as wide (sag.) as the brim in center, narrowing rapidly laterally, with pointed extremities, convex, elevated above the brim, anterior margin strongly arched forwards and posterior margin slightly arched backwards, the length (tr.) is much shorter than that of the brim. Border furrow wide and shallow, nearly transverse but slightly arcuated rearwards at center. Palpebral lobes elevated above the cheek, delimited by well-defined palpebral furrows, situated at a space between and opposite the middle of the basal glabellar lobes and the middle of the antero-lateral lobes, occupying a little more than 1 / 2 the glabellar length. Fixigena between the palpebral lobes narrow, moderately convex, sloping down to the axial furrow. Posterior limb subtriangular, exsagittally short and transversely long. Postero-lateral furrows very wide and deep, widening gradually outwards. Postero-lateral border narrow (exsag.), convex, ridge-like, as wide (tr.) as the basal glabellar width. Ocular ridge extremely faint, only as a weak depression exists in a continuous line of the palpebral lobe. Anterior facial sutures curving roundedly outwards, expanding laterally across the convex preocular areas, then strongly bending inwards to restrict the width of the anterior border; posterior sutures running directly outwards and slightly backwards to cut the postero-lateral border at a distance from the dorsal furrows equal to the glabellar basal width. Surface of cranidium with all furrows bearing pustules of different sizes, some of them on the glabella are larger but irrugularly distributed.

Occurrence: The upper part of the late Upper Cambrian Maotian Formation, at Pu'an of Sandu county, SE Guizhou (78F26-1, F2-7).

Family Kingstoniidae Kobayashi, 1933

Genus *Shuizuia* Lu et T. R. Zhou (gen. nov.)

Diagnosis: Kingstoniid trilobite with moderately convex and broad cranidium. Glabella wider than fixigena, faintly defined by shallow and broad dorsal furrows posteriorly, effaced in front. Palpebral lobe small, located at the mid-length of cranidium. Anterior sections of facial sutures slightly converging forwards, roundedly cutting the antero-lateral corners of cranidium; posterior sections directed obliquely backwards and outwards. Postero-lateral border well defined by broad postero-lateral furrow. Thorax composed of 7 segments. Axis tapering gradually rearwards. Pleural segment with shallow and broad pleural furrow and bluntly pointed extremity. Pygidium semicircular, wider than long. Axis triangular, defined by shallow and broad dorsal furrows. Articulating facet remarkably defined by antero-lateral furrow. Pleural fields separated by 5-6 pairs of very faintly pleural furrows.

Discussion: This genus closely resembles *Wanwanaspis* Kobayashi, 1966, in the broad glabella, narrow fixigena and small palpebral lobes, but the latter lacks clearly defined dorsal furrows both on the cranidium and pygidium, thus the glabella is indicated only by the pits on the posterior margin, the posterior furrows are absent, the pygidial axis is undifferentiated from the pleural fields, and the articulating facets are not clearly shown. *Kingstonia* Walcott, 1966, is another genus similar to *Shuizuia*, but the former has a narrower glabella, wider fixigenae and a rim-like border. *Shuizuia* is somewhat similar to some illaenids with small palpebral lobes, such as *Octillaenus* Salter, 1867, or *Ectillaenus* Salter, 1867, but *Shuizuia* has only 7 thoracic segments and less well defined glabella.

This new genus is alike to *Parakoldinia* Rozova, 1960, and a number of species formerly referred to *Kingstonia*, *Koldinia*, *Wanwanaspis* and *Wanwanoglobus* from Kazakhstan, Sayan Altai Basin, S. Siberia, Kulyumbe River section of NW Siberia Platform, S. Korea and N China of Upper Cambrian to Tremadocian in age are assigned to *Parakoldinia* by Shergold and Sdzuy (1984, p. 99). *Parakoldinia* differs from *Shuizuia* in the subovoid glabella instead of being contracted at the mid-length, in the less prominent dorsal furrows both on the cranidium and pygidium, and in the relatively narrower (tr.) postero-lateral limb.

Type Species: *Shuizuia laevigata* Lu et T. R. Zhou (sp. nov.) The generic name is derived from a tribe named "Shui nationality" living in the Shuizu Autonomous Region of SE Guizhou.

Geological and Geographical Distributions: Late Upper Cambrian of SE Guizhou.

Shuizuia laevigata Lu et T. R. Zhou (gen. et sp. nov.)

(Pl. 6, Figs. 11, 14; pl. 7, Figs. 1–4)

Diagnosis: *Shuizuia* with broad cranidium and pygidium.

Description: Exoskeleton elongate elliptical in outline, with a length ratio of cephalon, thorax and pygidium about 1.3:1:1.4. Cranidium moderately convex, semicircular, with a length and width ratio about 1:1.3. Glabella poorly defined posteriorly, effaced anteriorly, with a width at base about 1 / 2 the maximum width (tr.) of the cranidium. Glabellar furrows absent. Dorsal furrows slightly constricted in the mid-length, running slightly divergent forwards, about 1 / 2–1 / 3 the sagittal length of the cranidium and obliterated anteriorly. Occipital ring narrow (sag.), separated from the glabella by a very shallow and broad occipital furrow. Fixigena as wide (tr.) as 1 / 3 of the glabellar width at the base. Palpebral lobes small, situated at the mid-length of cranidium. Anterior sections of the facial sutures subparallel, cutting the antero-lateral corners of cranidium in a rounded angle; posterior sections obliquely extending backwards to enclose the wide (exsag.) subtriangular posterior limbs. Hypostoma subquadrate in outline, with an entire posterior margin. Median body convex, subcircular, defined by median furrow which is very faint sagittally. Posterior lobe crescentic, convex. Macular depressions long and broad. Lateral border flat, separated from median body by broad border furrow. Anterior wing triangular, extending directly outwards. Thorax with 7 segments. Axis slightly tapering backwards, defined by deep axial furrows, wider (tr.) than pleural field, which widens backwards. Pleural segment bearing a rounded, oblique terminal end; pleural furrow very shallow and broad. Pygidium semicircular. Axis conical, narrow, defined by shallow axial furrows, reaching the inner margin of doublure, poorly divided into 5 or 6 rings. Pleural furrow almost obliterated. Doublure relatively wide.

Occurrence: *Mictosaukia guizhouensis* Assemblage-Zone of the Upper Cambrian Maotian Formation at Pu'an Sandu, SE Guizhou (78F26-1, 78F2-7, 82Fg-1, 82Fg-10).

Shuizuia incerta Lu et T. R. Zhou (gen. et sp. nov.)

(Pl. 7, Figs. 5–8)

Diagnosis: *Shuizuia* with comparatively narrower cranidium and pygidium.

Description: Cranidium semielliptical, rounded in front, with the length and width subequal. Glabella well defined in posterior portion by wide dorsal furrows which disappear at about 3 / 4 the length of cranidium, with a breadth about one-half the basal width of cranidium, convex in front and sloping down anteriorly. Palpebral lobe small, about 1 / 7 the length of cranidium, situated at the mid-length of cranidium. Occipital ring narrow (sag. and exsag.); occipital furrow very shallow and broad. Postero-lateral limb subtriangular, less than 1 / 2 the basal width of glabella. Anterior sections of facial sutures subparallel, roundedly cutting the antero-lateral corners of cranidium.

Pygidium semicircular. Axis a little narrower than pleura in front, tapering rapidly backwards, reaching inner margin of doublure, separated by shallow transverse furrows into 4 rings and a terminal piece. Pleural fields marked by 5 pairs of very shallow pleural furrows.

Remarks: This species differs from the type species, *Shuizuia laevigata*, in the relatively narrower cranidium, the deeper dorsal furrows and the wider pygidial axis.

Occurrence: *Mictosaukia guizhouensis* Assemblage-Zone of the Upper Cambrian Maotian Formation, at Pu'an of the Sandu County, SE Guizhou (82Fg-10).

Superfamily Leiostegiacea Bradley, 1925
Family Leiostegiidae Bradley, 1925
Subfamily Leiostegiinae Bradley, 1925
Genus *Leiostegium* Raymond, 1913, sensu Walcott, 1925
Subgenus *Leiostegium* Raymond, 1913, sensu Walcott, 1925
Leiostegium (Leiostegium) constrictum Peng

(Pl. 9, Fig. 4)

1984 *Leiostegium* (*Leiostegium*) *constrictum* Peng, Cambrian-Ordovician Boundary (1), p. 342, pl. 10, figs. 3–7, 8a.

Remarks: The cranidium illustrated here from SE Guizhou is identical in all details with the type specimens described by Peng from the latest Upper Cambrian and earliest Lower Ordovician of NW Hunan. The chief characters of this form differing from other species of the subgenus *Leiostegium* (*Leiostegium*) are confined to the constricted glabella, and the anterior border which is narrow and strongly convex in the middle portion and expanded on the lateral portion.

Occurrence: Lower Ordovician Guotang Formation, at Shizilu of Sandu (82Fg-65).

Leiostegium (Leiostegium) punctatum (Zhou et Zhang)

(Pl. 9, Figs. 5–6)

1978 *Leiostegium* (*Alloleiostegium*) *punctatum* Zhou et Zhang, p. 13, pl. 2, figs. 1–8

1984 *Leiostegium* (*Leiostegium*) *punctatum*, Peng, p. 344, pl. 10, figs. 9, 10.

1985 *Leiostegium* (*Leiostegium*) *punctatum*, Zhou et Zhang, p. 82, pl. 21, figs. 5, 6; pl. 25, fig. 11.

Remarks: This species is represented by only one cranidium in our collection. It agrees with the type specimens described by Zhou and Zhang from the Lower Ordovician Yehli Formation at Wushan, Lulong of Hebei in the gradually tapering glabella with four pairs of short and obsolete lateral glabellar furrows, the position and size of the palpebral lobes, the narrow and upturned anterior border and the densely punctuated cranidial surface. This species is the type species of the subgenus *Alloleiostegium* Zhou and Zhang,

1978. After a carefully examination of new material collected from North and Northeast China, Zhou and Zhang (1985, p. 81) considered *Alloleiostegium* as synonymous of *Leiostegium*.

Occurrence: Lower Ordovician Guotang Formation, at Shizilu (78F26-12).

Leiostegium (Leiostegium) sp.

(Pl. 9, Figs. 7–8, 9a; Pl. 15, Fig. lla)

Four imperfectly preserved cranidia, trapezoidal in outline, bear a few unusual large and numerous small granules on the surface. The glabella is subparallel-sided in anterior half of the cranidial length and slightly diverging rearwards to the posterior margin. There are three pairs of shallow glabellar furrows of which the anterior pair is horizontally extended, opposite the anterior end of palpebral lobes inwards to 1 / 3 the glabellar width; the second pair is slightly oblique inwards and backwards from the dorsal furrows about 2 / 5 the glabellar width; posterior ones strongly oblique and long, but not reaching the occipital furrow. The fixigena is about 1 / 3 the glabellar width between the palpebral area. Anterior border very narrow and upturned in front of glabella; anterior border furrow narrow in middle, widening and deepening laterally. Eye-ridge faintly defined. Occipital furrow deep and broad, gently arched posteriorly; occipital ring wide in middle, narrowing laterally.

The very coarse tubercles and abundant small granules on the surface of cranidium are peculiar, and may be regarded as a characteristic feature to distinguish this form from other species of *Leiostegium*.

The cranidia of this form may be compared favorably with that of *L.* (*L.*) *constricta* Peng (1983, pl. 2, figs. 5–6; 1984, p. 342, pl. 10, figs. 3–7, 8a) in the outline, in the shape and proportion of glabella, in the location of palpebral lobes and in the nature of anterior border, including a pair of pits which lie behind anterior border furrow close to the antero-lateral corners of the glabella. However, in the latter the palpebral lobe and the ocular ridge are poorly defined, the glabellar furrows appear to be entirely absent. The present cranidia are distinguished from *L.* (*L.*) *punctatum* (Zhou et Zhang, 1978, p. 13, pl. 2, figs. 1–8; Peng, 1984, p. 344, pl. 10, figs. 9–10) by having a wider (tr.) constricted glabella and shallower glabellar furrows. However, the present specimens are small in size. They may represent some young forms of certain species. Therefore, it would be better to leave the nomenclature over till some adult specimens are known.

Occurrence: Lower part of the Guotang Formation, Lower Ordovician, Pu'an of Sandu, Guizhou (82Fg-16, 82Fg-21, 82Fg-24, 78Fg-12).

Subfamily Pagodiinae Kobayashi, 1935

Genus *Pagodia* Walcott, 1905

Pagodia(?) sp.

(Pl. 9, Fig. 10)

A badly and imperfectly preserved cranidium is questionably assigned to *Pagodia*. It shows, as far as it can be seen, some similarities to *Pagodia depressa* (Walcott) (Walcott, 1913, p. 176, pl. 16, fig. 6; Kobayashi, 1935, p. 268) from the late Upper Cambrian Fengshan Formation of North China Faunal Province. The chief features identical with *depressa* are the outline of the cranidium and glabella, the relative size and location of the palpebral lobes, the relative width of the glabella and fixigena, the low convexity of the glabellar height and the length (sag.) of the frontal border, but the bad state of preservation does not permit a precious specific determination. However the occurrence of this genus together with trilobites of Southeast China Faunal Province reveals an important evidence of the mixed nature of the transitional characters in SE Guizhou.

Occurrence: *Mictosaukia guizhouensis* Assemblage-Zone of late Upper Cambrian Maotian Formation, at Pu'an of Sandu County, SE Guizhou (82Fg-9).

Superfamily Dikelocephalacea Miller, 1889

Family Saukiidae Ulrich and Resser, 1930

Genus *Eosaukia* Lu, 1954

1954 *Eosaukia* Lu, Acta Palaeont. Sinica, vol. 2, p. 143.

1957 *Eosaukia*, Kobayashi, Jour. Fac. Sci, Univ. Tokyo, vol. 10, pt. 3, p. 375.

Type species: *Eosaukia latilimbata* Lu, 1954, from the Upper Cambrian of Sandu (Santu), SE Guizhou, by original designation.

Remarks: *Ptychaspis walcotti* Mansuy (Mansuy, 1915, p. 22, pl. 2, figs. 6a–b; pl. 3, figs. 10a-z) from the late Upper Cambrian of Penkai, Dongquan and Changpoung in the borderland of SE Yunnan and N. Vietnam, and from the Fengshan Formation of N and NE China (Sun, 1924, p. 68, pl. 5, fig. 2; Kobayashi, 1931, p. 183, pl. 22, figs. 18, 19) was questionably referred to *Eosaukia* by Kobayashi (1956, p. 281, pl. 40, figs. 1–4) and subsequently assigned to *Calvinalla* by Kobayashi (1933a, p. 254; 1933b, p. 129), by Lu (1957, p. 285, pl. 148, fig. 5) and by Lu et al. (1965, p. 451, pl. 88, figs. 13-16). Shergold included this species in his genus *Mictosaukia* (Shergold, 1975, p. 143). Another new species described as *Eosaukia* (?) *buravasi* by Kobayashi (1957, p. 376, pl. 5, figs. 1-10, 13-20) from the Upper Cambrian of Tarutao Island, SW Thailand, is splitted by Shergold (1975, p. 143) into two forms and assigned to two different genera: *Mictosaukia* (Kobayashi, 1957, p. 375–376, pl. 5, figs. 1–5, 7–10) and *Lophosaukia* (?) (op. cited, pl. 5, figs. 6, 11, 13–20). The specimens referred by Shergold to *Mictosaukia buravasi* differ from *Eosaukia latilimbata* Lu in the shorter glabella, the forwardly arched instead of straight anterior glabellar margin, the more posteriorly located palpebral lobes and in the shorter (exsag.) postero-lateral limb. The characters are closer to those of *Mictosaukia perplexa* Shergold (1975, p. 144, pl. 24, figs. 1–10) than to those of *Eosaukia*. The spacimens described by Shergold as *Lophosaukia* (?) sp. seem hardly to be as a member of this genus, because the palpebral lobes are much smaller than those of *Lophosaukia* and they are located in front instead of behind the mid-length of glabella, and they are devoid of a triangular anterior cranidial border which is an important feature of the genus *Lophosaukia*. The generic assignment of them remains a question open to discuss.

Geological and Geographical Distributions: Late Upper Cambrian of SE Guizhou.

Eosaukia latilimbata Lu

(Pl.10, Figs. 1–3)

1954 *Eosaukia latilimbata* Lu, Acta Palaeont. Sinica. vol. 2, no. 2, p. 128, pl. 1, figs. 5, 6.
1957 *Eosaukia latilimbata*, Lu, Index Fossils of China, Invertebrates (3), p. 284, pl. 147, fig. 11 (refigured).
1965 *Eosaukia latilimbata*, Lu et al., Trilobites of China, p. 443, pl. 86, fig. 14 (refigured).
1978 *Eosaukia latilimbata*, Yin et Li, in Palaeont. Atlas of SW China, Guizhou (1), p. 515, pl. 171, fig. 7 (refigured).

Remarks: Two cranidia and one pygidium in the present collection agree in all details with the type specimen of this species. The anterior pits on the anterolateral corners of glabella are distinctly marked. The eye-ridges are represented by shallow depressions directed inwards and slightly forwards from anterior end of palpebral lobes to the dorsal furrows opposite the anterior pair of glabellar furrows. The granules on the cranidial surface are large and densely distributed. The occipital node is small and situated rear the posterior margin of the occipital ring. The pygidium is fusiform, with a ratio of the length and the greatest width about 1:2. The axis is as wide as the pleural lobe anteriorly and rapidly tapering rearward to a rounded terminal piece, and it separated by deep and broad ring furrows into 5–6 rings. The pleural lobes are triangular in outline, with 5–6 pairs of pleural furrows and interpleural grooves. The marginal border is gently convex and the border furrow is very shallow.

Occurrence: Late Upper Cambrian *Mictosaukia guizhouensis* Assemblage-Zone, Maotian Formation, at Pu'an of Sandu (78F2-6, 78F2-7, 82Fg-9)

Eosaukia rectangula Lu et T. R. Zhou (sp. nov.)

(Pl. 10, Figs. 4–10)

Diagnosis: An *Eosaukia* with elongated rectangular glabella, fine granulations and narrow pygidial axis.

Description: Carapace elongated ovate. Cephalon semicircular, slightly convex. Cranidium roughly subtrapezoidal in outline, with a ratio of length, anterior width (measured the widest width (tr.) between the lateral sides) and posterior width (measured between extremities of postero-lateral border) about 1.8: 1.3:2.3. Glabella elongated rectangular, straight in front, with a ratio of length and width about 3:2. Three pairs of glabellar furrows: posterior pair complete, connected with each other in a straight line about 1 / 3 the glabellar width, laterally the furrows stretched obliquely outwards at an angle of 50°–60° to the dorsal furrows; second and anterior pairs discontinuous, less oblique and weakly defined. Dorsal furrows subparallel, bearing a pair of deep pits at the antero-lateral corners of glabella. No brim. Anterior border furrow straight in front of glabella, diverging from the anterior pits towards the anterolateral extremities of the anterior border which is gently convex, slightly upturned and subtrapezoidal outlined, with broadly arched anterior margin and a length (sag.) about 1 / 3 of the glabellar length. Occipital furrow shallow, broad, straight, bending forwards and bifurcated on lateral sides into a pair of short, broad furrows running obliquely backwards and outwards. Occipital ring a little longer (sag.) in middle, posterior margin slightly arched rearwards. Palpebral lobes about 1 / 3 of the glabellar length, situated opposite the mid-length of glabella. Fixigena about 1 / 3 of the glabellar width, slightly upsloping to the eye. Preocular areas triangular. Posterior limb elongated (tr.) subtriangular. Postero-lateral border about 1 / 2 the length (sag.) of occipital ring, widening laterally to the narrow, rounded lateral extremities, separated from the posterior limb by shallow and narrow furrow. Anterior facial sutures diverging forwards at an angle about 30° to the cephalic median line, cutting the lateral sides of anterior border in an obtuse to right angle; posterior sutures longer, oblique, cutting the postero-lateral border at a distance from the dorsal furrow equal to 7 / 8 of the glabellar basal width. Librigena wide, broadest between the posterior end of eye and genal angle; lateral and posterior borders of medium width, producing a very long genal spine which extends to the eighth thoracic segment; genal angle less than 90°.

Thorax composed of eleven segments. Axis convex from side to side, nearly equal width (tr.) from first to third rings, then gradually and regularly tapering rearwards. The width from first to fourth rings equal to the pleuron, while the succeeding rings decreasing in their width, the last or the eleventh ring has a width about 1 / 1.5 of the pleural width of same segment. The rings decreasing very slowly backwards in their length, the ratio of the length and width of first three rings is about 1:4.3, and that of the eleventh ring is about 1:2.5. Dorsal furrows broad, deeply incised throughout. Pleural segment with deep diagonal pleural furrow not reaching the lateral margin, articulating facet more than 1 / 2 the pleural width (tr.), lateral end rounded, no pleural spine.

Pygidium subfusiform to elliptical, broadly rounded in front and in posterior margin, acutely rounded in both extremities, with a length and width ratio about 1:2.5. Axis long, narrower than pleural lobe in frontal portion, conical in outline, reaching the marginal border, convex from side to side, composed of 3–4 axial rings and a triangular terminal piece. Dorsal furrows deep in anterior two-thirds of their length, shallowing posteriorly. Pleural lobe consisting 3 segments. First segment deeply incised by wide pleural furrow of about 3 / 4 its length (tr.), third pleural furrow very short. Interpleural grooves present, as long as the pleural furrows. Marginal border narrow and flat. Marginal furrow obscure.

Remarks: This species is distinguished from the type species, *Eosaukia latilimbata* Lu, by the longer glabella, the narrow pygidial axis with only 3–4 instead of 5–6 axial rings, 3–4 instead of 5–6 pleural segments and by the fine and condensed granulas on the cranidial surface.

Occurrence: *Mictosaukia guizhouensis* Assemblage-Zone of the Upper Cambrian Maotian Formation and the Sandu Formation, at Pu'an and Shizilu of Sandu county (78F2-6, 78F2-7, 82Fg-57, 82Fg-9).

Genus *Mictosaukia* Shergold, 1975

Mictosaukia guizhouensis Lu et T. R. Zhou (sp. nov.)

(Pl. 11, Figs. 1–4)

Diagnosis: A *Mictosaukia* with comparatively longer palpebral lobes, wider fixigenae and coarsely granulated surface ornamentation.

Description: Glabella subrectangular, basal lobe wider than the elongated rectangular anterior main lobe, broadly rounded in front. Dorsal furrows wide, deeply incised, parallel in 4 / 5 of the length from front, expanded outward on the lateral sides of the basal lobe of glabella. Posterior glabellar furrows complete, deeply incised, straight in central part, bending forwards at 1 / 4 its length into the dorsal furrows; second pair shallow and oblique, visible at the midlength of the main lobe; anterior glabellar furrows very short and shallow. Anterior border uniform in width (sag.), slightly arched forwards, narrowing and pointing laterally. Anterior border furrow very deep and wide, curving in a broad convex line to meet the dorsal furrows and slanting backward to the antero-lateral corners of glabella. Occipital furrow broad (sag.) and deep, straight in middle, bending forwards on sides. Occipital ring straight and wider (sag.) in central portion, narrowing obliquely forwards laterally. Palpebral lobe large and long, about 3 / 4 of the glabellar length, situated behind the glabellar midlength, slightly curving outwards. Fixigena narrower than 1 / 3 the width of glabella in front of basal lobe. Ocular furrows prominent, deep and narrow, very slightly curving outwards. Posterior limb very narrow (exsag.), a little shorter than the basal width of glabella; postero-lateral furrow deep, narrow in the inner part, widening outwards. Posterolateral border very narrow, gradually widening laterally. Anterior sections of facial sutures subparallel, obtusely rounded to cut the anterior border; posterior sections running directly outwards and very slightly backwards to cut the postero-lateral border at a distance equal to the basal width of glabella. Surface crowded with very coarse pustules.

Remarks: The new species bears some features similar to those of the new subgenus *Mictosaukioidia* described below, e. g. the configuration and size of the anterior border, the narrow glabella and the coarsely granulated surface ornaments, but the palpelbral lobes of this species are longer and situated behind instead of at the midlength of glabella, and the elongate (tr.) instead of triangular postero-lateral limb. The shape and the ratio of the length and width of the cranidium as well as the glabella are varied among the specimens due to the compression of different directions, Pl. 11, figs. 1 and 2 showing lateral compression, and Pl. 11, figs. 3 and 4 showing longitudinal compression, but the proportions of different structures in such specimens are fundamentally unchanged.

Except the type species, *Mictosaukia orientalis* (Resser et Endo, in Endo) has smooth cranidial surface, many species of this genus bear granulated surface ornaments, but the size of granules are different, for examples, *M. luanhensis* Zhou et Zhang (1978, p. 8, pl. 1, figs. 18-21; 1985, p. 104, pl. 7, figs. 8,9; pl. 26, figs. 10-14; pl. 28, fig. 11), *M. maculata* Qian (1985, p. 147, pl. 5, fig. 8a; text-fig. 2), *M. walcotti* (Mansuy) (Mansuy, 1915, 1916), *M. bella* (Walcott) (Walcott, 1913, p. 180, pl. 17, fig. 9), *M. collisto* (Walcott) (Walcott, 1913, p. 183, pl. 16, figs. 14, 14a) have very coarse granules on the surface of cranidium resembling this species, wheras the granules of other species are finer and greater in number. *M. guizhouensis* differs from the group with coarse granutaled surface ornaments chiefly in the narrower glabella, straight and less rounded glabellar front, and in the subparallel anterior sections of facial sutures.

Occurrence: Late Upper Cambrain *Mictosaukia guizhouensis* Assemblage-Zone of the Maotian Formation at Pu'an of Sandu (79F1-76=82Fg-1, 82Fg-9, 82Fg-10, 78F2-5, 78F2-6, 78F2-7).

Subgenus *Mictosaukioidia* Lu et T. R. Zhou (subgen. nov.)

Type species: *Mictosaukia* (*Mictosaukioidia*) *transita* Lu et T. R. Zhou (sp. nov.)

Diagnosis: A subgenus of *Mictosaukia* resembling the typical *Mictosaukia* in having a narrow (tr.) glabella broadly rounded in front and a similar anterior border, differing in the shorter palpebral lobes, the longer (exsag.) postero-lateral limb, the subparallel instead of divergent anterior sections of facial sutures, the long and stout genal spine and in the lateral and posterior marginal furrows continuously extending into the genal spine. The last feature together with the size and position of palpebral lobe and the non-spinose occipital ring resembling *Tellerina*, but differing from the latter by the wider (tr.) fixigenae, the relatively narrower anterior border and by the anterior sections of facial sutures being convex laterally or subparallel instead of strongly divergent in *Tellerina*.

Other species: "*Calvinella*" *elongata* Xiang (Xiang in Sun et Xiang, 1979, p. 10, pl. 4, figs. 11–13) from the late Upper Cambrian Baoshan Formation of Baoshan, western Yunnan may be assigned to *Mictosaukia* (*Mictosaukioidia*).

Remarks: This subgenus possesses some characters similar to those of and is morphologically transitional between *Mictosaukia* and *Tellerina*, so the specific name is given for the type species. The strongly and coarsely pustulated surface ornamentation of the subgenus also appears in some species of *Mictosaukia*, such as *M. luanheensis* Zhou et Zhang (1978, p. 8, pl. 1, figs. 18-20; 1985, p. 104, pl. 7, figs. 8, 9; pl. 26, figs. 10–14; pl. 28, fig. 11), *M. striata* (Resser et Endo) (Resser et Endo, in Endo et Resser, 1937, p. 189, pl. 56, figs. 24, 25; Peng, 1984, p. 350, pl. 8, figs. 1–3, 4a; Zhou et Zhang, 1985, p. 104, pl. 1, fig. 12; pl. 4, fig. 13; pl. 7, figs. 1–3; pl. 11, fig. 4; pl. 12, figs. 4–7; pl. 28, figs. 9, 10; Qian, 1985a, p. 81, pl. 16, figs. 3–6; pl. 17, figs. 1–4) and *M. maculata* Qian (1985b, p. 147, pl. 5, fig. 8; text-fig. 2), but the surface of cranidiam is smooth in the type species, *Mictosaukia orientalis* (Resser et Endo in Endo). From *Eosaukia* Lu, this subgenus can be differentiated by the arched instead of straight preglabellar furrow, the wide fixigenae between the palpebral lobes, the outwardly convex instead of strong divergent anterior sections of facial sutures and by the very deep and broad posterior pair of glabellar furrows and occipital furrow.

Geological and Geographical Distributions: Late Upper Cambrian of SE Guizhou and western Yunnan.

Mictosaukia (*Mictosaukioidia*) *transita* Lu et T.R. Zhou (subgen. et sp. nov.)

(Pl. 11, Figs. 5–10)

Description: Glabella subrectangular, broadly rounded in front, convex from side to side, anterior main lobe subsquare, parallel-sided, basal lobe expanded laterally, the length ratio of anterior main lobe to the posterior lobe is about 1:3, the ratio of the length of glabella, the width of basal lobe and the main lobe is 4:3.6:3. Anterior main lobe provided with two pairs of large, strongly convex nodes on the antero-lateral sides, and with two pairs of lateral glabellar furrows; the anterior pair is very weak, hardly discernible, the second pair is longer, faintly defined, shallowing and narrowing obliquely inwards, missing at 1 / 3 of the glabellar width; posterior ones complete, broad and deeply incised, straight in center, bending forwards laterally at 1 / 4 its length. Dorsal furrows deep and broad. Occipital ring uniform in width (sag.) in the central portion, narrowing and bending obliquely forwards from 1/4 the total length (tr.) of the ring, provided with a large median node in the posterior margin. Occipital furrow broad and deep, very slightly arched forwards in central portion, bending obliquely forwards laterally at 1 / 4 its length (tr.) into the dorsal furrows. Anterior border flattened at top, uniform in width, limited on both sides by oblique anterior marginal furrow which is very deep and broad in front of glabella. Fixigena wide, about 1 / 3 of the glabellar width between the palpebral lobes, flattenedly convex, gently upsloping from dorsal furrows. Posterior limb triangular, separated from the narrow postero-lateral border by very broad and deep postero-lateral furrow which is twice the width (tr.) of the occipital furrow. Surface of cranidium densely covered with pustules of median to large size.

Remarks: The species differs from *Mictosaukia guizhouensis* (sp. nov.) described in this paper in the shorter glabella with subsquare instead of elongated rectangular anterior main lobe, the wider fixigenae, the longer (exsag.) posterior limb and in the smaller palpebral lobes which are located at the midlength instead of behind the midlength of glabella. "*Calvinalla*" *elongata* Xiang (Xiang in Sun et Xiang, 1979, p. 10, pl. 4, figs. 11–13) from western Yunnan is here assigned to *Mictosaukia* (*Mictosaukioidia*). The Yunnanese species is similar to *transita* in the elongated rectangular glabella with deep and complete posterior glabellar furrows, the size of palpebral lobes, the triangular postero-lateral limb and in the wide fixigenae, but it differs from the latter in the longer frontal lobe of glabella and in the thick and broad (sag. and exsag.) anterior border.

Occurrence: *Mictosaukia guizhouensis* Assemblage-Zone of the late Upper Cambrian Maotian Formation, at Pu'an, Sandu County (Fl-79-87=82Fg-1, 82Fg-9, 82Fg-10, 78F2-6, 78F2-7).

Genus *Linguisaukia* Peng, 1984

Type species: *Linguisaukia spinata* Peng, 1984, from the Zhuitun Formation, Upper Cambrian, at Qianzhou, Jishou County, western Hunan, by original designation.

Supplements to the Generic Diagnosis: Since no complete dorsal shield, thorax and hypostoma of this genus have been known, the discovery of new material from SE Guizhou gives detailed features of them and a supplementary note of the generic diagnosis is given as follows.

Heteropygous Saukiid trilobite. Carapace elongated ovate, with a length ratio of cephalon, thorax and pygidium about 1:1.5:0.7. Hypostoma elongated elliptical excluding the triangular anterolateral wings, posterior furrows oblique, straight and meeting in center. Thorax composed of eleven segments; axis tapering regularly and gradually rearwards, each ring provided with a median node on the posterior margin; articulating geniculation of pleural segment at a position in the frontal margin about 1 / 4 the pleural width (tr.) from the deeply incised dorsal furrows, pleural extremities rounded, devoid of sharp spines. Surface of carapace ornamented with pustules.

Linguisaukia spinata Peng
(Pl. 12, Figs. 1-7; Pl. 13, Figs. 1–6)

1984 *Lingusaukia spinata* Peng, p. 347, pl. 6, figs. 4–8.

Supplementary Notes on the Specific Characters: The new material of this species collected from SE Guizhou contains three dorsal shields lacking of librigenae (pl. 12, figs. 4, 5; pl. 13, fig. 2), one twist carapace (pl. 13, fig. 1) in which the cephalon is overturned to show the ventral view of cephalon and the dorsal view of hypostoma, five cranidia, two librigenae and one hypostoma. They attribute some details of the cephalic doublure, the sutures of the ventral side, the position of hypostoma, etc.

Cephalon semicircular, with a small, transverse rectangular projection forming by cranidial tongue and anterior extension of librigena. Genal spines of the small specimen (pl. 12, fig. 1; pl. 13, fig. 5) very long, the length ratio of the carapace and the genal spines is about 3:5 (pl. 12, fig. 1). They extend in an outward arc backwards, the distance between their posterior ends equal to 1 / 2 the breadth of cephalon. The doublure of the librigena is narrow, almost uniform in width from postero-lateral border through genal angle forwards and continuously extends to the cephalic tongue, separated with each other by a median suture underneath the central part of the tongue. The hypostoma of the small specimen differs from the large hypostoma (pl. 12, fig. 6) in having a broad postero-lateral and posterior border which bear 7 pairs of short spines. The large hypostoma (pl. 12, fig. 6) is subelliptical in shape, central body convex, subpentagonal, rounded in front, a pair of straight, deep, narrow lateral furrows separated the central body from the posterior lobe, and meet with each other in the center at an obtuse angle not more than 100°; posterior lobe convex, narrowing antero-laterally and ending in a sharp acute angle about 30°–40° opposite the posterior end of antero-lateral wings which are roughly elongated biequilateral triangular in outline; marginal border narrow, smooth, non-dentition. Thorax composed of eleven segments, gently convex from side to side (pl. 12, figs. 4, 5; pl. 13, fig. 2), more convex in the small specimen (pl. 12, fig. 1), gradually and regularly tapering rearwards, the width (tr.) of the first ring a little narrower than the pleural width (tr.), the width (tr.) of the last or eleventh ring is about 1 / 2 the pleural width, each ring bears a median node on the posterior margin; pleural segment gently upsloping from the dorsal furrow at about 1 / 4 the pleural length, then geniculating with the beginning of articulating facet outwards and gently sloping down and ending in a rounded extremity. The thoracic pleural extremities in the small specimen (pl. 12, fig. 1) are more sharply angulated, but not enough to form pleural spine.

For other specific features, the reader may refer to the description of the original specimens of the type species given by S. C. Peng.

Occurrence: The Upper Cambrian Maotian Formation, at Pu'an of Sandu, SE Guizhou (79F-76, 7-11, 78F2-6).

Linguisaukia affinis Peng

(Pl. 13, Fig. 7)

1983 *Linguisaukia affinis* Peng, pl. 1, fig. 8.
1984 *Linguisaukia affinis* Peng, p. 348, pl. 7, figs. 3–7.

Description: An external mold of pygidium is assigned to *Linguisaukia affinis*, however no cephalon and other parts have been found in our collection. The pygidium is broad fusiform, with a length of about 12 mm and a width of 25 mm. Axis conical, narrow, convex from side to side, about 6 mm wide in front, narrowing rapidly rearwards, reaching and pointing to the posterior margin, composed of 5 axial rings divided by slightly wavy, transverse furrows. Pleural field roughly equal-laterally triangular in outline, consisting of 4 pleural segments, pleural furrows broad, shallowing outwards and backwards towards the marginal border; interpleural grooves faintly defined; articulating furrow deep and broad, articulating segment about half the width (exsag.) of the first pleural segment. Border wide, deeply sinuated at posterior margin. Surface marked by numerous fine terrace lines parallel to the postero-lateral margin. Border furrow indistinct.

Remarks: This species is distinguished from *Linguisaukia spinata* Peng (1984, pl.6 fig. 5), the type species, by the narrower pygidial axis, the narrower and deeper marginal sinus and by the broader pleural field.

Occurrence: Late Upper Cambrian Sandu Formation, Shizilu of Sandu county, SE Guizhou (82Fg-1, 82Fg-57, Fl-78-88 (=82Fg-8)).

Superfamily Remopleuridacea Hawle et Corda, 1874
Family Remopleurididae Hawle et Corda, 1874
Subfamily Macropyginae Kobayashi, 1937
Genus *Macropyge* Stubblefield in Stubblefield et Bulman, 1927
Subgenus *Promacropyge* Lu, 1964
Macropyge (Promacropyge) carinata Lu

(Pl. 14, Figs. 1–2)

1964 *Promacropyge carinata* Lu, in Index Fossils of South China, p. 34, pl. 6, figs. 2, 3.
1965 *Promacropyge carinata*, Lu et al., Trilobites of China, part. 2, p. 464, pl. 91, figs. 22, 23 (refigured)
1978 *Promacropyge carinata*, Yin and Li, in Palaeont Atlas. SW China, Guizhou (I). p. 522, pl. 173, fig. 9 (refigured).

Remarks: The new material adds some features which have not been described previously. The subparallel-sided glabella (pl. 14, fig. 2) has prominent, deep, subcircular posterior pair of glabellar furrows which become very narrow and shallow, and extending directly outwards to the dorsal furrows: the middle pair also very shallow and narrow, extending slightly oblique backward from the dorsal furrows almost to the axial line; anterior pair very short, inconspicuous, running horizontally inwards opposite the anterior end of palpebral lobes. Bacculae absent. Occipital furrow broadly sinuous in middle, stretching slightly obique forwards to the dorsal furrows opposite the posterior ends of palpebral lobes. Anterior sections of facial sutures slightly diverging forwards from the anterior ends of palpebral lobes, cutting the lateral sides of anterior border in a rounded angle. Surface of glabella (pl. 14, fig. 2), doublures of librigenae and thorax with fine striations parallel to their lateral margins, that of pygidium with horizontal striations (pl. 14, fig. 1).

Occurrence: Late Upper Cambrian Sandu Formation at Shizilu of Sandu county (82Fg-57) and Lower Ordovician (Tremadocian) Guotang Formation at Pu'an of Sandu county (82Fg-24).

Macropyge (Promacropyge) flexa Peng

(Pl. 14, Figs. 3–5)

1984 *Macropyge* (*Promacropyge*) *flexa* Peng, Cambrian-Ordovician Boundary of China (I), p. 356, pl. 2, fig. 6.

Description: The type specimen of this species described by Peng is a cranidium from the late Upper Cambrian Shengjiawan Formation of Cili county, NW Hunan. The discovery of a nearly complete dorsal shield and detached cranidium and pygidium from Sandu of SE Guizhou shares some details not known previously.

Carapace gently convex. Cranidium subsemicircular, with a ratio of length and basal width about 2:3. Glabella bell-shaped, with ill-defined sausage-shaped bacculae located near its postero-lateral corners. Three pairs of very shallow and long glabellar furrows: preoccipital pair subcircular in shape, depressed in middle, narrowing ontwards and slightly forwards to meet the dorsal furrows, and directed inwards and slightly backwards near the axial line; middle pair parallel to and almost as long as the posterior ones; anterior pair short, running inwards at the narrowest, sinuous antero-lateral portion of glabella opposite the anterior end of palpebral lobes. Occipital furrow deep, broadly sinuous in middle and arched forwards laterally to form a wavy line; occipital ring narrower in middle, gradually widening outwards, posterior margin straight. Anterior border narrow, upturned, flattenedly convex, separated from the glabella by deep and broad, gently arched marginal furrow. Palpebral lobe flat and thick, with a ratio of its length to the glabellar length about 1:1.3; palpebral area semicircular, flat, a little more than 1 / 3 the glabellar width between palpebral lobes. Postero-lateral limb narrow (exsag.), nearly uniform in width. Posterior border wider (exsag.) than postero-lateral limb, less than 1 / 2 the width (tr.) of occipital ring. Anterior sections of facial sutures slightly diverging forwards for a short distance, and then curved inwards and forwards to cut the anterior border; posterior section running almost horizontally outwards from the posterior end of palpebral lobe, and abruptly turning obliquely to cut the posterior border.

Thorax composed of 9 segments, narrowing rapidly rearwards, the width (except the pleural spines) ratio of the first and 9th segments is about 5:1. Axis gently convex, wider than the pleural lobe, about equal width (tr.) from the first to third rings, then rapidly narrowing from the fourth onwards posteriorly, confined by deep dorsal furrows laterally, the width ratio of the first ring to the ninth ring is about 2:1. Pleural segments flat, running outwards and slightly obliqne backwards; pleural furrow broad, stretching obliquely outwards and extending to the extremities of the pleural spines, following the curvature of the latter, the length of the pleural spines decreasing from the first pleura rearwards.

Pygidium elongated subconical, posterior marginal spine about 2.8 times the length of axis behind the latter. Axis conical, strongly elevated above the pleural lobes, consisting of 4–5 rings and a rounded terminal piece, first and second ring furrows distinct, others shallow and obsolete. Pleural field about 2 / 3 the width of axis, gently convex, with first two pleurae remarkably defined by broad and shallow pleural furrows not reaching the lateral margin. Posterior axial ridge narrow and flat.

Remarks: The constricted antero-lateral portion of glabella, the presence of ill-defined bacculae and the wider thoracic axis are the characters to distinguish this species from *Macropyge* (*Promacropyge*) *carinata*, the type species of the subgenus. *Macropyge* (*Promacropyge*) *pingius* Peng (1984, p. 357, pl. 2, figs. 1–5) from the late Upper Cambrian *Lotagnostus punctatus-Hedinaspis regalis* Assemblage-Zone of NW Hunan differs from *M.* (*P*). *flexa* in the obsolete glabellar furrows, and in the small number of the pygidial axial rings which are 4–5 in *flexa* but 2–3 in *pingius*.

Occurrence: Late Upper Cambrian Sandu Formation at Shizilu of Sandu County, SE Guizhou (82Fg-

51, 82Fg-57, 78-F27-14).

Macropyge (Promacropyge) ciliensis Peng

(Pl. 13, Figs. 8–9)

1984 *Macropyge* (*Promacropyge*) *ciliensis* Peng, Cambrian-Ordovician Boundary of China (I). p.358, pl. 2, figs. 7–10.

Remarks: Two cranidia in the present collection are available for examination. They agree in all details with *M*. (*P*). *ciliensis* described by Peng from the late Upper Cambrian *Lotagnostus punctatus-Hedinaspis regalis* Assemblage-Zone of NW Hunan, and differs from *M*. (*P*.)*carinata* Lu, *M*.(*P*.) *flexa* Peng and *M*. (*P*.) *pingius* Peng in the presence of a prominent preglabellar field and a wider palpebral area. The cranidium depicted on pl. 13, fig. 8 is suffered from vertical compression, so the glabella and occipital ring are pressed into irregular wrinkles.

Occurrence: Early Ordovician (Tremadocian) Guotang Formation at Pu'an of Sandu County, SE Guizou (82Fg-14, 78Fg-12).

Superfamily Asaphacea Burmeister, 1843
Family Asaphidae Burmeister, 1843
Subfamily Isotelinae Angelin, 1854
Genus *Golasaphus* Shergold, 1972

Golasaphus carinatus Lu et T. R. Zhou (sp. nov)

(Pl. 14, Figs. 6–10)

Description: Cranidium subrectangular, angulate anteriorly. Glabella gently convex, constricted near to palpebral lobes, rounded anteriorly, with a faint median keel and a marked axial node situated about 1 / 3 the way along the axial line from the rear. Occipital furrow shallow, separated a very narrow occipital band from glabella. A pair of low occipital lobes on the postero-lateral flanks of glabella in front of occipital band is traceable (pl. 14, fig. 6). Anterior border furrow very broad (sag.) and shallow. Anterior border slightly upturned, and slightly elevated above the preglabellar field, pointed anteriorly. Marginal furrow (= preglabellar furrow) wide and extremely shallow. Exsagittally the marginal furrow is separated from the dorsal furrows by a pair of convex subtriangular ridges on either side of the glabella. Fixigena narrow (tr.). Palpebral lobe semicircular, situated close to and at the mid-length of glabella. Anterior sections of facial sutures running forwards and outwards at an angle about 25° to the cranidial axis from the anterior ends of palpebral lobes, then curving inwards in a round angle across anterior border to meet with each other sagittally. Posterior sections sigmoidal, enclosing a wide (tr.) roughly subrectangular to subtriangular postero-lateral limb. Posterior border furrow deep, extending horizontally outwards. Posterior border equally-wide, moderately convex.

Pygidium semicircular. Axis highly convex, occupying 2 / 3 the total pygidial length, consisting of 5 rings and a terminal piece. Pleural field wider than axis. Pleural furrows very shallow, almost indiscernible. Pygidial border furrow broad and widening posteriorly, with concentric lines on doublure.

Remarks: This species resembles the type species, *G. momedahensis* Shergold (1971, pp. 75-80, pl. 10, figs. 1–8; pl. 11, figs. 1–9; text-figs. 22–25) from the Upper Cambrian of western Queensland, but it is differentiated from the latter by having a more constricted glabella, an even glabellar front and by a more anteriorly situated glabellar node. The pygidium of the new species is relatively longer (sag.) and its axis appears to be more convex. It can be easily distinguished from *Golasaphus simus* Shergold (1975, p. 214, pl. 55, figs. 1-8; text-fig. 71) in the comparatively wider (sag.) preglabellar field, the constricted glabella, the more anteriorly position of glabellar median node and in wider pygidium. *Golasaphus triqustrus* Shergold (1975, pp. 212-214, pl. 56, figs. 1-9; text-fig. 70) differs from the new species in the longer, drawn-out and triangular preglabellar field, strongly forwardly divergent anterior facial sutunes, the more posteriorly position of the axial node, and in the triangular outlined pygidium with broader (tr.) axis.

Occurrence: Lower part of the Goutang Formation, Lower Ordovician; Pu'an of Sandu, Guizhou (82Fg-24).

Subfamily Niobinae Jaanusson, 1959
Genus *Niobella* Reed, 1931

Niobella yangjiawanensis Chien
(Pl. 14, Fig. 11)

1961 *Niobella yangjiawanensis* Chien, Acta Palaeont. Sinica, vol. 9, no. 2, p. 105, pl. 3, figs. 1, 2; text-fig. 4.
1965 *Niobella yangjiawanensis*, Lu et al., Trilobites of China, (2), p. 502, pl. 102, fig. 3 (refigured).
1978 *Niobella yangjiawanensis*, Yin et Li, Palaeont. Atlas of SW China, Guizhou (1), p. 533, pl. 176, fig. 8 (refigured).

Remarks: A weathered cranidium with cylindrical glabella slightly contracted at middle. Occipital furrow faintly defined. Fixgena about 1 / 4 the basal glabellar width. Border slightly concave, about 1 / 4 the length of glabella, ogival anteriorly. Palpebral lobe medium-sized, situated at the mid-length of glabella. Postero-lateral limb nearly uniform in breadth (exsag.), separated from the posterior limb by broad postero-lateral furrow of considerable depth.

Occurrence: Late Upper Cambrian Sandu Formation, Shizilu of Sandu county, SE Guizhou (82Fg-57).

Family Nileidae Angelin, 1854
Genus *Troedssonia* Poletaeva, 1960

Comments: *Troedssonia*, type species *Symphysurus* (?) *wimani* Troedsson, 1937, was differentiated from *Symphysurus* by Poletaeva chiefly by the glabella which widens rapidly to the frontal portion, by the comparatively smaller palpebral lobes situated at the midlength of cranidium and by the posterior sections of facial sutures which are almost connected with the posterior end of palpebral lobes. Besides the type species, the following European and Asian species are assigned to this genus: *Symphysurus incipiens* Brögger, 1882; *S. breviceps* Angelin, 1878; *S. angustatus* Sars et Boeck, 1888; *S. fungiformis* Fedjanina, 1961; *S.* (*Troedssonia*) *elongata* Liu, 1977 (=*S.* (*T.*) *isoletus* Liu, 1977) (Peng, 1984, p. 368, pl. 13, figs. 5, 6); *S.* (*Troedssonia*) *lubrica* Lu et Chien, 1978 (Lu et Chien, 1978, p. 543, pl. 178, fig. 7; Lu et Qian, 1983, p. 67, pl. 11, figs. 3–5); *S.* (*Troedssonia*) *yanshanensis* Kuo et Duan, 1978 (p. 455, pl. 3, figs. 13–15)); *S.* (*Troedssonia*) *sulcatus* Kuo et Duan, 1978 (p. 456, pl. 3, figs. 11-12); *Troedssonia hunjiangensis* Kuo et Duan, 1982 (Kuo et al., 1982, p. 23, pl. 2, figs. 18–21; Zhou et Zhang, 1984, p. 135, pl. 13, figs. 8–11; Chen et al., 1985, p. 82, pl. 19, fig. 7b, non pl. 18, fig. 10; pl. 19, figs. 4–7a; pl. 20, figs. 1, 2, 3a, 4; Qian, 1986, p. 295, pl. 78, figs. 1–7a, 8; text-fig. 114); *Troedssonia laevigata* Kuo et Duan, 1982 (Kuo et Duan, 1982, p. 23, pl. 2, figs. 22, 23; Zhou et Zhang, 1984, p. 136, pl. 8, figs. 1–4; pl. 14, figs. 9–13; pl. 6, figs. 1–5; Chen et al., 1985, p. 82, pl. 18, fig. 10; pl. 19, figs. 4–7, non 7b; pl. 20, figs. 1, 2, 3b, 4, 5c, Qian 1986, p. 297, pl. 79, figs; 1–8; text-fig. 115); *Troedssonia subquadrata* Peng, 1984 (p. 369, pl. 12, figs. 8a, b); *Troedssonia longiforma* Peng, 1984 (p. 370, pl. 8, fig. 4b; pl. 12, fig. 9); and *Symphysurus* (*Troedssonia*) *yumenensis* Lu, Zhou et Zhou, 1986, (p. 110, pl.12, figs. 8–10). Because *Symphysurus breviceps* and *Symphysurus angustatus* possessed small posterior limbs, proportionately large palpebral lobes and a stout and more regularly expanded glabella, Zhou et Zhang (1984, p. 134) retained these two species in the genus *Symphysurus* (s. s.). Among the Chinese species, *S.* (*Troedssonia*) *hunjiangensis* Kuo et Duan, *S.* (*Troedssonia*) *laevigata* Kuo et Duan and the type species of *Troedssonia*, *Symphysurus* (?) *wimani* Troedsson, were transferred to the genus *Platypeltoides* Pribyl in Prantl and Pribyl, 1949, by Qian (1986, p. 294). In regard to whether *Troedssonia* is synonymous of *Platypeltoides*, Peng (1984, p. 367) is the opinion that the former is not congeneric with the latter, because the latter bears only 7 thoracic segments, while the former possesses a thorax composed of 8 segments. His opinion is cited in this paper. Except the species *Symphysurus* (?) *wimani* bearing 8 thoracic segments (Peng, 1984, p, 367, pl. 14, fig. 9), *Symphysurus* (*Troedssonia*) *lubricus* Lu et Chien (1978, p. 543, pl. 178, fig. 7; 1983, p. 67, pl. 11, figs. 3–5) also possesses 8 thoracic segments. The constancy of 8 thoracic segments in *Troedssonia* and 7 in *Platypeltoides* forms a basis at first to distinguish them into two separate genera. Moreover, *Troedssonia* as typified by *wimani* differs from *Platypeltoides* as typified by *croftii* Callaway, 1877, in the larger palpebral lobes and in the longer (exsag.) postero-lateral limbs. The hypostoma of both genera are different, the postero-lateral margin of the hypostoma of *Platypeltoides croftii* is rounded (Owens et al., 1982, pl. 3, c), while that of *Troedssonia lubrica* provides a pair of small triangular or spinose wings (Lu et Chien, 1983, pl. 11, fig. 3). The Upper Cambrian species *Troedssonia hunjiangensis* Kuo et Duan and *Troedssonia laevigata* Kuo et Duan of the *Richardsonella-Platypeltoides* Assemblage-Zone from the Hunjiang region of Jilin, NE China are excluded from *Troedssonia* and subsequently transferred to the genus *Platypeltoides* by Qian (1986, p. 260). These species may be distinguished from the typical from, *P.*

croftii, by the asaphid-type anterior sections of facial sutures, by the doublure which is connected with the librigenae instead of being free from the librigenae as shown by Owens et al. (1982, pl. 2, k; pl. 3, b), by the possession of an ogival arched anterior border, by the less expanded anterior end of glabella and by the almost effaced dorsal furrows. The differences, especially the first two are important and may be considered as generic or subgeneric level to exclude *hunjiangensis* and *laevigata* from *Platypeltoides* as pointed out by Qian (1986, p. 294). Other Chinese species of *Troedssonia* bear characters neither agreeable with those of *Platypeltoides* nor with those of *hunjiangensis* and *laevigata*. A new generic name *Hunjiangites* is here proposed for these two species with *Symphysurus* (*Troedssonia*) *hunjiangensis* Kuo et Duan, 1982, as the type species, and is provisionally placed into the Subfamily Isotelinae Angelin, 1854, of the Family Asaphidae Burmeister, 1843.

Troedssonia wimani (Troedsson)

(Pl. 15, Figs. 1–4)

1937 *Symphysurus* (?) *wimani* Troedsson, Palaeont. Sinica, n.s. (B) 2, Whole Ser. 106, p. 44, pl. 4, figs. 3–4, 5, 6(?)
1960 *Symphysurus* (*Troedssonia*) *wimani*, Poletaeva in Khalfin, L. L. (ed), 1960, Trudy SNIIGGIMS, vol. 19, p. 418.
1965 *Symphysurus* (*Troedssonia*) *wimani*, Lu et al., Trilobites of China, 1, p. 532, pl. 3, figs. 1-4 (refigured).
1980 *Symphysurus* (*Troedssonia*) *wimani*, Lu et Lin, Acta Palaeont. Sinica, 19(2), p. 128, pl. 3, figs. 3, 4.
1984 *Symphysurus* (*Troedssonia*) *wimani*, Lu and Lin, Cambrian-Ordovician Boundary (1), p. 116, pl. 15, figs. 4-13.
1984 *Platypeltoides wimani*, Apollonov and Chugaeva, pp. 10, 11, pl. 16, figs. 4, 8, 12; pl. 18, figs. 12, 14–19; pl. 21, figs. 8,9.
1984 *Troedssonia wimani*, Peng, Cambrian-Ordovician Boundary (1), p. 367, pl. 10, fig. 8b; pl. 13, figs. 3, 4; pl. 14, fig. 9.

Remarks: This species has a wide-spread geographic distribution in the Lower Ordovician of Eastern and Central Asia, and it occurred with typical trilobites of the Southeast China Faunal Province, North China Faunal Province and the transitional regions between them. Therefore it is usually used as a guide fossil for the correlation of different faunal provinces. In our collection, a small pygidium with a thoracic segment attached measuring 5 mm wide and 3 mm long (pl. 15, fig. 4) is in association with the large specimens and is regarded as a young form of this species.

Occurrence: Early Lower Ordovician Guotang Formation (Tremadocian), at Shizilu and Pu'an of Sandu County, SE Guizhou (82Fg-64, 82Fg-24).

Troedssonia elongata Liu

(Pl. 14, Fig. 12)

1977 *Symphysurus* (*Troedssonia*) *elongata* Liu, in Zhou, Liu, Meng and Sun, p. 223, pl. 67, fig. 1.
1977 *Symphysurus* (*Troedssonia*) *isotelus* Liu, Ibid, p. 223, pl. 67, fig. 2.
1982 *Symphysurus* (*Troedssonia*) *elongata*, Liu p. 333, pl. 224, fig. 15 (refigured).
1982 *Symphysurus* (*Troedssonia*) *isotelus*, Liu, p. 333, pl. 224, fig. 16 (refigured).
1984 *Troedssonia elongata* Peng, p. 368, pl. 8, figs. 5, 6.

Description: Glabella subrectangular, constricted at mid-length, with three pairs of shallow, pit-like glabellar furrows which are isolated from axial furrows. Palpebral lobes medium-sized, lying close to glabella and slightly behind its mid-length. Anterior sections of facial sutures slightly divergent forwards as far as to the antero-lateral corner of glabella, then curved inwards; posterior sections oblique outwards, enclosing the long (tr.), triangular posterior limbs. Posterior border furrow deep; posterior border very narrow, wire-like.

Remarks: A single cranidium is identical in all aspects with *Troedssonia isotelus* Liu (in Zhou and others, 1977, p. 223, pl. 67, fig. 1) which is considered to be a junior synonym of *T. elongata* Liu (loc, cit., fig. 2) by Peng (1984, p. 368) after reexamining Liu's specimens. Thus, the present specimen is here identified with the latter species.

Occurrence: Lower part of Guotang Formation, Lower Ordovician, Pu'an of Sandu, Guizhou.

Superfamily Ceratopygacea Linnarsson, 1869
Family Ceratopygidae Linnarsson, 1869
Genus *Diceratopyge* Troedsson, 1937
(=*Paraceratopyge* Poletaeva, 1945)

***Diceratopyge mobergi* Troedsson**

(Pl. 15, Figs. 5–6)

1937 *Diceratopyge mobergi* Troedsson, p. 39, pl. 2, figs. 11; pl. 3, figs. 6–7; pl. 5, fig. 2.
1965 *Diceratopyge mobergi*, Lu et al., p. 546, pl. 114, figs. 1–2 (refigured)
1980 *Diceratopyge mobergi*, Lu et Lin, p. 128, pl. 3, figs. 1–4.
1984 *Diceratopyge mobergi*, Lu et Lin, p. 118, pl. 16, figs. 6–8.

Remarks: In our collection, a complete but strongly oblique distorted dorsal shield and a longitudinally pressed pygidium, both preserved in shale, are available for identification. They agree with all apects to the type specimens of this species from Eastern Tianshan, the pressed-deformation caused the pygidium shorter and broader than the type specimens, and the diagonal compression dislocated the cephalon to overlap the anterior segments of thorax. The librigena has a long genal spine which extends directly posteriorly and slightly laterally to a position behind the posterior margin of the pygidium. The anterior facial sutures meet with each other at the center of frontal margin of cephalon to form a single suture, thus it separates the librigenae as in some asaphids, (known as the isoteliform type sutures). The pleural furrows of thoracic segments are wide and deep in anterior portion, so as to make the anterior marginal portion of the segments a narrow and convex ridge which is parallel to the anterior margin; the posterior portion of furrows decreases in depth backwards to the posterior margin of segments without any sharp boundary. The comparison of this form to other species have been given in a recent paper by Lu and Lin (1984, pp. 118–119).

Occurrence: The upper part of the late Upper Cambrian Maotian Formation, at Pu'an of Sandu County, SE Guizhou (78F-16).

***Diceratopyge agenor* Lu et Chien**

(Pl. 15, Fig. 7)

1978 *Diceratopyge agenor* Lu et Chien, in "Fossil Atlas of Southwest China", Fasc. 1 (Guizhou), p. 549, pl. 174, fig. 14.
1983 *Diceratopyge agenor* Lu et Chien, Palaeont. Cathayana, no. 1, p. 68, pl. 10, fig. 12; pl. 11, fig. 2.

Remarks: The cephalon in the present collection from the Lower Ordovician Guotang Formation bears all detailed features agreeable with those of the type specimens described by Lu and Chien in 1978 (see 1983, p. 68–69) from the Upper Cambrian Sandu Formation. Because of the different state of preservation, some features of the present specimen are not shown in the type specimens and other species of the genus *Diceratopyge*, e.g. there are three pairs of broad and shallow muscular impressions on the glabella: the two anterior pairs are long and horizontal, separated with each other by a longitudinal ridge and not reaching the dorsal furrows; the posterior pair is obscure, located between the middle pair and the occipital furrow. The occipital furrow is strongly oblique backwards and inwards from the dorsal furrows and arched backwards in center. The anterior border is narrow and convex, and the brim is broad (sag,) and flattened. The doublure as shown on the weathered surface of cephalon is as broad (sag.) as the brim and is separated by a longitudinal median suture into two parts. About 18–20 concentric terrace lines paralled to the anterior and lateral margins of the anterior border, brim and librigenae are well-marked. The panderian organ on the librigenae is situated at the midway between the posterior facial suture and the genal spine in front of the posterior furrow. *D. agenor* differs from the type species of *Diceratopyge*, *D. mobergi* Troedsson (1937, p. 39, pl. 2, fig. 11; pl. 3, figs. 6, 7; pl. 5, fig. 2) from the Upper Cambrian Torsagtagh Formation of Eastern Tianshan, Xinjing and *D. troedssoni* Sdzuy (1955, p. 21, pl. 4, figs. 93, 94; text-figs. 15–17) from the Tremadocian Leimitz Schiefer of Germany in the shorter and slightly contracted glabella, the wider (sag.) brim, narrower anterior border and narrower (tr.) postero-lateral limb. Furthermore the anterior border furrow is angulated forwards mesially and the glabella is strongly curved forwards in *D. troedssoni*, while the anterior border furrow is broadly arched forwards and the glabella is gently curved in *agenor* and *mobergi.*

Occurrence: Lower Ordovician Guotang Formation, at Pu'an of Sandu (78Fg-12).

***Diceratopyge sanduensis* Lu et T. R Zhou (sp. nov.)**

(Pl. 15, Figs. 8–10)

Diagnosis: A *Diceratopyge* with short and wide glabella, deep occipital furrow and short pygidial axis.

Description: Cephalon semicircular in outline. Glabella subquadrate, rounded anteriorly, well defined by axial and preglabellar furrows, slightly constricted at its mid-length. Glabellar furrows faint, represented by pits which are isolated from axial furrows. Occipital ring equally-wide, bending rearwards. Palpebral lobes small, semicircular, situated slightly rearwards behind mid-length of cranidium. Fixigena narrow (tr. exsag.). Anterior sections of facial sutures long, strongly diverging outwards from anterior ends of palpebral lobes, as far as to cranidial margin at an angle slightly more than 50° with sagittal line; posterior sections occurring horizontally, then obliquely backwards to enclose a narrow (exsag.) and long (tr.) posterior limb. Posterior border furrow deep and transverse. Thorax with 6 segments. Axis much narrower than pleural fields, with a width about 1 / 5 total width of thorax. Pleural segments with rounded panderian protuberances in the outer end of pleural doublures. Pleural furrows broad (exsag). Pygidium semicircular. Axis occupying 2 / 3 total pygidial length, with 6 rings and a terminal piece. Pleural furrows well defined, running outwards and merging into pleural field before reaching margin. A pair of delicate spines extending rearwards and outwards derived from the second pleural segment.

Remarks: The new species is closely similar to *Diceratopyge mobergi* Troedsson from the Upper Cambrian Torsuq tagh Formation in western Quruq tagh, Xinjiang (Troedsson, 1937, p. 39, pl. 2, fig. 11; pl. 3, figs. 6–7; pl. 5, fig. 2), but it can be distinguished by: (1) in the new species, the glabella is relatively wide (tr.) but longer in the latter, (2) the anterior sections of facial suturs are more divergent forwards, (3) the occipital furrow is deeper, (4) the thorax is relatively wide (tr.) and short (sag.), (5) the pygidial axis is shorter (sag.) than the latter. The new species differs from *D. agenor* Lu et Chien (in Yin and Li, 1978, p. 549, pl. 179, fig. 14; Lu et Qian, 1983, p. 68, pl. 10, fig. 12; pl. 11, fig. 2) in having a shorter glabella, more posteriorly situated palpebral lobes, narrower posterior limbs and a deeper occipital furrow. *Diceratopge guizhouensis* Lu et Qian (1983, p. 70, pl. 11, fig. 1) differs from the new species in (1) very shallow dorsal and preglabellar furrows, (2) short (sag.) preglabellar field, (3) smaller palpebral lobes which are located more forwardly and closer to the dorsal furrows, and (4) in the longer glabella. The shape of *D. sanduensis* is similar to that of *D. obesa* Peng (1984, p. 371, pl. 5, figs. la-b, 2,3), but the latter has shorter (sag.) preglabellar field and anterior border, the fixigena is relatively narrower, the pygidium is narrower, the pygidial axis is wider, with only 4 instead of 7 rings includiug the terminal piece, and the pygidial border is well defined. *D. elongata* Peng (1984, p. 372, pl. 5, figs. 4–6) is distinguished from *sanduensis* by the much longer glabella with a longitudinal keel, the longer and wider pygidial axis and by the remarkable pygidial border.

Occurrence: Lowermost part of the Lower Ordovician Guotang Formation; Shizilu, Sandu, Guizhou (78F2-4, 82Fg-6, 82Fg-21).

Diceratopyge sp.

(Pl. 17, Figs. 10–11)

An external mold of cranidium is suffered from deformation. The glabella is contracted at middle as that of *Diceratopyge sanduensis* (sp. nov.) (pl. 15, figs. 8–10), but the anterior border and the preglabellar field are narrower (tr.) and the palpebral lobes are larger. The last feature also serves to differentiate from other species of *Diceratopyge*. The preglabellar field is a little wider (sag. and exsag.) than that of *D. mobergi* Troedsson, the type species of this genus. The surface is densely granulated which are not seen in other species of *Diceratopyge*, except *D. sanduensis* (sp. nov.) which bears some granules.

Occurrence: *Mictosaukia guizhouensis* Assemblage-Zone, late Upper Cambrian Maotian Formation, Pu'an of Sandu (82Fg-8).

Suborder Harpina Whittington, 1959
Family Palaeoharpidae Lu et Chien, 1977*
Genus *Palaeoharpes* Lu et Chien, 1977**

Palaeoharpes spectabilis Lu et T. R. Zhou (sp. nov.)

(Pl. 15, Figs. 11, 12)

Diagnosis: A *Palaeoharpes* with irregularly arranged radiating punctures on the brim, rim and cheek

*,** The family Palaeoharpidae and the genus *Palaeoharpes* were erected in 1977 (Lu et Qian, in Zhou, Liu, Meng and Sun, in "Fossil Atlas of Central-South China" Part 1, p. 243), one year earlier than the publication of "Fossil Atlas of Southwest China, Guizhou (1)" (Lu et Chien in Yin et Li, 1978, pp. 564-565). The English version of the diagnostic features and discussions of them were given in "Palaeontologia Cathayana" no. 1, pp. 70–71, 1983.

roll, fine reticulation on the preglabellar field and cheek, and entirely coalesed anterior and posterior sections of facial sutures.

Description: Cephalon semicircular in outline, moderately convex, with a ratio of length and width at base about 3:5. Glabella narrow, strongly convex laterally and anteriorly, very gradually tapering forwards, with a length a little less than 1 / 2 the cephalic length. Three pairs of lateral glabellar furrows: two anterior pairs very short, pit-like, close to the dorsal furrows; posterior pair long, running strongly oblique inwards and backwards and reaching the occipital furrow. Occipital furrow straight; occipital ring uniform in length (sag.), narrowing laterally. External rim narrow, thread-like, upturned. Brim flat and broad. Cheek roll flat, broad in middle portion, very gradually narrowing postero-laterally. Preglabellar area very slightly concave. Eye-tubercles large, circular, elevated, situated opposite the anterior lateral glabellar furrows. Ocular ridges narrow but distinct, extending horizontally inwards from the eyes to the dorsal furrows. No trace of anterior and posterior sections of facial sutures observable, they seem completely coalesed with each other and probably united in an antero-lateral direction from the eye tubercle to the brim and the external rim. Alae semi-elliptical in outline, elevated, well defined by prominent alar furrows, situated at the inner posterior corners of cheek lobes. Genal prolongations short and broad, ending in a broadly rounded genal angle. Surface of brim and cheek roll ornamented with coarse irregularly arranged radiating punctures; surface of preglabellar field and cheek lobes marked with fine reticulating net-works.

Remarks: The new species is closely similar to *Palaeoharpes primigenius* Lu et Chien (Lu et Chien, in Yin and Li, 1978, p. 565, pl. 183, fig. 13; Lu et Chien, 1983, p. 72, pl. 12, fig. 9), the type species of the genus, from the Upper Cambrian Sandu Formation at Pu'an of Sandu, SE Guizhou. The main differences are confined to the separation of coalition of the anterior and posterior sections of facial sutures and to the convexity of cheek roll. In *primigenuis*, both sections of the facial sutures are separated, parallel to each other and concurring antero-laterally from the eye tubercles to the margin, the cheek roll is elevated in the anterior part; while in *spectabilis*, both sections of facial sutures are completely united, and the cheek roll is flattened. The progressive coalition and finally the disappearance of the facial sutures are probably due to the advancement in the evolution from the Upper Cambrian *Palaeoharpes primigenius* to the Lower Ordovician *P. spectabilis*.

The surface markings on the cephalon are very conspicuous in *spectabilis*, but they can not be seen in *primigenius*. *Palaeoharpes danzhaiensis* Lu et Chien (Lu et Chien in Yin and Li, 1978, p. 566, pl. 163, fig. 10; Lu et Chien, 1983, p. 72, pl. 12, fig. 10) from the Upper Cambrian of Danzhi county, SE Guizhou has a noticeable separation of the anterior and posterior sections of facial sutures like that of *primigenius* and unlike that of *spectabilis*. Besides, *P. danzhaiensis* differs from the latter two species in having narrow brim and cheek roll, broader preglabellar area and in more angulated genal prolongations. *Palaeoharpes simplex* Liu (Liu in Zhou et al., 1977, p. 243, pl. 73, fig. 10) from the Lower Ordovician Panjiazui Formation of Jiuxi, Taoyuan of NW Hunan differs from our species in the narrower glabella, the broader preglabellar area, the shorter (sag. and exsag.) brim and in the longer genal prolongations.

Occurrence: Lower Ordovician Guotang Formation, at Pu'an of Sandu (78F26-12, 82Fg-24).

Suborder Cheirurina Harrington et Leanza, 1957
Family Encrinuridae Angelin, 1854
Subfamily Pseudokoldinioidiinae Lu et T. R. Zhou (subf. nov.)

Diagnosis: Proparian trilobites with subparallel-sided posterior portion and forwardly expanded anterior portion of glabella, four pairs of short glabellar furrows, medium-sized palpebral lobe, very narrow and convex anterior border, strong genal spines, broad semielliptical pygidium with well-segmented axis and pleural lobes provided with 2–3 pairs of short lateral spines.

Remarks: The new subfamily is typified by the genus *Pseudokoldinioidia* Endo, 1944, and is differentiated from other subfamilies of the family Encrinuridac by the medium-sized palpebral lobes which are small or blind in most of other subfamilies and by the same number of the pygidial axial rings and the pleural segments.Except in Staurocephalinae, the number of the axial rings are considerably greater than that of the pleural segments in other subfamilies.

Besides *Pseudokoldiniodia* Endo, the genus *Encrinuella* Reed, 1915, from the Llanvirnian of the Northern States of Burma and from western Yunnan may be included in the Pseudokoldinioidiinae.

Geological and Geographical Distributions: Late Upper Cambrian to late Lower Ordovician (Llanvirnian), North, Northeast and Southwest China and Burma.

Genus *Pseudokoldinioidia* Endo, 1944, emend. herein

Endo (1944, p. 72) erected this genus based on a single cranidium. As stated by Endo, it is characterized by the triangular cranidium, strongly forwardly expanded glabella with rounded front, three pairs of glabellar furrows, wide fixigenae and very narrow, convex anterior border. This genus includes five species:

1. *Pseudokoldinioidia granulosa* Endo (Endo, 1944, p. 72, pl. 9, fig. 6; Lu et al., 1965, p. 640, pl. 131, fig. 24 (refigured), the type species, from the late Upper Cambrian of Lashushan, Laohushan of Liaoning Province, NE China.
2. *Pseudokoldinioidia bifurcata* Qian (1985, p. 152, pl. 1, fig. 9; pl. 2, figs. 1, 7, 9, 10; pl. 5, fig. 8b) from the late Upper Cambrian Tangcun Formation of Jingxian, southern Anhui.
3. *Pseudokoldinioidia perpetis* (Zhou et Zhang) (Zhou et Zhang, 1985, p. 93, pl. 4, figs. 6, 7; pl. 13, figs. 5–7; pl. 18, figs. 8–12; pl. 21, figs. 1–4; pl. 22, figs. 13-15; pl. 23, figs. 13–15; pl. 24, figs. 5–10; pl. 25, figs. 4–6; pl. 27, fig. 1–2; pl. 28, fig. 12) which is originally described from the Tremadocian of Nei Mongol, S. Jilin, N. Hebei, N. Shanxi and N. Anhui as a species of *Missisquoia*.
4. *Pseudokoldinioidia expansa* Lu et T. R. Zhou (sp. nov.) from the late Upper Cambrian of SE Guizhou.
5. *Pseudokoldinioidia encrinuroides* Lu et T. R. Zhou (sp. nov.) from the early Lower Ordovician (Tremadocian) Guotang Formation of SE Guizhou.
6. *Pseudokoldinioidia* sp. from the late Upper Cambrian of SE Guizhou.

Except the type species *Pseudokoldinioidia granulosa* represented by only one cranidium, all other species of this genus have their cranidia in association with the pygidia, the librigenae and the hypostoma. The following emended generic diagnosis of this genus is given according the finding of the mentioned new materials:

Medium-sized proparian trilobites with subparallel-sided or slightly medially contracted glabella which expands forwards and extends to the anterior border forrow, and rounded in front. Four pairs of short glabellar furrows of which the two anterior pairs are weakly defined and the two posterior pairs are deeply incised. Anterior border convex, narrow, defined by deep border furrow from the glabella and fixigenae. Occipital ring convex, wide in middle (sag.), narrowing towards both sides; occipital furrow deep and wide. Palpelbral lobes medium-sized, convex, about 1 / 3 to 2 / 5 the length of glabella. Eye-ridges obsolete. Fixigenae about 1 / 2 to greater than 1 / 2 the glabellar width between the palpebral lobes. Posterolateral limb subtriangular to elongate (tr.) striped triangular. Posterior border longer (tr.) than the basal width of glabella, widening laterally and bending posteriorly to produce a pointed genal spine, separated from the postero-lateral limb by a wide (exsag.) and deep posterior border furrow. Anterior sections of facial sutures converging forwards; posterior sections long, running sigmoidally and obliquely outwards from the palpebral lobe, strongly bending backwards to cut the posterior border from the librigena in front of the genal angle. Librigena narrower than fixigena, convex, narrowing toward both extremities; border wide, border furrow deep and wide. Pygidium semielliptical, axis long, divided by 6–7 deep ring furrows, pleural lobe separated by prominent pleural and interpleural furrows, with 2–4 anterior pair of pleurae forming serrations on lateral margins. Surface granulated. Associated hypostoma elongated subquadrate in outline, antero-lateral wings small, triangulur; lateral border narrow. Median body oval, anterior lobe circular, convex, posterior lobe flattenedly convex, median furrow shallow in middle, deepening laterally and posteriorly. Posterior border projected backwards in middle.

Remarks: The systematic position of this genus is dubious before the discovery of the proparian nature of facial sutures. In some papers (Lu et al., 1965, pp. 635–640; Qian, 1986, p. 152), whatever family is the assignment of this genus is uncertain. When Zhou and Zhang (1985, p. 93) tentatively placed *perpetis* into the genus *Missisquoia* (*perpetis* is regarded as a member of *Pseudokoldinioidia* in the present paper), this species falls naturally into the family Missisquoiidae Hupé, 1955.

This genus occurs in the late Upper Cambrian and early Tremadocian strata in Eastern Asia. It represents obviously a primitive encrinurid group with some primitive characters such as the larger eyes as compared with those of the later forms with small eyes in the Middle Ordovician to Silurian time, the glabellae are generally not so strongly expanded as some Ordovician and Silurian genera with globose anterior glabellar lobes, the powerful proparian genal spines reveal its affinities to some typical encrinurid genera with long genal spines in later geologic age, the deep pitted or short glabellar furrows indicate an encrinurid character but the apodemes of both the occipital furrow and the glabellar furrows are very rudimentary or not completely developed.

Pseudokoldinioidia is closely similar to *Missisquoia* Shaw, 1951, in the subtriangular to subsemicircular cranidium, in the presence of short and deep glabellar furrows, in the narrow and convex anterior border, but *Pseudokoldinioidia* has proparian facial sutures instead of opisthoparian ones in *Missisquoia* and has a more forwardly expanded glabella. *Encrinurella* Reed, 1915, from the Llanvirnian of Burma and western Yunnan differs from *Pseudokoldinioidia* in the small eyes, the gonatoparian facial sutures, the absence of genal spines and in the pliomerid-like pygidium. *Parakoldinioidia* Endo, 1937, is distinguished from *Pseudokoldinioidia* by the subrectangular glabella which is constricted in the middle and with a notch in front, and by the triangular-shaped and depressed anterior border. Like *Pseudokoldinioidia*, the genus *Tangshanaspis* Zhou and Zhang, 1978 (Zhou et Zhang, 1985, p. 95) also has four pairs of glabellar furrows, but in the latter the glabella is subrectangular and carinated, with an anterior medial notch and the fixigenae are wider.

Geological and Geographical Distributions: Late Upper Cambrian of Northeast and Southwest China, early Tremadocian of North, Northeast, South and Southwest China.

Pseudokoldinioidia expansa Lu et T. R. Zhou (sp. nov.)

(Pl. 16, Figs. 1–6)

Diagnosis: A *Pseudokoldinioidia* with broad cranidium, short and broad glabella, straight occipital furrow in middle, powerful genal spines and short pygidial axis.

Description: Cephalon roughly semicircular, except the long genal spines. Cranidium trapezoidal in outline, rounded in front, slightly longer than 1 / 2 its breadth between the genal spines. Glabella subparallel-sided from the base to 1 / 2 its length, then strongly expanded forwards and reaching the anterior border furrow; with a greatest width at 9 / 10 of the glabellar breadth, the ratio of breadth of the anterior portion and the basal portion is about 4.5: 3.5, convex from side to side, sloping downwards anteriorly. Four pairs of lateral glabellar furrows: anterior pair very shallow and short, weakly defined, closed to the dorsal furrows; next pair also very shallow, a little longer than the anterior one, slightly oblique backwards and inwards; third and posterior pairs short, deeply incised on both sides of glabella, and make an angle of about 40°-45° to the dorsal furrows. Anterior border narrow and convex, uniform in width throughout both the extremities, regularly arched anteriorly; border furrow deep, as wide as the anterior border (sag. and exsag.). Occipital ring convex, uniform in breadth in middle portion (sag.), narrowing and bending forwards on sides; occipital furrow broad (tr.) and wide (sag and exsag.), straight in middle, slightly oblique forwards and outwards laterally. Fixigenae moderately convex, a little less than 1 / 2 the glabellar width between the palpebral lobes, narrowing rapidly anteriorly in front of palpebral lobes. Postero-lateral limb narrow, elongated (tr.) triangular, with a ratio of width to the base of glabella about 4:3.5. Posterior border convex, very narrow in inner portion, rapidly widening laterally and strongly bending backwards to form a strong genal spine, separated from the postero-lateral limb by a very broad (exsag.) and deep posterior border furrow which is about twice the breadth (sag.) of the occipital furrow. Palpebral lobe crescentic, medium-sized, about 2 / 5 the glabellar length, elevated above the fixigena, situated at the position a little behind the mid-length of glabella, well defined by deep palpebral furrow. Anterior sections of facial sutures short, converging forwards to cut the lateral margin of anterior border at an angle about 60° to the cranidial axis; posterior sections long, sigmoidal, running in a broad concave curve strongly outwards and reaching genal area, then rapidly bending rearwards to cut the postero-lateral limb in front of the genal spines. Librigenae (pl. 16, figs. 4 left, 5) narrow, elongated subtriangular, pointed in both extremities, with broad lateral border defined by deep and wide border furrow.

Pygidium fusiform, broad, acutely rounded on both extremities, with a length slightly greater than 1 / 2 its width. Axis strongly convex, as wide as the pleural lobe in front, tapering rapidly rearwards, reaching the ill-defined posterior marginal furrow, with rounded posterior end, separated by 5 broad, deep, forwardly curved transverse furrows into 5 rings, a half ring and a terminal piece. Pleural lobe triangular, moderately convex in inner portion, strongly sloping downwards laterally from the midwidth, separated by 4–5 pronounced interpleural furrows extending to the lateral margin; pleural furrow deep and wide, anteior 2-3 pleurae terminated by short lateral spines. Postero-lateral marginal border inconspicuous, can be recognized only from the incomplete pleural furrows not reaching the lateral margin.

Remarks: The new species differs from *P. granulosa* Endo (1944, p. 72, pl. 9, fig. 6), the type species of this genus, in the broader cranidium, shorter glabella and smaller granules on the surface. Endo stated that *granulosa* bears only three pairs of glabellar furrows, but the type specimen is too badly preserved, whether the anterior pair of the four pairs in other species is present or absent in this genus cannot be ascertained until the new material is available. *P. bifurcata* Qian (1986, p. 152, pl. 1, fig. 9; pl. 2, figs. 1, 7, 9, 10; pl. 5, fig. 8a) from the late Upper Cambrian Tangcun Formation of S Anhui is easily distinguished from our new species by the bifurcated posterior pair of glabellar furrows, wider fixigenae, a small number and coarser granules on the surface of glabella and fixigenae, and in the narrower pygidium which is subtriangular instead of fusiform in outline in our species. Like *P. granulosa*, Qian said that *bifurcata* also provides three pairs of glabellar furrows, but the cranidia of the latter are also not well preserved. *Pseudokoldinioidia perpetis* (Zhou et Zhang) (1985, p. 93, pl. 4, figs. 6, 7; pl. 13, figs. 5–7; pl. 18, figs. 8–12; pl. 21, figs. 1–4; pl. 22, figs. 13–15; pl. 23, figs. 13–15; pl. 24, figs. 5–10; pl. 25, figs. 4–6; pl. 27, figs. 1–2; pl. 28, fig. 12) from the Tremadocian of North, Northeast and South China bears four pairs of glabellar furrows like those of *P. expansa*, but the former has a longer and less forwardly expanded glabella, a posteriorly arched occipital furrow, a shorter pygidium with a greater number of axial rings and pleurae (8 rings plus one terminal lobe, and 8 pairs of pleurae). The hypostoma detached from cranidia and pygidia as assigned to *perpetis* by Zhou and Zhang (pl. 25, fig. 6) is different from those of many encrinurids, it remains a question whether this belongs really to the family Encrinuridae or this is a primitive hypostoma of the encrinurids of the Tremadocian age.

Occurrence: *Mictosaukia guizhouensis* Assemblage-Zone of the late Upper Cambrian Maotian Formation, at Pu'an, Sandu County (78F2-6, 82Fg-1).

Pseudokoldinioidia encrinuroides Lu et T. R. Zhou (sp. nov.)

(Pl. 17, Figs. 1–6; Pl. 9, Fig. 9)

Diagnosis: A *Pseudokoldinioidia* with subquadrate to subtrapezoidal cranidium, elongated, strongly forwardly expanded glabella, wide fixigenae, pygidium with narrow axis and broad pleural lobes, surface ornamented with coarse and fine granules.

Description: Cranidium roughly subquadrate to subtrapezoidal in outline, with a ratio of length and basal width about 1:1.4, gently arched anteriorly. Glabella narrow, long, with a ratio of length and basal width about 1.7:1, strongly convex and steeply sloping downwards in front, rapidly widening from 2 / 3 the length forwards, with a ratio of the greatest width between the antero-lateral corners and the basal width about 3:2. Four pairs of very weak, pitted glabellar furrows on both sides of glabella. Anterior border very narrow and convex, thread-like; anterior border furrow very narrow and deep. Occipital ring convex, crescentic; occipital furrow deep and broad, arched backwards. Fixigenae as wide as the glabella between the palpebral lobes, gently convex. Dorsal furrows very deep and broad. Postero-lateral limb subtriangular, flat. Posterior border very narrow in inner portion, widening rapidly outwards to the short genal spine. Posterior border furrow as wide as or a little wider than the occipital furrow. Palpebral lobe medium-sized, crescentic, strongly convex, situated at the midlength of cranidium, eye ridge conspicuous, running directly inwards and slightly forwards to the dorsal furrows opposite the antero-lateral side of glabella behind the anterior glabellar furrow. Anterior sections of facial sutures short, converging anteriorly to cut the anterior border at an angle about 60°–70° to the axial line; posterior sections long, extending obliquely backwards at an angle about 45° to the axial line to cut the posterior border in front of genal spines. Genal spine short. Surface of cranidium ornamented with large, scatteredly distributed pustules and numerous fine granules between them.

Pygidium broad fusiform, acutely rounded on both extremities, with a length about 1 / 2 of its greatest width. Axis strongly convex, subtriangular in shape, narrowing regularly backwards, reaching the border furrow, less than 1 / 3 the pygidial width, with a ratio of length and width about 1.5:2.5, divided by 7–8 deep and wide ring furrows into 8–9 axial rings and a rounded terminal piece. Pleural lobe wider than the axis, triangular, consisting of 7–8 pleural segments which are well marked by deep and broad pleural furrows and separated by distinct interpleural grooves. First 2–3 segments extending laterally to form short, free lateral spines, the rest of other segments behind these are fused with each other. Clearly there are no marginal border outside the first 2–3 pleural segments, and the marginal border is weakly developed behind these if ever. The fusion of the anterior pleural segments may not take place simultaneously on both sides in a single pygidium, e.g. the pygidium dipected on pl. 17, fig. 6 has well fused anterior pleural segments surrounded by ill-defined marginal border on the right pleural lobe, whereas on the left, at least the outer portion of the first and second pleural segments are freely separated, demonstrated by the presence of 2 free pleural spines. The retarded differentiation of the pleural segment on right side is probably not caused by some pathologic accidents, but might be due to the unsymmetrical development of organs in some few individuals of animals and plants. However the separation of the last thoracic segment from the first or the anterior pygidial segment, as did in other trilobites, is noticeable in the small immature pygidium (pl. 17, fig. 3) with 3 pairs of free pleural spines. The pygidium of a fully adult (pl. 17, fig. 4) has only 2 pairs of free spines. Anyhow the development of the pygidium shows one of the primitive characters of the genus *Pseudokoldinioidia* in the trilobite evolution.

Remarks: The present species may be differentiated from other species of *Pseudokoldinioidia* by its narrow and strongly expanded glabella with weakly defined lateral glabellar furrows. Moreover, it differs from *Pseudokoldinioidia perpetis* (Zhou et Zhang) from the Tremadocian in the narrower cranidium, wider (exsag.) postero-lateral limb, broader pygidium with comparatively narrower axis, and by the coarse pustules on the surface of cranidium. The last feature of *encrinuroides* is similar to that of *P. bifurcata* Qian from the late Upper Cambrian of Jilin Province, but *bifurcata* has a broader glabella characterized by very deep and broad glabellar furrows. From *P. expansa* described in foregoing paragraphs, *encrinuroides* differs in the backwardly arched occipital furrow, the wider (exsag.) postero-lateral limb and in the coarsely granulated surface ornamentation.

Occurrence: Lower part of the Guotang Formation, early Tremadocian, at Pu'an of Sandu County (78Fg-12, 78Fg-16).

Pseudokoldinioidia sp.

(Pl. 16, Figs. 7, 8)

A cranidium associated with a pygidium bear many characters similar to those of *P. expansa* (sp. nov.) described in this paper, but the glabella is relatively broader and less expanded anteriorly, the postero-lateral limb is wider (exsag.), the pygidial axis is not so rapidly tapered rearwards and the granules on both the cranidium and pygidium are larger and scatteredly distributed. Because of the bad preservation of the specimens, the specific reference can not be made.

Occurrence: *Mictosaukia guizhouensis* Assemblage-Zone of late Upper Cambrian Maotian Formation, at Pu'an of Sandu County (82Fg-10).

Suborder Calymenina Swinnerton, 1915
Family Calymenidae Milne Edwards, 1840
Subfamily Bathycheilinae Pribyl, 1953
(=Pharostomatinae Hupé, 1953= ? Eulominae Venek, 1964)
Genus *Pharostomina* Sdzuy, 1955
(=*Colpocoryphoides* Harrington et Leanza, 1957)

Pharostomina sanduensis Zhou

(Pl. 11, Figs. 11–13)

1981 *Pharostomina sanduensis* Zhou, Acta Palaeont. Sinica, v. 20, no. 3, p. 243, pl. 1, figs. 10, 11.

The original diagnostic description given by the junior author in Chinese is here translated into English as follows: "Cranidium subtrapezoidal. Glabella broad conical, bluntly rounded in anterior margin. Three pairs of glabellar furrows: two anterior pairs shallow and short, running obliquely rearwards; posterior pair very deep and broad, close to the dorsal furrows, Fixigena narrow. Anterior border narrow and short (tr.), upturned, middle part convex posteriorly. Anterior border furrow broad. Brim convex on sides, concave posteriorly in central portion to form a wide depression in front of glabella connecting with the anterior border furrow. The depression is twice the length (sag.) of the anterior border. Palpebral lobes small; eye ridge indistinct. Postero-lateral limb wide, triangular; posterior border furrow wide, horizontally stretching outwards; posterior border narrow. Anterior sections of facial sutures converging anteriorly, strongly curving inwards to cut the anterior border; posterior sections obliquely cutting the posterior border. Occipital furrow wide, straight, rather shallow in middle portion, deepening laterally. Occipital ring wide in middle. Surface of cranidium ornamented with coarse granules".

Remarks: The genus *Pharostomina* has a wide-spread geographic distribution in the Tremadocian and Arenigian of Germany, South and Central Americas and SW China. The present species more resembles *P. ferentaria* Sdzuy (1955, p. 34, pl. 6, figs. 77–79, 81–88; pl. 2, fig. 80; text-fig. 33) than the type species, *P.*

oepikii Sdzuy (1955, p. 31, pl. 6, figs. 62–75; pl. 2, fig. 99; text-fig. 31, 32), both from the Tremadocian Leimitz Schiefer of Germany, but the very coarse granulated cranidium of it is quite different from the fine granulations of the German forms. The South American species *Colpocoryphoides trapezoidalis* Harrington et Leanza (1957, p. 223, figs. 12, 1–6) from the Upper Tremadocian and Arenigian of Argentina differs from *P. sanduensis* in having wider cranidium and finer and a great number of surface granules. The last feature also serves sufficiently to differentiate the present species from the Mexican Tremadocian form *Pharostomina mexicana* Robison and Pantoja (1968, p. 783, pl. 99, figs. 7–12).

Occurrence: Lower Ordovician (Tremadocian) Guotang Formation, at Pu'an and Shizilu of Sandu County (82Fg-12, 82Fg-24, 82Fg-65).

Family Uncertain

Genus *Protarchaegonus* Sdzuy, 1955

Protarchaegonus sanduensis Zhou

(Pl. 17, Figs. 7–9)

1981 *Protarchaegonus sanduensis* Zhou, Acta Palaeont. Sinica, v. 20, no. 3, p. 245, pl. 1, figs. 12, 13.

Zhou's description of this species in Chinese is here translated into English as follows: "Glabella bluntly conical, convex, rounded in anterior margin, Dorsal furrows deep. Three pairs of glabellar furrows: anterior pair very short; middle pair longer, slightly oblique rearwards: posterior ones longest, oblique, terminal part running directly posteriorly. Fixigena narrow, about 1 / 4 the glabellar width between the palpebral lobes; eye ridge prominent. starting from the palpebral lobes oblique forwards and inward. Palpelbral lobe small, semicircular, located a little behind the midlength of cranidium. Anterior border convex, wide in middle portion, gradually narrowing laterally. Brim convex, with a width (sag.) about twice that of the anterior border. Anterior sections of facial sutures long, diverging anteriorly to make an angle about 40° to the axial line; posterior sections running obliquely outwards and backward to cut the postero-lateral border. Occipital furrow deep, straight in middle, oblique forwards laterally. Occipital ring narrow, uniform in breadth. Postero-lateral limb triangular, postero-lateral furrow deep and short. Posterior border widening laterally, slightly bending backwards, extremities pointed. Pygidium transversely broad elliptical. Axis conical, with pointed terminal end, divided into 5 rings. Pleural lobe with 4 segments, pleural furrow deep, interpleural furrow well-defined, both extending to the postero-lateral margin".

Remarks: Three cranidia in the presant collection are available for examination. This species is distinguished from the type species, *P. moroffi* Sdzuy (1955, p. 40, pl. 7, figs. 2–7; text-fig. 36) from the Lower Tremadocian of Germany by the longer cranidium, longer and narrower glabella, the prominent eye-ridge, the wider (sag.) brim, and by the wider (tr.) postero-lateral limb and postero-lateral border.

Occurrence: Lower Ordovician (Tremadoc) Guotang Formation, Pu'an of Sandu (78F26-12).

REFERENCES

Apollonov, M. K., Chugaeva M. N. and Dubinina S. V., 1981. The Cambro-Ordovician boundary in the Malyi Karatau Range, South Kazakhstan. Short Papers 2nd Internat. Symp. Cambrian System, 1981. Open File Report 81–743.

________, ________ and ________, 1983. The Cambrian and Ordovician trilobites from the Batyrbay section of the Malyi Karatau Range. Lower Palaeozoic Stratigr. and Paleont. of Kazakhstan. Nauka Kazakh. SSR (in Russian).

________, ________ and ________, 1984. Trilobites and conodonts from the Batyrbay section (Uppermost Cambrian-Lower Ordovician) in Malyi Karatau Range. Nauka Kazakh. SSR (in Russian).

Chen Junyuan et al., 1985. Study on Cambrian-Ordovician Boundary strata and its biota in Dayangcha, Hunjiang, Jilin, China. China Prospect Publishing House, Beijing.

Chen Yunyuan et al., 1986. Aspects of Cambrian-Ordovician Boundary in Dayangcha, China Prospect Publishing House, Beijing.

Chien Yiyuan (Qian Yiyuan), 1961. Cambrian trilobites from Sandu and Duyun, Southern Kweichou. Acta Palaeont. Sinica, 9 (2).

Endo, R., 1931. The Cambrian Period. In Iwanami Lecture Series, Section of Geology, Paleontology, Mineralogy and Petrology, 11(2) (in Japanese).

________ 1944. Restudies on the Cambrian formations and fossils in southern "Manchukuo". Bull. Cent. Nat. Mus."Manchukuo", 7, 1-100.

________ and Resser, S. E., 1937. The Sinian and Cambrian formations and fossils of southern "Manchukuo". Bull Manchurian Sci. Mus., 1.

Ergaliev, G. Kh., 1983. Upper Cambrian and Lower Ordovician trilobites of Great Karatau and Ulutau. Lower Palaeozoic Stratigr. and Paleont. of Kazakhstan. Nauka Kazakh. SSR (in Russian).

Harrington, H. J. and Leanza, A. F., 1957. Ordovician trilobites of Argentina. Spec. Publ. Dept. Geol., Univ. Kansas, 1.

Kobayashi, T., 1931. Studies on the stratigraphy and palaeontology of the Cambro-Ordovician. Jap. J. Geol. Geogr., 8(3), 131–189.

________, 1933. Faunal study of the Wanwanian (Basal Ordovician) Series with special notes on the Ribeiridae and the Ellesmereoceroids. Journn. Fac. Sci., Imp. Univ. Tokyo, Sect. 2, 3(7).

__________, 1935. Cambrian formations and faunas of South Chosen, Journn. Fac. Sci., Imp. Univ. Tokyo, Sect. 2, 4(2).

__________, 1955. The Ordovician fossils of the Mackay Group in British Columbia, Western Canada, with a note on the Early Ordovician palaeogeography. Journn. Fac. Sci., Imp. Univ. Tokyo, Sect. 2, 9(3), 355–493.

__________, 1956. Miscellaneous notes on the Cambrian-Ordovician geology and palaeontology, 26, Supplementary notes on the Cambrian trilobites from southeastern Asia. Trans. Proc. Palaeont. Soc. Japan, N. S. 24, 277–284.

__________, 1957. Upper Cambrian fossils from peninsular Thailand. Journn. Fac. Sci., Univ. Tokyo, 10(3)

__________, 1966. The Cambro-Ordovician formations and faunas of S. Korea. The Chosen Group of N. Korea and NE. China. Journn. Fac. Sci. Univ. Tokyo. Sect. 2. 16 (2).

Kuo Hongjun and Duan Jiye, 1978. Cambrian and Early Ordovician trilobites from northeastern Hebei and western Liaoning. Acta Palaeont. Sinica, 17(4), 445–460.

Kuo Hongjun, Duan Jiye and An Sulan, 1982. Cambrian-Ordovician boundary in the North China Platform with description of trilobites. Papers for 4th Internat. Symposium on Ordovician System. Jour. Changchun College of Geology, 1982, (3), 9–28.

Li Shanji and Yin Gongzheng, 1978. Class Trilobita. In: Palaeont. Atlas of Southwest China. Guizhou (1), 385-595. Geol. Publ. House, Beijing (in Chinese).

Lu Yanhao,1954. Upper Cambrian trilobites from Sandu, southeastern Kueichou. Acta Palaeont. Sinica, 2(2), 117–152.

__________, 1957. "Trilobita", In: Index Fossils of China. Invertebrates(3). Science Press. Beijing (in Chinese).

__________, 1964. "Trilobita" In: Wang Yu (ed.), Index Fossils of South China. Sci. Press, Beijing (in Chinese).

__________, 1984. Report on the Cambrian-Ordovician boundary work in China. In: Stratigr. & Palaeont. Systemic Boundaries in China. Cambrian-Ordovician Boundary (1), 1–8. Anhui Sci. Techn. Publ. House, Hefei.

__________ and Lin Huanling., 1980. Cambro-Ordovician boundary in western Zhejiang and the trilobites contained therein. Acta Palaeont. Sinica, 19(2), 118–135.

__________ and __________, 1984. Late Cambrian and Earliest Ordovician trilobites of Jiangshan-Changshan area, Zhejiang. In: Stratigr. & Palaeont. Systemic Boundaries in China. Cambrian-Ordovician Boundary (1), 45–164. Anhui Sci. Techn. Publ. House, Hefei.

__________, Lin Huanling, Han Nairen, Li Luozhao and Ju Tianyin, 1984. On the Cambrian-Ordovician Boundary of the Jiangshan-Changshan area. W. Zhejiang. In: Stratigr. & Palaeont. Systemic Boundaries in China, Cambrian-Ordovician Boundary (1). Auhui Sci. Techn. Publ. House. Hefei.

__________ and Mu Enzhi, 1980. Boundaries of the Ordovician System in China. In: Scientific papers on geology for international exchange, prepared for the 26th International Geological Congress, Stratigraphy and Palaeontology, 4, 15–25. Geol. Publ. House, Beijing (in Chinese).

__________ and Qian Yiyuan, 1983. Cambro-Ordovician trilobites from eastern Guizhou. Palaeont. Cathayana, 1, 1–106.

__________, Zhang Wentang, Zhu Zhaoling, Qian Yiyuan and Xiang Liweng, 1965. Trilobites of China. Science Press, Beijing (in Chinese).

__________, Zhou Zhiqiang and Zhou Zhiyi, 1986. New materials of *Onychopyge* faunas, with a discussion on the evolution of *Onychopyge* (Trilobita). Bull. Nanjing Inst. Geol. Palaeont., 7, 114–126.

__________, Zhou Zhiyi and Zhou Zhiqiang, 1981. Cambrian–Ordovician boundary and their related trilobites in the Hangula region, W. Nei Monggol. Bull. Chinese Acad. Geol. Sci., Ser. 10, 2(1), 1–22.

__________, Zhu Zhaoling, Qian Yiyuan, Lin Huanling, Zhou Zhiyi, Yuan Kexing, 1974. Bio-environmental control hypothesis and its application to the Cambrian biostratigraphy and palaeozoogeography. Mem. Nanjing Inst. Geol. Palaeont., 5 (in Chinese).

__________, __________, __________, Zhou Zhiyi, Chen Junyuan, Liu Gengwu, Yu Wen, Chen Xu and Xu Hankui, 1976. Ordovician biostratigraphy and Palaeozoogeography of China. Mem. Nanjing Inst. Geol. Palaeont., 7 (in Chinese).

Lndvigsen, R., 1982. Upper Cambrian and Lower Ordovician trilobite biostratigraphy of the Rabbitkettle Formation, western district of Mackenzie. Life Sci. Contrib., Roy. Ontario Mus., 134.

Mansuy, H., 1915. Faunas Cambrian du Haut–Tonkin. Mem. Serv. Geol. l'Indochine, 5(1).

Owens, R. M., Fortey, R. A., Cope, J. W. C., Rushton, A. W. A. and Bassett, M. G., 1982. Tremadoc faunas from the Carmarthen district, South Wales. Geol. Mag., 119(1), 1–38.

Palmer, A. R., 1955. Upper Cambrian Agnostidae of the Eureka district, Nevada. Journn. Paleont., 29(1), 86–101.

__________, 1968. Cambrian trilobites of East-Central Alaska. U. S. Geol. Surv. Prof. Paper, 559–B.

Peng Shanchi, 1984. Cambrian-Ordovician boundary in the Cill-Taoyuan border area, northwestern Hunan with descriptions of relative trilobites. In: Stratigr. & Palaeont. Systemic Boundaries in China, Cambrian-Ordovician Boundaries (1), 285-405. Anhui Sci. Techn. Publ. House, Hefei.

Petrunina, Z. E., Poletaeva, O. K., Semenova, V. S. & Fedianina, E. S., 1960. Arthropoda. In: Khalfin, L. L. (ed.), Biostratigraphy of the Palaeozoic of Syano-Altai folded region. Lower Palaeozoic, 1, 409-430. Trypy Sniiggims, 19 (in Russian).

Qian Yiyuan,1985. Late Late Cambrian trilobites from the Tangcun Formation of Jingxian, southern Anhui. Palaeontologia Cathayana, 2, 137–153.

Rasetti, F., 1954. Early Ordovician trilobite faunule from Quebec and Newfoundland. Jour. Palaeont., 28, 581–587.

Raymond, P. E., 1924. New Upper Cambrian and Lower Ordovician trilobites from Vermont. Proc. Boston Soc. Nat. Hist., 37(4), 339–466.

__________, 1937. Upper Cambrian and Lower Ordovician Trilobita and Ostracoda from Vermont. Bull. Geol. Soc. Amer., 48, 1079–1146.

Robison, R. A. and Pantoja-Alor, J., 1968. Tremedocian trilobites from the Nochixtlan Region, Oaxaca, Mexic. Jour. Paleont., 42(3), 767–800.

Sdzuy, K., 1955. Die Fauna der Leimitz-Schiefer(Tremadoc). Abh. Senck. Naturf. Ges., 492.

Shaw, A. B., 1951. Palaeontology of northwestern Vermont, 1. New Late Cambrian trilobites. J. Paleont., 25, 97–114.

Shergold, J. H., 1971. Late Upper Cambrian trilobites from the Gola Beds, western Queensland. Bull. Bur. Min. Res. Geol. Geophys.

Australia, 112, 1–126.

__________, 1975. Late Cambrian and Early Ordovician trilobites from the Burke River Structural Belt, western Queensland, Australia. Bull. Bur. Min. Res. Geol. Geophys. Australia, 153 (1-2).

__________, 1977. Classification of the trilobite *Pseudagnostus*. Palaeontology, 20(1), 69–100.

__________, 1982. Idamean (Late Cambrian) trilobites, Burke River Structural Belt, western Queensland. Bull. Bur. Min Res. Geol. Geophys. Australia, 187, 1–69.

__________ and Sdzuy, K. 1984. Cambrian and early Tremadocian trilobites from Sultan Däg, central Turkey. Senckenbergiana Lethaea, 65(1/3) 51–135.

Sun Yunzhu (Sun, Y. C.), 1924. Contribution to the Cambrian faunas of North China. Palaeont. Sinica, Ser. B, 1(4).

__________ and Xiang Liwen, 1979. Late Upper Cambrian trilobite fauna from western Yunnan. Mem. Chinese Academy of Geology, 1(1), 1–17.

Troedsson, G. T., 1937. On the Cambro-Ordovician faunas of western Quruq tagh, eastern Tienshan. Palaeont. Sinica, n. s. B, 2.

Walcott, C. D., 1913. Cambrian faunas of China. In: Research in China. 3. Carnegic Inst. Washington.

Winston, D. and Nicholls, H., 1967. Late Cambrian and Early Ordovician faunas from the Wiberms Formation of Central Texas. Jour. Palaeont., 41(l), 66–98.

Yin GongZheng and Li Shanji, 1978. Trilobita In: Fossil Atlas of Southwest China, Guizhou (l), 385-595. Geol. Publ. House, Beijing-(in Chinese).

Yin Gongzheng, Gong Lianzan, Cai Ying and Jiao Huiliang, 1984. On the Cambrian-Ordovician boundary in Guizhou of China. In: Papers for 27th International Geological Congress, 1980, Moscow, 25-34. Geol. Publ. House, Beijing (in Chinese).

Zhou Tianmei, Liu Yiren, Meng Xiansong and Sun Zhenhua, 1977. Class Trilobita, In: Palaeontological Atlas of Central-South China. 1. 104–266, Geol. Publ. House, Beijing (in Chinese).

Zhou Tianrong, 1981. New materials of early Tremadocian trilobites from Sandu and Pu'an, Guizhou, Acta Palaeont. Sinica, 20(3), 241–246.

Zhou Zhiyi and Zhang Jinling, 1978. Cambrian-Ordovician Boundary of the Tangshan area with descriptions of the related trilobite fauna. Acta. Palaeont. Sinica, 17(1).

Zhou Zhiyi and Zhang Jinling, 1985. Uppermost Cambrian and Lowest Ordovician trilobites of North and Northeast China. In: Stratigr. & Palaeont. of Systemic Boundaries in China. Cambrian-Ordovician Boundary(2), 63–194. Anhui Sci. Techn. Publ. House, Hefei.

EXPLANATION OF PLATES

All the specimens are housed in Nanjing Institute of Geology and Palaeontology, Academia Sinica

Plate I

1–3. *Geragnostus* (*Micragnostus*) *elongatus* Chien
  1. Cephalon, ×7, Coll. no. 82Fg-57, Cat. no. 104291.
  2. Pygidium, ×8, Coll. no. 78F27-14, Cat. no. 104292.
     Late Upper Cambrian Sandu Formation, Shizilu of Sandu.
  3. Pygidium, ×18, Coll. no. 78Fg-12, Cat. no. 104293.
     Lower Ordovician (Tremadocian) Guotang Formation, Pu'an of Sandu.

4–6. *Geragnostus* (*Micragnostus*) sp.
  4. Cephalon, ×15, Coll. no. 82Fg-24, Cat. no. 104294.
  5. Pygidium, ×15, Coll. no. 82Fg-24, Cat. no. 104295.
     Lower Ordovician (Tremadocian) Guotang Formation, Pu'an of Sandu.
  6. Pygidium, ×15, Coll. no. 82Fg-1, Cat. no. 104296.
     Mictosaukia guizhouensis Assemblage-Zone, late Upper Cambrian Maotian Formation.

7–11. *Neoagnostus shiziluensis* Lu et T. R. Zhou (sp. nov.)
  7. Cephalon, Holotype, ×20, Coll. no. 78F2-6, Cat. no. 104297.
  8. Cephalon, ×12, Coll. no. 82Fg-1, Cat. no. 104298.
  9. Pygidium, ×20, Coll. no. 78F2-6, Cat. no. 104299.
  10. Pygidium, ×35, Coll. no. 82Fg-1, Cat. no. 104300.
     Lower part of late Upper Cambrian Maotian Formation, Pu'an of Sandu.
  11. Incomplete exoskeleton, ×6.5, Coll. no. 82Fg-5, Cat. no. 104301.
     Middle part of late Upper Cambrian Sandu Formation, Shizilu of Sandu.

12–14. *Neoagnostus bilobus* (Shaw)
  12–14. Cephala, ×8, ×8, ×8, Coll. no. 78F26-12. Cat. nos. 104302, 104303, 104304.
     Lower Ordovician (Tremadocian) Guotang Formation, Pu'an of Sandu.

Plate 1

Plate 2

1–2. *Neoagnostus bilobus* (Shaw)
  1. Pygidium, ×12, Coll. no. 78F26-12, Cat. no 104305.
  2. Pygidium, ×7, Coll. no. 82Fg-24, Cat. no. 104306.
  Lower Ordovician (Tremadocian) Guotang Formation, Pu'an of Sandu.

3–5. *Rhaptagnostus changshanensis* Lu et Lin
  3. Exosketon, ×6.5, Coll no. 78F2-14, Cat. no. 104307.
  4. Cephalon, ×6, Coll. no. 82Fg-1, Cat. no. 104308.
  5. Pygidium, ×7, Coll. no. 78F2-7, Cat. no. 104309.
  Late Upper Cambrian Maotian Formation, Pu'an of Sandu.

6–8. *Leiagnostus* (?) *insoletus* Lu et T.R. Zhou (sp. nov.)
  6. Cephalon, ×7, Holotype, Coll. no. 82Fg-57, Cat. no. 104310.
  7. Cephalon, ×7, Coll. no. 82Fg-50, Cat. no. 104311.
  8. Pygidium, ×7, Coll. no. 82Fg-57, Cat. no. 104312.
  Late Upper Cambrian Maotian Formation, Pu'an of Sandu.

9–10. *Euloma* (*Archaeuloma*) *elongatum* Lu et T.R. Zhou (sp. nov.)
  9. Cranidium, ×7, Coll. no. 78F2-7(=82Fg-1), Cat. no. 104313.
  10. Cranidium, ×7, Holotype, Coll. no. 78F2-7(=82Fg-1), Cat. no. 104314.
  Upper part of late Upper Cambrian Maotian Formation, Pu'an of Sandu.

11–12. *Euloma* (*Archaeuloma*) *sanduensis* Lu et T.R. Zhou (sp. nov.)
  11. Exoskeleton, ×4.6, Holotype, Coll. no. 82Fg-1, Cat. no. 104315.
  12. Cranidium, ×6, Coll. no. 78F2-6, Cat. no. 104316.
  Lower part of late Upper Cambrian Maotian Formation, Pu'an of Sandu.

Plate 2

Plate 3

1. *Leiagnostus* sp.
   Exoskeleton, ×7, Coll. no. 82Fg-64, Cat. no. 104317.
   Lower Ordovician (Tremadocian) Guotang Formation, Shizilu of Sandu.
2. *Euloma* (*Archaeuloma*) sp.
   Cranidium, ×6.5, Coll. no. 82Fg-51, Cat. no. 104318.
   Late Upper Cambrian Sandu Formation, Shizilu of Sandu.

3–4. *Proteuloma subquadratum* (Lu et Qian)
   3. Cranidium and thorax, ×5, Coll. no. 82F-21, Cat. no. 104319.
   4. Exoskeleton, ×4, Coll. no. 78F26-12, Cat. no. 104320.
   Lower Ordovician (Tremadocian) Guotang Formation, Pu'an of Sandu.

5. *Proteuloma* sp.
   Cranidium, ×6, Coll. no. 82Fg-66, (see also Pl. 15, Fig. 11b). Cat. no. 104321.
   Lower Ordovician (Tremadocian) Guotang Formation, Shizilu of Sandu.

6–7. *Pareuloma* cf. *spinosa* Palmer
   6. Cranidium, ×5.5, Coll. no. 82Fg-63, Cat. no. 104322.
   7. Incomplete cranidium, ×6, Coll. no. 82Fg-57, Cat. no. 104323.
   Upper part of late Upper Cambrian Sandu Formation, Shizilu of Sandu.

8–10. *Shumardia puanensis* Lu et T.R. Zhou (sp. nov.)
   8. Exoskeleton, ×15, Holotype, Coll. no. 82Fg-21, Cat. no. 104324.
   Lower Ordovician (Tremadocian) Guotang Formation, Pu'an of Sandu.
   9. Cranidium, ×24, Coll. no. F5-102, Cat. no. 104325.
   10. Exoskeleton, ×16, Coll. no. Fg-15, Cat. no. 104326.
   Lower Ordovician (Tremadocian) Guotang Formation, Limuzhai of Sandu.

11–12. *Akoldinioidia hewenensis* (Zhou)
   11. Exoskeleton, ×18, Coll. no. 78F2-7, Cat. no. 104327.
   12. Cephalon, ×14, Coll. no. 78F2-7, Cat. no. 104328.
   Mictosaukia guizhonensis Assemblage-Zone, late Upper Cambrian Maotian Formation, Pu'an of Sandu.

Plate 3

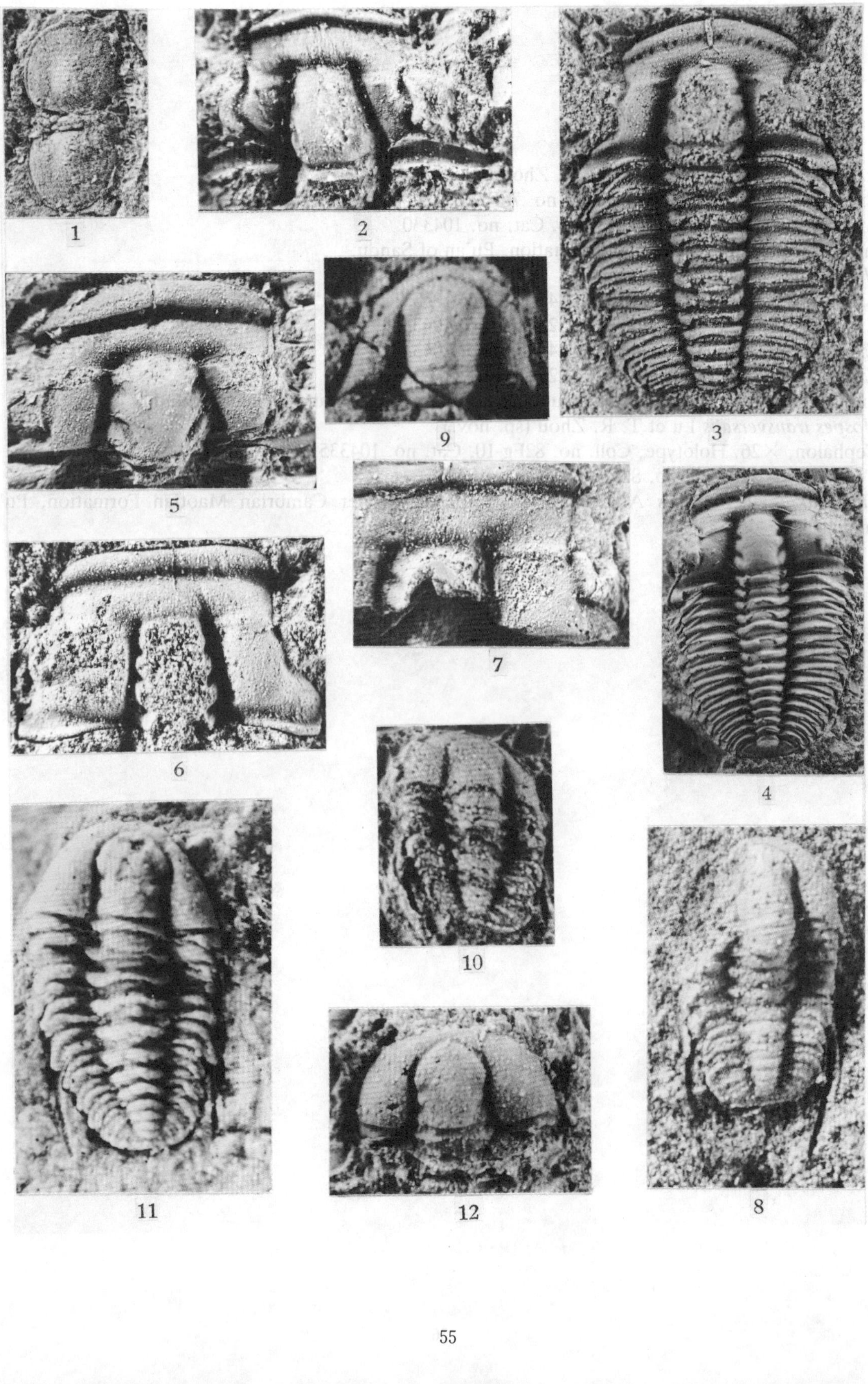

Plate 4

1–2. *Akoldinioidia sanduensis* Lu et T.R. Zhou (sp. nov.)

1. Exoskeleton, ×15, Holotype, Coll. no. 78F2-6, Cat. no. 104329.
2. Exoskeleton, ×15, Coll. no. 78F2-6, Cat. no. 104330.

Late Upper Cambrian Maotian Formation, Pu'an of Sandu.

3–6. *Hospes guizhouensis* Zhou

3. Exoskeleton, ×24, Coll. no. 82Fg-24, Cat. no. 104331.
4. Exoskeleton, ×21, Coll. no. 82Fg-22, Cat. no. 104332.
5. Exoskeleton, ×25, Coll. no. 82Fg-24, Cat. no. 104333.
6. Exoskeleton, ×23, Coll. no. 82Fg-22, Cat. no. 104334.

Lower Ordovician (Tremadocian) Guotang Formation, Pu'an and Shizilu of Sandu.

7–8. *Hospes transversalis* Lu et T. R. Zhou (sp. nov.)

7. Cephalon, ×26, Holotype, Coll. no. 82Fg-10, Cat. no. 104335.
8. Cephalon, ×32, Coll. no. 82Fg-1, Cat. no. 104336.

Mictosaukia guizhouensis Assemblage-Zone of late Upper Cambrian Maotian Formation, Pu'an of Sandu.

Plate 4

1

2

3

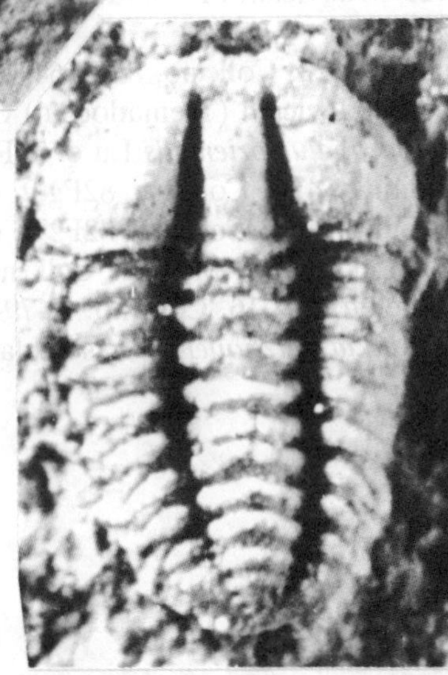
6

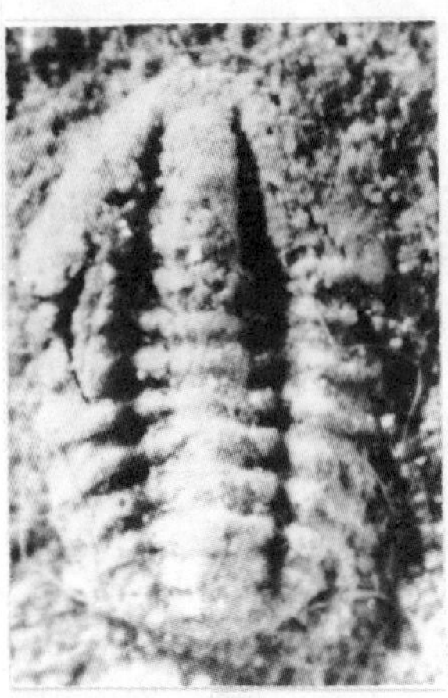
4

5

7

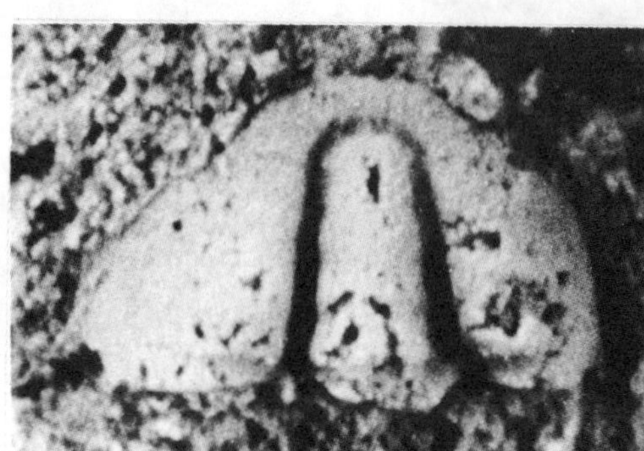
8

Plate 5

1–2. *Hospes guizhouensis* Zhou
  1. Exoskeleton, ×22, Coll. no. 82Fg-24, Cat. no. 104337.
  2. A slab with 4 individuals, ×23, Coll. no. 82Fg-16, Cat. no. 104338.
   Lower Ordovician(Tremadocian) Guotang Formation, Pu'an and Shizilu of Sandu.

3. *Hospes transversalis* Lu et T.R. Zhou (sp. nov.)
  Cephalon, ×22, Coll. no. 82Fg-10, Cat. no. 104339.
  Mictosaukia guizhouensis Assemblage-Zone of late Upper Cambrian Maotian Formation, Pu'an of Sandu.

4. *Hospes sanduensis* Zhou
  Cephalon, ×20, Coll. no. 82Fg-21, Cat. no. 104340.
  Lower Ordovician (Tremadocian) Guotang Formation, Shizilu of Sandu.

5. *Hospes* sp.
  Cephalon, ×22, Coll. no. 82Fg-64, Cat. no. 104469.
  Lower Ordovician (Tremadocian) Guotang Formation, Shizilu of Sandu.

6–9. *Puanocephalus orientalis* Lu et T.R. Zhou (gen. et sp. nov.)
  6. Pygidium, ×16, Coll. no. 82Fg-9, Cat. no. 104341.
  7. Pygidium, ×17, Coll. no. 82F2-6, Cat. no. 104342.
  8. Cephalon, ×17, Holotype, Coll. no. 78F2-6, Cat. no. 104343.
  9. Exoskeleton, ×16, Coll. no. Fl-79-87, Cat. no. 104344.
  Mictosaukia guizhouensis Assemblage-Zone, Upper Cambrian Maotian Formation, Pu'an of Sandu.

Plate 5

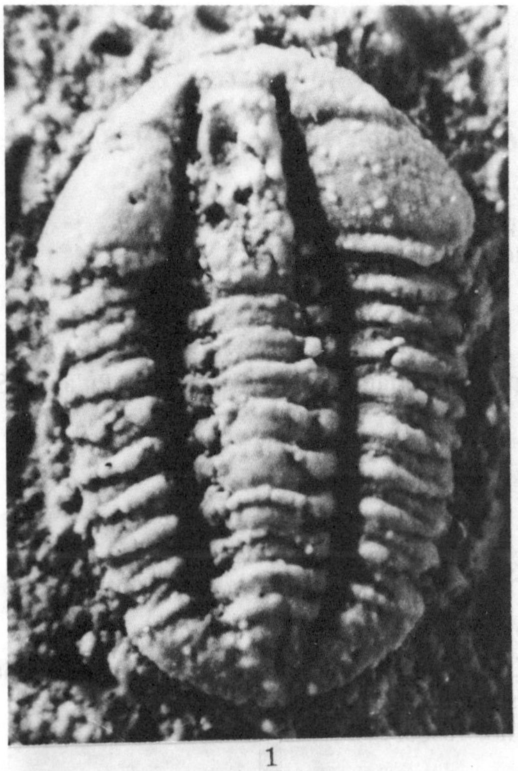

1

2

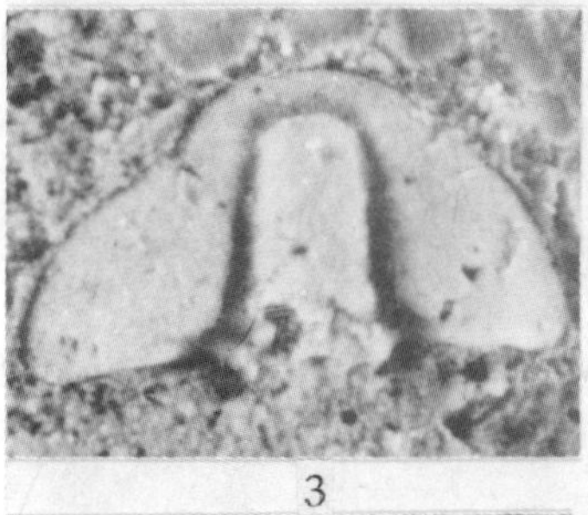

3

4

6

5

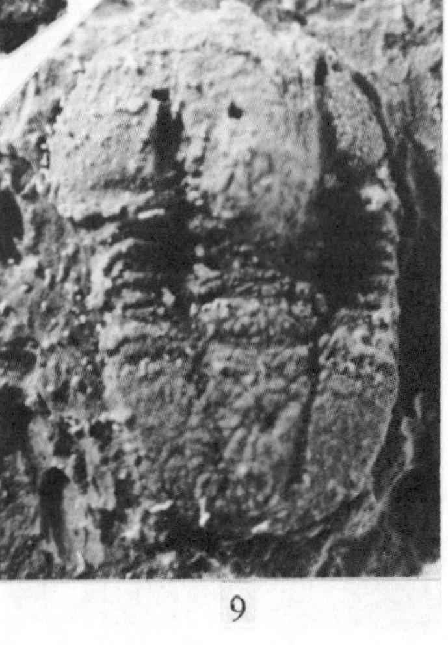

9

7

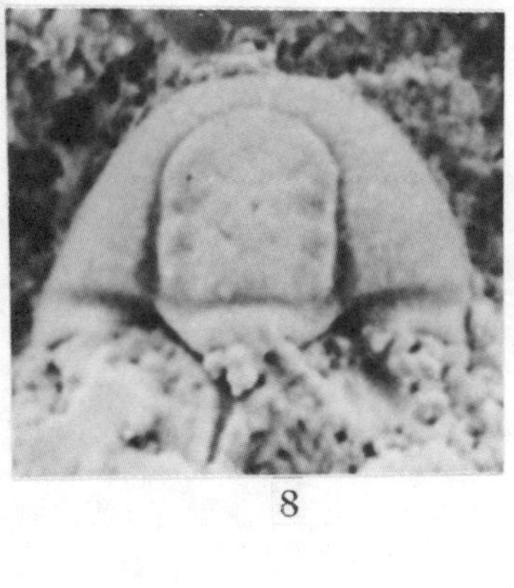

8

Plate 6

1–4. *Puanocephalus orientalis* Lu et T.R. Zhou (gen. et sp. nov.)
1. Cephalon and thorax, ×35, Coll. no. 82Fg-1, Cat. no. 104345.
2. Cephalon, ×30, Coll. no. 82Fg-1, Cat. no. 104346.
3. Cephalon, ×7, Coll. no. Fl-79-87, Cat. no. 104347.
4. Cephalon, ×7, Coll. no. 78F2-6, Cat. no. 104348.
Mictosaukia guizhouensis Assemblage-Zone, late Upper Cambrian Maotian Formation, Pu'an of Sandu.

5. *Proteuloma subquadratum* (Lu et Chien)
Cephalon, ×4.5, Coll. no. 78F27-12, Cat. no. 104349.

6–7. *Euloma* (*Archaeuloma*) *elongatum* Lu et T.R. Zhou (sp. nov.)
6. Cranidium, ×6, Coll. no. 82Fg-65, Cat no. 104350.
7. Cephalon, ×6, Coll. no. 82Fg-51, Cat. no. 104351.
Late Upper Cambrian Maotian Formation, Pu'an of Sandu.

8. *Lorrettina* (?) sp.
Cranidium, ×13, Coll. no. 82Fg-10, Cat. no. 104352.
Mictosaukia guizhouensis Assemblage-Zone, Upper Cambrian Maotian Formation, Pu'an of Sandu.

9–10. *Lorrettina* (*Puanella*) *intermedia* Lu et T.R. Zhou (subgen. et sp. nov.)
9. Cranidium, ×13, Holotype, Coll. no. 78F26-1, Cat. no. 104353.
10. Cranidium, ×6, Coll. no. F2-6, Cat. no. 104354.
Mictosaukia guizhouensis Assemblage-Zone, Upper Cambrian Maotian Formation, Pu'an of Sandu.

11–14. *Shuizuia laevigata* Lu et T.R. Zhou (gen. et sp. nov.)
11. Exoskeleton, ×5, Holotype, Coll. no. 78F26-1, Cat. no. 104355.
12. Pygidium, ×20, Coll. no. 82Fg-1, Cat. no. 104356.
13. Cranidium, ×7, Coll. no. 78F2-7, Cat. no. 104357.
14. Hypostoma, ×14, Coll. no. 78F2-7, Cat. no. 104358.
Mictosaukia guizhouensis Assemblage-Zone, late Upper Cambrian Maotian Formation, Pu'an of Sandu.

Plate 6

Plate 7

1–4. *Shuizuia laevigata* Lu et T.R. Zhou (gen. et sp. nov.)
   1. Cranidium, ×13, Coll. no. 82Fg-1, Cat. no. 104359.
   2. Pygidium, ×4, Coll. no. 82Fg-10, Cat. no. 104360.
   3. Cranidium, ×7, Coll. no. 78F2-7, Cat. no. 104361.
   4. Cranidium, ×17, Coll. no. 82Fg-10, Cat. no. 104362.

   Mictosaukia guizhouensis Assemblage-Zone, late Upper Cambrian Maotian Formation, Pu'an of Sandu.

5–8. *Shuizuia incerta* Lu et T.R. Zhou (gen. et sp. nov.)
   5. Cranidium, ×25, Coll. no. 82Fg-10, Cat. no. 104363.
   6. Pygidium, ×16, Coll. no. 82Fg-10, Cat. no. 104364.
   7. Pygidium, ×21, Coll. no. 82Fg-10, Cat. no. 104365.
   8. Cranidium, ×20, Coll. no. 82Fg-10, Cat. no. 104366.

   Mictosankia guizhouensis Assemblage-Zone, late Upper Cambrian Maotian Formation. Pu'an of Sandu.

9. *Parabolinella* aff. *latilimbata* Lu et Chien

   Cranidium, × 5. External mould (see also Pl. 17, Fig. 12, Latex cast), Coll. no. 82Fg-64, Cat. no. 104367.

   Lower Ordovician (Tremadocian) Guotang Formation, Pu'an of Sandu.

Plate 7

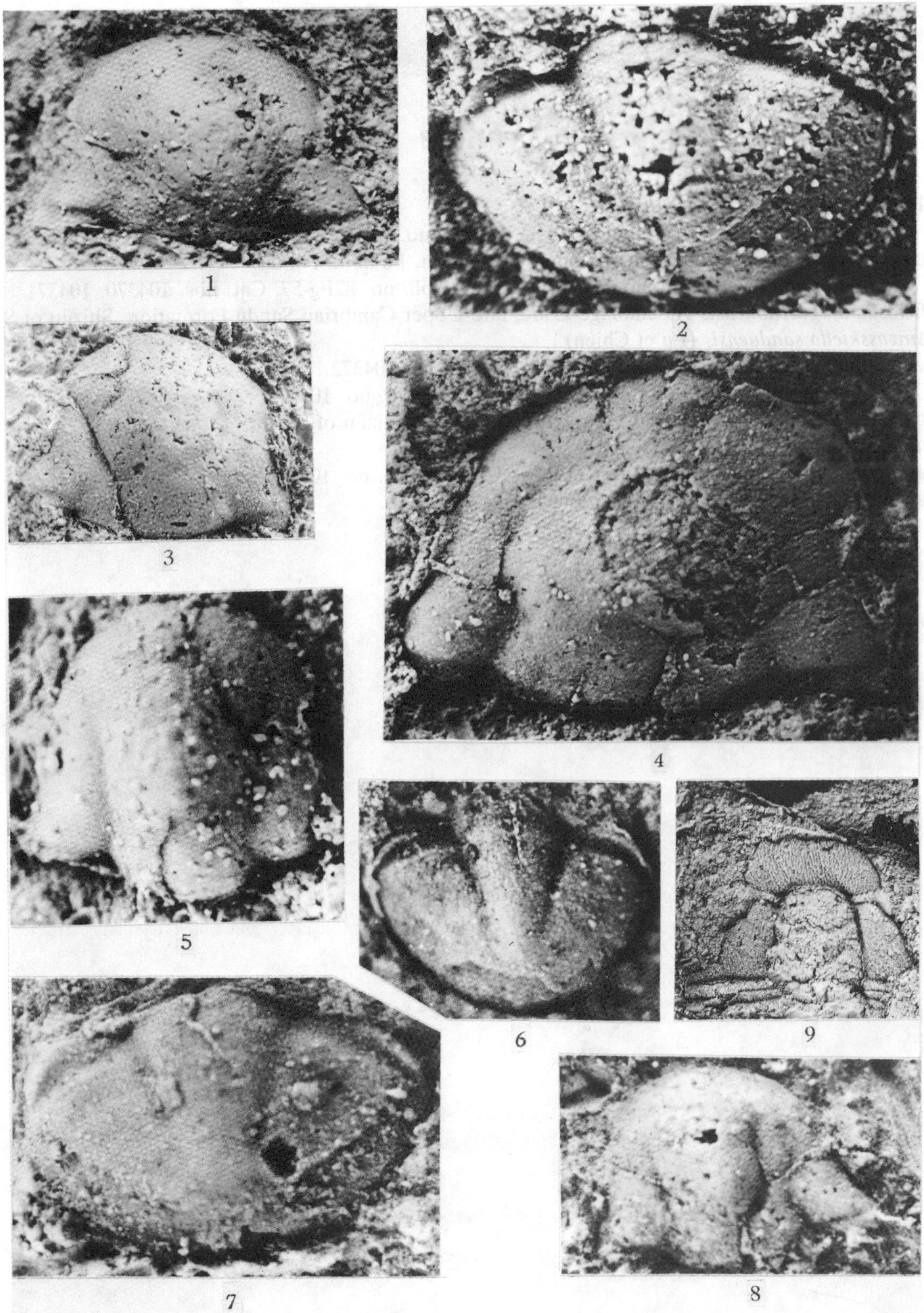

Plate 8

1–4. *Acrocephalina convexa* Lu et Chien
   1,4. Cranidium and its external mould, ×3, ×4. Coll. no. 78F26-12. Cat. nos. 104368, 104369.
   Lower Ordovician (Tremadocian) Guotang Formation, Pu'an of Sandu.
   2,3. Cranidium and its external mould, ×4.7, ×4.7. Coll. no. 82Fg-57, Cat. nos. 104370, 104371.
   Mictosaukia guizhouensis Assemblage-Zone, late Upper Cambrian Sandu Formation, Shizilu of Sandu.

5–6. *Amzasskiella sanduensis* (Lu et Chien)
   5. Incomplete cranidium, ×4. Coll. no. 82Fg-66, Cat. no. 104372.
   6. Cranidium and part of thorax, ×4. Coll. no. 78F2-4. Cat. no. 104373.
   Lower Ordovician (Tremadocian) Guotang Formation, Shizilu of Sandu.

7. *Hedinaspis shengi* Lu
   Nearly complete exoskeleton, ×2.6, Coll. no. 82Fg-57, Cat. no. 104374.
   Late Upper Cambrian Sandu Formation, Shizilu of Sandu.

Plate 8

Plate 9

1–3. *Taoyuania xenisma* Liu
  1. Cranidium. ×2. Coll. no. 78Fg-12, Cat. no. 104375.
  2. Cranidium, ×2. Coll. no. 78Fg-12, Cat. no. 104376.
  3. Pygidium, ×2.5. Coll. no. 82Fg-24, Cat. no. 104377.
  Lower Ordovician (Tremadocian) Guotang Formation, Pu'an of Sandu.

4. *Leiostegium* (*Leiostegium*) *constrictum* Peng.
  Cranidium, ×7. Coll. no. 82Fg-65, Cat. no. 104378.
  Lower Ordovician (Tremadocian) Guotang Formation, Shizilu of Sandu.

5–6. *Leiostegium* (*Leiostegium*) *punctatum* (Zhou et Zhang)
  5. Cranidium, ×4.5. Dorsal view. Coll. no. 78F26-12. Cat. no. 104379.
  6. Same as 5, ×4.5, Left lateral view
  Lower Ordovician (Tremadocian) Guotang Formation, Pu'an of Sandu.

7–8. 9a. *Leiostegium* (*Leiostegium*) sp.
  7. Cranidium, ×27. Coll. no 82Fg-24. Cat. no. 104380. (see also Pl. 15. Fig.11a)
  8. Cranidium, ×20. Coll. no. 82Fg-21, Cat. no. 104381.
  9a. Incomplete cranidium (lower right), ×30. Coll. no. 78Fg-12. Cat. no. 104382.
  Lower Ordovician (Tremadocian) Guotang Formation, Pu'an of Sandu.

9. *Pseudokoldinioidia encrinuroides* Lu et T. R. Zhou (sp. nov.)
  Cranidium, X30. Coll. no. 78Fg-12. Cat. no. 104383.

10. *Pagodia* sp.
  Strongly weathered cranidium, ×21. Coll. no. 82Fg-9, Cat. no. 104384.
  Mictosaukia guizhouensis Assemblage-Zone, late Upper Cambrian Maotian Formation, Pu'an of Sandu.

Plate 9

Plate 10

1–3. *Eosaukia latilimbata* Lu
1. Cranidium, ×17. Coll. no. 78F2-6, Cat. no. 104385.
2. Cranidium, ×7. Coll. no. 78F2-7, Cat no. 104386.
3. Pygidium, ×5. Coll. no. 82Fg-9, Cat. no. 104387.

Mictosaukia guizhouensis Assemblage-Zone, late Upper Cambrian Maotian Formation, Pu'an of Sandu.

4–10. *Eosaukia rectangula* Lu et T.R. Zhou (sp. nov.)
4. Exoskeleton, ×4.5. Holotype, Coll. no. 78F2-6, Cat. no. 104388.
5. Cranidium, ×6, Coll. no. 78F2-7, Cat. no. 104389.
6. Cranidium, ×5. Coll. no. 78F2-7. Cat no. 104390.
7. Cranidium, ×3.75. Coll. no. 82Fg-1, Cat. no. 104391.
8. Exoskeleton, ×1.5. Coll. no. 82Fg-9, Cat. no. 104392
9. Exoskeleton, × 3.7. Coll. no. 82Fg-57, Cat. no. 104393.
10. Librigena, ×3. Coll. no. 82Fg-9, Cat. no. 104394.

Mictosaukia guizhouensis Assemblage-Zone, late Upper Cambrian Maotian and Sandu Formations, Pu'an and Shizilu of Sandu.

Plate 10

Plate 11

1–4. *Mictosaukia guizhouensis* Lu et T.R. Zhou (sp. nov.)
1. Cranidium, ×7. Holotype, Coll. no. 78F2-5, Cat. no. 104395.
2. Cranidium, ×7. Coll. no. 79F1-76 (=82Fg-1). Cat. no. 104396.
3. Cranidium, ×2. Coll. no. 79F1-76(=82Fg-1). Cat. no. 104397.
4. Small cranidium, ×8. Coll. no. 78F2-6. Cat. no. 104398.

Mictosaukia guizhouensis Assemblage-Zone, late Upper Cambrian Maotian Formation, Pu'an of Sandu.

5–10. *Mictosaukia* (*Mictosaukioidia*) *transita* Lu et T.R. Zhou (subgen. et sp. nov.)
5. Cranidium, ×7. Coll. no. 78F2-7, Cat. no. 104399.
6. Cranidium, ×3, Coll. no. 82Fg-1, Cat. no. 104400.
7. External mould of fig. 6, with a pygidium of *Linguisaukia affinis* Peng on right side. Cat. no. 104401.
8. Librigena, ×2.5. Coll. no. 82Fg-9, Cat. no. 104402.
9. Cranidium, ×4. Coll. no. 82Fg-10, Cat. no. 104403.
10. Cranidium, ×7. Holotype. Coll. no. 79F1-87(=82Fg-1). Cat. no. 104404.

Mictosaukia guizhouensis Assemblage-Zone, late Upper Cambrian Maotian Formation, Pu'an of Sandu.

11–13. *Pharostomina sanduensis* Zhou
11. Cranidium, ×4.5, Coll. no. 82Fg-24. Cat. no. 104405.
12. Incomplete cranidium, ×7, Coll. no. 82Fg-65, Cat. no. 104406.
13. Small cranidium, ×16. Coll. no. 82Fg-12. Cat. no. 104407.

Lower Ordovician (Tremadocian) Guotang Formation, Shizilu of Sandu.

Plate 11

Plate 12

1–7. *Linguisaukia spinata* Peng

1. Exuviation specimen showing the inverted cephalon with hypostoma and doublure in original position, and the rotated thorax (11 segments) and pygidium. The lateral view of occipital spine in front of the first thoracic segment is probably a part from the cranidium on the ecdysis of the animal, ×4. Coll. no. 78F2-6. Cat. no. 104407.
2. Cranidium, ×5. Coll. no. 82Fg-1. Cat. no. 104408.
3. Cranidium, ×7. Coll. no. 82Fg-1. Cat. no. 104409.
4. Exoskeleton, ×2. Coll. no. 79F1-76. Cat. no. 104410.
5. Exoskeleton, ×5. Coll. no. 82Fg-10. Cat. no. 104411.
6. Hypostoma,×5. Coll. no. 78F2-6, Cat. no. 104412.
7. Pygidium, ×5. Coll. no. 78F2-10, Cat. no. 104413.

Mictosaukia guizhouensis Assemblage-Zone, late Upper Cambrian Maotian Formation, Pu'an of Sandu.

Plate 12

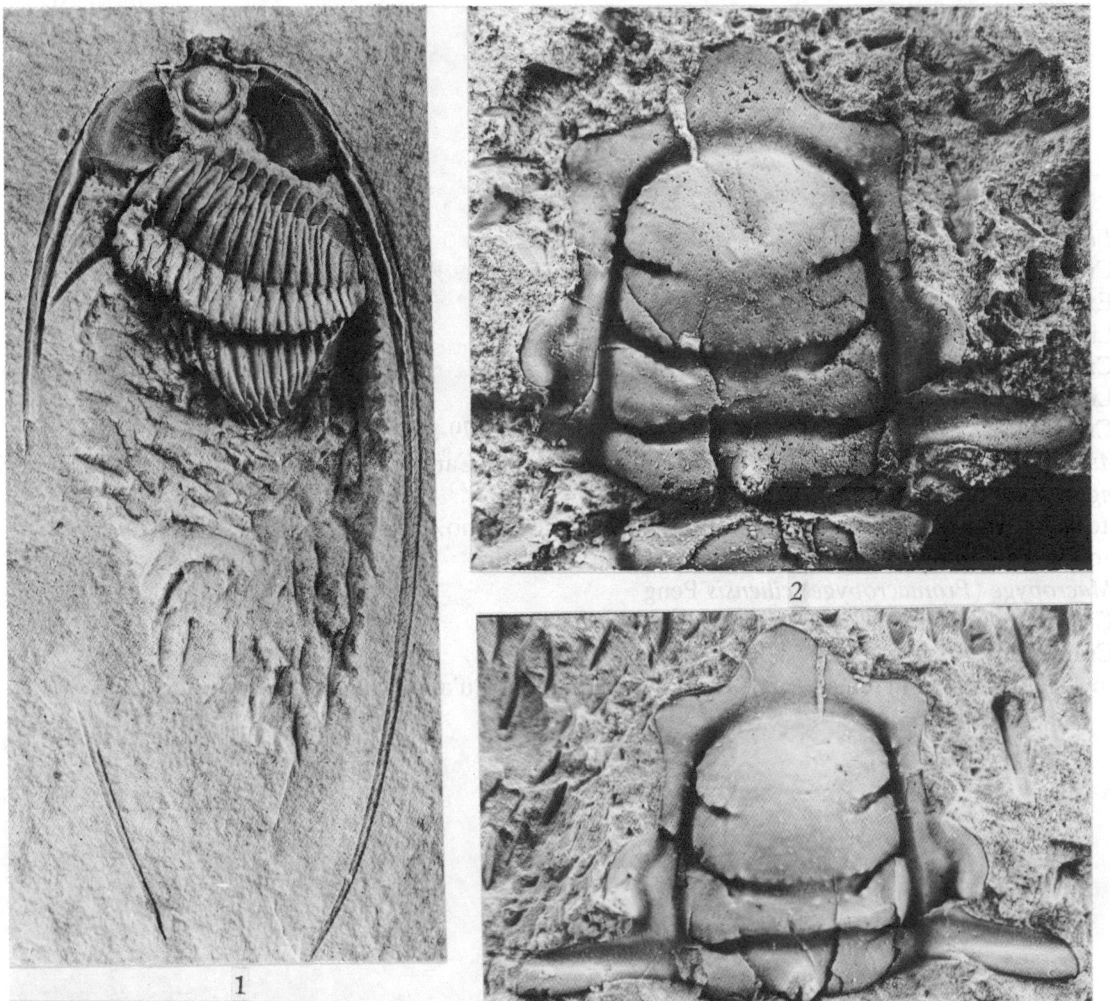

Plate 13

1–6. *Linguisaukia spinata* Peng
1. Cranidium, ×4.5. Coll. no. 78F2-6. Cat. no. 104414.
2. Exoskeleton, ×4. Coll. no. 82Fg-10, Cat. no. 104415.
3. Librigena, ×4. Coll. no. 78F2-6, Cat. no. 104416.
4. Cranidium, ×7. Coll. no. 82Fg-1. Cat. no. 104417.
5. Librigena, ×4.5. Coll. no. 82Fg-1. Cat. no. 104418.
6. Cranidium, ×4.5. strongly lateral pressed. Coll. no. F1-79-85(=82Fg-8). Cat. no. 104419.

Mictosaukia guizhouensis Assemblage-Zone, late Upper Cambrian Maotian Formation, Pu'an of Sandu.

7. *Linguisaukia affinis* Peng
External mould of pygidium, ×3. Coll. no. 82Fg-57. Cat. no. 104420.
Late Upper Cambrian Sandu Formation, Shizilu of Sandu.

8–9. *Macropyge* (*Promacropyge*) *ciliensis* Peng
8. Cranidium, ×11. Coll. no. 78Fg-12. Cat. no. 104421.
9. Cranidium, ×5.5. Coll. no. 82Fg-14. Cat. no. 104422.

Lower Ordovician (Tremadocian) Guotang Formation, Pu'an of Sandu.

Plate 13

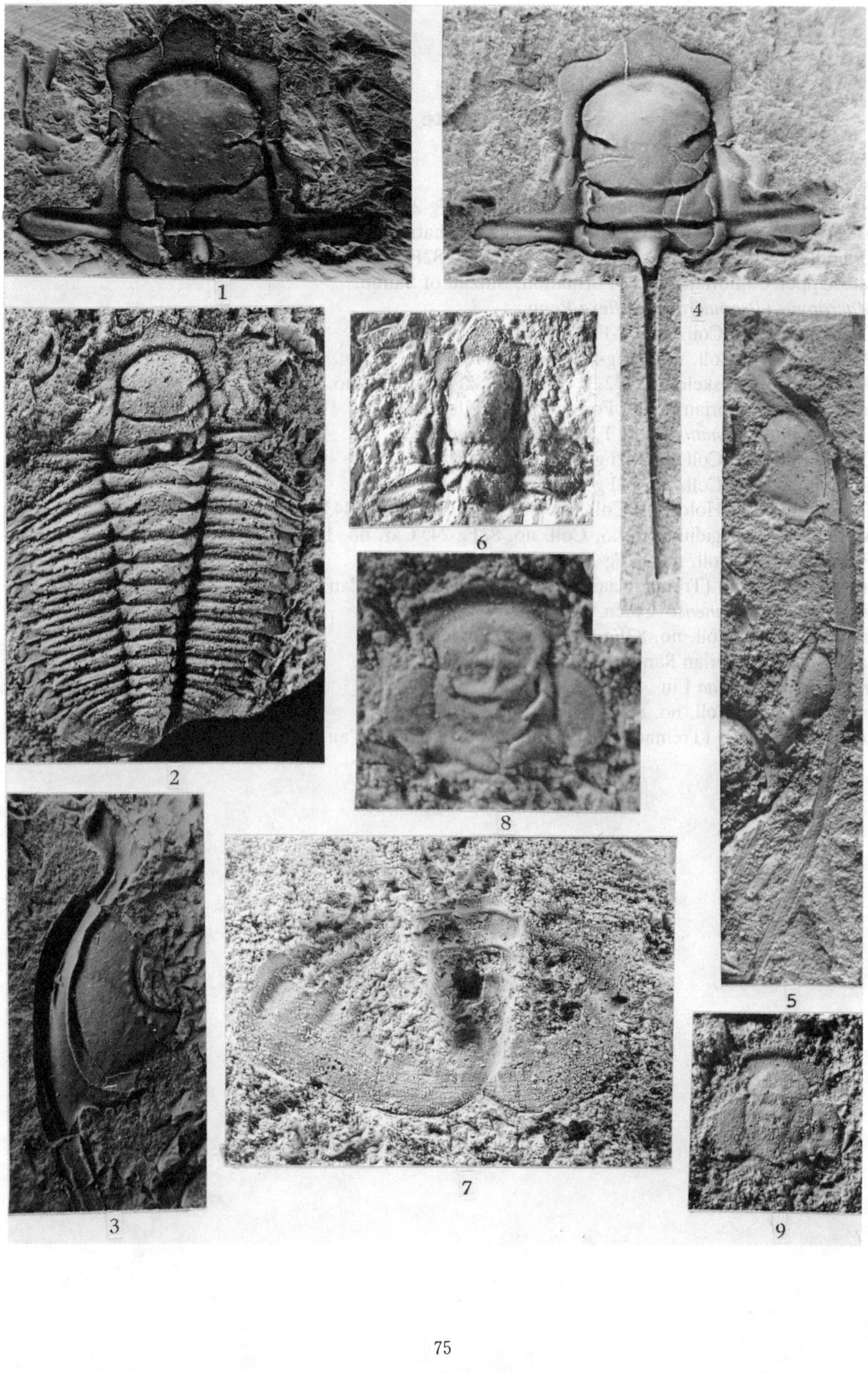

Plate 14

1–2. *Macropyge* (*Promacropyge*) *carinata* Lu
  1. Nearly complete exoskeleton, ×4, Coll. no. 82Fg-24. Cat. no. 104423.
     Lower Ordovician (Tremadocian), Guotang Formation, Pu'an of Sandu
  2. Cranidium with a part of thorax, ×4, Coll. no. 82Fg-57. Cat. 104424.
     Late Upper Cambrian Sandu Formation, Shizilu of Sandu.

3–5. *Macropyge* (*Promacropyge*) *flexa* Peng
  3. Cranidium, ×6, Coll. no. 78-F27-14. Cat. no. 104425.
  4. Pygidium, ×4. Coll. no. 82Fg-57. Cat. no. 104426.
  5. Latex cast of exoskeleton, ×2.3. Coll. no. 82Fg-51. Cat. no. 104427.
     Late Upper Cambrian Sandu Formation, Shizilu of Sandu.

6–10. *Golasaphus carinatus* Lu et T.R. Zhou (sp. nov.)
  6. Cranidium, ×5. Coll. no. 82Fg-24, Cat. no. 104428.
  7. Cranidium, ×4. Coll. no. 82Fg-24, Cat. no. 104429.
  8. Cranidium, ×4. Holotype. Coll. no. 82Fg-24. Cat. no. 104430.
  9. Medium-sized pygidium, ×4.5. Coll. no. 82Fg-24. Cat. no. 104431.
  10. Pygidium, ×3. Coll. no. 82Fg-24. Cat. no. 104432.
     Lower Ordovician (Tremadocian) Guotang Formation, Pu'an of Sandu.

11. *Niobella yangjiawanensis* Chien
    Cranidium, ×4. Coll. no. 82Fg-57, Cat. no. 104433.
    Late Upper Cambrian Sandu Formation, Shizilu of Sandu.

12. *Troedssonia elongata* Liu
    Cranidium, ×7. Coll. no. 82Fg-17. Cat. no. 104434.
    Lower Ordovician (Tremadocian) Guotang Formation, Pu'an of Sandu.

Plate 14

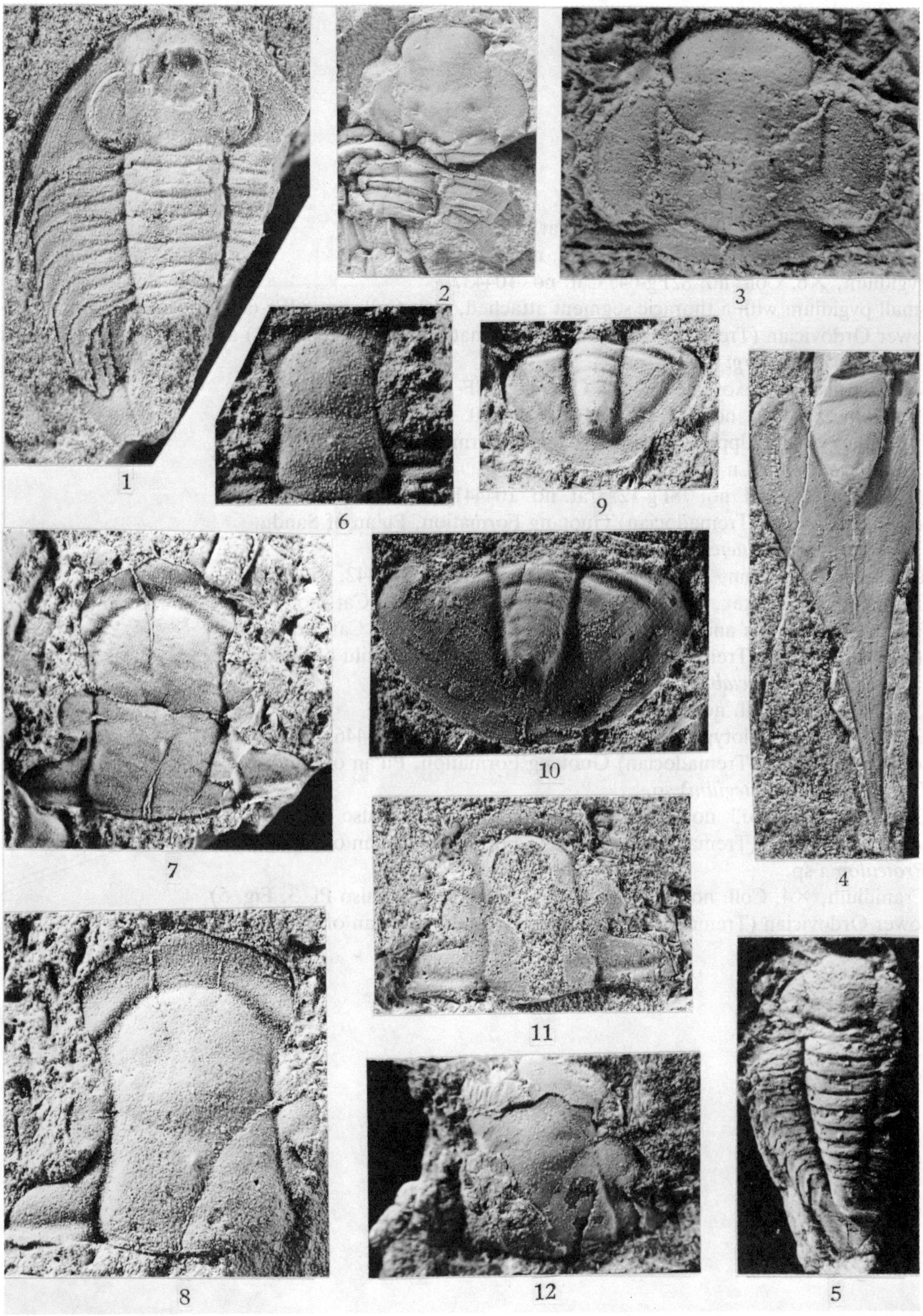

Plate 15

1–4. *Troedssonia wimani* (Troedsson)
1. Cranidium, ×6.5, Coll. no. 82Fg-64, Cat. no. 104435.
2. Cranidium, ×4. Coll. no. 82Fg-24, Cat. no. 104436.
3. Pygidium, ×6. Coll. no. 82Fg-64, Cat. no. 104437.
4. Small pygidium with a thoracic segment attached, ×7. Coll. no. 82Fg-64, Cat. no. 104438.

Lower Ordovician (Tremadocian) Guotang Formation, Shizilu (82Fg-64) and Pu'an (82Fg-24), Sandu.

5–6. *Diceratopyge mobergi* Troedsson
5. Strongly distorted exoskeleton, ×5 Coll. no. 78F-16. Cat. no. 104439.
6. Pygidium, ×5 Coll. no. 78F-16, Cat. no. 104440.

Upper part of late Upper Cambrian Maotian Formation, Pu'an of Sandu.

7. *Diceratopyge agenor* Lu et Chien

Cephalon, ×7.5, Coll. no. 78Fg-12. Cat. no. 104441.

Lower Ordovician (Tremadocian) Guotang Formation, Pu'an of Sandu.

8–10. *Diceratopyge sanduensis* Lu et T.R. Zhou (sp. nov.)
8. Thorax and pygidium, ×5. Coll. no. 82Fg-6, Cat. no. 104442.
9. Cranidium and thorax, ×3.7. Holotype, Coll. no. 82F2-4. Cat. no. 104443.
10. Fragmentary thorax and pygidium, ×4. Coll. no. 82Fg-21. Cat. no. 104444.

Lower Ordovician (Tremadocian) Guotang Formation, Shizilu of Sandu.

11–12. *Palaeoharpes spectabilis* Lu et T.R. Zhou (sp. nov.)
11. Cephalon, ×4. Coll. no. 82Fg-24, Cat. no. 104445.
12. Cephalon, ×8. Holotype, Coll. no. 78F26-12. Cat. no. 104446.

Lower Ordovician (Tremadocian) Guotang Formation, Pu'an of Sandu.

11a. *Leiostegium* (*Leiostegium*) sp.

Cranidium, ×4, Coll. no. 82Fg-24, Cat. no. 104447. (see also Pl. 9, fig. 7,8,9a)

Lower Ordovician (Tremadocian) Guotang Formation, Pu'an of Sandu.

11b. *Proteuloma* sp.

Cranidium, ×4, Coll. no. 82Fg-24, Cat. no. 104448. (see also Pl. 3, Fig. 5).

Lower Ordovician (Tremadocian) Guotang Formation, Pu'an of Sandu.

Plate 15

Plate 16

1–6. *Pseudokoldinioidia expansa* Lu et T.R. Zhou (sp. nov.)
1. Cranidium, ×20, Coll. no. 78F2-6. Cat. no. 104449.
2. Pygidium, ×12, Coll. no. 82Fg-1. Cat. no. 104450.
3. Cephalon showing the librigena and genal spine, ×20, Coll. no. 78F2-6. Cat. no. 104451.
4. Cranidium, ×36, Holotype, Coll. no. 82Fg-1. Cat. no. 104452.
5. Cranidium, ×20, Coll. no. 78F2-6. Cat. no. 104453.
6. Pygidium, ×12, Coll. no. 82Fg-1. Cat. no. 104454.

Mictosaukia guizhouensis Assemblage-Zone, late Upper Cambrian Maotian Formation, Pu'an of Sandu.

7–8. *Pseudokoldinioidia* sp.
7. Cranidium, ×20. Coll. no. 82Fg-10, Cat. no. 104455.
8. Pygidium, ×35, Coll. no. 82Fg-10, Cat. no. 104456.

Mictosaukia guizhouensis Assemblage-Zone, late Upper Cambrian Maotian Formation, Pu'an of Sandu.

Plate 16

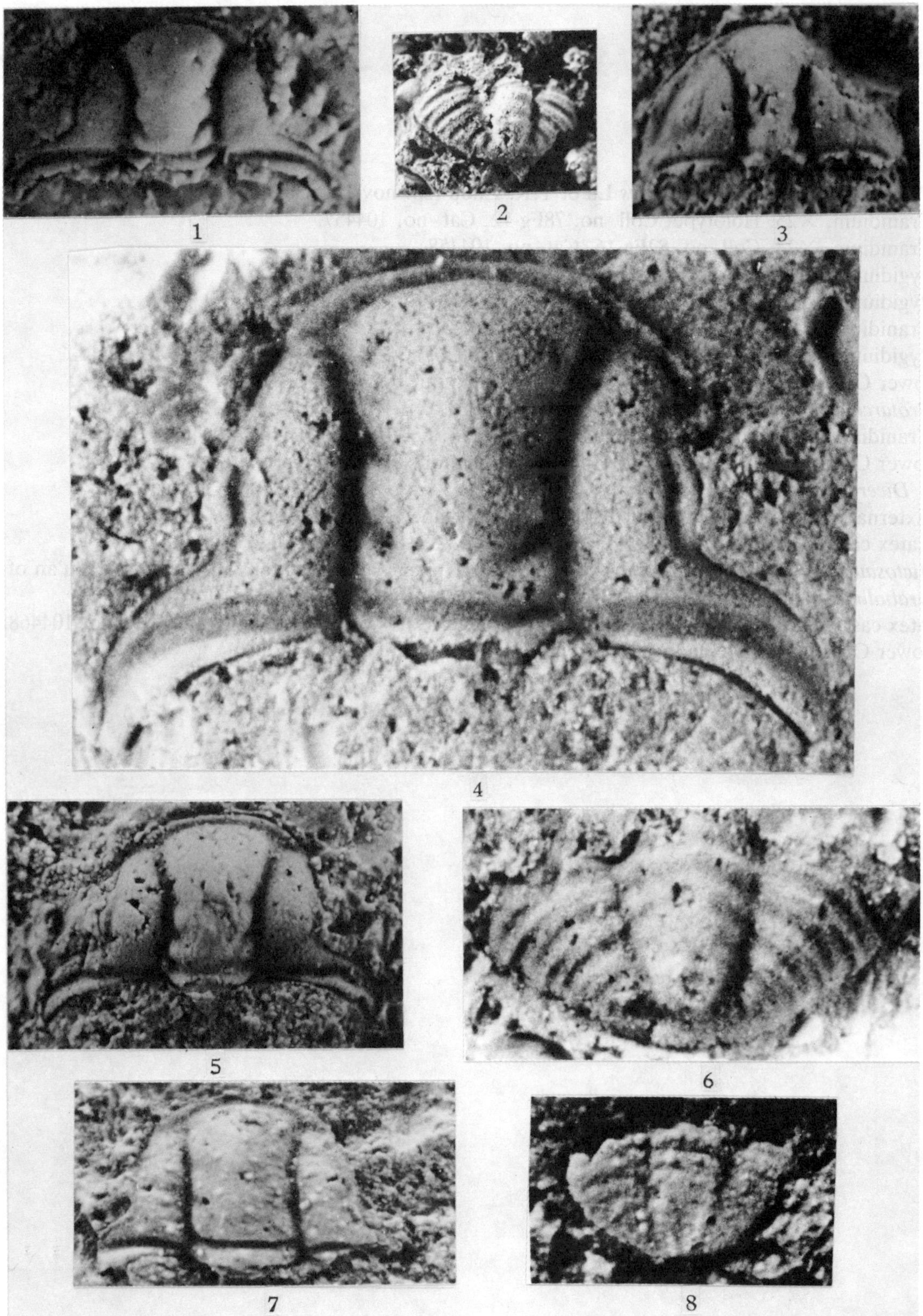

Plate 17

1–6. *Pseudokoldinioidia encrinuroides* Lu et T.R. Zhou (sp. nov.)

1. Cranidium, ×25, Holotype. Coll. no. 78Fg-12, Cat. no. 104457.
2. Cranidium, ×24, Coll. no. 82Fg-16, Cat. no. 104458.
3. Pygidium, ×22, Coll. no. 78Fg-12, Cat no. 104459.
4. Pygidium, ×27, Coll. no. 78Fg-12, Cat. no. 104460.
5. Cranidium, ×20, Coll. no. 78Fg-12, Cat. no. 104461.
6. Pygidium, ×27, Coll. no. 78Fg-12, Cat no. 104462.

Lower Ordovician (Tremadocian) Guotang Formation, Pu'an of Sandu.

7–9. *Protarchaeogonus sanduensis* Zhou

Cranidia, ×20, ×20, ×20, Coll. no. 78Fg-12. Cat. no. 104463, 104464, 104465.

Lower Ordovician (Tremadocian) Guotang Formation, Pu'an of Sandu.

10–11. *Diceratopyge* sp.

10. External mould of cranidium of fig. 11, ×3. Coll. no. 82Fg-8, Cat. no. 104466.
11. Latex cast of cranidium of fig. 10, ×3. Coll. no. 82Fg-8, Cat. no. 104467.

Mictosaukia guizhouensis Assemblage-Zone, late Upper Cambrian Maotian Formation, Pu'an of Sandu.

12. *Parabolinella* aff. *latilimbata* Lu et Chien

Latex cast of external mould of cranidium (Pl. 7, fig. 9), ×5, Coll. no. 82Fg-64. Cat. no. 104468.

Lower Ordovician (Tremadocian) Guotang Formation, Pu'an of Sandu.

Plate 17

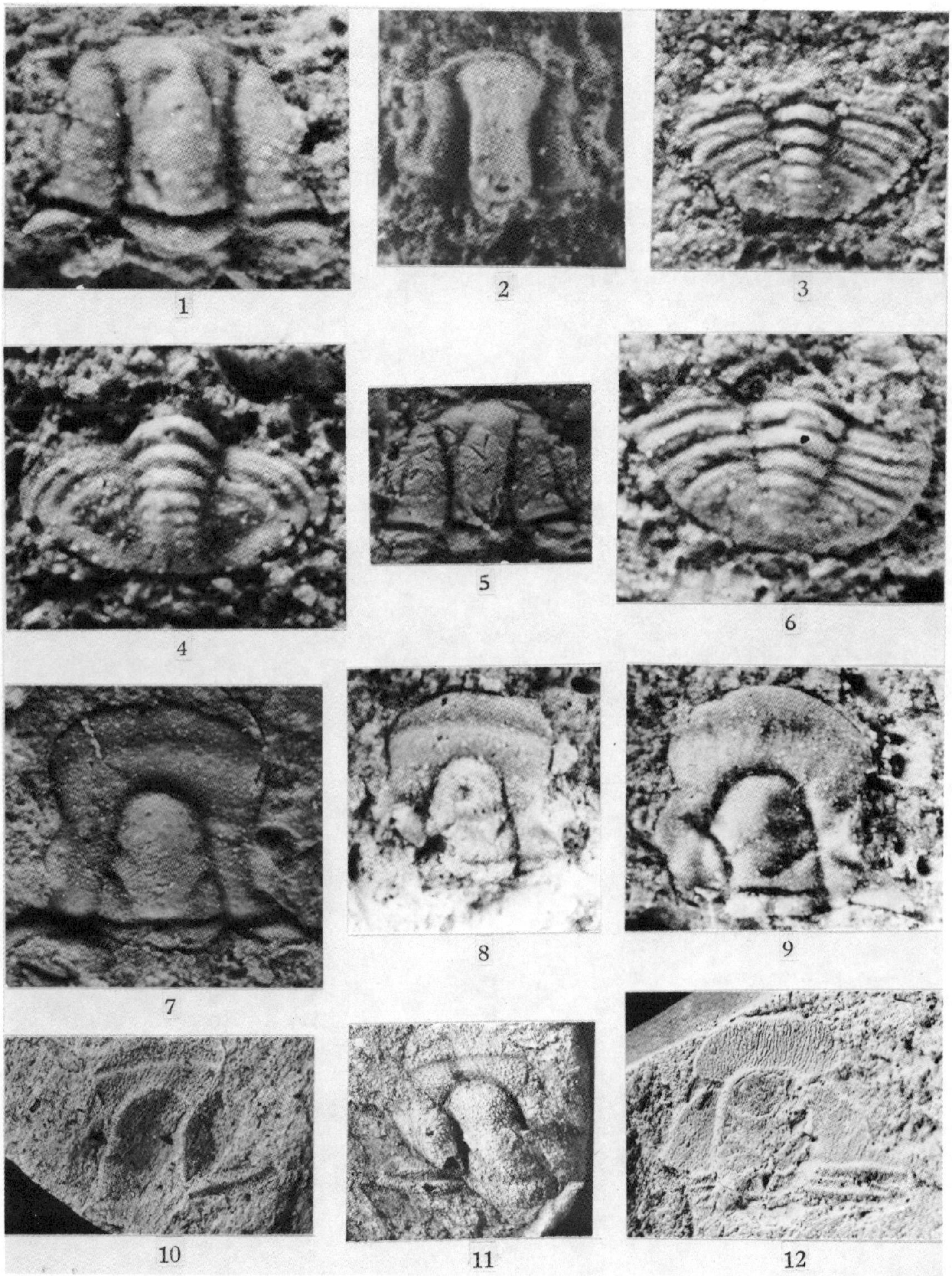

PALAEONTOLOGIA CATHAYANA Volume 5, 1990, p.85–138

DEVONIAN STRATIGRAPHY, PALAEOGEOGRAPHY AND MINERAL RESOURCES IN HUNAN

Yu Changmin and Xu Hankui
(Nanjing Institute of Geology and Palaeontology, Academia Sinica)

Peng Ji and Xiao Songtao
(Geological Institute of the Colour-Metal Company of Hunan)

Liu Zhuhan
(Institute of Coal Mining of Hunan)

Abstract

The Devonian rocks in Hunan are distributed in the shallow sea basin extending from SW to NE. The Jiangnan Oldland, trending NE-SW, separated the basin into northwestern Hunan and central Hunan. Based on the differences of lithology and fossil content, the Devonian of Hunan may be distinguished into the littoral facies including the outer littoral facies and the inner littoral facies, and the shallow marine facies, which can be further differentiated into the reef platformal facies zone and the inter-reef depressional facies zone.

The framework of subdivision of Devonian in each facies is outlined, with a special discussion on the lithostratigraphic and biostratigraphic characters of the Chitzechiao Formation and the Shetienchiao Formation and their correlatives, and the biostratigraphic characters of the Hsikuangshan Formation. Some remarks are made on the age of the Shaotung Formation and the Devonian-Carboniferous boundary in Hunan.

The Devonian paleogeography of Hunan is discussed with special emphasis on the development of Devonian reef complex, which can be generalized into 4 phases, ranging from the late Middle Devonian to the end of the Late Devonian Frasnian stage.

Two kinds of ore deposit associations can be recognized. The iron deposit of Ningxiang type is known to occur mostly in the transitional terrain between the outer and inner littoral facies zones. The siderite and its sister ore deposits have been proved to occur mostly at the transitional terrain between the reef platformal facies zone and the inter-reef depressional facies zone in the shallow marine facies. It is for the first time to point out that the Devonian paleogeography and reef complex are the main controlling factors and genetic mechanisms affecting the distributional pattern of the iron ore deposits of the Ningxiang type and also the formation of the siderite and its sister ore deposits in Hunan.

PREFACE

Hunan is one of the famous provinces of China for the well development of the Devonian strata. In particular, the Upper Devonian of Hunan including the Shetienchiao and the Hsikuangshan Formations had been considered as the representative of the Upper Devonian of China, which had been called "the Hunan Series" (Tian, C. C., 1938). The Devonian of Hunan has various types of facies with well exposed sections in different facies zones, and contains many biotas, such as benthic and pelagic, invertebrates and vertebrates, megaflora, spores and algae as well, giving rise to many biofacies, especially the development of the reef complex. Therefore, Hunan is an ideal province for studying the Devonian stratigraphy, palaeoecology and reef complex. Devonian is one of the important periods in which a great variety of metallic and non-metallic mineral resources are formed. There are well-known oolitic and pisolitic hematite ore deposits in Hunan, or so-called the Ningxiang type; in addition, many kinds of coloured metal deposits, such as siderite, pyrite and Cu, Pb, Zb, Hg-metal sulfide ore deposits, and Au, Ca, W. Mn-ore deposits as well, have been found there. A study on the Devonian of Hunan is therefore of great economical value.

It was early in 1938 that the well-known Chinese geologist, C. C. Tian and his colleagues studied the Devonian of Hunan and subdivided it into 4 formations, namely, the Tiaomachien Formation, the Chitzechiao Formation, the Shetienchiao Formation and the Hsikuangshan Formation, 8 members and 19 brachioped zones (Tian, 1938). Tian's work has laid a solid foundation for later working on the Devonian in this area. Sun, Y. T. (1935, 1958), Wang Y. L. (1938), Sze, H. C. (1937), Chi, Y. S. (1942), Liu, K. C. (1942) and others have also made contributions to the Devonian faunas and floras and the correlation and subdivision of Devonian in this area. (Text-fig.1)

During the last three decades, regional mapping, reconnaisance and exploration of mineral resources in Hunan have been conducted by the Geological Bureau and the Prospecting Corporation of Hunan. Meanwhile, researches into regional stratigraphic work carried out by those specialists from different institutions and colleges have accumulated large amounts of new data and materials on the Devonian stratigraphy and palaeontology. These have been summarized by Wang Yu and Yu Changmin (1963) and recently by Zhao Ruxuan and others (1978), who have made a general statement and discussion on the distribution and subdivision of Devonian in Hunan.

In 1977, under the sponsorship of the Nanjing Institute of Geology and Palaeontology, Academia Sinica, a research team was organized for studying the Devonian strata and the iron-bearing deposits. The work was carried out in cooperation with the Geological Institute of Metallurgical Industry of Hunan, the Regional Geological Surveying Team of Hunan and the Institute of Geology of Hunan. Those who joined in the field work were: Yu Changmin, Xu Hankui, Wang Chenyuan, Cai Chongyang, Lu Lichang, Mu Daochen and others from the Nanjing Institute of Geology and Palaeontology; Peng ji from the Geological

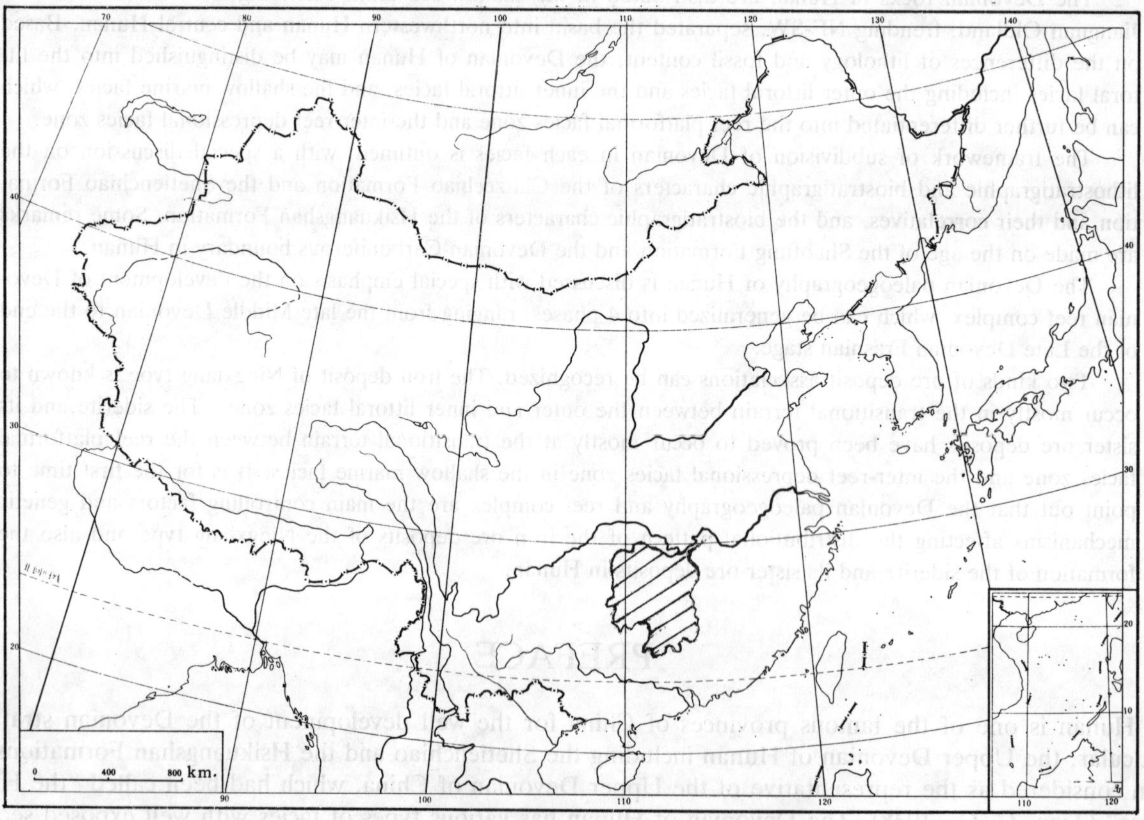

Text-fig.1. Showing the situation of Hunan in China.

Institute of the Metallurgical Industry; Xia Zhifen from the Institute of Geology; Yu Risheng and Liu Zuhan from the Regional Geological Surveying Team.

In laboratory, fossil identification and mineral and rock determination were made by the following colleagues in the respective fields: corals–Yu Chang-min and Jiang Shuigen; brachiopods–Xu Hankui and Liu Zuhan; tentaculites and ammonoids–Ruan Yiping; conodonts–Wang Chengyuan and Wang Zhihao; bryozoans–Hu Zhaoxuan; minerals and rocks–Peng Ji; map-compiling and drawing–Xiao Songtao.

The result was written by Yu Chang-min, Xu Han-kui, Pen Ji and Liu Zu-han in a paper entitled "The Devonian strata of Hunan and its characteristics of ore deposits", which was submitted to the Symposium on the siderite ore deposits sponsored jointly by the Chinese Academy of Sciences and Nanjing Institute of Geology and Paleontology in 1979 in Changsha, Hunan. Papers presented to that symposium were published only in Chinese. Owing to the limited circulation, there are not many people even in China knowing about this collected works, to say nothing of those outside this country. It was during the last few years when some of the present writers were talking over this subject that most colleagues both in China and abroad expressed their interest and desire for an English version of that paper. Therefore, we decide to rewrite it in English by adding some materials or new data from this area.

This paper involves the zonation of lithofacies in Hunan, the division of strata, the biostratigraphic characters, the development of the reef complex and its possible effect on the formation of ore deposits, including metal and non-metal, coloured metal and iron ore deposits, particularly presenting a model to interprete the Devonian palaeogeography and reef complex as the main possible controlling factors and genetic mechanism affecting the distributional pattern of the iron ore deposits and the formation of the siderite ore association.

STRATIGRAPHY AND SUBDIVISION OF DEVONIAN IN HUNAN

Regional Distribution and Facies Pattern of the Devonian in Hunan.

The Devonian in Hunan is distributed in the shallow sea basin extending from SW to NE. On the north, it is truncated by the oldland, while its southwestern portion conjoins with the sea basin of Guangdong and Guangxi; the eastern portion is in connection with western Jiangxi, where the sediments are mainly of littoral facies. The Jiangnan Oldland, trending NE to SW, separated the basin into two different depositional areas: NW Hunan and central Hunan.

The Devonian in northwestern Hunan is more or less similar to that in the western part of Hubei, where only the Upper Devonian is probably developed and can be mainly divided into two parts. The lower part is called the Huangchiateng Formation, yielding *Cyrtospirifer, Tenticospirifer, Atrypa, Pseudozaphrentis difficile* and *Leptophloeum rhombicum*. The Yungtaikuan Formation consisting mainly of quartzose sandstone and being formerly assigned to the Middle Devonian may be the approximate correlative bed of the Huangchiateng Formation, which can be roughly correlated with the Frasnian. They are different from each other in lithofacies. The upper part of the Upper Devonian sequence is called the Hsiehchingsu Formation, yielding *Yunnanella* and *Leptophloeum rhombicum*, corresponding at least to part of the Famennian. All of these are referable to littoral facies, consisting chiefly of terrigenous clastics with oolitic hematite ore deposits.

The Devonian in eastern Hunan also consists mostly of terrigenous clastics, which may be referable to littoral facies. However, the main oolitic hematite ore beds are younger than those in northwestern and central Hunan. They were assigned to the Frasnian by Liao (1954), and later revised to be correlated with Famennian. The recent finding of megafossils from the hematite beds and from the argillaceous limestone intercalations has proved that they may be correlated with the Shaotung Formation and its equivalents in central Hunan, being approximately the Sturian in age. The important megafossils are; *Caninia* sp., *Zaphrentites* sp., *Hunanoproductus hunanensis*, *Tenticospirifer vilis*, etc.

In central and southern Hunan, the main Devonian depositional area, most of the Devonian stratigraphical units of Hunan have been named after this area since 1938 (Tien, 1938). Therefore, while discussing the Devonian biostratigraphic divisions of Hunan, we have to focus our attention on the Devonian biostratigraphic division of central Hunan.

The Devonian of central Hunan has been studied for more than five decades. However, the earlier stratigraphic work was mainly based on the geological route survey and on the observation of the isolated outcrops. As both geological mapping and mineral resources exploration have been under way in Hunan, a number of sections have been measured and the framework of the Devonian subdivisions has been achieved. The following is a scheme of the Devonian subdivision taken traditionally as the standard in Hunan:

Upper Devonian

Shaotung Formation

Hsikuangshan Formation

Shetienchiao Formation

Late Middle Devonian

Chitzechiao Formation

Early and late Middle Devonian, probably being Lower and Middle Devonian somewhere in southern Hunan

Tiaomachien Formation

It is obvious that the above framework of Devonian subdivisions does not include and summarize all of the Devonian strata developed in Hunan, especially with complicated and different lithofacies or with abundant and diversified biotas. In the present paper, an attempt is made to give a comprehensive analysis of lithofacies and biotic features, especially to consider the development of the reef and reef complex as a cause to the diversification and differentiation of the facies. Based on the above consideration, the Devonian strata in Hunan can be allocated to the following two basic facies and can be further subdivided into four facies zones or belts:

Littoral Facies

The inner littoral facies belt, occurring close to and around the old land, consists mainly of terrigenous clastics with a few plants, fishes, bivalves and fragments of inarticulate brachiopod.

The outer littoral facies belt is located between the inner littoral facies belt and the shallow marine facies. It consists primarily of detrital sediments with some mudstones and marls. The characteristic feature of the biota is the dominance of brachiopods, together with bivalves, gastropods, trilobites, ostracods and bryozoans. Generally, the coral is very rare in this facies belt, except in the marginal portion of the reef platform and in the transitional belt between the outer littoral facies belt and the shallow marine facies zone, where the corals seem to be more adaptable.

Shallow Marine Facies

The reef platformal facies belt of shallow marine facies was formed in the time from late Middle Devonian to early Late Devonian. The measured sections in Maanshan of Shaodong district and in Qukoupu of Shaoyang district are typical for the Devonian reef complex. The Chitzechiao Formation at its type locality, however, may serve as the representative of the biostratigraphic division. Apart from the basal part, which is composed of sandy shales, argillaceous limestones and marls, the Chitzechiao Formation at its type locality consists almost exclusively of carbonates, including biogenic limestone, bioclastic limestone, dolomitic limestone and dolomitites. It should be pointed out that the reef platformal facies belt is a general term combining various parts of reef complex with various lithofacies and biotas. It can be recognized to include the massive biogenic and dolomitic limestones with massive and laminated stromatoporoids and algae in the reef complex; abundant disarticulated brachiopods, fragmentary rugose corals, broken branch of tabulate corals, abraded, laminated and small colonial rugose corals in the biocalcirudites on the marginal slope of the reef platform; and also well-bedded limestone and dolomitic limestone with abundant branching stromatoporoids *Amphipora* and *Paramphipora*, all indicating a restricted and quiet environment in the back reef facies; whereas close to the former region where the dolomitites are well developed, the fossil is very seldom. Further precise study can lead us to distinguish more and more facies belts and microfacies zones, but it is beyond the topic of the present paper.

The inter-reef depressional facies belt is located between the reef platforms or surrounding them to form an elongated and narrow belt. The rock contents of this facies belt mostly contain greyish black marls and argillaceous limestones with crystalline ferric pyrite, and intercalated somewhere with carbonaceous shales for the late Middle Devonian. In the late Middle Devonian, there are siliceous beds, siliceous shale and siliceous marls intercalated with carbonaceous shale, sometimes with siliceous nodules. All of these constitute the upper part of Chizechiao Formation and the lower part of Shetienchiao Formation. The benthic biotas in this facies zone are remarkably on the decrease, instead there are abundant tentaculites associated with ammonoids and conodonts, which recall some similarities to the pelagic facies, also called the Nandan facies in Guangxi and Guizhou, indicating a stagnant and much reductive environment. However, there exist some differences between the typical Nandan facies in South China and the depressional facies zone in Hunan. The tentaculites occurring in Hunan are those mostly with slightly thicker shells as compared with the thin-shelled nowakiids in the Nandan facies in South China. The development of the depressional facies lasted for a short time, ranging from late Middle Devonian to early Late Devonian in central Hunan, whereas the Nandan facies was developed from late Early Devonian to Earliest Carboniferous in South China. This may be due to the paleogeographical situation in which the central Hunan was located close to the old land during the Devonian period.

The Lithostratigraphic Subdivision of the Chitzechiao Formation and the Shetienchiao Formation and their Correlatives

Based on the foregoing analysis of the facies zonation, it is appropriate to present the framework of Devonian lithostratigraphic division and biostratigraphic character for each of the facies belt so as to understand the development of the Devonian stratigraphy and to interprete the Devonian Paleogeography and the mechanic controlling factors to the formation of ore deposits. However, owing to the limited space, the Devonian subdivisions of littoral facies are not discussed.

The Chitzechiao Formation

The stratotype of the Chitzechiao Formation is located at the Wanlao Hill of Qiziqiao (Chitzechiao) town of Xiangxiang district. The section is exposed along the hill near a cement plant and measured by Messrs. Hu Zhaoxuan and Deng Zhanqiu of Nanjing Institute of Geology and Paleontology. By courtesy of them, the record is quoted as follows (Text-fig.2):

The overlying bed: The Lungkouchung Formation with alternation of grayish yellow medium-to thick-bedded quartzose sandstone and argillaceous siltstone, with two-metre-thick mudstone and marls in the basal part.

——————Conformable——————

The Chitzechiao Formation

11. Greyish white and gray medium-bedded limestone, occasionally intercalated with brownish gray thin-bedded argillaceous limestone, containing rich corals and stromatoporoids: *Thamnopora* sp., *Disphyllum* sp., *Hexagonaria* sp. 22.4m
10. Dark, blackish gray, thick-bedded limestone, containing stromatoporoids. 100m
9. Blackish gray and dark gray thick-bedded limestone 43.9m
8. Grayish black medium-bedded argillaceous limestone with brachiopods, numerous stromatoporoids and gastropods. 10.7m
7. Dark gray or grayish black thick-bedded limestone. 30.1m
6. Predominantly grayish white, gray and blackish gray thick-bedded limestone, with some chert concretions on the rock surface, intercalated with several thin-bedded dolomitic limestone yielding stromatoporoids. 132.3m
5. Dark gray-black thick-bedded coarse crystalline limestone, containing chert concretions with *Hexagonaria* sp., *Sociophyllum* sp. and some stromatoporoids. 22.8m
4. Dark gray to gray thin-bedded argillaceous limestone, containing abundant *Endophyllum* sp. 5.6m
3. Gray-black to dark gray calcareous shale. 5.8m
2. Grayish black thin-bedded impure limestone, containing *Stringocephalus* sp. 3m
1. Alternation of grayish black, grayish green calcareous shale and dark gray thin-bedded argillaceous limestone. 5–10m

——————Conformable——————

The underlying bed: Greenish sandstone and purple sandy shale as the top of the Tiaomachien Formation.

Lithologically, the Chitzechiao Formation can be subdivided into two parts: the lower part (member), about 20m thick, is composed of calcareous and sandy shales, and argillaceous limestones; the upper part (member), about 360m thick, consists mainly of massive and thick-bedded biogenic limestone, bioclastic limestone and dolomitic limestone, somewhere with dolomitites.

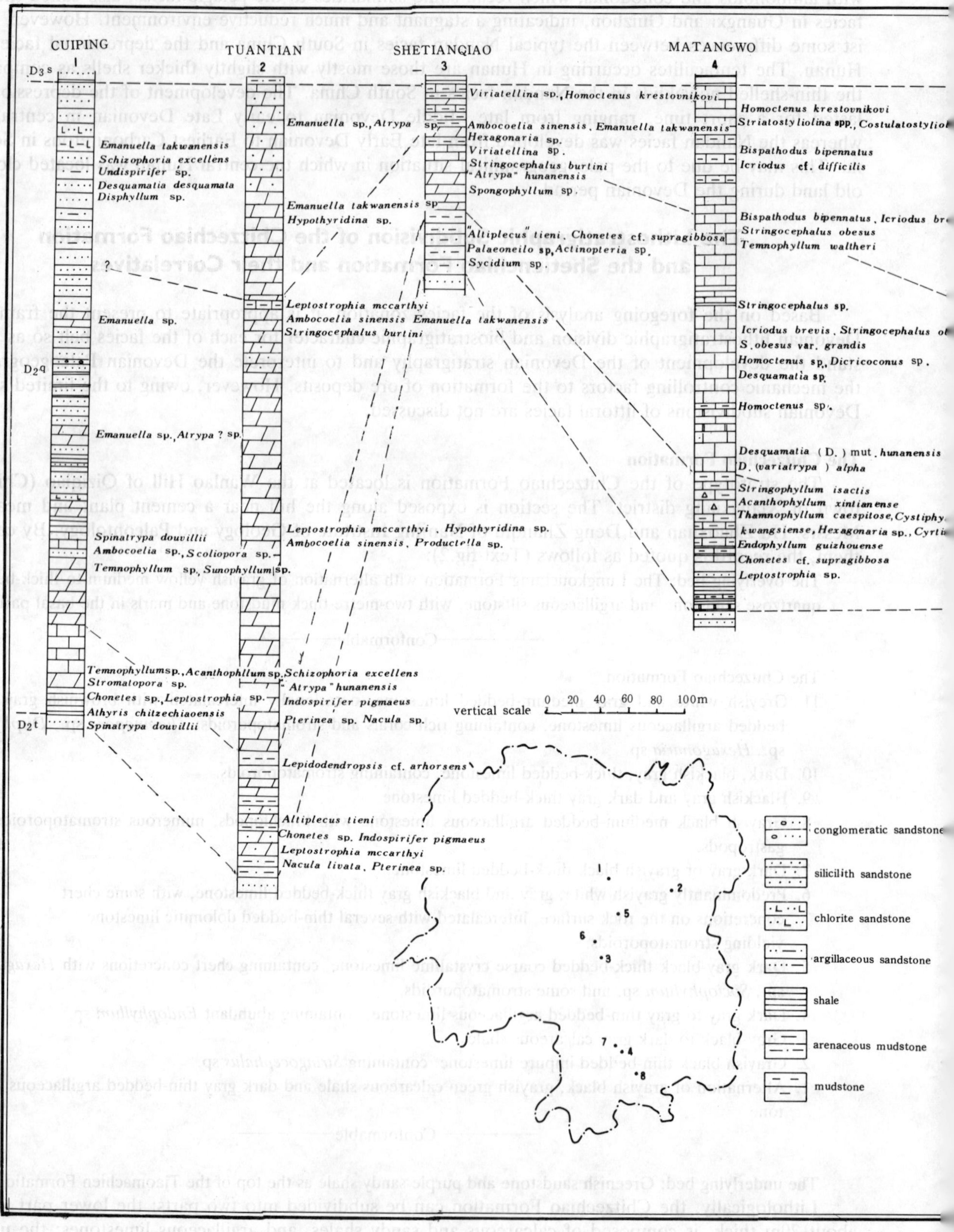

CUIPING
TUANTIAN
SHETIANQIAO
MATANGWO
1
2
3
4
D3s
D2q
D2t
Emanuella takwanensis
Schizophoria excellens
Undispirifer sp.
Desquamatia desquamata
Disphyllum sp.
Emanuella sp.
Emanuella sp., Atrypa ? sp.
Spinatrypa douvillii
Ambocoelia sp., Scoliopora sp.
Temnophyllum sp., Sunophyllum sp.
Temnophyllum sp., Acanthophllum sp.
Stromatopora sp.
Chonetes sp., Leptostrophia sp.
Athyris chitzechiaoensis
Spinatrypa douvillii
Emanuella sp., Atrypa sp.
Emanuella takwanensis sp.
Hypothyridina sp.
Leptostrophia mccarthyi
Ambocoelia sinensis Emanuella takwanensis
Stringocephalus burtini
Leptostrophia mccarthyi Hypothyridina sp.
Ambocoelia sinensis, Productella sp.
Schizophoria excellens
"Atrypa" hunanensis
Indospirifer pigmaeus
Pterinea sp. Nacula sp.
Lepidodendropsis cf. arhorsens
Altiplecus tieni
Chonetes sp. Indospirifer pigmaeus
Leptostrophia mccarthyi
Nacula livata, Pterinea sp.
Viriatellina sp., Homoctenus krestovnikovi
Ambocoelia sinensis, Emanuella takwanensis
Hexagonaria sp.
Viriatellina sp.
Stringocephalus burtini
"Atrypa" hunanensis
Spongophyllum sp.
"Altiplecus" tieni Chonetes cf. supragibbosa
Palaeoneilo sp., Actinopteria sp.
Sycidium sp.
Homoctenus krestovnikovi
Striatostyliolina spp., Costulatostyliol
Bispathodus bipennatus
Icriodus cf. difficilis
Bispathodus bipennatus, Icriodus bre
Stringocephalus obesus
Temnophyllum waltheri
Stringocephalus sp.
Icriodus brevis, Stringocephalus ob
S. obesus var. grandis
Homoctenus sp., Dicricoconus sp.
Desquamatia sp.
Homoctenus sp.
Desquamatia (D.) mut. hunanensis
D (variatrypa) alpha
Stringophyllum isactis
Acanthophyllum xintianense
Thamnophyllum caespitose, Cystiphy
kwangsiense, Hexagonaria sp., Cyrti
Endophyllum guizhouense
Chonetes cf. supragibbosa
Leptostrophia sp.
vertical scale
0 20 40 60 80 100m
1
2
3
4
5
6
7
8
conglomeratic sandstone
silicilith sandstone
chlorite sandstone
argillaceous sandstone
shale
arenaceous mudstone
mudstone

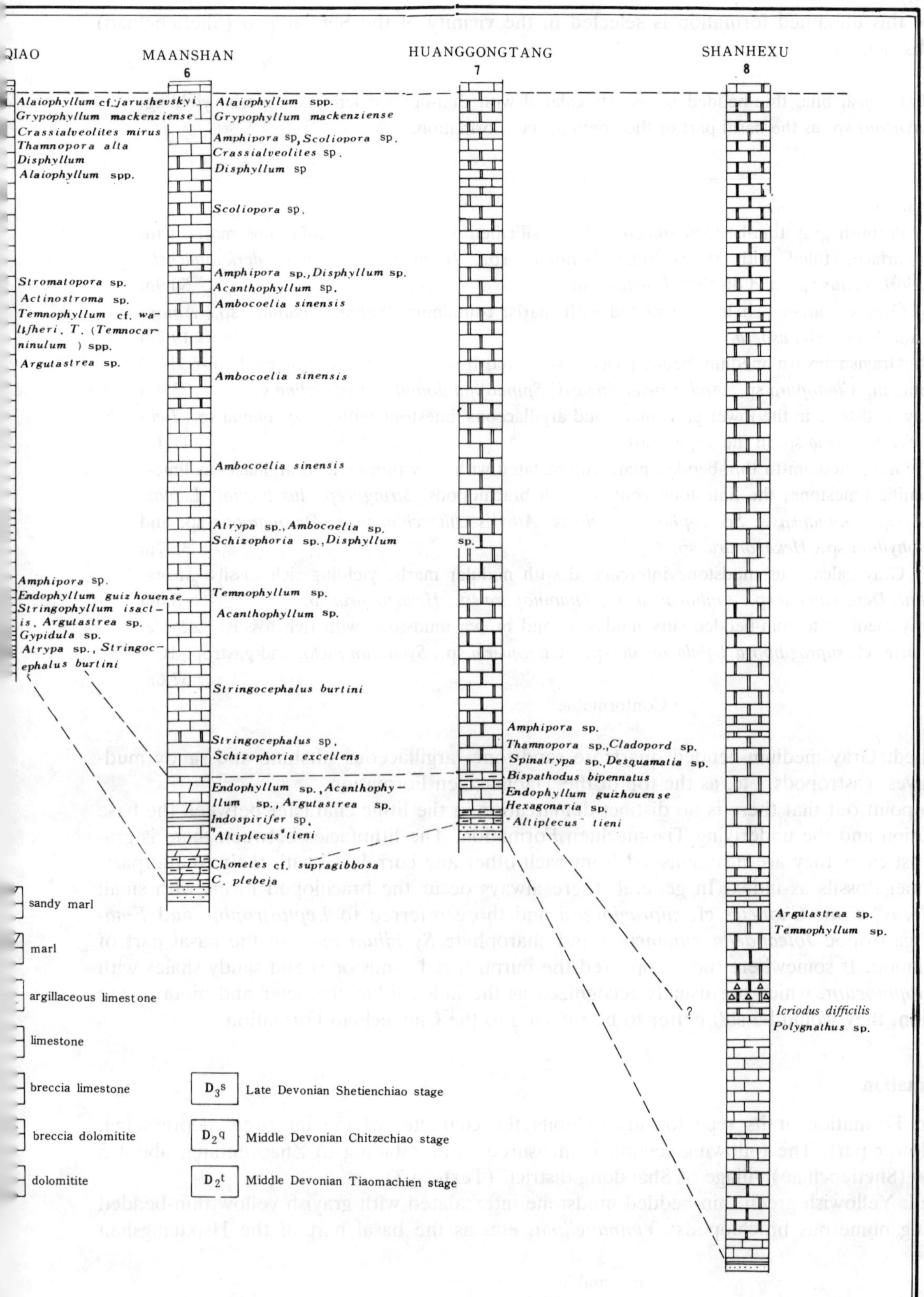

Text-fig. 2. Columnar correlation of the Middle Devonian Chitzechiao Formation in Hunan.

(2) The unnamed formation (in the inter-reef depressional facies zone correlating approximately with the Chitzechiao Formation)

The stratotype of this unnamed formation is selected in the vicinity of the Shetianqiao (Shetienchiao) village of Shaodong district.

The overlying bed: Grayish blue thin-bedded marls intercalated with thin-bedded lenticular marls, yielding: *Chontes* sp., *Leiorhynchus* sp. as the basal part of the Shetienchiao Formation.

———Conformable———

The unnamed formation

7. (ADS 452-466) Purplish gray thin-bedded siliceous shale, silicified and siliceous-argillaceous marls, with smooth rock surface, filled with tentaculites: *Styliolina* sp., *Homoctenus ultimus derkaouaensis*; brachiopod: *Leiorhynchus* sp., and bivalve: *Buchiola* sp. 50.9m
6. (ADS 509-510) Gray calcareous shale intercalated with marls, containing *"Spongophyllum"* sp., *Ambocoelia sinensis* and *Spinatrypa aspera.* 19.7m
5. (ADS 505-508) Grayish brown medium-bedded calcareous mudstone intercalated with medium-bedded limestone, containing *Cladopora* sp., *Ambocoelia sinensis, Spinatrypa douvillii, Viriatellina* sp. 23.4m
4. (ADS 504) Gray mudstone in the lower part, marls and argillaceous limestone with *Desquamatia* sp., *Leptostrophia* sp., *Schizophoria* sp. in the upper part. 11.1m
3. (ADS 502-503) Gray, medium-to thin-bedded marls intercalated with very thin-bedded argillaceous limestone and dolomitic limestone; the limestone contains rich brachiopods: *Stringocephalus burtini, Leptostrophia* sp., "*Atrypa*" *hunanensis*, *Schizophoria excellens, Athyris chitzechiaoensis*, *Desquamatia* sp. and corals: *Spongophyllum* sp., *Hexagonaria* sp. 35.7m
2. (ADS 500-501) Gray calcareous mudstone intercalated with nodular marls, yielding rich fossils: *Stringocephalus burtini*, *Desquamatia* sp., *Ambocoelia* sp., *Thamnopora* sp., *Hexagonaria* sp. 8.6m
1. (ADS 499) Gray medium-to thin-bedded silty mudstone and brown mudstone with rich fossils: "*Altiplecus*" *tieni*, *Chonetes* cf. *supragibbosa*,? *Palaeoneilo* sp., *Actinopteria* sp., *Sycidium melo*, and gastropods. 37.6m

———Conformable———

The underlying bed: Gray medium-bedded quartzose sandstone, argillaceous siltstone and sandy mudstone with a few bivalves, gastropods, etc. as the top of the Tiaomachien Formation.

It is necessary to point out that there is no distinct demarcation in the lithic character between the base of Chitzechiao Formation and the underlying Tiaomachien Formation. The lithofacies between them is gradually changed. In most cases they are distinguished from each other and correlated with their counterparts mainly by using the megafossils as index. In general, there always occur the brachiopod forms with small shell, such as "*Altiplecus*" *tieni*, *Chonetes* cf. *supragibbosa* and those referred to *Leptostrophia* and *Emanuella*, together with gastropod *Joleaudella hunanensis* and charophyte *Sycidium melo* in the basal part of the Chitzechiao Formation. If somewhere there appeared the purplish red sandstones and sandy shales with *Bothriolepis* and *Leptophloeum*, which are usually recognized as the index lithic character and biota of the Tiaomachien Formation, they would be still better to be referred to the Chitzechiao Formation.

The Shetienchiao Formation

The Shetienchiao Formation at its type locality exhibits the character of the inter-reef depressional facies, especially its lower part. The following section is measured from Zifatang to Zhaorentang, about 5 km NE of Shetianqiao (Shetienchiao) village of Shaodong district. (Text-fig.3)

The overlying bed: Yellowish green thin-bedded mudstone intercalated with grayish yellow thin-bedded nodular marls, yielding numerous brachiopods: *Yunnanellina*, etc. as the basal part of the Hsikuangshan Formation.

———Conformable———

The Shetienchiao Formation

18. (ADS 485–487) Grayish green, yellowish green thin-bedded mudstone with brachiopods: *Desquamatia* sp., *Productella* sp., *Cyrtospirifer hunanensis* (sp. nov.), *C.* sp., *Tenticospirifer* sp. 25m
17. (ADS 484) Grayish green, yellowish green thin-bedded mudstone with brachiopods: *Spinatrypa* sp., *Des-*

quamatia sp., *Gypidula* sp., *Hypothyridina* sp., *Tenticospirifer* sp., and rugose corals: *Hunanophrentis uniforme*, *Disphyllum geinitzi*. 15.6m

16. (ADS 481-493) Grayish yellow limestone and thin-bedded nodular limestone, with rich brachiopods: *Spinatrypa* sp., *Desquamatia* sp., *Productella* sp., *Cyrtospirifer hunanensis* (sp. nov), *Hypothyridina* sp., *Chonetes* sp., *Tenticospirifer* sp. 51.2m
15. (ADS 480) Yellowish green calcareous mudstone, with thin-bedded crystalline limestone in the basal part. 28.6m
14. (ADS 477-479) Yellowish green calcareous mudstone intercalated with thin-bedded nodular marls, containing rich brachiopods: *Productella* sp., "*Camarotoechia*" sp., *Spinatrypa* sp., *Iowatrypa* sp., *Desquamatia* (*Seratrypa*) sp., *Hypothyridina* sp. 24m
13. (ADS 475-476) Yellowish green thin-bedded calcareous mudstone and thin-bedded nodular marls with brachiopods: *Chonetes* sp., "*Camarotoechia*" sp., *Cyrtospirifer hunanensis* (sp. nov.), *Spinatrypa* cf. *varistriata*. 50.2m
12. (ADS 471-474) Yellowish green thin-bedded calcareous mudstone intercalated with brownish yellow thin-bedded limestone, containing plentiful brachiopods with small shell: *Chonetes* sp., "*Camarotoechia*" sp., *Cyrtospirifer hunanensis* (sp. nov.), *Dalmanella*? sp., *Productella* sp. 50.6m
11. Yellowish green thin-bedded calcareous mudstone with a few brachiopods: *Cyrtospirifer hunanensis* (sp. nov.), "*Atrypa*" sp., "*Camarotoechia*" sp. 39.3m
10. Grayish black medium-bedded fine crystalline limestone intercalated with medium-bedded bioclastic limestone, yielding ammonoid: *Manticoceras* sp.; conodonts: *Palmatolepis gigas*, *Ancyrodella ioides*, *Ancyrognathus triangularis*, *Tripodellus* sp., *Ozarkodina elegans*, *Ligonodina faciformis*, *Hindeodella germana*; bivalve: *Buchiola vetrostriata*. 1.7m
9. Yellowish green thin-bedded silty marls or marls, strongly weathered, with argillaceous siltystone seen under microscope, with bivalves: *Buchiola retrostriata* (Buch), *B*. sp. A, *B*. sp. B. 86.4m
8. Dark gray to yellowish green thin-bedded marls intercalated with argillaceous limestone or nodular limestone, with brachiopod: *Leiorhynchus* sp.; bivalve: *Leptodus* sp.; ostracod: *Richterina* sp. 56m
7. Dark gray thin-to medium-bedded argillaceous limestone and silty marls, with *Buchiola gigantea* (sp. nov.), *B*. sp.; *Manticoceras* sp.; *Richterina* sp., *Perimarginia* sp. 64.4m
6. Dark gray to grayish yellow thin-bedded marls, silty marls intercalated with lenticular limestone, yielding *Buchiola* sp., *Pterinopectinella* sp.; *Richterina* sp., *Perimarginia* sp. 38m
5. Dary gray to grayish black thin-bedded marls intercalated with sandy-argillaceous limestone or lenticular limestone, with *Buchiola retrostriata*; *Leiorhynchus*? sp.; *Richterina* sp., *Perimarginia* sp. 66.5m
4. Alternation of dark gray to grayish black thin-to thick-bedded argillaceous limestone and silty marls, with *Buchiola* sp.; *Manticoceras*? sp., *Richterina* sp. 39.2m
3. Interbeddings of grayish yellow, brown to yellowish green thin-bedded marls and silty marls, with *Chonetes* sp.; *Buchiola* sp.; *Manticoceras* sp. 40.1m
2. Dark gray to grayish black thin-bedded marls intercalated with argillaceous limestone or lenticular marls with *Buchiola* sp., *Lingula* sp.; *Styliolina* sp. 14.4m
1. Grayish blue thin-bedded marls, silty marls intercalated with thin-bedded lenticular marls, yielding *Chontes* sp., *Leiorhynchus* sp. 103m

The underlying bed: Purplish gray thin-bedded siliceous shale, silicified and siliceous-argillaceous marls, with smooth rock surface, filled with tentaculites: *Styliolina* sp., *Homoctenus ultimus derkaouaensis*; brachiopod: *Leiorhynchus* sp., and bivalve: *Buchiola* sp., as the top of the unnamed formation approximately correlating with the Chitzechiao Formation in the reef platformal faces zone.

The Shetienchiao Formation mentioned above can be roughly subdivided into two parts: The lower part (member) from Bed 1 to Bed 10 about 508m thick, is mainly composed of dark gray marls and limestones intercalated with yellowish green argillaceous limestones, predominated by pelagic biotas, such as conodonts, tentaculites and ammonoids, together with a few benthic forms, mainly brachiopods, bivalves or occasionally corals. The upper part (member) from Bed 11 to Bed 18, about 294m thick, consists of yellowish green calcareous mudstone and marls, yielding abundant brachiopods and bivalves together with some solitary rugose corals. The difference between these two parts is not only in lithology but also in biotas, reflecting a change of the environment.

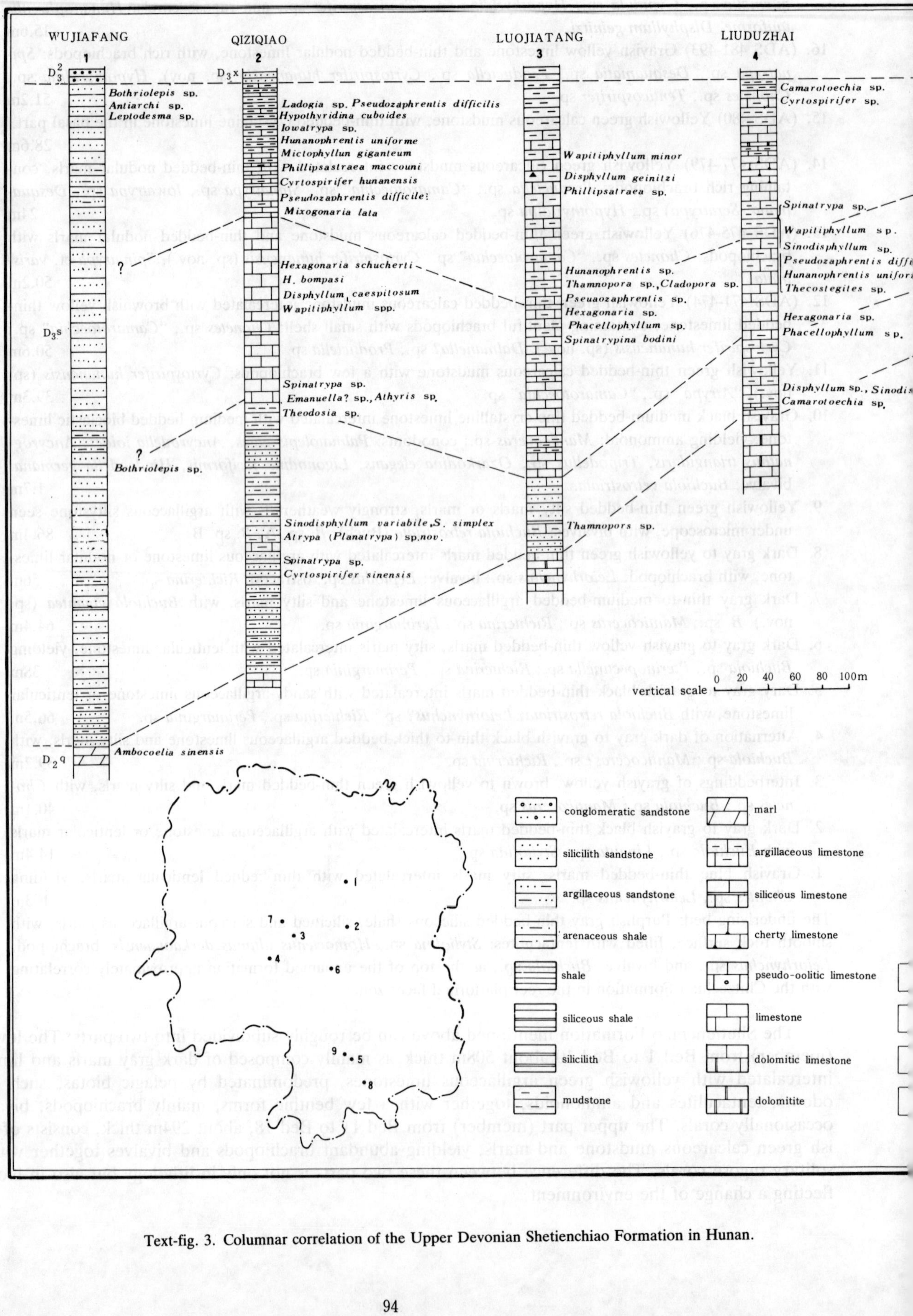

Text-fig. 3. Columnar correlation of the Upper Devonian Shetienchiao Formation in Hunan.

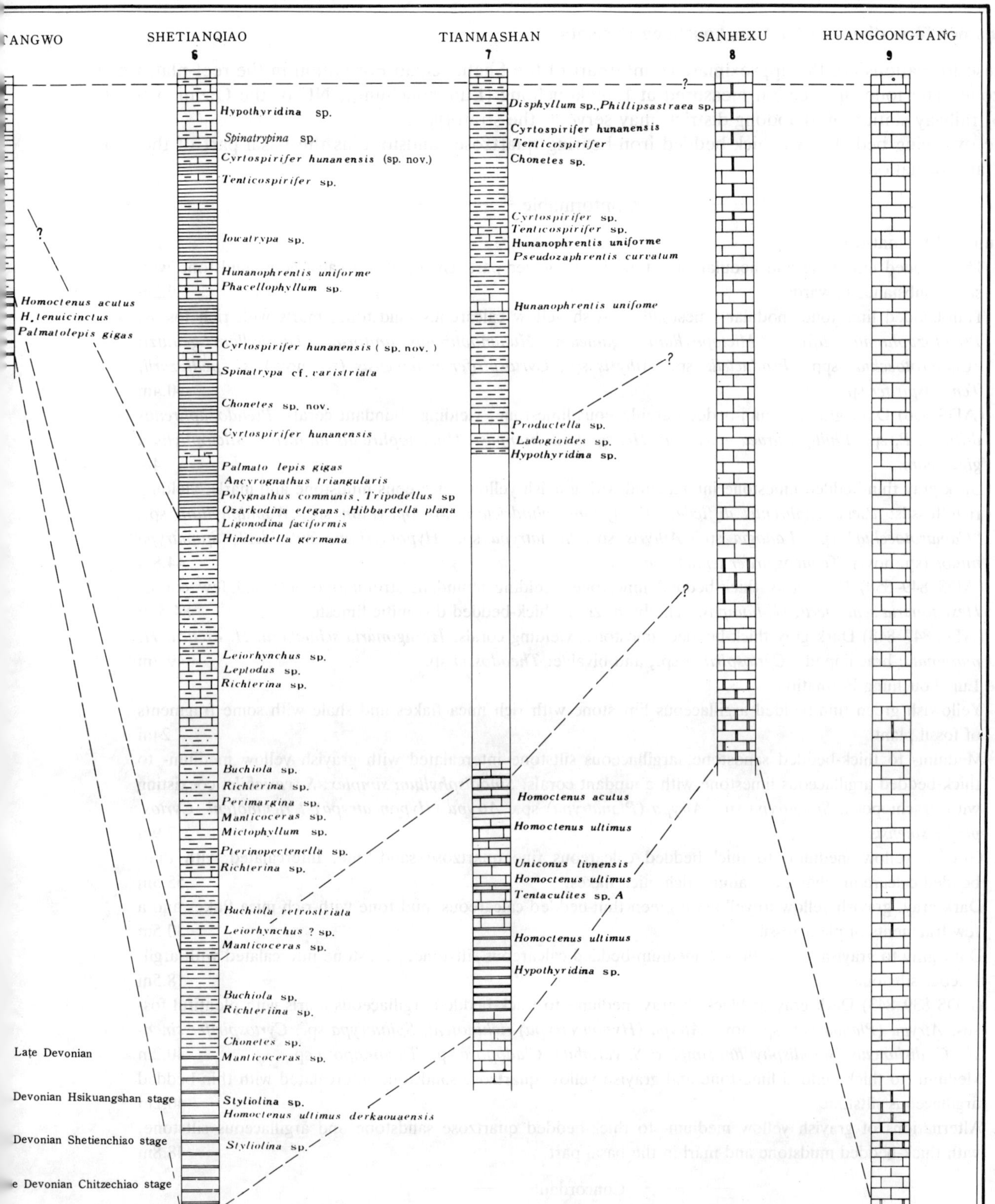

TANGWO
SHETIANQIAO
6
TIANMASHAN
7
SANHEXU
8
HUANGGONGTANG
9
Hypothyridina sp.
Spinatrypina sp.
Cyrtospirifer hunanensis (sp. nov.)
Tenticospirifer sp.
Iowatrypa sp.
Hunanophrentis uniforme
Phacellophyllum sp.
Cyrtospirifer hunanensis (sp. nov.)
Spinatrypa cf. varistriata
Chonetes sp. nov.
Cyrtospirifer hunanensis
Palmato lepis gigas
Ancyrognathus triangularis
Polygnathus communis, Tripodellus sp
Ozarkodina elegans, Hibbardella plana
Ligonodina faciformis
Hindeodella germana
Leiorhynchus sp.
Leptodus sp.
Richterina sp.
Buchiola sp.
Richterina sp.
Perimarkina sp.
Manticoceras sp.
Mictophyllum sp.
Pterinopectenella sp.
Richterina sp.
Buchiola retrostriata
Leiorhynchus ? sp.
Manticoceras sp.
Buchiola sp.
Richterina sp.
Chonetes sp.
Manticoceras sp.
Styliolina sp.
Homoctenus ultimus derkaouaensis
Styliolina sp.
Disphyllum sp., Phillipsastraea sp.
Cyrtospirifer hunanensis
Tenticospirifer
Chonetes sp.
Cyrtospirifer sp.
Tenticospirifer sp.
Hunanophrentis uniforme
Pseudozaphrentis curvatum
Hunanophrentis unifome
Productella sp.
Ladogioides sp.
Hypothyridina sp.
Homoctenus acutus
Homoctenus ultimus
Uniconus livnensis
Homoctenus ultimus
Tentaculites sp. A
Homoctenus ultimus
Hypothyridina sp.
Homoctenus acutus
H. tenuicinctus
Palmatolepis gigas
Late Devonian
Devonian Hsikuangshan stage
Devonian Shetienchiao stage
e Devonian Chitzechiao stage

The unnamed formation and the Lungkouchung Formation

These formations are the approximate counterpart of the Shetienchiao Formation in the reef platformal facies zone. The following section measured at Lungkouchong (Lungkouchung), NE of the Qiziqiao (Chitzechiao) railway station of Shaodong district, may serve as the stratotype.

The overlying bed: Brown thick-bedded iron-bearing quartzose sandstone as the basal part of the Hsikuangshan Formation.

————Conformable————

The unnamed formation:

14. Thin-bedded limestone and nodular limestone in the lower part, changed into calcareous mudstone with sandy substance upwards. 9.2m
13. Thin-bedded limestone, nodular limestone, grayish yellow calcareous mudstone, marls with rich fossils: *Pseudozaphrentis difficile, Mictophyllum giganteum, Hunanophrentis uniforme, Disphyllum geinitzi*; "*Camarotoechia*" spp., *Productella* sp., *Athyris* sp., *Cyrtospirifer hunanensis* (sp. nov.), *C. pellizzarii, Tenticospirifer* sp. 40.4m
12. (ADS 856) Dark gray medium-bedded argillaceous limestone, yielding abundant corals: *Pseudozaphrentis difficile, P.* sp., *Phillipsastraea maccouni, Hexagonaria bompasi, Hunanophrentis uniforme, Mictophyllum giganteum.* 4m
11. Dark gray thin-bedded limestone intercalated with grayish yellow calcareous mudstone and marls, yielding rich fossils: *Pseudozaphrentis difficile*, *Disphyllum cylindricum*, *Mictophyllum* sp., *Wapitiphyllum* sp.; "*Camarotoechia*" sp., *Ladogia* sp., *Athyris* sp., *Spinatrypa* sp., *Hypothyridina cuboides*?, *Spinatrypa minor* (sp. nov.), *Tenticospirifer tenticulum.* 44.8m
10. (ADS 849–850) Dark gray thick-bedded limestone, yielding abundant stromatoporoids and few corals: *Hexagonaria schucherti, H. bompasi.* The basal 2m is thick-bedded dolomitic limestone. 84.5 m
9. (ADS 847–848) Dark gray thick-bedded limestone, yielding corals: *Hexagonaria schucherti, H. arctica, H. marginata;* brachiopods: *Cyrtospirifier* sp.; and bivalve: *Theodossia* sp. 49.7m

The Lungkouchung Formation

8. Yellowish green thin-bedded argillaceous limestone with rich mica flakes and shale with some fragments of fossil plant. 24m
7. Medium- to thick-bedded sandstone, argillaceous siltstone intercalated with grayish yellow medium- to thick-bedded argillaceous limestone with abundant corals: *Sinodisphyllum simplex*, *S. variabile*, coexisting with brachiopods: *Spinatrypa* sp., *Atrypa* (*Planatrypa*) sp., *Atrypa* (*Hyponeatrypa*) ? *richthofeni*, *Cyrtospirifer sinensis.* 7.9m
6. Grayish yellow medium- to thick-bedded calcareous fine quartzose sandstone, intercalated with thin-bedded calcareous shale, containing rich mica flakes. 55.5m
5. Dark gray, grayish yellow to yellowish green thin-bedded calcareous mudstone with rich mica flakes and a few fragments of plant fossil. 30.5m
4. Dark gray to grayish yellow thin-to medium-bedded calcareous siltstone, sandstone intercalated with argillaceous siltstone. 18.5m
3. (ADS 839–843) Dark gray or blackish gray medium- to thick-bedded argillaceous marls with plentiful fossils: *Atrypa* (*Planatrypa*) sp. nov., *Atrypa* (*Hyponeatrypa*) *richthofeni, Spinatrypa* sp., *Cyrtospirifer sinensis, C. disjunctus; Sinodisphyllum simplex, S. variabile, Cladopora* sp., *Thamnopora* sp. 40.2m
2. Medium- to thick-bedded limestone and grayish yellow quartzose sandstone intercalated with thin-bedded argillaceous siltstone. 40.2m
1. Alternations of grayish yellow medium- to thick-bedded quartzose sandstone and argillaceous siltstone, with thick-bedded mudstone and marl in the basal part. 48.8m

————Concordant————

The underlying bed: Thick-bedded limestone as the top part of the Chitzechiao Formation.

Lithologically, the section can be divided into the following parts:

The unnamed formation:

The upper member (from Bed 11 to the top), about 98m thick, consists mainly of marls and calcareous mudstones with abundant corals of high diversity, brachiopods and bivalves.

The lower member (from Bed 9 to Bed 10), about 134m thick, is composed of thick-bedded limestones, with massive rugose corals together with massive stromatoporoids.

The Lungkouchung Formation (from Bed 1 to Bed 8), about 265m thick, was named from the Lungkouchung member by Tian (1983). It is mainly composed of sandstone with some intercalations of mudstones and marls. Fossils are rare, but occasionally some fragments of plants and a few specimens of *Cyrtospirifer* were found at the adjacent locality.

The Biostratigraphic Character of the Chitzechiao Formation and its Correlatives in Hunan

Rugose corals

Devonian rugose corals first appeared near the basal part of the Chitzechiao Formation (about 40m from the base in the type section) and rapidly increased in the number of genera and species, including: *Endophyllum guizhouense, E. yunnanense, Stringophyllum isactis, Sunophyllum* sp., *Sociophyllum* sp., *Temnophyllum* (*Truncicarinulum*) spp., *Cystiphylloides kwangsiense*, *C.* spp. They occur mainly in the marls or argillaceous limestones; slightly above this bed, there appear some massive corals in the limestone and argillaceous limestone bed, including: *Argutastrea quadrigemina, A.* spp., *Disphyllum* sp., *Hexagonaria* sp. These corals can be recognized as the *Endophyllum guizhouense* assemblage zone, which can be correlated with the *E. assemblage* zone widely distributed in Guangxi, Yunnan, Guangdong, Guizhou of South China (Yu and others, 1978). Corals in this assemblage zone are also widely distributed in central and southern Hunan, not only being in the reef platformal facies zone, but also discernible somewhere in the inter-reef depressional zone, for example, in Matangwo of Xingtian district.

The large colonial skeleton of forms of *Endophyllum* together with the skeleton of other taxa always form the main constituent part of biostrome, which can be taken as the indices of the initial phase of the Devonian reef development in Hunan.

Rugose corals in the middle part of the Chitzechiao Formation are rare. This may be due to the facies controlling, as the lithofacies in this part was changed into the reef facies with massive or thick-bedded limestone and dolomitic limestone. *Temnophyllum waltheri, Temnophyllum* sp. and *Temnophyllum* (*Truncicarinulum*) spp. constitute the *Temnophyllum waltheri* assemblage zone, although the forms of *Temnophyllum* and its subgenus *Truncicarinulum* occur throughout the Givetian and range into the Frasnian.

The upper part of the Chitzechiao Formation is characterized by the occurrence of *Alaiophyllum* cf. *jarushevskyi*, *A.* spp., *Grypophyllum mackenziense, Temnophyllum leei,* which can be recognized as the *Grypophyllum mackenziense* assemblage zone instead of the *Alaiophyllum xiangxiangense - Grypophyllum mackenziense* zone, which has been recognized by Jiang Shuigen (1980) in an unofficial informative paper. In contrary to Jiang's opinion, we prefer considering this zone as the upper part of the Middle Devonian Chitzechiao Formation. *Grypophyllum mackenziense* originally reported in W. Canada has been first recognized by Pedder as an index fossil occurring in the upper *hermann-cristatus* conodont zone (Pedder, 1963, 1973). The same coral species has been identified by the senior author of the present paper from the same level in Guiling and other localities in Guangxi and in Shaodong, Xinghua, Lianyuan, Shaoyang of Hunan.

Viewed as a whole, the rugose corals in the Chitzechiao Formation and its equivalents in Hunan can be recognized as the following assemblage zones:

3. *Grypophyllum mackenziense* Zone
2. *Temnophyllum waltheri* Zone
1. *Endophyllum guizhouense* Zone

The upper two coral assemblages have been found only in the reef platformal facies Zone.

Stromatoporoids:

There can be seen many massive and laminated stromatoporoids in the limestone bed of the Chitzechiao Formation. The described forms include: *Hammatostroma tschussovense, H.* spp., *Atelodictyon dahekouense, Stromatoporella solitaria, Actinostroma clathratum, A.* spp., *Gerronostroma* spp., *Idiostroma radiata, Stromatopora carteri, Clathrocoilona abeona*, etc. In the dolomitic limestone there often occur

Amphipora rudis, A. spp., but the other taxa are rare.

Tabulate corals:

The common forms are *Thamnopora raritabulata, T. yanetae*, *T. tumefacta* and forms assigned to *Pachyfavosites.* They occur in coexistence with *Crassialveolitella multipora, Scoliopora* sp., *Natalophyllum* sp. in the lower part of this formation. In the dolomitic limestone bed, however, only a few forms of *Thamnopora, Cladopora* still exist together with *Amphipora* and *Paramphipora*, indicating that they are tolerant of the environment with a high content of salinity.

Brachiopods:

They are common forms in the Devonian of Hunan. However, they have been found rarely in the reef platformal facies zone, despite of some forms of *Stringocephalus* and *Atrypa* being discernible in the lower part. The brachiopods are frequently seen in the inter-reef depressional zone, especially in the marl beds in the outer littoral facies zone. From the Shetianqiao section, one of the typical sections in depressional facies zones, the following divisions can be recognized:

The upper part with rare forms of *Ambocoelia sinensis, Emanuella takwanensis* and *Spinatrypa* sp. etc.; The middle part abundant in *Stringocephalus burtini, S. grandis, Desquamatia magna, "Atrypa" hunanensis* and *Athyris chitzechiaoensis* , etc.; The lower part characterized by the predominance of small forms, such as *"Altiplecus" tieni,"A."* sp., *Chonetes* cf. *supragibbosa*? *Indospirifer* sp., etc. These are widely distributed in Hunan.

Tian (1938) subdivided the brachiopods in the Chitzechiao Formation into three zones (in ascending order):

3. *Meristella flayelliformis* Zone
2. *M. kutsingensis* and *Stringocephalus burtini* Zone
1. *Stringocephalus burtini* mut. *alpha & beta* Zone

The above subdivision was based on the study of the materials from the locality near Changsha, where the Chitzechiao Formation is about 560m thick and falls in the outer littoral zone. However, the forms identified by Tian as *Meristella* are not really the case, because the internal characters of them are quite different from those of the typical *Meristella,* which ranges from Silurian to Early Devonian. These forms in Hunan may be assigned to *Athyris.* Moreover, there are no adequate evidences to support Tian's opinion of separating the *Stringocephalus burtini* zone from the zone with its subspecies; on the contrary, the species and subspecies are always found to be associated with each other.

Tentaculites:

Occurring only in the middle and upper parts of the Chitzechiao equivalents in the inter-reef depressional facies zone, with such common forms as *Striatostyliolina phillippovae, Viriatellina assaensis, V."tinuta"* and other allied new forms. They are forms with thick shell and have been found in the Middle Devonian Givetian equivalents in Ural region. (Lyashanko, 1959). Other biotas in this formation include: bivalves: *Nacula livata, Tolmaia* sp., *Pterinea consinulis*; charophyte: *Sycidium melo*; plants, fish remains and gastropods with *Joleaudella hunanensis* as one of the index fossils in the basal part of this formation.

Conodonts:

It should be pointed out that in the Shetianqiao section as in many other sections in the depressional facies zone, the siliceous shale, siliceous beds and carbonaceous shale with *Buchiola* sp., *Richterina* sp., *Leiorhynchus* sp., *Chonetes* sp., *Styliolina* sp., *Homoctenus ultimus derkaouaensis* were formerly considered as the basal part of the Upper Devonian Shetienchiao Formation. However, recent finding of the conodonts in this bed in the Dinghechong section, about 3 km north of Qukoupu in Xingshao district reveals that the conodonts are index elements of late Middle Devonian. The conodonts include: *Polygnathus varcus, P. linguiformis, P* cf. *pseudofoliatus, P.* sp., *Icriodus brevis, I.* sp., *Hindeodella* sp., *Lonchodina* sp., *Neoprioniodus*, etc. Above it in the marls and argillaceous limestone, there are conodonts including: *Polygnathus asymmetricus, P. asymmetricus ovalis, P. communis, P.* sp., *Ancyrodella rotundiloba, A. buckeyensis, A.*

rugosa, etc. (from Jing Yuanjia, 1981, in an unofficial informative paper). Therefore, the siliceous beds in the depressional zone widely spread in central Hunan may be assigned to the upper part of the Chitzechiao equivalents. Conodont collections from the sections in Xingtian and Linwu have been identified by Wang Chengyuan and Wang Zhihao as: *Icriodus brevis, I. difficilis, Bispathodus bipennatus, Polygnathus* sp. They are typical forms of the *varcus* zone. Also in the Dongbiankou section of Xingshao district and the Shetianqiao section of Shaodong district, there are conodonts found in the upper part of the Chitzechiao equivalent including: *Polygnathus timorensis, P. xylus xylus, P. xylus ensensis* and many other forms (from Jing Yuanjia, 1980, in an unofficial informative paper). All of these prove that the existence of the *varcus* zone in the equivalents of the Chitzechiao Formation. In addition to the existence of *Stringocephalus*, there is no doubt that the age of the Chitzechiao Formation is late Middle Devonian, approximately correlating with the Givetian.

The Biostratigraphic Character of the Shetienchiao Formation and its Correlatives

Rugose corals:

It was early in 1958 that Sun Yunzhu published a monograph describing the rugose corals from the Upper Devonian of Hunan. He recognized the *Pseudozaphrentis* coral zone in the upper and the *Sinodisphyllum* coral zone in the lower and correlated them with the European Frasnian coral fauna.

Sun's material was mainly from the unnamed formation in Qiziqiao of Xiangxiang district, only a few species being collected from the Shetienchiao Formation at its type locality. Based on a thorough consideration of the vertical distribution of the rugose corals in the Qiziqiao section and a correlation with those in other localities in Hunan, the Shetienchiao correlatives in Hunan can be recognized as three rugose coral assemblage zones, they are in descending order as follows:

3. *Phillipsastraea maccouni* Assemblage Zone
2. *Hexagonaria schucherti* Assemblage Zone
1. *Sinodisphyllum variabile* Assemblage Zone

Rugose corals in the *Sinodisphyllum variabile* Zone are in low diversity, that is, only *Sinodisphyllum variabile, S. simplex, S.* spp. and, occasionally, a few species of *Pseudozaphrentis* are found. However, the individial corals are numerous and well preserved in their original growing place in the Lungkouchung Formation.

Rugose corals are abundant in the *Hexagonaria schuchert* Zone including: *Hexagonaria schucherti, H. orientale, H. philomena, H. longiseptatum, Disphyllum caespitosum, D. cylindricum, D. irregulare, "Peneckiella"* sp., *Wapitiphyllum* spp. and species of *Pseudozaphrentis*, etc. The colonial corals are more in number than the solitary forms. Especially, the massive forms are dominant, occurring together with algae, stromatoporoids and tabulate corals and forming the biostrome in the lower member of the unnamed formation.

Corals were most developed in the upper member of the unnamed formation in Qiziqiao section. They are in high diversity and constitute the *Phillipsastraea maccouni* zone, including *Phillipsastraea maccouni, P.* spp., "*Trapezophyllum*" sp., "*Haplothecia*" sp., *Disphyllum geinitzi, D. jevalanense*, "*Peneckiella*" *regularis, Mixogonaria lata, Mictophyllum giganteum, M. sinense, M. modicum, M. curvatum, M.* spp., *Pseudozaphrentis difficile, P. curvatum, P. conicum, Hunanophrentis uniforme, H. zaphrentoides*, etc. Both of the colonial forms and the massive forms in this zone are less in dimension; in contrast, the solitary forms are largely developed into various external forms. The coral fauna is quite similar to that of late Devonian Frasnian in Mackenzie River area in W. Canada (Smith, 1945; McLean and Pedder, 1984). Some of colonial corals, such as forms of *Hexagonaria* and *Phillipsastraea* recall some similarities to those in Frasnian of Belgium (Hsien, 1977; Coen-Aubert, 1976).

Rugose corals in the inter-reef depressional facies zone are very rare, with only a few species: *Hunanophrentis uniforme, Pseudozaphrentis* sp., and *Disphyllum* sp. in the upper member of the Shetienchiao Formation at its type locality.

Tabulate corals:

Comparatively rare in this formation. There are some species of *Thamnopora, Cladopora, Alveolitella, Syringopora*, etc. found in the reef platformal zone.

Stromatoporoids:

They are also rare in this formation but *Clathrodictyon* sp., *Hammatostroma delicatulum, H. katovense, Stromatopora* sp. have been found in the limestone bed; however, in the dolomitic limestone, forms of *Amphipora* and *Paramphipora* are still abundantly seen.

Brachiopods:

Tian (1938) recognized the following five brachiopod zones in this formation:

Shetienchiao member
- 5. *Atrypa hunanensis* Zone
- 4. *Camarotoechia shetienchiaoensis*-*Cyrtospirifer sinensis* Zone
- 3. *Hypothyridina cuboides* Zone

Lungkouchung member
- 2. *Cyrtospirifer sinensis* Zone
- 1. *Atrypa douvillii lungkouchungensis* Zone

In fact, so far as the brachiopods in the Lungkouchung Formation are concerned, the individual specimens are really abundant in the marl intercalation, however, the generic and specific diversity is low. The dominant form is *Cyrtospirifer sinensis*, being somewhere coexistent with *Atrypa* (*Planatrypa*) sp., *A.* (*Hyponeatrypa*)? *richthofeni, Spinatrypa*. Moreover, the form described by Tian as *Atrypa douvillii lungkouchungensis* has not been found again in this formation, indicating at least that this species is rarely present. Therefore, we prefer choosing *Cyrtospirifer sinensis* as the leading form of the assemblage zone to represent the Lungkouchung brachiopod fauna. As regards Tian's Zone 3 to Zone 5, they are based on material collected from different sections in different localities. However, *Desquamatia hunanensis* (= *Atrypa hunanensis*) is rarely found; "*Camarotoechia*" *shetienchiaoensis* is in fact a common species in most sections. Although *Hypothyridina cuboides*, the index fossil for the middle Frasnian of Europe, was recognized by Tian as the zone fossil of the middle part of his Shetienchiao member, it was not described in his monograph, nor has it been hitherto found in China. Instead, *Hypothyridina hunanensis, H. linglingensis*, "*Theodossia*" sp., "*Ladogioides*" sp., *Cyrtospirifer* sp., *Tenticospirifer* sp. are very common in the middle part of the Shetienchiao Formation. It seems, therefore, appropriate to revise the brachiopod fauna of the Shetienchiao Formation as the following three assemblage zones:

3. "*Camarotoechia*" *shetienchiaoensis* Assemblage Zone
2. *Hypothyridina hunanensis* -"*Ladogioides*" Assemblage Zone
1. *Cyrtospirifer sinensis* Assemblage Zone

Ammonoids, bivalves and ostracods:

The ammonoids are abundant in the lower member of the Shetianchiao Formation, widely spread in the depressional facies zone in central and southern Hunan. Prof. Ou Yuanren (1973) described in an unofficial paper *Manticoceras subordiforme* Au and *M. crassum* Au from this part, proposed to erect a subzone for this fauna and correlated it with the Frasnian *Manticoceras cordatum* Zone in West Europe. In association with the ammonoid are always bivalves: *Buchiola retrostriata, B. gigantea* and ostracods: *Richterina* sp., *Perimarginia* sp. which form an association as frequently reported in the Upper Devonian in South China.

Conodonts:

They have been found in the Shetianqiao section in association with the ammonoid fauna. The following species are identified by Wang Chengyuan and Wang Zhihao as: *Palmatolepis gigas, Ancyrodella ioides, Ancyrognathus triangularis, Tripodella* sp., *Ozarkodina elegans, Hindeodella germana.* At least a *gigas* zone of late Devonian Frasnian can be dated for this part of the Shetienchiao Formation.

Tentaculites:

Mainly occurring in the lower and middle parts of the Shetienchiao Formation; among them *Homoctenus* sp., *Ultimus derkaoensis* have been found in the early Late Devonian in Morocco, capable of serving as an index for correlation.

In general, the boundary between the Shetienchiao Formation and its overlying Hsikuangshan Formation in the benthic sequence is mainly demarcated by using megafossils. The first appearance of *Yunnanellina* and *Yunnanella* fauna together with algae and Receptaculitid can be recognized as the base of the Hsikuangshan Formation.

The Biostratigraphic Character of the Hsikuangshan Formation

The Hsikuangshan Formation in Hunan mostly falls in the littoral facies zone, composed mainly of terrigenous clastics with intercalations of marls and nodular limestones, with the exception of that in southern Hunan, where the approximate equivalent of Hsikuangshan Formation is mainly composed of carbonate rocks. Since the Hsikuangshan Formation has been studied detailedly by Tian (1983) and others, and the lithofacies is not differentiated as compared with that in the preceding stages, in this paper we intend to make some division and to reconsider the biostratigraphic character of the Hsikuangshan Formation based on the accumulated new material and collected data.(Text-fig. 4)

Brachiopods:

As brachiopods are the most common biota in the Hsikuangshan Formation, the zonation is important for regional correlation. Tian (1938) subdivided the Hsikuangshan Formation at its type locality - Xikuangshan (Hsikuangshan) of Xinhua district-into three members and seven brachiopod zones:

Makunao limestone Member

7. *Spirifer* (*Tenticospirifer*) *hsikuangshanensis* Zone
6. *Camarotoechia hsikuangshanensis* Zone
5. *Spirifer* (*Hunanospirifer*) *wangi* Zone
4. *Yunnanella supersynplicata* Zone
3. *Yunnanella synplicata* Zone

Tutzutang limestone Member

2. *Yunnanellina hanburyi* mut. *latiformis* and *Athyris gurdoni* Zone
1. *Yunnanellina hanburyi* and *Y. uniplicata* Zone

Changlungchieh shale Member

As the brachiopod *Atrypa* sp. and the coral *Disphyllum* sp. have been found in the lower part of the Changlungchieh member, it has proved the existence of the Frasnian biota in this part.

It is found that there exist some ambiguous points in Tian's framework of biostratigraphic zonation, which need to be clarified and revised. For example, the constituent element of Tian's brachiopod Zone 1 is almost similar to that in zone 2; and brachiopods in Zones 3-4 are very difficult to be distinguished from each other; whereas the brachiopods in Zones 5-7 are in fact always mixed together. Moreover, forms of *Yunnanella* have been found either in the basal part or in the upper part of the Makunao Member. It is therefore appropriate to give a revised zonation as follows:

Yunnanellina hanburyi-Cyrtiopsis davidsoni Assemblage

The brachiopods in this assemblage zone correspond to those in Tian's Zones 1 and 2, and in the upper part of the Changlungchieh shale. In this assemblage, *Yunnanellina* is the most common and distinct form widespread in Hunan; *Cyrtiopsis davidsoni*, which has been reported from the Frasnian of West Germany is also very common in Hunan, especially in Shaodong; *Cyrtospirifer* mostly with *C. pellizzarii*, *C. subarchiaci* and *Tenticospirifer*, such as *T. vilis* are small-sized forms dominant in this assemblage. In addition, there are still many species, such as *Athyris gurdoni*, *Productella lachrymosa* var. *asiatica* characterized by having very small shell. It can be seen somewhere in Xinhua that a few forms of *Yunnanella* are in association with those of *Yunnanellina* at the top of this assemblage zone.

Yunnanella grandis - Tenticospirifer gortani Assemblage

Brachiopods are flourishing with many species of *Yunnanella,* including *Y. grandis, Y. synplicata, Y. mesoplicata.* The forms assigned to *Tenticospirifer* are characterized by having large size, complex medium

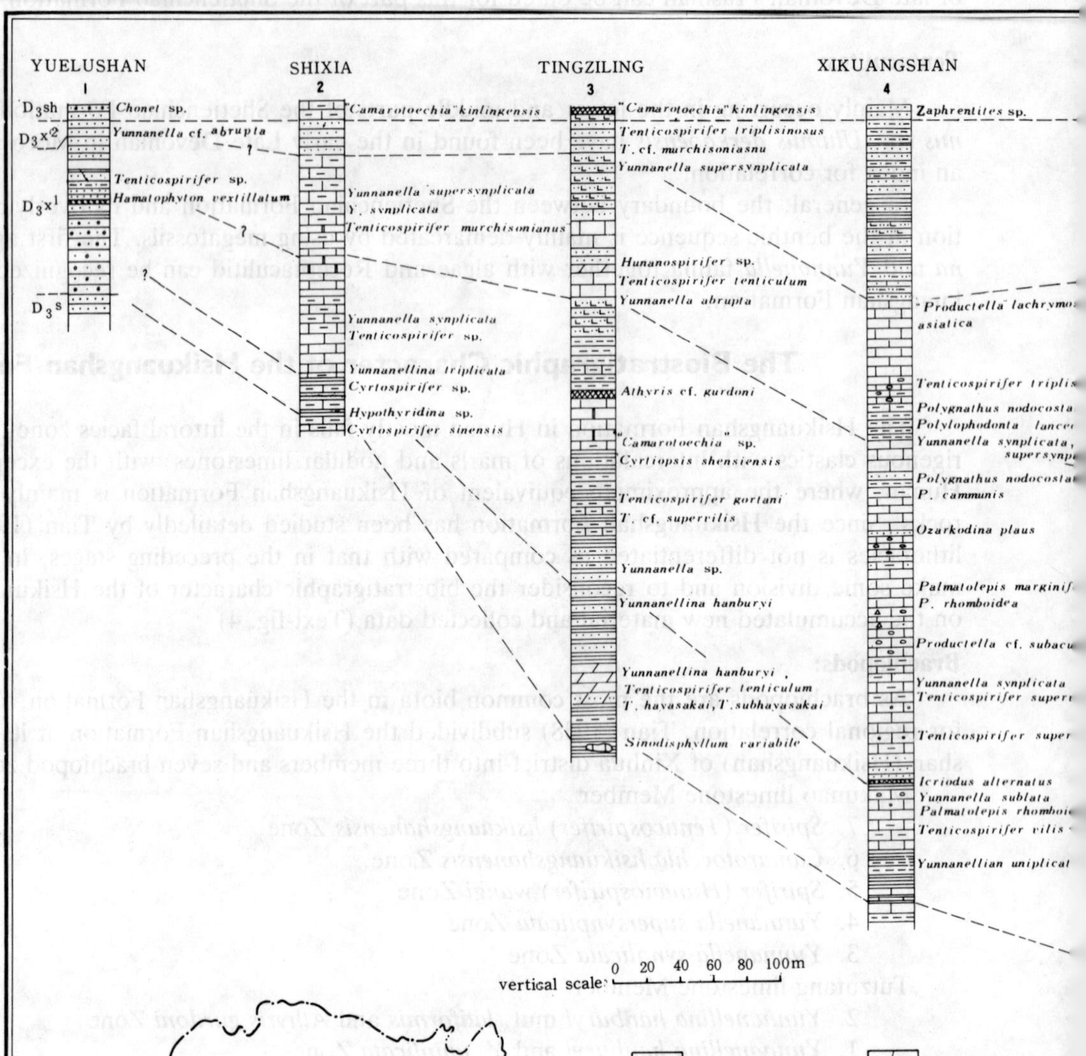

Text-fig. 4 Columnar correlation of the Upper Devonian Hsikuangshan Formation in Hunan.

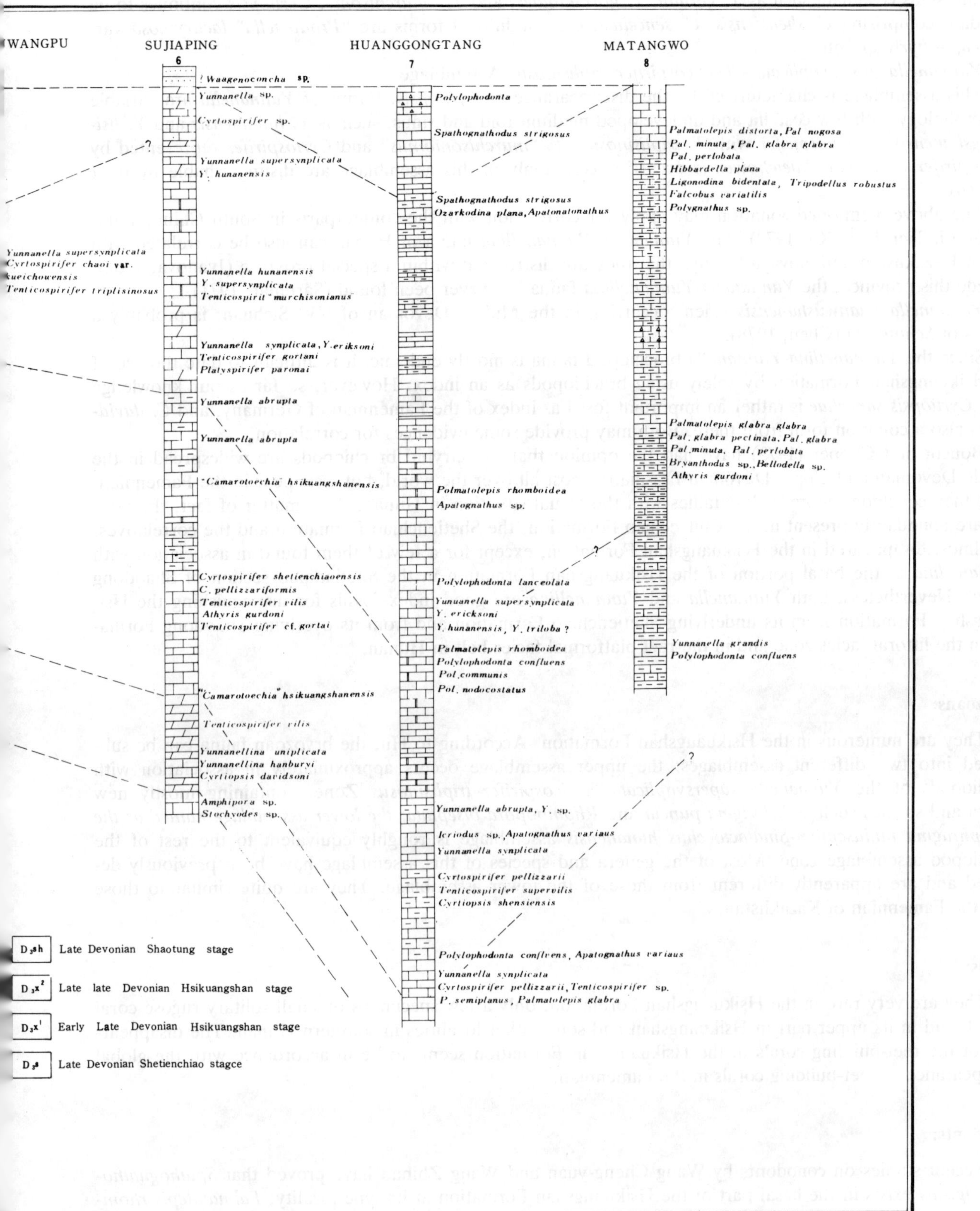
WANGPU
SUJIAPING
HUANGGONGTANG
MATANGWO
6
7
8
Waagenoconcha sp.
Yunnanella sp.
Cyrtospirifer sp.
Yunnanella supersynplicata
Y. hunanensis
Yunnanella supersynplicata
Cyrtospirifer chaoi var.
kueichowensis
Tenticospirifer triplisinosus
Yunnanella hunanensis
Y. supersynplicata
Tenticospirit "murchisomianus"
Yunnanella synplicata, Y. eriksoni
Tenticospirifer gortani
Platyspirifer paronai
Yunnanella abrupta
Yunnanella abrupta
"Camarotoechia" hsikuangshanensis
Cyrtospirifer shetienchiaoensis
C. pellizzariformis
Tenticospirifer vilis
Athyris gurdoni
Tenticospirifer cf. gortai
"Camarotoechia" hsikuangshanensis
Tenticospirifer vilis
Yunnanellina uniplicata
Yunnanellina hanburyi
Cyrtiopsis davidsoni
Amphipora sp.
Stachyodes sp.
Polylophodonta
Spathognathodus strigosus
Spathognathodus strigosus
Ozarkodina plana, Apatonatonathus
Polmatolepis rhomboidea
Apatognathus sp.
Polylophodonta lanceolata
Yunuanella supersynplicata
Y. ericksoni
Y. hunanensis, Y. triloba ?
Palmatolepis rhomboidea
Polylophodonta confluens
Pol. communis
Pol. nodocostatus
Yunnanella abrupta, Y. sp.
Icriodus sp. Apatognathus variaus
Yunnanella synplicata
Cyrtospirifer pellizzarii
Tenticospirifer supervilis
Cyrtiopsis shensiensis
Polylophodonta conflvens, Apatognathus variaus
Yunnanella synplicata
Cyrtospirifer pellizzarii, Tenticospirifer sp.
P. semiplanus, Palmatolepis glabra
Palmatolepis distorta, Pal nogosa
Pal. minuta, Pal. glabra glabra
Pal. perlobata
Hibbardella plana
Ligonodina bidentata, Tripodellus robustus
Falcobus variatilis
Polygnathus sp.
Palmatolepis glabra glabra
Pal. glabra pectinata, Pal. glabra
Pal. minuta, Pal. perlobata
Bryanthodus sp., Bellodella sp.
Athyris gurdoni
Yunnanella grandis
Polylophodonta confluens
?
D3sh Late Devonian Shaotung stage
D3x2 Late late Devonian Hsikuangshan stage
D3x1 Early Late Devonian Hsikuangshan stage
D3s Late Devonian Shetienchiao stagce

fold and sinus costellae, such as *T. gortani, T. gortanioides*, and *T. "murchisoni"*. *Cyrtiopsis* continues to be abundant comprising *C. shensiensis, C. senceliae*, etc.; additional forms are *"Productella" lachrymosa* var. *asiatica, Athyris* sp., etc.

Yunnanella supersynplicata - Tenticospirifer triplisinosus Assemblage.

This assemblage is characterized by the disappearance of *Cyrtiopsis*. Forms of *Yunnanella* are variable in morphology with few costella and undeveloped medium fold and sinus, such as *Y. hunanensis* and *Y. hsikuangshanensis; Tenticospirifer* with *T. triplisinosus, T. "murchisoniomus"* and *Cyrtospirifer* represented by *Cyrtospirifer chaoi* var. *kueichowensis* occurring commonly in this assemblage are distinguishable by their large size.

The above-mentioned zonation may apply for correlation with the counterparts in South China, Tsinling, South Tienshan (Xu, 1979). The *Yunnanella-Yunnanellina* fauna of Hunan can also be correlated with that of Kazakhstan and Novaya Zemlya that they are distributed within a special province (Ivanova, 1962). Outside this province, the *Yunnanella-Yunnanellina* fauna has never been found (Sartenar, 1971).

Yunnanella kuanwushanensis Chen, occurring in the Middle Devonian of NW Sichuan, is probably a species of *Schnurella* (Chen, 1978).

Since the *Yunnanellina-Yunnanella* brachiopod fauna is mostly endemic, it is difficult to date the age of the Hsikuangshan Formation by solely using brachiopods as an index. However, so far as our knowledge goes, *Cyrtiopsis sanceliae* is rather an important fossil as index of the Famennian of Germany, and *C. davidsoni* is also a common form from there. Both may provide some evidences for correlation.

Boucot and Cooper (1973) have had the opinion that the atrypid brachiopods are widespread in the Middle Devonian and Upper Devonian Frasnian almost all over the world, but disappear in the Famennian. What they concluded more or less tallies with the actual situation in Hunan. As a matter of fact, the atrypids are abundantly present in the Chitzechiao Formation, the Shetienchiao Formation and the correlatives, but almost disappeared in the Hsikuangshan Formation, except for a few of them found in association with *Yunnanellina* at the basal portion of the Hsikuangshan Formation in the Shetianqiao section of Shaodong district. Nevertheless, both *Yunnanella* and *Yunnanellina* are good index fossils for distinguishing the Hsikuangshan Formation from its underlying Shetienchiao Formation and from its overlying Shaodong Formation in the littoral facies zone and in the reef platformal facies belt in Hunan.

Bryozoans:

They are numerous in the Hsikuangshan Formation. According to Hu, the bryozoan fauna can be subdivided into two different assemblages: the upper assemblage occurs approximately in association with brachiopods of the *Yunnanella supersynplicata-Tenticospirifer triplisinosus* Zone, containing mostly new genera and species such as *Schulgina puncta* and *Rhombopora biseptata; the lower assemblage named as the Multiphragma multiseptum-Sinatactoechus hunanensis* assemblage is roughly equivalent to the rest of the brachiopod assemblage zone. Most of the genera and species of this assemblage have been previously described and are apparently different from those of the lower assemblage. They are quite similar to those from the Famennian of Kazakhstan.

Corals:

They are very rare in the Hsikuangshan Formation, only a few specimens of small solitary rugose coral being found in its upper part in Hsikuangshan and some other localities in southern Hunan. The disappearance of the reef-building corals in the Hsikuangshan Formation seems to be in accordance with the global disappearance of reef-building corals in the Famennian.

Conodonts:

Recent studies on conodonts by Wang Cheng-yuan and Wang Zhihao have proved that *Spathognathodus strigosus* exists in the basal part of the Hsikuangshan Formation at its type locality; *Palmatolepis rhomboidea* occurs in the bed with brachiopods of *Yunnanellina hanburyi-Cyrtiopsis davidsoni* Zone, indicating the existence of this index conodont which is widely distributed in Europe, North America and Australia; *Palmatolepis quadrantinodosa* is seen in the bed with brachiopods of *Yunnanella grandis-Tenticospirifer gor-*

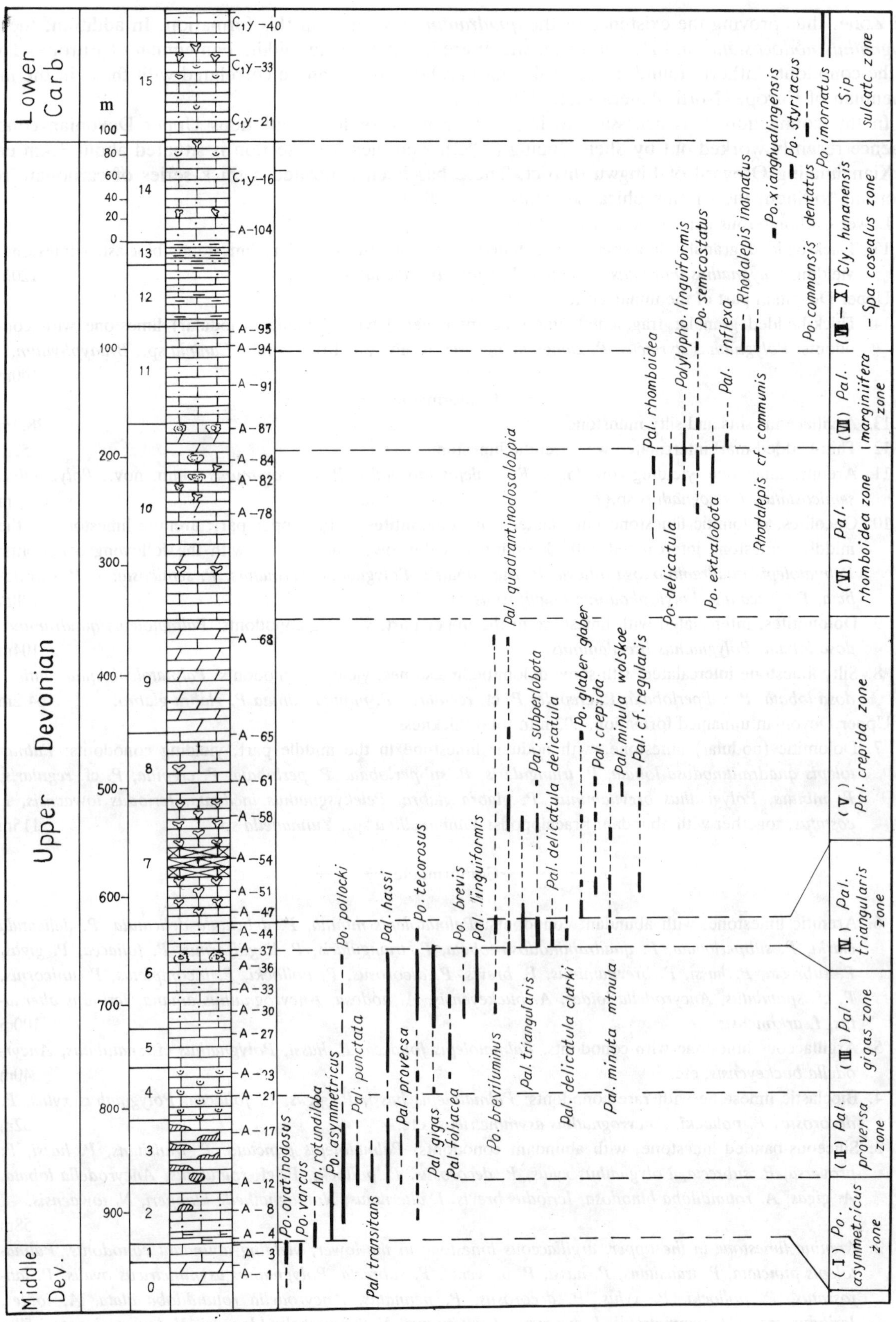

Text-fig. 5. Showing the stratigraphic distribution of main platformal conodonts in the Upper Devonian of Xianghualing, Lingwu (after Sheng Qiming, 1982).

dani Zone, thus proving the existence of the *quadrantinodosa* zone in this formation. In addition, there are *Polygnathus nodocostatus* and *P. lanceolata*, which are indexes of the middle Famennian of Europe. To sum up, the conodonts hitherto found in the Hsikuangshan Formation can be correlated with those in the middle Famennian of Europe, North America and Australia.

In this connection, it is necessary to introduce a more or less continuous Upper Devonian conodont sequence recently worked out by Shen Qiming (1982). The measured section is situated about 3 km east of the Xianghualing Oreyard of Lingwu district. There has been deposited a thick series of carbonate rocks. According to Shen, the stratigraphical sequence is as follows:

Lower Carboniferous part of the unnamed formation

15. Thick-bedded granular limestone, intercalated with dolomitic (nodular) limestone, bioclastic limestone, yielding *Polygnathus inornatus, P. anida, P. communis dentatus.* 120m

Upper Devonian part of the unnamed formation

14. Thick-bedded granular fragmental limestone intercalated with dolomitic (nodular) limestone with conodonts: *Polygnathus styriacus, P. xianghualingensis* sp. nov.; and rich corals: *Caninia* sp., *Bradyphyllum.* 100m

———Conformable———

13. Argillaceous silts and silty mudstone. 28.2m
12. Thin-bedded marl intercalated with arenitic limestone. 52m
11. Arenitic limestone, yielding conodonts: *Rhodalepis inornatus, R.* cf. *communis* comb. nov., *Polygnathus semicostatus, Mestognathus* sp.(?). 92m
10. Oncolites, dolomitic limestone intercalated with dolomitites in the upper part; arenitic limestone in the middle; limestone intercalated with dolomitites in the lower; upper part with the following conodonts: *Palmatolepis quadrantinodosa inflexa, P. rhomboidea, Polygnathus germanus, P. semicostatus, P. extralobata, P. delicatula, Polylophodonta linguiformis.* 188m
9. Dolomitites, intercalated with limestone in the upper part, yielding conodonts: *Palmatolepis quadrantinodosa lobata, Polygnathus brevilaminus.* 104m
8. Silty limestone intercalated with some dolomitic limestones, yielding conodonts: *Palmatolepis quadrantinodosa lobata, P. subperlobata, P. crepida, P.* cf. *regularis, P. minuta minuta P. glabra glabra.* 64.2m

Upper Devonian unnamed formation, 302m in total thickness.

7. Dolomitic (nodular) limestone, with nodular limestone in the middle part, yielding conodonts: *Palmatolepis quadrantinodosa lobata, P. triangularis, P. subperlobata, P. perlobata, P. crepida, P.* cf. *regularis, P. minuta, Polygnathus brevilaminus, P. glabra glabra, Pelekysgnathus inclinata, Icriodus iowaensis, I. costatus*, together with abundant brachiopods: *Yunnanellina* sp., *Yunnanella* sp. 115m

———Conformable———

6. Arenitic limestone, with abundant conodonts: *Palmatolepis minuta, P. delicatula delicatula, P. delicatula clarki, P. subperlobata, P. quadrantinodosa lobata, P. triangularis, P. linguiformis, P. foliacea, P. gigas, P. subrecta, P. hassi, P. brevilaminus, P. brevis, P. dicorosus, P. pollocki, P. incompletus, P. unicornis, P.* cf. *spatulatus, Ancyrodella ioides, A. buckeyensis, A. nodosa, Ancyrognathus amana, Icriodus alternatus, I. arkonensis.* 100m
5. Argillaceous limestone with conodonts: *Palmatolepis foliacea, P. hassi, Polygnathus* cf. *spatulatus, Ancyrodella buckeyensis*, etc. 40m
4. Bioclastic limestone with rare conodonts: *Palmatolepis hassi, P. gigas, P. foliacea, Polygnathus xylus, P. decorosus, P. pollocki, Ancyrognathus asymmetricus*, etc. 52m
3. Siliceous-banded limestone, with abundant conodonts: *Palmatolepis punctata, P. transitans, P. hassi, P. proversa, P. subrecta, Polygnathus xylus, P. decorosus, P. pollocki, P.* cf. *spatulatus, Ancyrodella lobata, A. gigas, A. rotundiloba binodosa, Icriodus brevis, I. alternatus, Nothognathella klapperi, N. iowaensis.* 58m
2. *Arenitic limestone in the upper; argillaceous limestone in the lower; yielding abundant conodonts: Palmatolepis punctata, P. transitans, P. hassi, P. proversa, P. subrecta, Polygnathus asymmetricus ovalis, P. latifossatus, P. pollocki, P. xylus, P. decorosus, P. pennatus, Ancyrodella rotundiloba alata, A. ioides, Icriodus brevis, I. symmetricus, I. expansus, I. alternatus, Nothognathella klapperi, N. iowaensis,* etc. 58m

—————Conformable—————

The underlying bed of unnamed formation

1b. Limestone with conodonts: *Polygnathus varcus, P. ovatinodosus, P. cristatus, P. latifossatus, P. xylus, Icriodus expansus, I. symmetricus, I. difficilis, I. alternatus, Ancyrodella rotundiloba rotundiloba.*

1a. *Massive limestone with conodonts at the top: Polygnathus ovatinodosus, Icriodus expansus.* 100m

Base not exposed

From the above-mentioned section, it can be seen that there exists an almost complete and continuous Upper Devonian conodont sequence in southern Hunan, which may serve as reference section for further correlation of the Devonian strata in Hunan. The Devonian section should be subdivided into several lithostratigraphic units. Moreover, *Yunnanella* and *Yunnanellina* first appeared in the bed with *Palmatolepis quadrantinodosa lobata* and other texa. This case is slightly different from that in the type locality of the Hsikuangshan Formation. Further detailed biostratigraphic work is called for in the future. (Text-fig.5)

A Brief Account of the Age of the Shaodong Formation

Traditionally, the Devonian-Carboniferous boundary has been drawn at the base of the Shaodong Formation in Hunan, which has been erected as the Shaodong Member at its type locality in Jieling of Shaodong district and can be correlated with the Etroeungtian in France (Yoh and Hou, 1962; Hou, 1965). While doing field work in Xintian of southern Hunan, we collected some brachiopods *Camarotoechia kinlingensis, Tenticospirifer* sp. together with some rugose corals *Ceriphyllum* and *Caninia*, which can be identified as from the Shaodong equivalent. It is interesting to note that the coexisting conodonts identified by Wang Chengyuan and Wang Zhihao are: *Polylophodonta* sp, *Spathognathodus strigosus, Apatognathus varians, Ozarkodina plana,* which are generally referred to typical Devonian conodonts. In the vicinity of Baiducun of southern Hunan, we have also collected a coral form assigned to *Caninia* which is similar to that in the Shaodong Formation. The Devonian conodonts *Palmatolepis distorta, P. glabra glabra* have also been identified by Wang Chengyuan as coexisting with this coral form. This led us to consider that the age of the Shaodong Formation may be of Devonian.

It was almost at the same time that Wu, Zhao and Jiang (1981) published what they had achieved on the rugose corals from the Shaodong Formation. In that paper Wu and others thought that the Shaodong Member should be changed into the Shaodong Formation with two coral assemblage zones: the *Caninia dorlodoti* Assemblage in the upper part and the *Ceriphyllum elegantulum* Assemblage in the lower part. They have also correlated the Shaodong Formation with the Etroungtian of West Europe, and treated it as the basal formation of the Lower Carboniferous. The corals in the upper assemblage are: *Caninia patula, C. shaodongensis, C. jielingensis, Zaphrentoides steroseptatus, Diphyphyllum antiquatum, Dematophyllum minor, D. hunanense, Syringopora ramulosa, S. zhongshanensis*, etc.

The lower assemblage includes the following coral species: *Zaphrentoides delanouei, Z. konincki, Complanophyllum compressum, C. minor, Caninia shaodongensis.*

However, the foraminifers from the upper part of the Shaodong Formation identified by Wang, K.L. (1978) include: *Septotournayella rauserae, Septaglomospiranella complanata, Glomospiranella rare, Vicinesphaera* sp., *Asterosphaera* sp. According to Wang, this fauna is equivalent to the Ctia Zone or *Septatournayella rauserae* Zone in the Russian Platform, Urals and Donbass Basin, and the Tn la in Dinant Basin. Therefore, from the foraminiferal correlation, the Shaodong Formation is appropriately assigned to the Late Devonian.

Wang and Ziegler (1982) have studied the conodont fauna from the Shaodong Formation at its type locality. In the lower part of this formation corresponding to the *Ceriphyllum elegantum* coral Assemblage Zone, Wang and Ziegler (1982) have identified the coexisting conodonts including: *Polygnathus semicostatus, Icriodus costatus* and *Bispathodus aculeatus aculeatus*, indicating that this part is Late Devonian in age.

The conodont fauna occurring in the upper part in association with the corals of *Caninia dorlodoti-Caninia patula* Assemblage and the brachiopods of *Mesoplica* and *Sphenospira* includes the stratigraphically important forms: *Bispathodus aculeatus plumulus, Clydagnathus cavusformis* and *Polygnathus communis communis*, which are very similar to those of Lower K-Zone in England. They came to the conclusion that "the Shaodong Member is Uppermost Devonian in age"; "The Shaodong Member is only a part of the Hsikuangshan Formation and belongs to the latest Famennian" (Wang and Ziegler, 1982).

Recent advancements in the study of Devonian-Carboniferous boundary in South China have made it

Table 1. Showing the relationship of the Devonian stratigraphic units in the facies zones in Hunan.

	Central and Southern Hunan: shallow marine facies, reef platformal	Central and Southern Hunan: shallow marine facies, interreef depressional	Central and Southern Hunan: littoral facies, outer littoral	Central and Southern Hunan: littoral facies, inner littoral	Eastern Hunan	Northwestern Hunan
Upper Devonian	Shaotung Fm.	Unnamed Fm. limestones	sandstones and siltstones	sandstones and siltstones	Fanxia Fm. iron ore bed	
Upper Devonian	Unnamed Fm. in S. Hunan with thick bedded limestones and nodular limestones	Unnamed Fm. in S. Hunan with thick bedded limestones and nodular limestones	Hsikuangshan Fm.: sandstones Mem. (Oujiachong Mem.) thick bedded limestones intercalated with marls (Makunao Mem.) (Nilangli iron ore bed) Marls and limestones (Tutzutang Mem.)	Unnamed Fm. or Gr. sandstones	Jingchong Fm. Upper Huolongquan Fm.	Hsiehchingsu Fm.
Upper Devonian	Unnamed Fm.: marls and limestones thick-bedded limestones	Shetienchiao Fm.: marls and calcareous mudstones Mem. limestones Mem.	Chanlongchie Fm.: Upper Mem. Lower Mem. Unnamed Fm. with marls and sandstones	Unnamed Fm. or Gr. sandstones (continued)	Middle and lower Huolongquan Fm.	Huangchiateng Fm.
Middle Devonian	Lungkouchung Fm.	Shetienchiao Fm. (limestones Mem., continued)	Unnamed Fm. with marls and sandstones (continued)		Mashiling Fm.	Yungtaikuan Fm.
Middle Devonian	Chitzechiao Fm.: thick bedded limestones marls and sandy shales	Unnamed Fm. with siliceous bed Unnamed Fm.: argillaceous limestones marls intercalated with siliceous shales limestones	argillaceous limestones and marls intercalated with sandstones	argillaceous limestones and marls intercalated with sandstones	Chitzechiao Fm. — ? —	
Middle Devonian	Unnamed Fm. siltstones, limestomes, marls	Unnamed Fm. siltstones, limestomes, marls	argillaceous limestones and marls intercalated with sandstones (continued)			
Lower Dev.	Tiaomachian Fm. sandstones, conglomeratic sandstones	Tiaomachian Fm. sandstones, conglomeratic sandstones	Tiaomachian Fm. sandstones, conglomeratic sandstones			

clear that some Devonian conodont elements are found to be coexistent with the coral *Cystophrentis*, which has been by tradition recognized as the index coral of the *Cystophrentis* Zone at the base of the Lower Carboniferous. Whether the Devonian-Carboniferous boundary should be drawn at the base or at the top or even through the *Cystophrentis* Zone calls for further detailed work and this is beyond the scope of this paper. The relationship of the Devonian stratigraphic units in the facies zone in Hunan is shown in Table 1.

GENERAL ACCOUNT OF THE DEVONIAN LITHOFACIES AND PALEOGEOGRAPHY IN HUNAN

As a result of the pre-Devonian epierogenic movement, most parts of Hunan uplifted to form the land and were subjected to a long period of denudation until the marine transgression approached Hunan in the Devonian time.

Before the deposition of Devonian rocks, there already existed a major difference in the relief of the basement between northwestern Hunan and central and southeastern Hunan. The former area is of platform nature, where the pre-Devonian sediments had not suffered from intensive folding and denudation. The Devonian rocks are, therefore, in disconformity with the underlying Silurian rocks. However, in central and southeastern Hunan, the pre-Devonian rocks are extensively folded and faulted and had suffered from strong erosion. The Devonian rocks lie unconformably upon the Ordovician and Silurian rocks in the region south of Yuechenling-Guangdimiao granite bodies and upon the Cambrian rocks in southeastern Hunan, where they form the elongated belts trending northeastward, approximately in accordance with the main direction of the Devonian marine transgression.

In consideration of the remarkable difference in characters of the Devonian sediments between northwestern Hunan and central and southern Hunan, and the general pattern of the facies, it is reasonable to presume that the "Jiangnan Old Land" had existed in the Devonian. It might be an uplifted area suffering from a long period of denudation. The scope of the old land changed from stage to stage, depending on the Devonian marine transgression and regression. It is presumed that the old landmass might have extended to its large scale in the Early Devonian and reached to the smallest in the late Middle Devonian when the Chitzechiao Formation was deposited.

It is necessary to point out again that owing to the limited evidence from the conodonts and ammonoids hitherto available, the time-determination and the time-correlation of Devonian sequence in different areas in Hunan are mainly based on the megafossils. Therefore, the meaning of the "stage" here used is not a strict chronostratigraphical term, but a general term used for a certain period when certain formation and its approximate correlatives were deposited

During the period from Early Devonian to Middle Devonian, northwestern Hunan was under the situation of denudation. It was since the beginning of late Middle Devonian that the marine transgression spread over this area. The terrigenous clastics with considerable thickness, mainly composed of pure quartzose sandstone and yielding rare plant fossils, are known as the Yungtaiguan Formation. The deposition of quartzose sandstones near the old land indicates that they were deposited after a long period of denudation and transportation. To the north and above or equivalent to it, the grain size of sediments becomes finer, intercalated with hematitic ironstones, shales with ferric and manganic, marls and limestones, which yield *Cyrtospirifer*, with rare solitary rugose coral *Pseudozaphrentis* and bivalves or sometimes with fragments of plant fossils, known as the Huangchiateng Formation of the littoral facies. Later on, there are deposits of intertidal and shallow subtidal sediments known as the Hsiehchingsu Formation, which is approximately part of the Famennian correlative. The lower part of the Hsiehchingsu Formation is composed of arenaceous, argillaceous rocks, intercalated with two layers of oolitic hematite and containing abundant brachiopods *Yunnanella*, bryozoans and ostracods. The upper part of the Hsiehchingsu Formation is mainly composed of quartzose sandstones. All of these indicate the shallow marine sedimentation. Then, it changed into a regressive phase, which is characterized by sediments mainly of quartzose sandstones, silts intercalated with some black shales. Approximately at the end of the Devonian, the marine water began to retreat from this area. It was only in some depressional regions that the clastics were received. As a whole, during the Devonian time, northwestern Hunan may be a platformal area with graded shoreline, covered with shallow marine and oscillated in transgression and regression. For this reason, it is determined as the littoral facies zone.

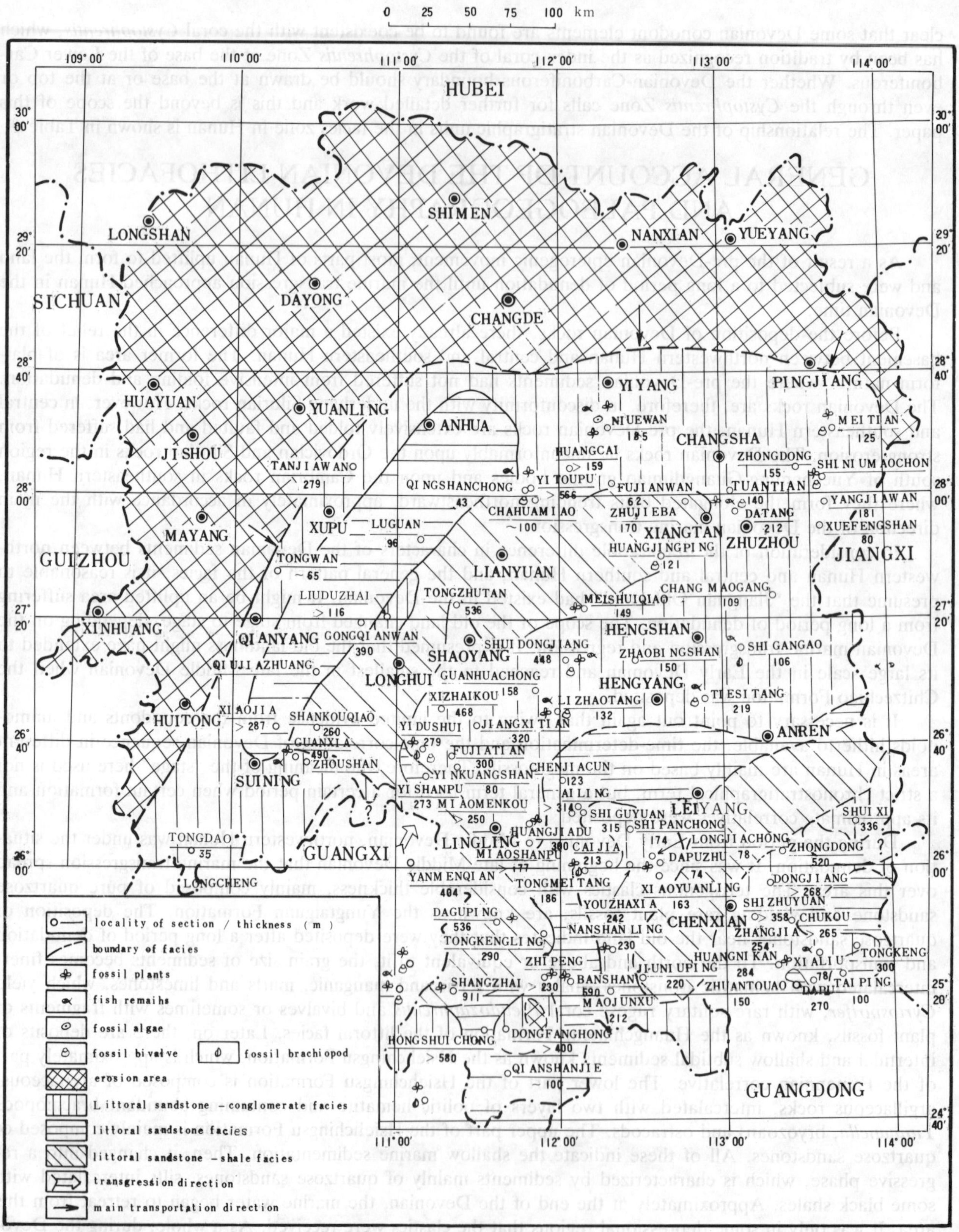

Text-fig. 6. Paleogeographic map of Early Devonian to Middle Devonian Tiaomachien "stage" in Hunan.

The Early Devonian to early Middle Devonian

Those regions covered by the marine transgression in this period were limited to some domains located to the east of Xingning, south of Shuidongjiang of Shaodong district and west of Changning district, where the terrigenous clastics of littoral facies were deposited. The lower part of the clastic series is mainly composed of conglomerates, sandy conglomerates, sandstones, silts and silty shales, which are different from place to place, mainly depending on the local geological background. The colour of the sediments is purple for the lower part, and changes into greenish yellow for the upper part. *Zosterophyllum* and *Taeniocrada* are the common fossil plants in the Lower Devonian, whereas *Protopteridium* and *Psilophyton* become dominant in the Middle Devonian. Among the fish remains, *Crossopterygii, Dipnoi, Antiarchi* are frequently seen, together with forms of bivalves. Toward the north, the grain size of sediments becomes coarser and their thickness is on the decrease, but the thickness of basal conglomerates is on the increase. This is in accordance with the main direction of the marine transgression, reflecting the uneven relief of the basement. This is also proved by the fact that there is a well developed sequence of the clastics of Early to Middle Devonian with the basal conglomerate bed attaining 40m thick at Shuidongjiang of Shaodong district, however, at Shetianqiao, only a few kilometers apart from Shuidongjiang, the whole clastic sequence vanishes from sight.

The late Middle Devonian probably including early Middle Devonian (the Tiaomachien Formation and its correlatives)

The Tiaomachien Formation tentatively recognized as the late Middle Devonian probably including early Middle Devonian clastics is widespread in Hunan, except in Miaoershan of Xingning district, Dongjiangling and Qichangling of Daoxian district, where the said formation is absent. The rock contents of the Tiaomachien Formation mainly are the purple quartzose sandstones, argillaceous sandstones, silty shales, thin-bedded pisolitic hematites, locally with copper-bearing variegated beds and basal conglomerates of various thicknesses. Plants, bivalves and fish remains are the main components of the fossil assemblage. Among the fossil plants, forms referred to *Lepidodendropsis, Taeniocrada, Protopteridium* and *Psilophyton* are common; *Bothriolepis sinensis* and *Hunanolepis tieni* are the well known forms of the pisces; bivalves include *Prothyris lanceolata*, and forms referred to *Palaeoneilo* and *Clossites*. Pisces and bivalves are mostly seen in the upper part of the formation; sometimes, they are coexistent with some brachiopods, such as forms referred to *Lingula* and *Chonetes*. Fossil and plants occur throughout the formation.

The facies change of the Tiaomachien Formation is mainly represented by the change of thickness and grain size of the sediments and by the presence of intercalations of iron-bearing horizons and the amount of terrigenous materials. These are as a whole, under the control of such factors as the distance from the old land, the relief of the sea floor and the magnitude of subsidence of the basin. Detailedly, in Jiangyong, Jianghua, Daoxian, Linwu, Guiyang and Lingling of southern Hunan, the sediments are 200-300m thick, containing fewer gravels; toward the north in the southern part of Shaoyang district, there appear the basal conglomerates; further northward to Tongwan of Huaihua district, Qingtangpu of Anhua district, Fujianwan of Ningxiang district, the thickness of sediments is reduced to about 100m, and the amount of gravels decreases again; whereas in Tanjiawan of Xupu district, Shaping of Taoyuan district, Niuewan of Yiyang district, Changsha and Liuyang districts close to the old land, the sediments become coarser, attaining to 200m in thickness inclusive of basal conglomerates.

In an elongated area from the northeastern corner of central Hunan, toward southwest through Liuyang, Henshan to Changning, the sediments are much coarser and thicker than those on the sides of the area. It might be a relatively uplifted area. In contrast, the area around Liuyang, Xingshao, Longhui might be a depressive area with considerable magnitude of subsidence, where the sediments are 200m to 500m thick and fine in grain size, comprising principally silts and pelites.

In Putangpu of Anhua district, Tongzhutan of Xingshao district, Liuduzai of Longhui district and in Wugang, the thickness of the Tiaomachien Formation changes remarkably within a short distance, but the characters of lithology and biota remain more or less the same. (Text-fig.6)

In Jiangyong, Daoxian and Guiyang, the Tiaomachien Formation generally bears one to three layers of thin-bedded pisolitic or oolitic hematite with no industrial value; northward to Shaodong, Shuangfeng, Changsha, the hematite beds change into ferruginous sandstones and then thin out.

The late Middle Devonian (Chitzechiao Formation) to the early Late Devonian (Shetianchiao Formation)

Generally, the marine transgression during this period continued to advance (Text-figs. 7,8). There appeared in Hunan the deposition of carbonate rocks with considerable thickness. In central and southern Hunan, the marine sediments are widespread with flourishing biota. Reefs or banks were gradually formed in the appropriate area by the extensive activities of frame-builders, such as stromatoporoids, corals and algae. The extent and distribution of the reef complex are controlled by the environmental factors like the water depth, the distance from the old land, from where the terrigenous material derived and the relief of sea floor, especially the existence of the submarine terrace formed by faulting. In turn, the development of the reef or any kind of build-up gave rise to the complicated relief of sea floor and the diversification and differentiation of lithofacies and biofacies.

As discussed in the preceding chapter, in central Hunan where the littoral facies is developed, it can be further subdivided into the outer littoral facies zone and the inner littoral facies zone. The extent of the inner littoral facies zone in the time when the Chitzechiao Formation and its approximate correlatives were deposited, covered the area north of the line along the Ningxiang-Changsha districts and Gugang of Liuyang district. Later on, in the Shetienchiao "stage", the extent slightly expanded southward to Anhua, Lianyuan, Shaoshan and their adjacent region.

The extent of the outer littoral facies zone in the Chitzechiao "stage" covered the area north of the line along Xianxi, Qingtangpu of Anhua district, Zhujiaduan of Lianyuan district, Shuangfeng and Zhuzhou districts. In the Shetienchiao "stage", the eastern part remained unchanged, but its western front shifted forward to Xinghua-Lianyuan.

The shallow marine facies is further subdivided into the reef platformal facies zone and the inter-reef depressional facies zone. In the Chitzechiao "stage", the reef platformal facies zone was located to the south of Xingshao-Liling. The sections measured in Maanshan of Shaodong district, in Qiziqiao of Xiangxiang district and in Huanggongdang of Xintian district may serve as typical for studying the reef complex (Text-fig. 9). In the Shetienchiao "stage" the extent of this facies zone was slightly narrower, at least in Qiziqiao and its adjacent region, the reef platformal facies has not yet been discernible.

The inter-reef depressional facies zone is scattered in distribution, and its extent remained unchanged from the Chitzechiao "stage" to the Shetienchiao "stage". In southern Hunan, it covered the regions of Xintian-Lingling-Daoxian in a horseshoe-like arrangement, surrounding the reef platformal facies zone. In central Hunan, there are several elongated depressional facies zone in NE trending, alternating with the reef platformal facies zones.

As regards the development of the Devonian reef in Hunan, it can be generally subdivided into the following phases:

The mud-mound and the biostrome-forming phase

At the beginning of the Middle Devonian Chitzechiao "stage", there were lots of argillo-calcareous sediments deposited on the warm and shallow platform in central Hunan. Many biotas, mainly of benthos were flourishing, including bivalves, brachiopods, bryozoans, crinoids, etc. Later on, corals were growing up including those solitary and small colonial rugose corals and branching and laminated tabulate corals.

The accumulated bio-detritus from the skeleton of these organisms usually built up to form the bank, although the strong and compact wave-resistant structure has not yet been formed. As the reef-building organisms continued growing, with the participation of stromatoporoids and algae, especially with the extensive development of *Endophyllum*, large colonial rugose coral skeletons had accumulated and combined together to form biostromes, which extended widely in central and southern Hunan and which can be taken as an indication of the initial phase of the Devonian reef-building process in Hunan.

The reef-forming phase mainly by stromatoporoids

In the middle and late Chitzechiao "stage", the extent of the marine transgression was enlarged. Because of the frequent activities of the reef-building organisms, there exist thick and massive reef limestone and dolomitites with considerable thickness overlying the *Endophyllum* bed in the reef platformal facies zone in central Hunan.

The reef limestone may attain to several hundred meters in total thickness, its peripheral part being

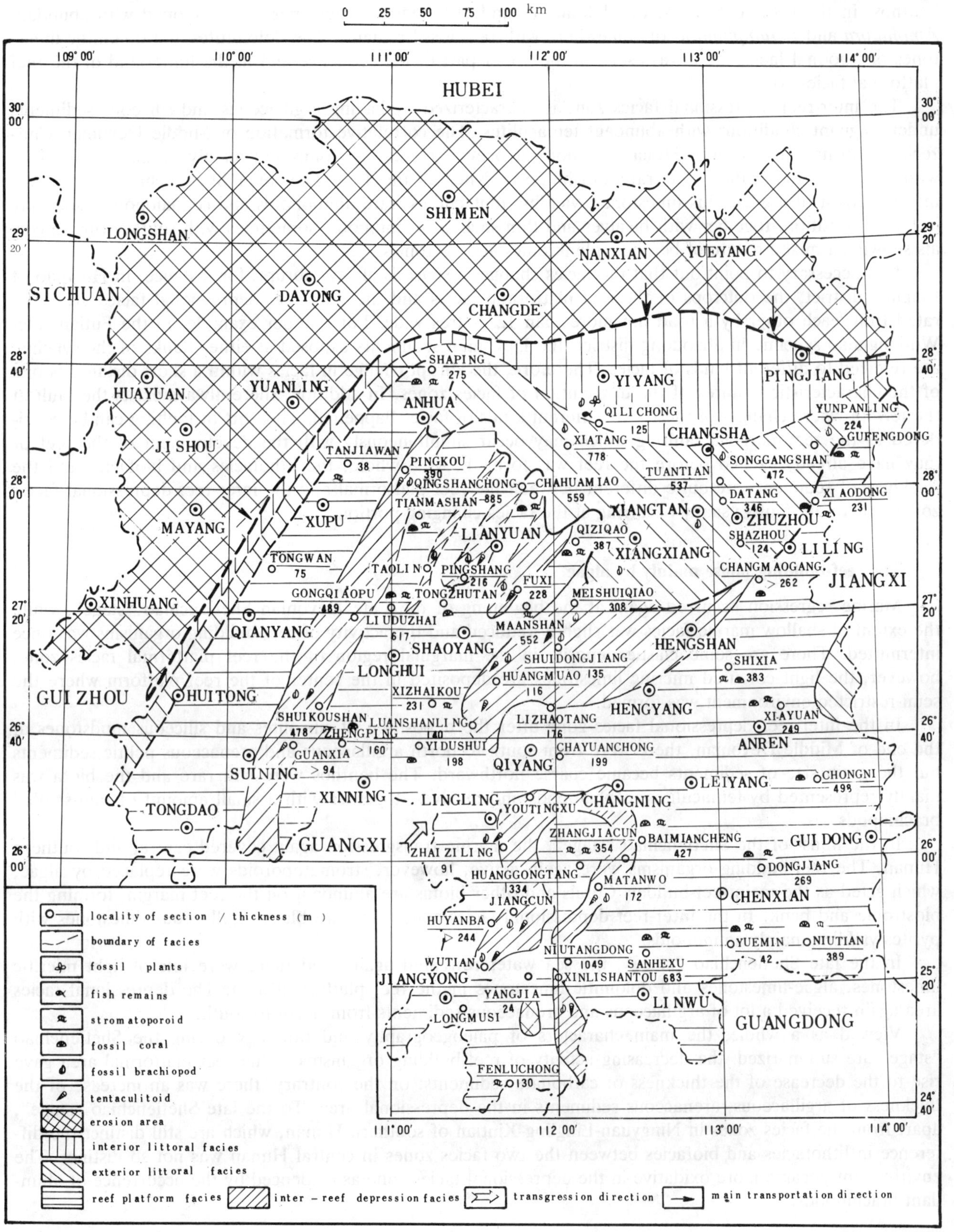

Text-fig.7. Paleogeographic map of Middle Devonian Chitzechiao "stage" in Hunan.

flanked by limestones, dolomitites and argillaceous limestones with abundant corals, brachiopods and other benthos. In the restricted facies, the dolomitites and well-bedded limestones are developed with abundant *Amphipora* and *Paramphipora*, or somewhere with rare fossil remains. The dolomitites and dolomitic limestones of lagoonal facies are always seen in the upper part of the sequence and in the hinterland of the reef platformal facies zone.

The inter-reef depressional facies zone is characterized by having argillaceous and siliceous sediments under stagnant conditions with abundant tentaculites. The correlative formation of Middle Devonian Chitzechiao Formation in central Hunan has not yet been named. It is composed mainly of shales, silicolites with abundant tentaculites and rare brachiopods, whereas the Late Devonian Shetienchiao correlatives mainly consist of argillaceous limestones, yielding tentaculites, rare ammonoids and cephalopods, especially in later deposition, together with bivalves and thin-shelled brachiopods, sometimes with solitary rugose corals showing a mixed nature of benthos, planktonics and nectonics.

It is necessary to point out that the facies change in space is always abrupt, for instance, in Huaqiao of Lingling district, the outcrops of the reef platformal facies and the inter-reef depressional facies are separated from each other by a road not more than 12 metres wide. For another example, in the Futiandang-Wudongmiao syncline in Shaodong district, the Devonian sediments exposed at either flanks of the syncline are referred to different kinds of facies. Therefore, the abrupt change in facies within a short distance is one of the characteristic features of the distribution of facies zones. In addition, there always exists the fault at the terrain where changes of these two different kinds of facies happened. This sort of faulting can be considered as the reef-marginal faulting which may occur simultaneously with the development of the reef or may have already existed before the reef was formed. As a whole, the lithofacies and biofacies and the thickness of sediments are quite different between the reef platformal and the inter-reef depressional facies zones. There are faults usually occurring at the facies changing portion.

The reef-building phase mainly by algae

Marine regression occurred again at the beginning of the Late Devonian Shetienchiao "stage", while the extent of shallow marine facies was slightly reduced and the activity of reef-building organisms was once intermitted. There are some clastic deposits in the marginal region of the reef platformal facies zone, however, the light coloured micritic limestones are deposited in the center of the reef platform where the semi-restricted environment was formed.

In the inter-reef depressional facies zone after the deposition of silicolites and siliceous mudstones at the end of Middle Devonian, there were from south to north argillaceous and arenaceous pelitic sediments but the grain size of sediments became coarse northward. The benthos were still rare and the biota was mainly represented by tentaculites, cephalopods and bivalves, together with a small amount of thin-shelled brachiopods.

In the midst of the Shetienchiao "stage", the marine transgression again covered central and southern Hunan. The reef-building organisms were again active, however, stromatoporoids were replaced by algae, which acted as the main reef-builder. Corals and other biotas are abundant on the reef margin, forming the biostrome and bank. In the inter-reef depressional facies zone there are black argillaceous limestones with pyrites yielding mainly ammonoids.

In the late Shetienchiao "stage", the sea water retreated again, and there were lots of light micritic limestones, algae-limestones and dolomitic limestones in the reef platformal area. The depressional facies area again received a lot of argillaceous and arenaceous sediments from north to south.

Viewed as a whole, the main characters of paleogeography and lithology during the Shetienchiao "stage" are summarized: the decreasing activity of reef-building organisms in the reef platformal area gave rise to the decrease of the thickness of carbonate sediments; on the contrary, there was an increase of the thickness of argillaceous, arenaceous sediments in the depressional area. To the late Shetienchiao "stage", apart from the facies zones in Ningyuan-Lingling-Xintian of southern Hunan, which are still distinct the difference in lithofacies and biofacies between the two facies zones in central Hunan was not so distinct. The environment became more oxidative in the depressional facies zone as evidenced by the occurrence of abundant brachiopods.

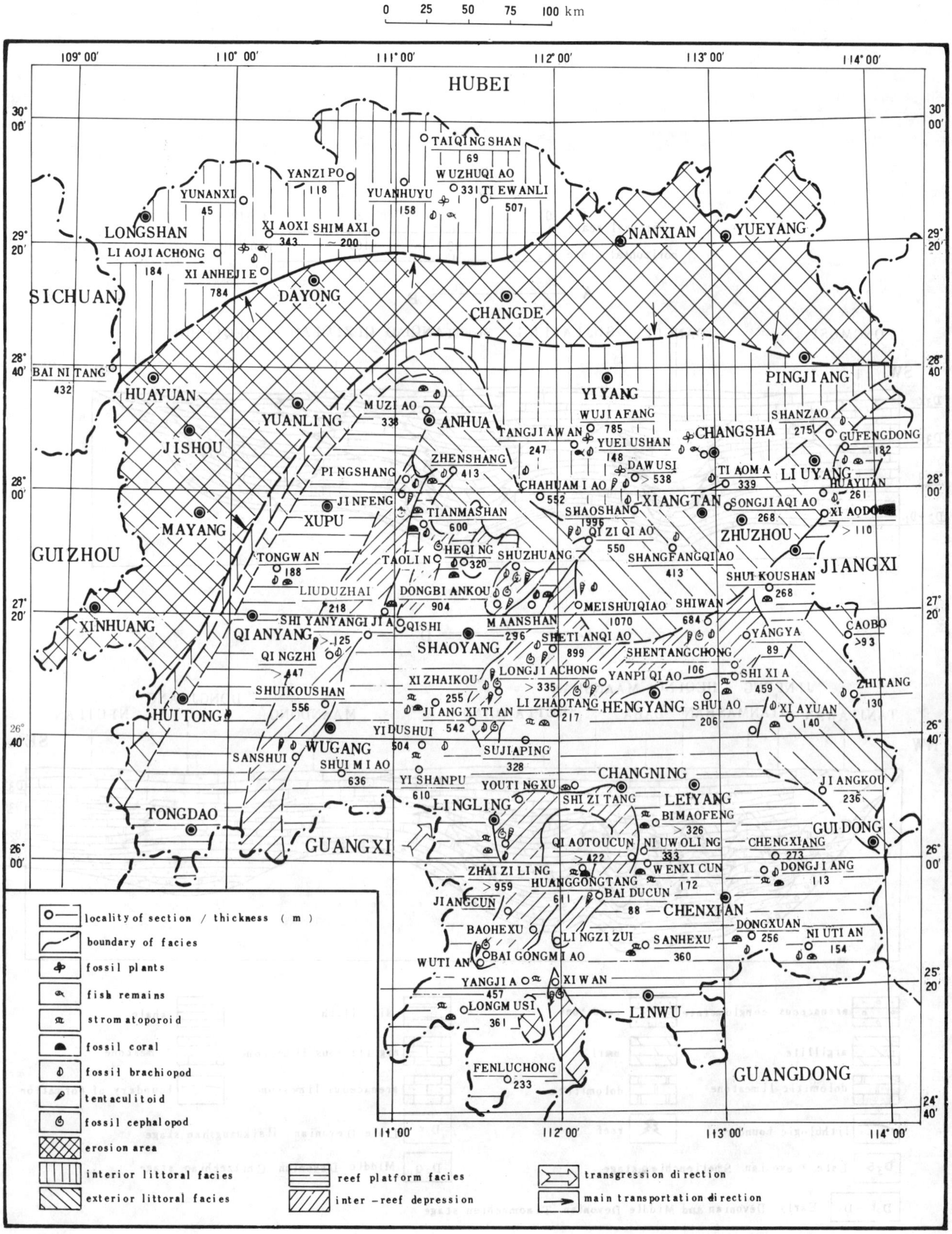

Text-fig. 8. Paleogeographic map of Late Devonian Shetienchiao "stage" in Hunan.

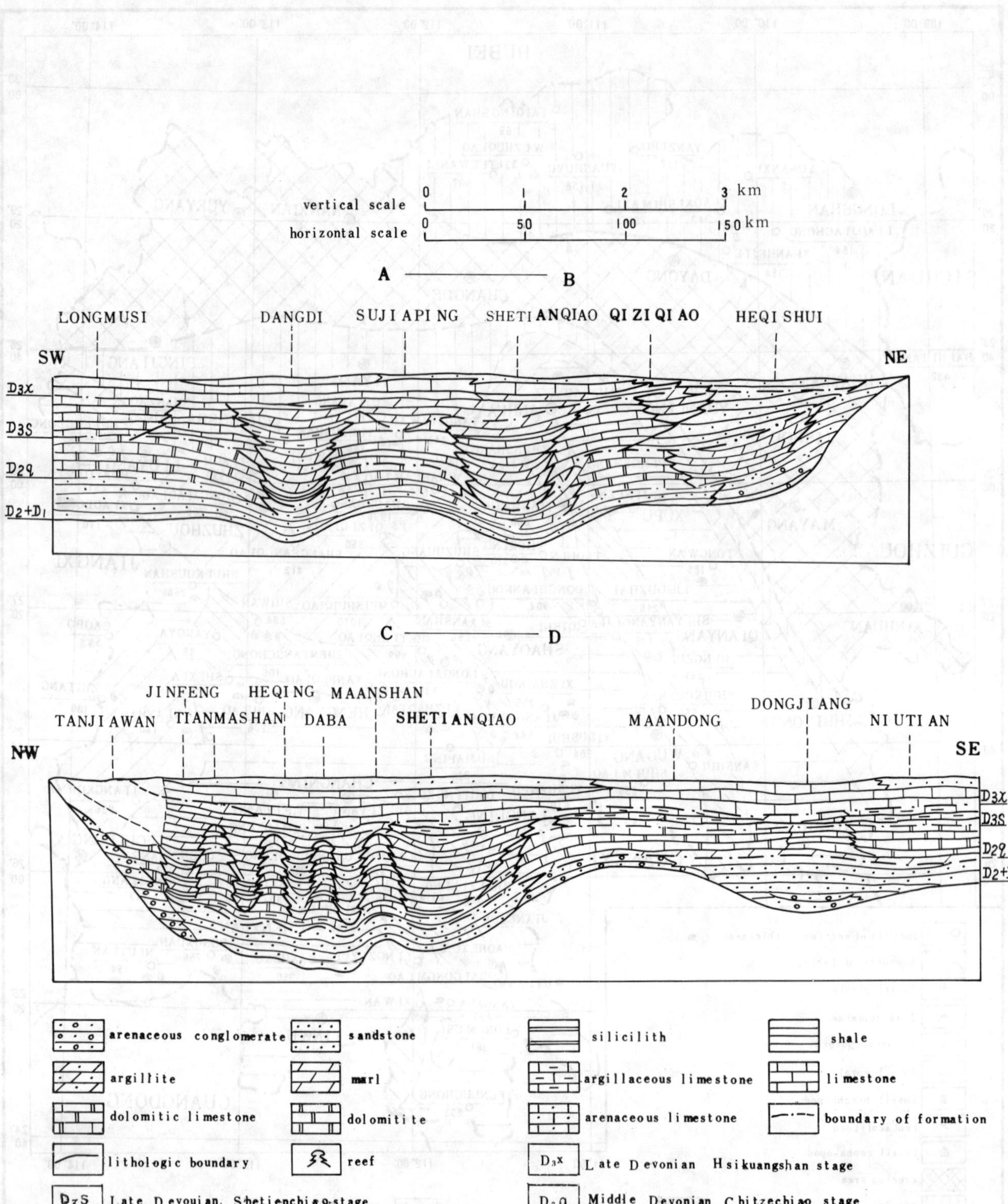

Text-fig. 9. Showing the lithofacies change in the Devonian of Hunan.

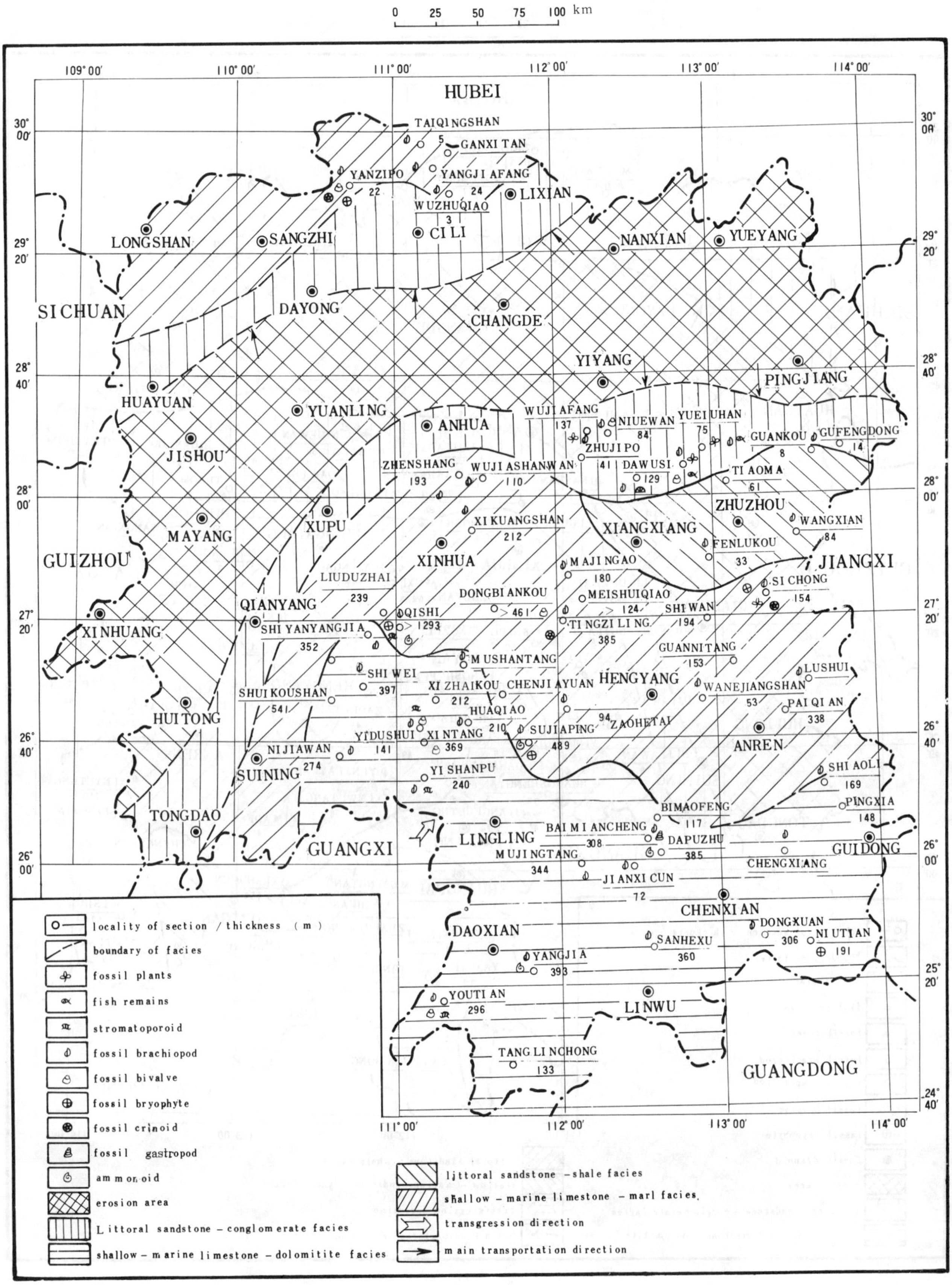

Text-fig. 10. Paleogeographic map of Late Devonian early Hsikuangshan "stage" in Hunan.

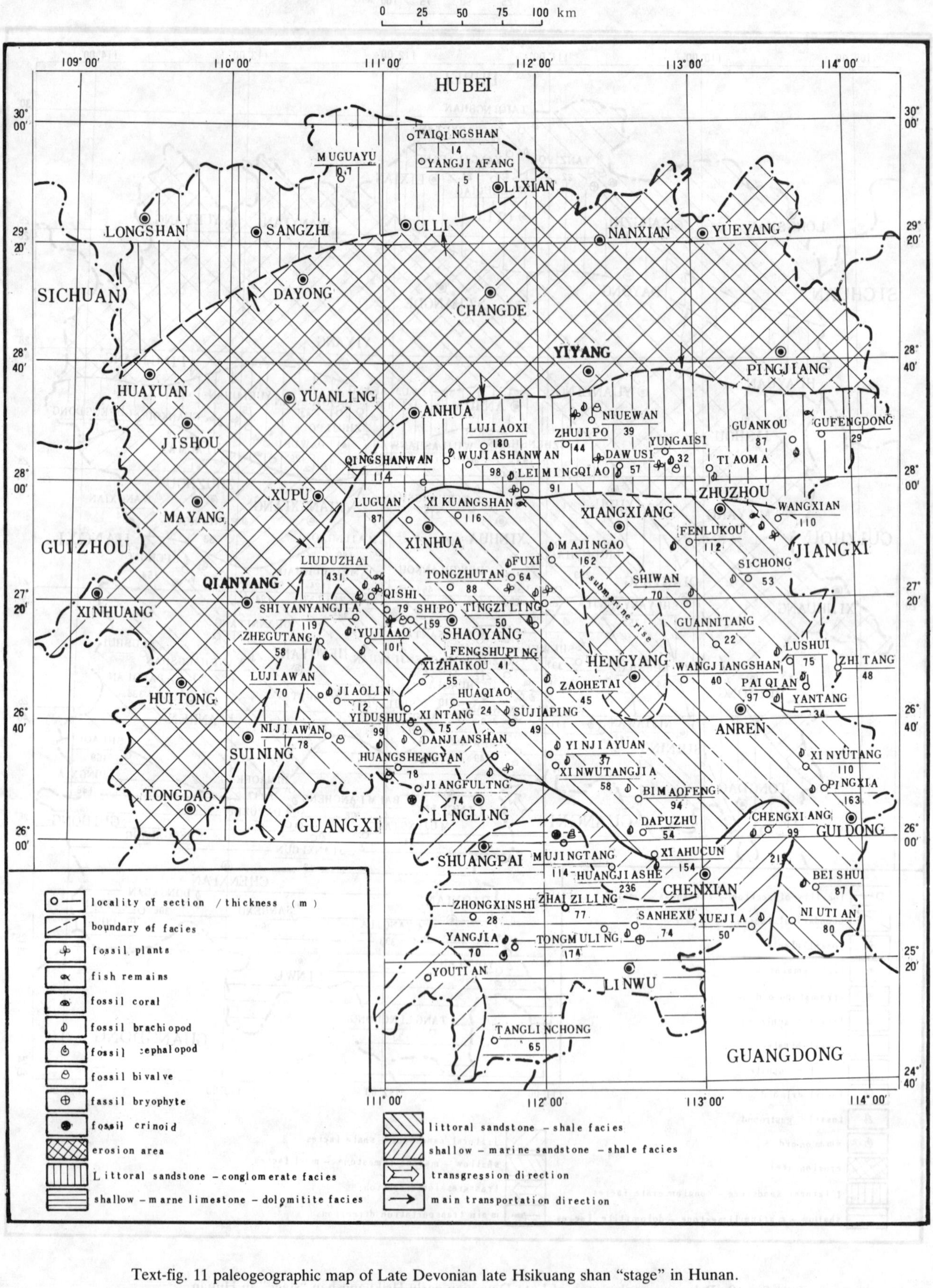

Text-fig. 11 paleogeographic map of Late Devonian late Hsikuang shan "stage" in Hunan.

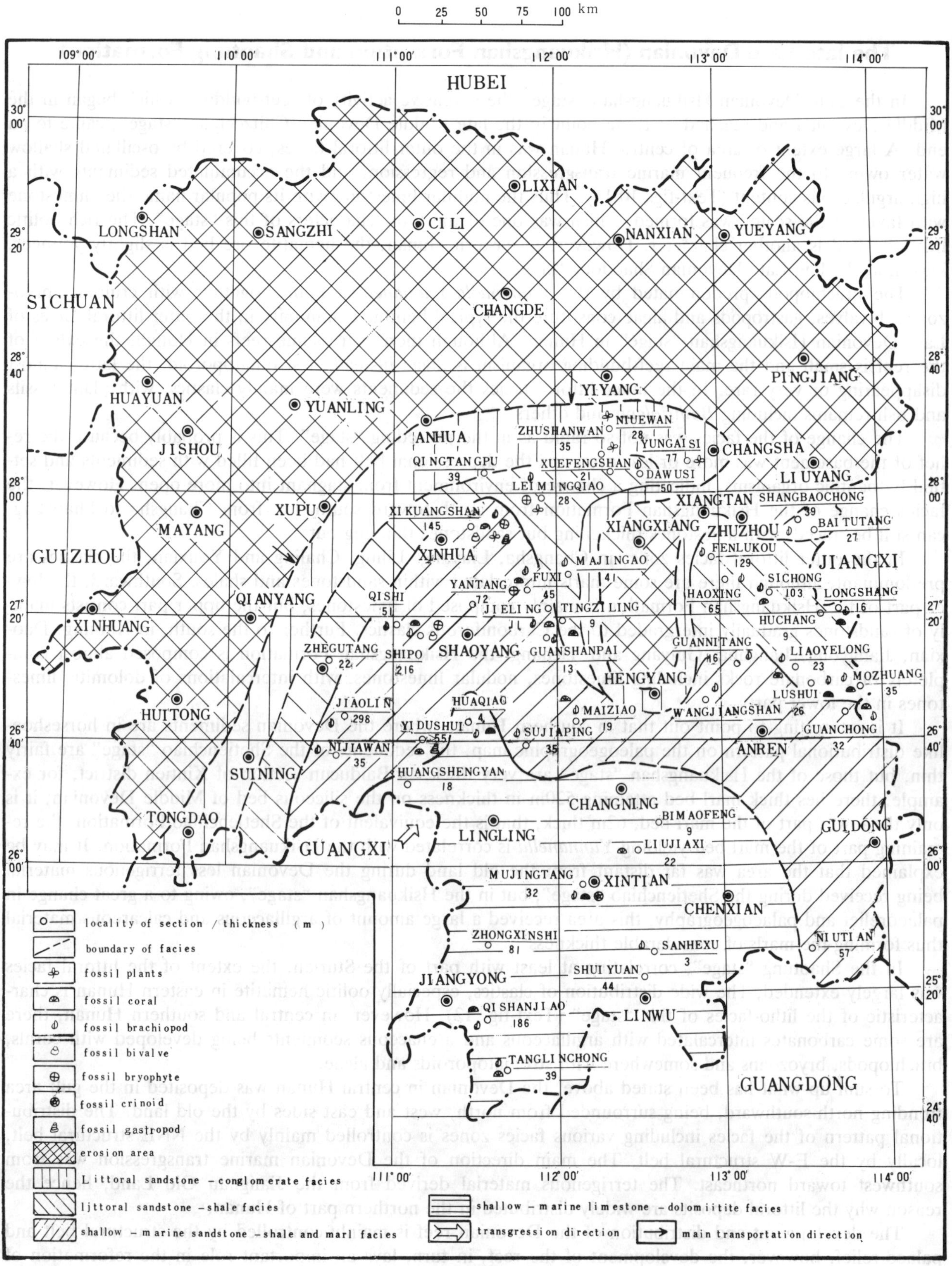

Text-fig. 12. Paleogeographic map of late Late Devonian Shaodong "stage" in Hunan.

The late Late Devonian (Hsikuangshan Formation and Shaotung Formation)

In the Late Devonian Hsikuangshan "stage" the extensive activity of reef-building, which began in the Middle Devonian and reached its acme point in the late Middle Devonian Chitzechiao "stage", came to an end. A large extent of area of central Hunan was in the outer littoral facies, covered by oscillated shallow water owing to the frequent marine transgression and regression, and the accumulated sediments with a high argillaceous content (Text-figs 10,11). Thus the nodular limestone with its popular name the "limestone with favus of the scalp" was formed, which was one of the lithic characters of this "stage". The rich oolitic hematite bed is another character. However, in eastern Hunan, the hematite ore bed is slightly younger, known as the late Late Devonian Shaodong "stage".

The brachiopods predominated by the *Yunnanella-Yunnanellina* fauna together with crinoids, bryozoans, bivalves, gastropods and algae constitute the typical benthic community in the outer littoral facies of Late Devonian Hsikuangshan "stage" in Hunan. Although there were some reefs in Hunan, the extent of the reef is small and the main reef-building organism is algae instead of stromatoporoid, the latter almost disappearing or very rare. In the inner littoral facies the sediments were mainly clastics, with plant fossils and fish remains including *Bothriolepis* and others.

The change of the facies was not so rapid as in the preceding "stage". This is probably because the relief of the basement was more or less even and the depressional belt had been filled with sediments and settled by benthic organisms, indicating a change of environment from stagnant into more open. However, the facies change of the Hsikuangshan Formation from north toward south, i.e. from Changsha to Shaoyang, can still be traced, with limestones thickening but sandstones thinning out.

In the outer littoral facies zone, in Changsha, Liuyang, Liling, Chaling and Youxian, the clastics are predominant, although there are some marls intercalated within sandstones and shales. Southward, the lower part of the Hsikuangshan Formation is mainly composed of limestones, but the upper part consists mainly of sandstones gradually intergraded to the Carboniferous clastics. Further to the south, in Xinning, Daoxian, Jiangyong, Jianghua, Guiyang and Yichang, the Hsikuangshan Formation is composed almost completely of carbonate rocks including limestones, nodular limestones, with intercalations of dolomitic limestones in the lower part.

It is interesting to point out that in southern Hunan, where the Devonian sediments are in horseshoe-like distributional pattern on the paleogeographic map, the sediments of the Shetienchiao "stage" are fairly thin, but those of the Hsikuangshan "stage" are very thick. In Baiducun section of Xintien district, for example, there lies thick marl bed attaining 520m in thickness on the siliceous bed of Middle Devonian; it is only the lower part of the marl bed, 62m thick, that is the equivalent of the Shetienchiao Formation, the remaining part of the marl bed yielding *Yunnanella* is correlated with the Hsikuangshan Formation. It may be explained that the area was far distant from the old land during the Devonian less terrigenous material being received during the Shetienchiao "stage"; but in the Hsikuangshan "stage", owing to a great change in palaeorelief and palaeogeography, this area received a large amount of argillaceous and calcareous material thus forming the marls of considerable thickness.

In the Shaotung "stage", correlating at least with part of the Sturian, the extent of the littoral facies was largely extended, The wide distribution of clastics, especially oolitic hematite in eastern Hunan is characteristic of the litho-facies of this "stage" (Text-fig. 12). However, in central and southern Hunan, there are some carbonates intercalated with argillaceous and arenaceous sediments being developed with corals, brachiopods, bryozoans and somewhere with stromatoporoids and algae.

To sum up what has been stated above, the Devonian in central Hunan was deposited in the gulf area trending north-southward, being surrounded from north, west and east sides by the old land. The distributional pattern of the facies including various facies zones is controlled mainly by the NNE structural belt, locally by the E-W structural belt. The main direction of the Devonian marine transgression was from southwest toward northeast. The terrigenous material derived from the Jiangnan Old Land, hence the reason why the littoral deposits are widely distributed in the northern part of Hunan.

The development and distribution of the Devonian reef is mainly controlled by the structural belt and palaeo-relief, however, the development of the reef, in turn, lays an important role in the reformation of the palaeo-relief and in the diversification and differentiation of lithofacies and biofacies, especially in the time when the Chitzechiao Formation and the Shetienchiao Formation were deposited.

CHARACTERISTICS OF THE DEVONIAN ORE-DEPOSITS IN HUNAN

The ore-bearing Horizons and the Character of Mineralization

Mineral resources are rich in the Devonian rocks in Hunan. The mineralization of Devonian in this province is characterized by the following aspects: the multiple ore-bearing horizons, the wide distribution, the various kinds of mineral resources and the diverse type of ore-deposits. From the Tiaomachien Formation at the base up to the Shaodong Formation on the top of Devonian, almost each of the Devonian formations in Hunan bears ore deposits or shows ore occurrence. Central and southern Hunan is well known for the occurrence of a great variety of coloured metal resources, whereas in northeastern and eastern Hunan is known as the famous area with iron deposits of the so-called Ningxiang type, a kind of oolitic or pisolitic hematite (ironstone). The mineral resources in Hunan include oolitic or pisolitic hematite, siderite, pyrite, and Cu, Pb, Zb, Hg – metal sulfide deposits, as well as Au, Ga, W, Mn ore deposits. There are not only typical stratified ore deposits, but also vein deposits with diverse types. At the first appearance, the mineral resources seem to be at random and in disorder but actually are controlled by certain horizons and facies.

The Devonian is the primary facies in the Late Palaeozoic cycle in Hunan, even in South China. There always exist conglomerates as the basement overlying various kinds of rocks in different ages, which indicates a long-term hiatus between Devonian and pre-Devonian. Concerning the upper limit of the Devonian ore-deposits in Hunan, the ore-deposits in the Shaotung Formation and its approximate equivalents are included, as discussed biostratigraphically in the preceding chapter. In fact, in some mining area in eastern Hunan, gradual and continuous sedimentation and mineralization can be seen through the Hsikuangshan Formation into the Shaotung Formation. The rock content of these two formations are mainly calcareous, argillaceous and arenaceous with high components of Fe, Mn, P, As, Cu, Zn, Sb, V, Mo, Ga, Ni, Co, Pb, etc., which are relevant to the elements absorbed by the clay and with abundant B, Ti, Cr, Sn, which are related to detrital sediments. However, there exists a remarkable change of lithofacies after the deposition of the Shaodong Formation and its approximate equivalents. The massive limestones and dolomitites with an increasing content of Sr and Ba forming the overlying beds indicate a change of the speed and magnitude of the marine transgression and probably a change of the climate, which might be a result from the evaporation of sea water.

Relationship between Lithofacies and Ore-deposit Association

Viewed from the character of lithofacies and paleogeographic pattern, the Devonian of Hunan may fall into two kinds of deposits; the littoral facies and the shallow marine facies. Accordingly, two kinds of ore-deposit association are distinct from each other. The littoral ore-deposit association occurs mainly in the area with terrigenous clastics which are intercalated with argillaceous limestones. Of this association, the ironstones of the Ningxiang type is most important. A small amount of gallium has been even found in the surrounding rocks close to the ore-body in a few mining areas. Besides, there are indications of mineralization of copper in the sandstones and gold in the conglomerates of the Tiaomachian Formation (Text-fig. 13). Ore deposits in association with shallow marine occur chiefly in the carbonate rocks, including siderite, pyrite, galena, sphaerite, stibnite and cinnabar. The above two associations are always kept apart in space, somewhere also in time intervals. It is evident that further subdivision of the lithofacies into facies zones makes it possible to understand the distribution of ore deposits. As the littoral facies is further subdivided into the outer littoral facies zone and the inner littoral facies zone, the iron deposit of Ningxiang type is known to occur mostly in the transitional terrane between these two zones. The siderite and its sister deposits have been proved to occur mostly at the transitional terrane between the reef platform facies zone and the inter-reef depressional facies zone in the shallow marine facies.

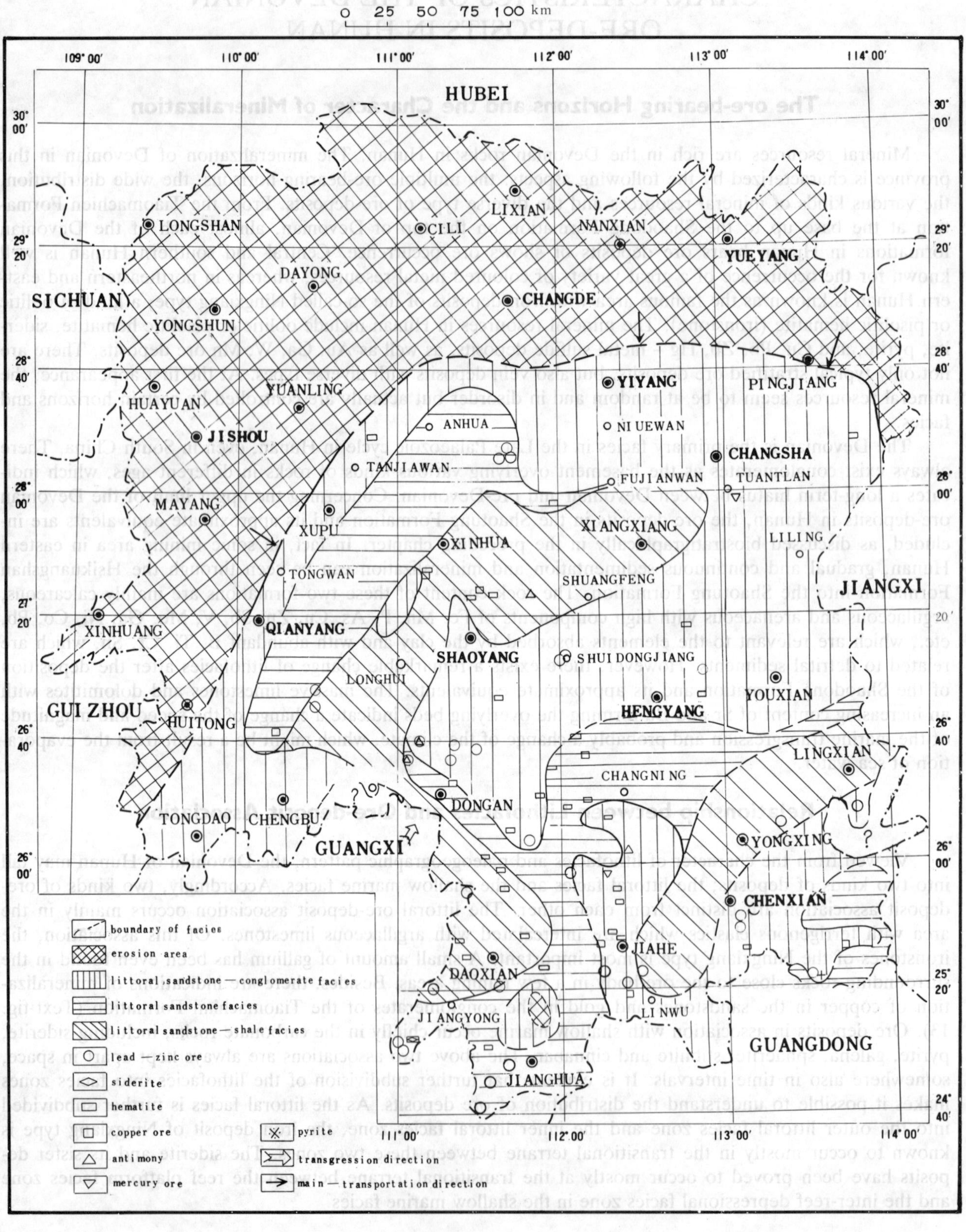

Text-fig. 13. Showing the distribution of lithofacies and mineral resources in Hunan Province in Early Middle Devonian.

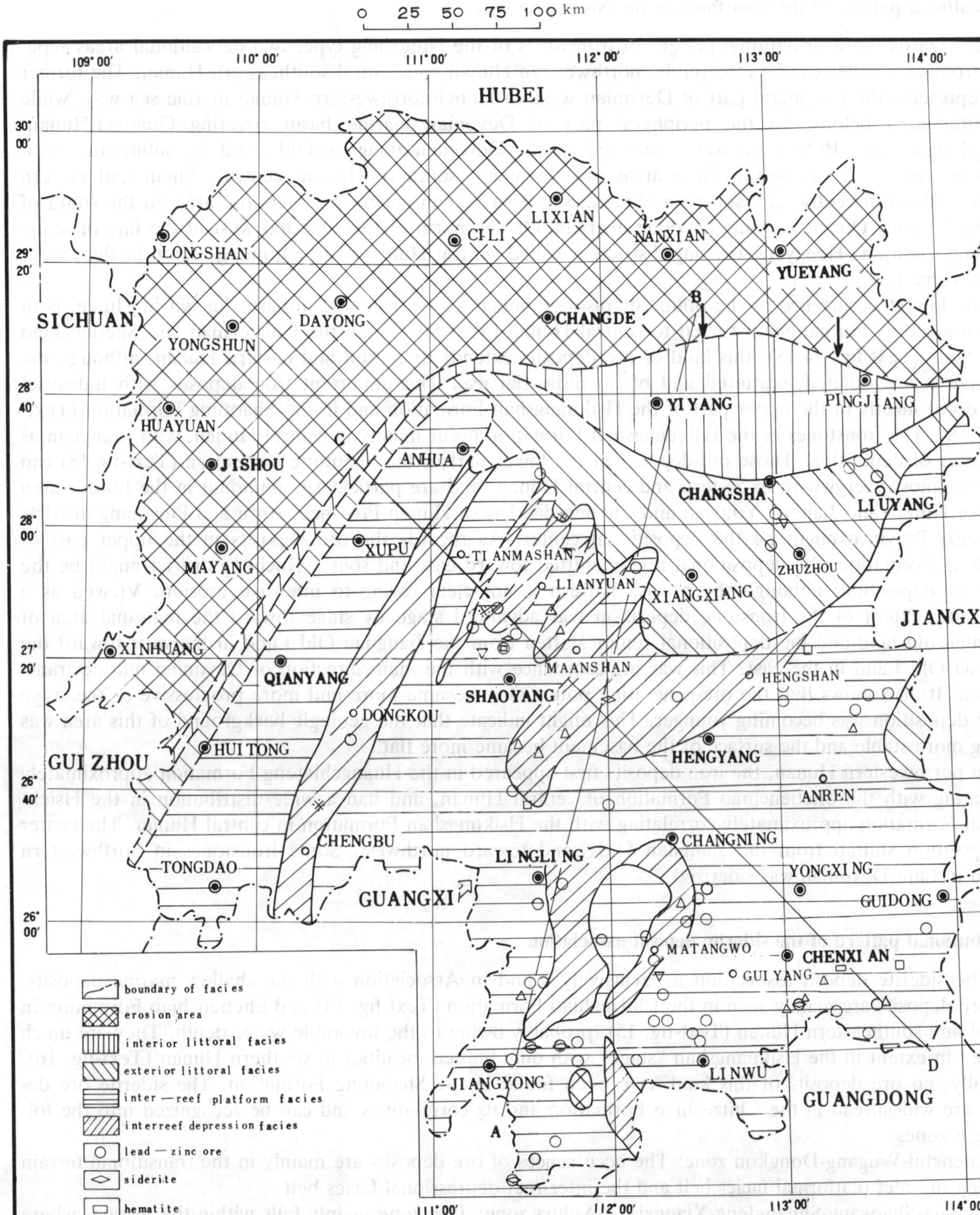

Text-fig. 14. Showing the distribution of lithofacies and mineral resources in Hunan Province during Middle Devonian Chitzechiao "stage".

Distributional pattern of the ironstones of the Ningxiang type.

As regards the distributional pattern of ironstones of the Ningxiang type, two depositional areas separated from each other can be discernible, northwestern Hunan and central southeastern Hunan. The former area represents the peripheral part of Devonian western Hubei-northwestern Hunan marine sea way, while the latter area belongs to the peripheral part of Devonian marine basin covering Guangxi-Hunan-Guangdong-Jiangxi. Both areas were separated by the Devonian Jiangnan Old Land (or submarine rise). For convenience, the following three areas are discussed: southern Hunan, central Hunan and eastern Hunan, instead of central and southern Hunan. Southern Hunan generally covers the area on the south of Huitong-Wugang-Qiyang-Leiyang-Guidong districts. On the north of this line and west of the line or along Pingjiang-Changsha-Hengyang-Leiyang districts is called central Hunan. The adjoining region in the east is eastern Hunan.

The lowest occurrence of the oolitic or pisolitic hematites (ironstones) hitherto known in Hunan is in the Tiaomachien Formation restricted to central Hunan (Text-fig 13). In the Chitzechiao and Shetienchiao Formations (Text-figs 14,15), this kind of ore deposits extends to central and eastern Hunan, although the occurrences are isolated, scattered and of no industrial use. The important iron deposits with industrial value occur mostly in the upper part of the Hsikuangshan Formation and in the Shaotung Formation (Text-fig. 17,18). The ironstones in the Hsikuangshan Formation occur mainly in central Hunan, with Lianyuan as the center of deposition. Those ore deposits in the lower part of the Shaotung Formation (Text-fig.18) can be seen almost everywhere in eastern and central Hunan, and are particularly abundant in the border area between Hunan and Jiangxi. Youxian and Chaling districts of Hunan Province, Anfu and Pingxiang districts of Jiangxi Province might be the depositional center. As regards the ore deposits in the upper part of Shaodong Formation, the depositional region shifted toward east and south. Rucheng district might be the center of deposition, although the precise pattern is not clear owing to intensive erosion. Viewed as a whole, the extent of the ironstone deposition was advanced stage by stage toward the marginal area of Devonian old land and the depositional center shifted from the Jiangnan Old Land in the west toward the Cathysia Old Land in the east. This was in accordance with the main direction of Devonian marine transgression. It also shows that the ore-generating potentiality became more and more progressive as the stage of ore deposition was becoming younger. This might indicate that the geologic background of this area was getting more stable and the surface of the basement became more flat.

In northwestern Hunan, the iron deposits first appeared in the Huangchiateng Formation approximately correlating with the Shetienchiao Formation of central Hunan, and had a wide distribution in the Hsiehchingsu Formation approximately correlating with the Hsikungshan Formation in central Hunan. The center of deposition shifted from the Jiangnan Old Land toward northwest. Some ironstones in northwestern Hunan in Late Devonian was siderized.

Distributional pattern of the siderite deposit association

The siderite deposit association is exclusively found in Association with the shallow marine deposits. The ore deposits are mainly seen in the Chitzechiao Formation (Text-fig. 14) and Shetienchiao Formation in central and southeastern Hunan (Text-fig. 15), probably owing to the favorable water depth. They are much reduced in extent in the Hsikuangshan "stage" with only limited localities in southern Hunan (Text-fig. 16). Basically, no ore deposits of this kind have been found in the Shaodong Formation. The siderite ore deposits are widespread in the Chitzechiao Formation and its correlatives and can be recognized into the following 8 zones:

Chengbu-Wugang-Dongkou zone: The occurrences of ore deposits are mainly in the transitional terrain between the reef platformal facies belt and the inter-reef depressional facies belt.

Xinhua-Shaoyang-Shuangfeng-Xiangxiang-Anhua zone: This zone mainly falls within the terrain, where the reef platformal deposits and the inter-reef depressional deposits made an abrupt change in facies. It seems that this area is favorable for ore deposition, as the occurrences of the ore deposit are always abundant and widely extended.

Xiangtan-Liuyang zone: The ore deposits are mainly formed in the outer littoral facies zone, probably referred to the back reef facies under the lagoonal environment.

Hengdong-Hengyang zone: The ore deposits occur in the transitional zone between the reef platformal

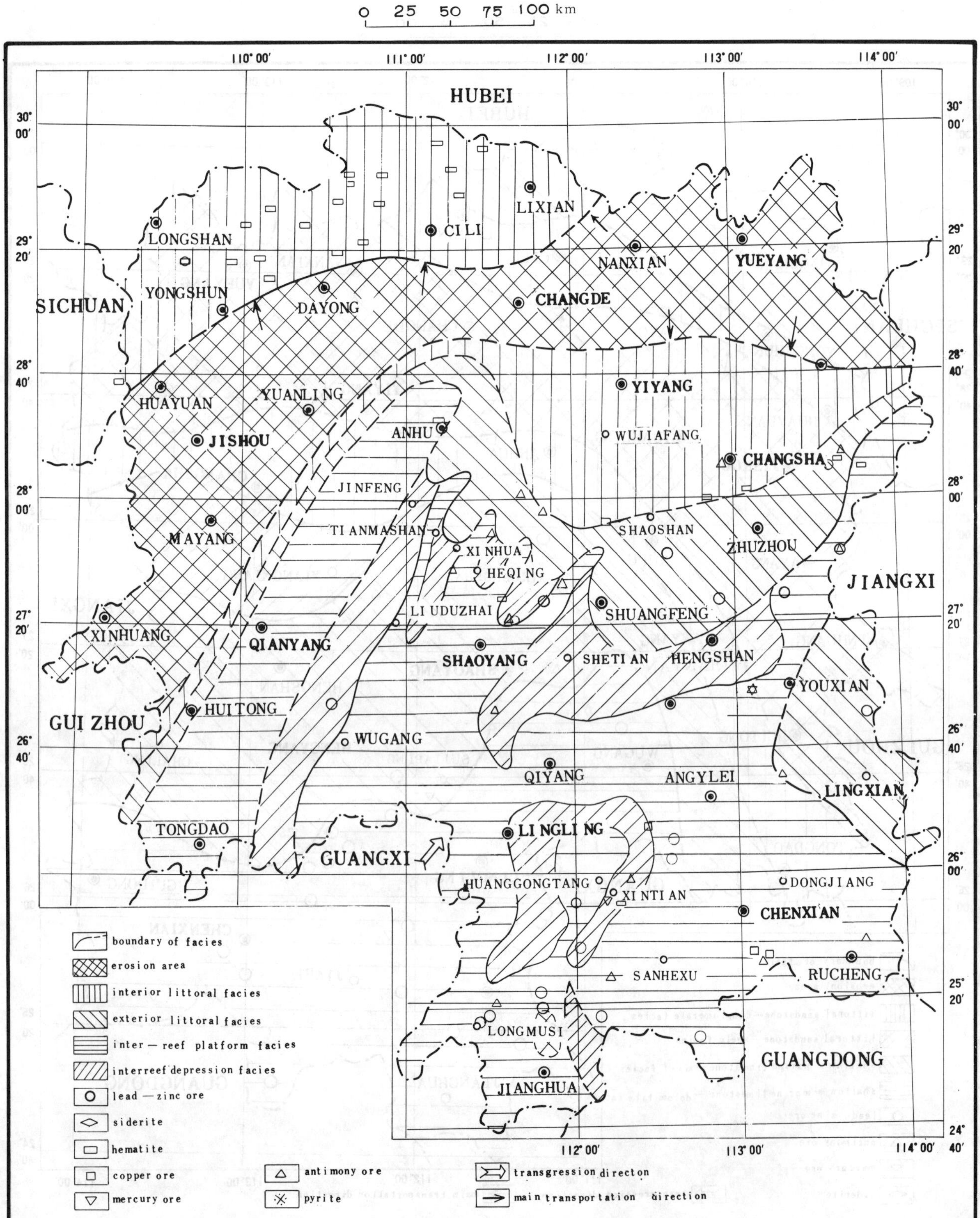

Text-fig. 15. Showing the distribution of lithofacies and mineral resources in Hunan Province during Late Devonian Shetienchiao "stage".

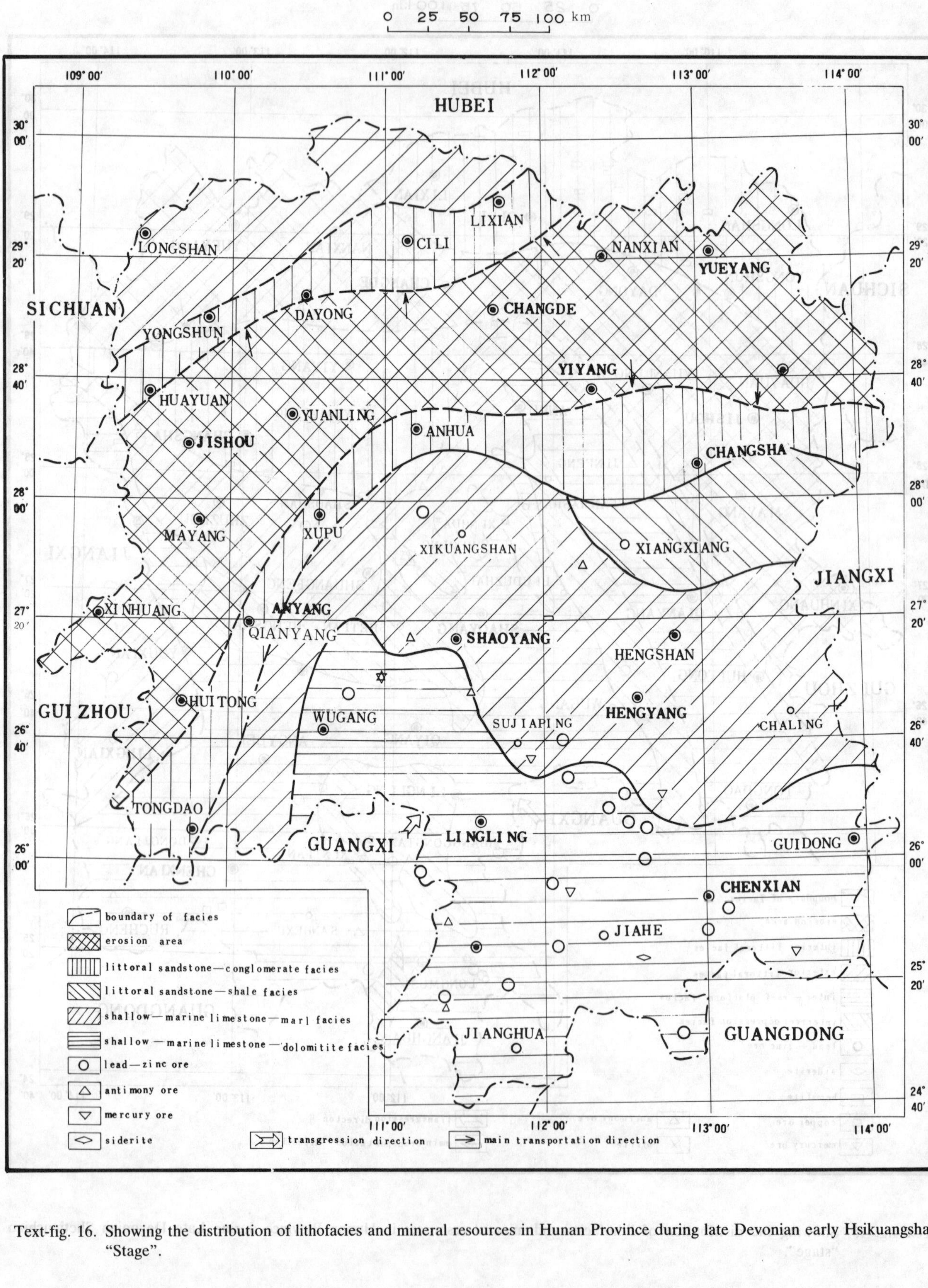

Text-fig. 16. Showing the distribution of lithofacies and mineral resources in Hunan Province during late Devonian early Hsikuangshan "Stage".

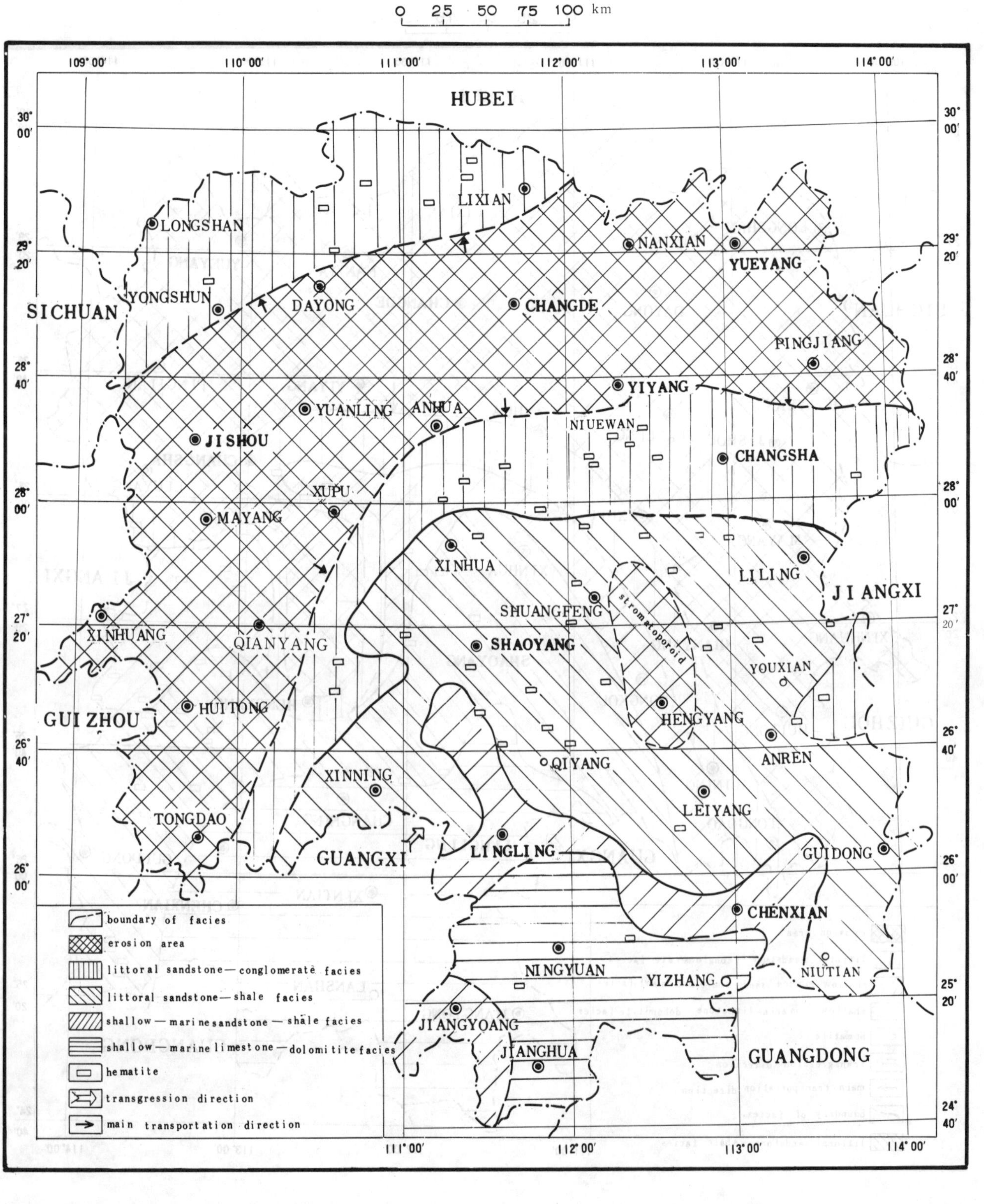

Text-fig. 17. Showing the distribution of lithofacies and mineral resources in Hunan Province during late Devonian Late Hsikuangshan "Stage".

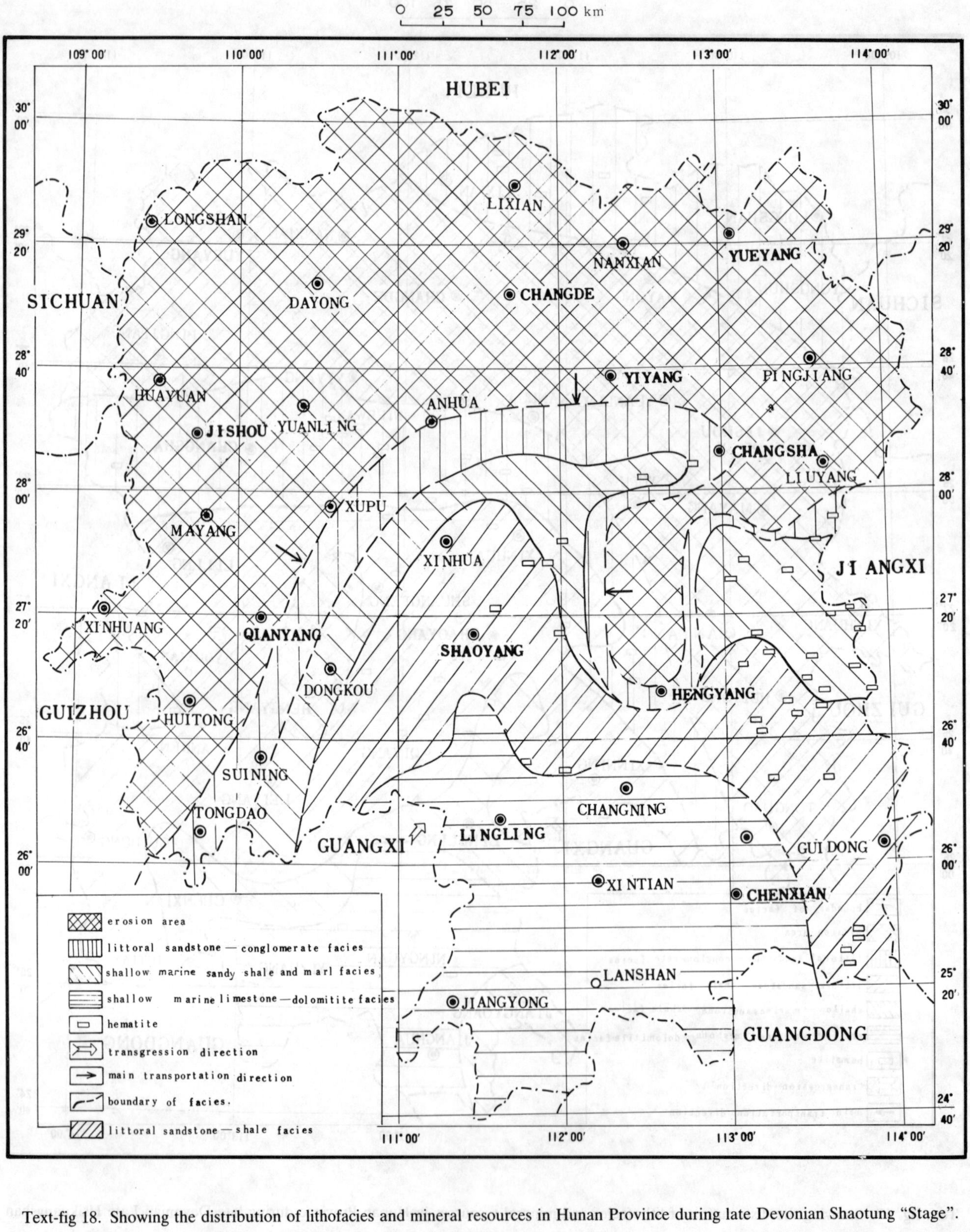

Text-fig 18. Showing the distribution of lithofacies and mineral resources in Hunan Province during late Devonian Shaotung "Stage".

facies belt and the inter-reef depressional belt.

Qiyang-Shaoyang zone: The ore deposits are mostly found in the peripheral part of the inter-reef depressional zone, but some are in the transitional part of the reef and inter-reef zones.

Lingling-Daoxian-Jiangyong-Lanshan-Linwu-Jiahe-Xingtian-Changning zone: The occurrences of the ore deposit are more or less in semicircular arrangement encircling the inter-reef zone and are mostly in the transitional zone.

Chaling-Youxian-Lianhua (of Jiangxi Province) zone: The ore deposits occur in the peripheral part of the reef platformal facies zone and in the transitional part between the reef platform and the outer littoral facies, probably under the lagoonal facies.

Cunxian-Yizhang zone: The occurrences of ore deposit in this zone are considerably concentrated and extended to some mining areas in northern Guangdong. They are mainly in the impure limestones occurring in the transitional zone between the clastic deposits and the carbonate sediments. It seems that they are not favorable for the extreme facies. There are no ore deposits found in the carbonaceous shale or siliceous rock, which are typical for the depressional facies; therefore, it is appropriate to assume that the ore deposits are formed under the restricted or semirestricted environment. It is not impossible that some of the ore deposits in this zone might be in connection with the lagoonal environment in the reef platformal facies.

The distribution and zonation of the ore deposits in the Shetianchiao Formation are more or less similar to those in the Chitzechiao Formation as stated above. However, the occurrences of ore deposits are less in extent. Among the eight zones, only four zones, viz. (ii), (iii), (vi), (viii), still bear industrial importance. This is probably related to the factors that the border of the littoral facies zone was shifted slightly southward and the development of the reef became declining toward the upper part of the Shetianchiao Formation.

Some Remarks on Metallogeny

Derivation of ore-forming materials

In consideration of the characters and distribution of the ore deposit associations in Hunan, it is appropriate to refer them to the stratabound ore deposits including not only the stratiform ore deposits such as oolitic and pisolitic hematites, but also the vein-like ore with more or less complex structure.

The ironstone of the Ningxiang type is the typical stratiform ore deposit. The constitutive material of the ore beds is more or less similar to that of the country rock. Both of the ore-forming and the rock-forming materials are land-derived, and they were synchronous deposits.

Although some of the siderite, pyrite, lead, zinc, antimonial and mercurial sulfide deposits in Hunan are vein-like, penetrating in some cases the country rock, yet they are still considered by the present writers as the stratabound ore deposit. They might be reformed during the post-depositional process. Traditionally, the siderite ore deposit association in Hunan has long been known as of hydro-thermal origin, the ore-forming material being derived from the magmatism, although the convincing evidence has not yet been available. Even though in southern Hunan, the coloured metal resources are plentiful and much research work has been done on the specialization of the magmatogenetic deposit, the positive result hitherto got is too less. On the contrary, it can be seen that most of ore deposits occurred in certain horizons in the marine sequence, for example, certain horizons in the Chitzechiao Formation are important for ore deposits. Apart from the vein-like ore deposits of siderite, pyrite, galena, sphalerite and stibnite in central and southern Hunan, there also exist stratiform ore body, residual stratification, biotic structure and other indications of sedimentation. In the lead-zinc mining area in Fankou of Guangdong Province, close to southern Hunan, the isotopic data reveal that the lead-zinc ores coexisting with siderites are synchronous to their country rock. Therefore, it is preferable to suggest that the siderite ore association is of sedimentary origin. However, its active chemical properties make it easy to be affected by the magmatic or metamorphic fluid or even the heated underground water and are easy to be replaceable, changeable and reformable as well.

Similar to the littoral ore deposit association, the ore deposits in the shallow marine facies in Hunan have their major distributional pattern in the border area surrounding the old land. So far no Devonian magmatism has been known in this area, the material provided for mineralization should be, therefore, considered as being derived from the old land.

This is favorable by the fact that there are high amounts of elements which constitute the Devonian ore-forming elements existing in the Banxi group of Proterozoic and in rocks of Sinian, Cambrian and Ordovician ages, which constitute the basement of the old land.

In this connexion, there is another problem regarding the form of transportation of the ore-forming material. As known from the traditional theory, the ore-forming elements are transported in the form of a solution or in the colloid form. The former viewpoint has been rejected by a large number of up-to-date available data and the latter has not been proved to have common significance. Recent hydro-chemical analysis and research on the geochemical element reveals that the main transportation form of metallic elements in the sea water is clay absorption and mechanical suspension. Thus, at the beginning of sedimentation, the ore-forming elements were mingled with terrigenous sediments. The sedimentation itself only provided the material background for mineralization. There might have been a process in which the metallic elements were entering the water and separating themselves from the carrier. It seems that a suitable environment is needed for completing the whole process. Since the source of material and the transporting form of the ore-forming material in the littoral facies are considered to be similar to those in the shallow marine facies, the distinction of these ore-deposits has to be interpreted as the result of differentiation under physical and chemical conditions in different facies. These will be discussed below.

Controlling factors and genetic mechanism affecting ironstones of the Ningxiang type

It is noted that the location of the ore deposit of the Ningxiang type is closely related to the sedimentary cycle. The ore bed is mainly located at the transitional part between the detrital and the chemical sediments in the litho-procession (i.e. spatial pattern) and at the changing level of these two kinds of deposits in the litho-succession (i. e. time development).

By analysing the distribution of the ore-bearing sequence, 4 kinds of sediments can be distinguished in ascending order as follows:

The pure quartzose sandstone, which is located just beneath the iron ore bed. It represents the products resulting from the mechanical differentiation in the inner littoral facies zone and plays a pioneer role in the course of mineralization.

The iron ore bed, of which the constituent components are hematite and chamosite accompanied with siderite, collophane etc. The oolitic texture is well developed, indicating that they are mainly formed during the diagenetic stage; however, there are also some indices of scouring and transportation.

The chlorites or the chloritic argillites lies immediately above the iron ore bed. The existence of the oolite or pseudoolith points to the diagenetic character.

The micritic limestones or microcrystalline dolomitites are mostly argillaceous and located at the upper part of the sequence. There are rare fossil fragments of benthonic organism. The pellet or the organic framework structure is also rarely seen, indicating that the sedimentation was mainly chemical or microbiogenic. As a whole, the ore-bearing sequence frequently seen in eastern Hunan is characterized by having a simple composition of constituent minerals and a high degree of differentiation. It is hard to find the sand being coexistent with carbonates.

The above-stated pattern of the ore-bearing sequence is usually found in the marine transgressive sequence. As regards the distributional pattern in space, a distinction can be made among clastic sediments, clastic argillaceous limestones, carbonates by tracing them to the shoreline where clastics are dominant in the shelf area. The distributional pattern of different facies is so distinctive that it may reflect the tectonic background of this area being comparatively stable and the erosional surface of the sedimentary basement being not quite elevated in the Devonian time in Hunan.

Based on the foregoing account, it is found that the transitional terrain between the inner littoral facies zone characterized by abundant clastic sediments and the outer littoral facies zone with dominant argillaceous limestones is advantageous to iron ore deposits. Lithologically, the favorable condition for ore deposition is in the changing part between the clastic rock complex and the argillaceous limestones in the central part of the littoral facies zone. Out of this zone where the carbonates are dominant, the ore beds are thin, narrow and low-graded; toward the inner zone where the clastic rocks are dominant, the iron ore deposits are developed, whereas both thickness and grade are greatly variable, which makes it difficult to find rich ore.

Viewed from the evolutionary history of the sedimentary iron deposits, it has been commonly assumed that the iron oolites with colloid texture as an important type of iron deposits can only be formed after the appearance of organic material and free-stated oxygen in the earth crust. However, in the Devonian time there were plenty of oxygen and electrolyte. Under such circumstances, iron could not be carried and concentrated in the form of real or colloid solution. It is probable therefore that iron might be transported under the form of suspension or being absorbed by the clay. The iron oolite might have resulted from diagenesis although it is small and scattered in amount. The iron ore deposit might have been formed through other ways of process.

According to the geochemical properties of iron and the geochemical conditions of marine ooze, it has been generally agreed that the activity of bacteria and the decay of organic material would change the nature of the void water from alkaline and oxidative to acid and reductive, and could dissolve much more iron in it. When the void water was gradually pressed out by compaction, the iron would be carried up to the surface of the sea bottom. If the circumstance there was oxidative, the iron solution would be transformed into ferrous hydroxide or ferric oxide colloid (hematite); if the circumstance was weakly reductive, part or whole of the iron would be combined with silicon and aluminium to form iron silicate colloid (chamosite). These kinds of colloid might precipitate around a core which might be composed of dust, or fragment or any kind of particles to form oolite. If the circumstance was strongly reductive, the hematite or chamosite might be replaced by siderite or pyrite. In morphological character, the oolite is variable and its zonation can always be discernible. These are in correspondence with the environmental condition. One of the important factors is hydrodynamism.

As compared with the country rock, there is evidently less amount of organic material existing in the ore bed. It seems that this phenomenon is in contradiction with the above-stated assumption. In fact, however, the iron ore bed is redeposited after the panning of the formed oolite. In the mining area in Huchang and Daping, Hunan, the chlorite slate with carbon is situated above and beneath the iron ore bed, yielding large amounts of pseudo-ooliths, which are variable in size and generally round in outline. These pseudo-ooliths are quite similar to oolites but without concentric bands and with a low content of iron. From the arrangement of sand grains and pyrites in the rocks, it is assumed that the pseudo-ooliths were originally formed during the late diagenetic stage, free from being transported and panned.

The size of oolite, its morphological character, the degree of sorting and mixing, the vertical and horizontal gradation, the ratio of colloid and the type of matrix, all of these reflect the degree of panning. As a whole, oolites were formed at the diagenetic stage when the environment was reductive, but they were concentrated at the syngenetic stage when the environment was oxidative. This process was a result from the small-magnitude regression. The pioneer lithofacies (quartzose sandstone) beneath the ore bed is nothing more than an indication for this regression. The large volume of the oolite, the heavy specific gravity and the small surficial area of the oolite all played a decisive role in the preservation of iron during the process of natural panning.

To sum up, the process of mineralization includes two necessary steps: first, oolites were formed through hydrolization and diagenesis from the terrigenous sediments in the outer littoral facies zone during the large-scale transgression. Secondly, the iron oolite was concentrated by panning and mechanical differentiation in the inner littoral facies zone during the course of small-scale regression. As a rule, the ironstones occur at the changing level in the regressive cycle within a transgressive megacycle. These procedures might repeat for several times, but usually once or twice at most.The characteristics and kinds of mineral and their assemblages in the iron ore reflect the environmental conditions and evolutionary history of the iron ore. They are expressed as in Table 2.

It can be seen that there are many factors in connection with the environmental conditions and time-dependence, such as the late or post Proterozoic deposits, the warm climate around the tropical area in the past, the long-term break, the tectonically stable area, the persistent but slightly oscillatory marine transgression, the proper restricted basin, etc. Besides, the diagenesis needs some preconditions, such as source material from the eroded old land, suitable marine transgressive cycle, slow sedimentation, and relief of the basement.

It is, therefore, important for us to know about the planation of the basement and the stability of the regional geologic background while making ore-estimation in a large extent of area. It is also important to know about the regional lithological and paleogeographical pattern while making ore-valuation in certain mining area.

Table 2. Showing the genetic mode of the iron ore deposit of the Ningxiang type.

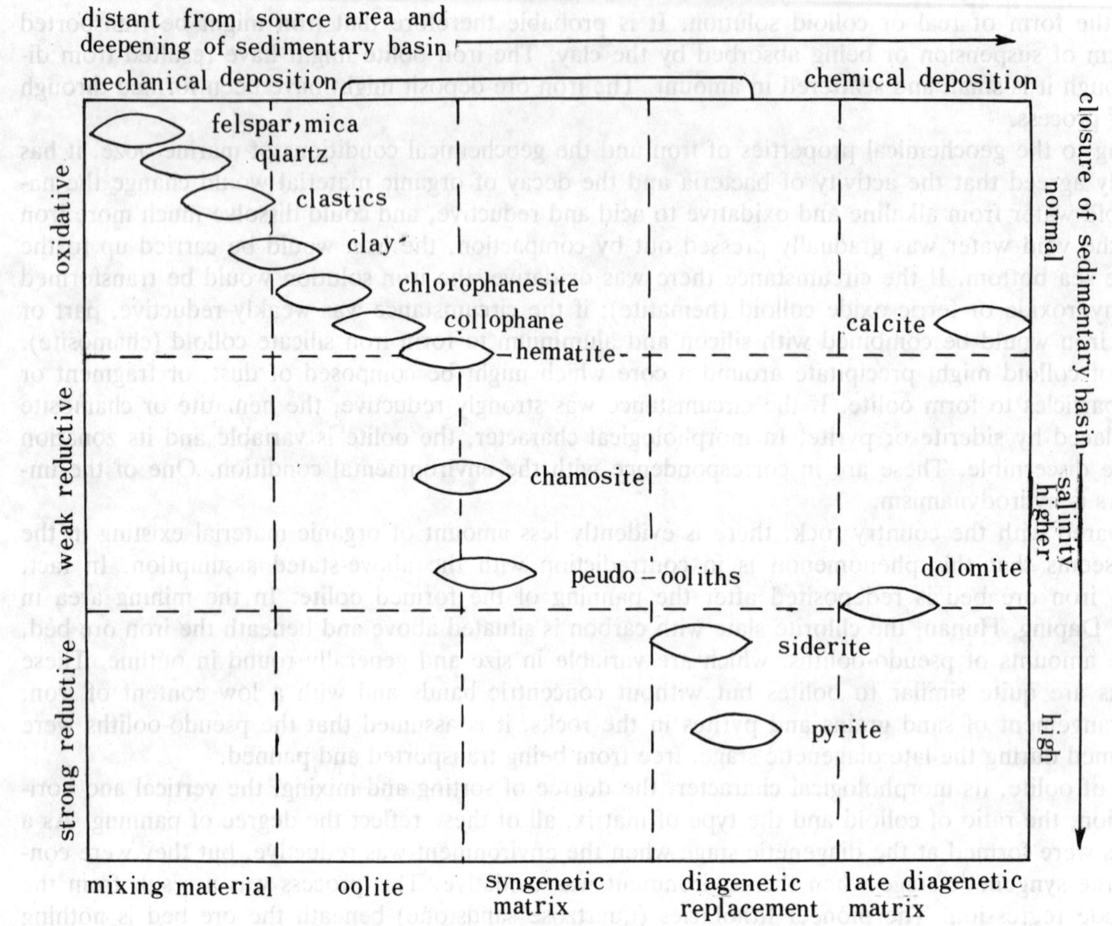

Genetic Mechanism of the Siderite ore Association.

In recent years, quite a lot of siderite ore deposits have been discovered in the carbonate formation in many provinces of South China including Hunan. The age of the ore-bearing formation is Devonian (as in Hunan), Carboniferous, Permian and Triassic. Despite the different ages of the ore-bearing formations, they bear the common characters which may be summarized as follows: (1) lying unconformably on the older rock indicating a long-term break before deposition; (2) in the border area close to the old land after long period of erosion; (3) with distinct lithofacies and great thickness within the mining area, but difficult to correlate with the circumscribed lithofacies; (4) located mainly at the facies changing portion from dolomitites or dolomitic limestones to marls, no ore being found in pure limestones nor in marls; (5) with distinct sequence or succession, and with calcirudites and bioclastics in the lower, framestones in the middle, dolomitic limestones or dolomitites in the upper, occasionally with barite or gypsum in the upper; (6) with large amounts of manganese coexisting with iron, both increasing synchronously; (7) with a high content of manganese and sometimes magnesium in the siderite, appearing isomorphic (generally Mn > 1%, but much higher in Hunan in particular); (8) the siderite ore deposit may change vertically or laterally into sulfide ore deposits of Fe, Cb, Pb, Zn, Hg, etc. to form the sister ore deposits, sometimes with high contents of Cd, Ge, In, Ga. These characters having puzzled the people can be now explained in the light of mineralization from the reef complex.

The distribution of the reef may be controlled by the basement tectonic and the basement relief. However, the development of reef complex to form the wave-resistant structure had, in turn, an effect on the relief form. In addition, the persistent activity of the edge fault caused the relief intensity. They changed hydrodynamic and hydrochemical conditions and locally built up restricted or semi-restricted environments.

It should be stressed that the acidic and reductive conditions are important for the hydrolization of the terrigenous material and that iron and manganese are easy to be dissolved in such marine condition (Yeh, 1963). The organic material accumulated and preserved, decomposed and dissolved in the restricted environment would cause the water component to be more acidic and reductive, and evaporation would increase the water salinity. All of these factors would have an effect on dissolving a great deal of metallic elements and were advantageous to the ore formation. This is perhaps why ore formation in the marine condition is always lithologically found in the dolomitic facies. Because metallic elements are easy to be dissolved in the solution with high salinity under warm and dry weather. Thiede and Cameron (1978) have also pointed out that the concentration of heavy metallic elements is relevant to evaporites. The existence of large amounts of electrolyte in the above-stated environment neutralizes the electrostatic charge of the suspended material from terrigenous sediments. This is probably why there are abundant argillaceous, rather than arenaceous sediments in the Late Devonian Shetianchiao and Hsikuangshan Formations in the inter-reef depressional facies in Hunan. It is interesting to know that the region where the inter-reef depressional facies and the lagoonal facies are developed is advantageous to the deposition of terrigenous materials, known as the suitable environment for the dissolution and separation of metallic elements.

It is impossible to assume that the dissolution and precipitation of ore-forming elements would happen simultaneously under the same condition and in the same place, unless a change in physical and chemical conditions should happen. The precipitation of the ore-forming elements generally takes place at the facies changing portion between the reef platformal and the inter-reef depressional facies, because this terrain is the center of the change in physical and chemical conditions. Only in rare case, the ore-forming materials could precipitate in their original place, owing to the enrichment by evaporation or by pouring of fresh water. From the mineralized map it can be seen that the location of ore body is mainly close to the reef area, while the ore-bearing zone is in the centre of the basin. But in the peripheral part of the basin, the ore body is mainly close to the inter-reef depressional facies and the lagoonal facies.

It is important to emphasize that the function of organisms and bacteria in the process of mineralization can not be neglected. For example, in the mining area in Huqing, central Hunan, it can be seen that stromatoporoids and corals in the reef have been replaced by galena, sphalerite, siderite and pyrite. There still remain some relic textures of organisms. All of these indicate that organisms play an vital role in the absorption of metallic elements and provide organic materials for speeding up diagenesis. The major function performed by bacteria in resolving terrigenous material, dissolving metallic elements, separating out mineralizer (H_2S) and precipitating metallic elements is absolutely necessary. As a whole, they are in connection with the reef environment.

Despite physical action, chemical action, biogenetic mineralization and bacterial action, they are all relevant to and referred to different aspects of the reef ore deposition.

The development of the reef is generally controlled by the factors: e.g. paleoclimate and the scope of marine transgression. There is a remarkable difference in the characters of sediments and ore deposits between Middle Devonian and Upper Devonian in Hunan as shown in the following table (Table 3):

Table 3. Showing the sedimentary characters and iron deposits in the Middle and Upper Devonian of Hunan.

Sedimentary character \ Stratigraphic unit		Chitzechiao Fm.	Shetianchiao Fm.
Reef Platform	Biofacies	mainly stromatoporoids, with algae, corals	mainly algae, with corals
	Lithology	abundant dolomitites	rare dolomitites
Inter-reef depression	Biofacies	rare benthonic fauna	no benthonic fauna, abundant tentaculites
	Lithology	mainly marls	carbonaceous shales, silicolites, marls
Ore deposits		mainly Fe, Mn, Pb, Zn, rare Sb, Hg	mainly Sb, Hg, rare Pb, Zn, less Fe, Mn

From the above table, it is assumed that the salinity level of the water body was lower but more reductive in the inter-reef depressional facies zone in the time when the Chitzechiao and Shetianchiao Formations and their equivalents were deposited. This might be caused by the gradual sea regression and the climatic change from dry-hot to warm-wet.

In regard to the horizontal pattern of ore deposits in the Devonian reef complex in Hunan, it is generally presumed that there occur successively siderite, galena and sphaerite, stibnite and cinnabar, pyrite in the terrain from the reef platform to the inter-reef depressional facies zone. Concretely speaking, the pyrite has always been found in the zone close to the inter-reef facies; the stibnite mostly in the silicified limestones is presumed to be close to the inter-reef depressional facies. The siderite occurring along with galena and sphalerite is close to the reef platform. The distributional pattern of these ore deposits is schematically expressed in Table 4

Table 4. Schematically showing the distributional pattern of the siderite and its sister ore deposit.

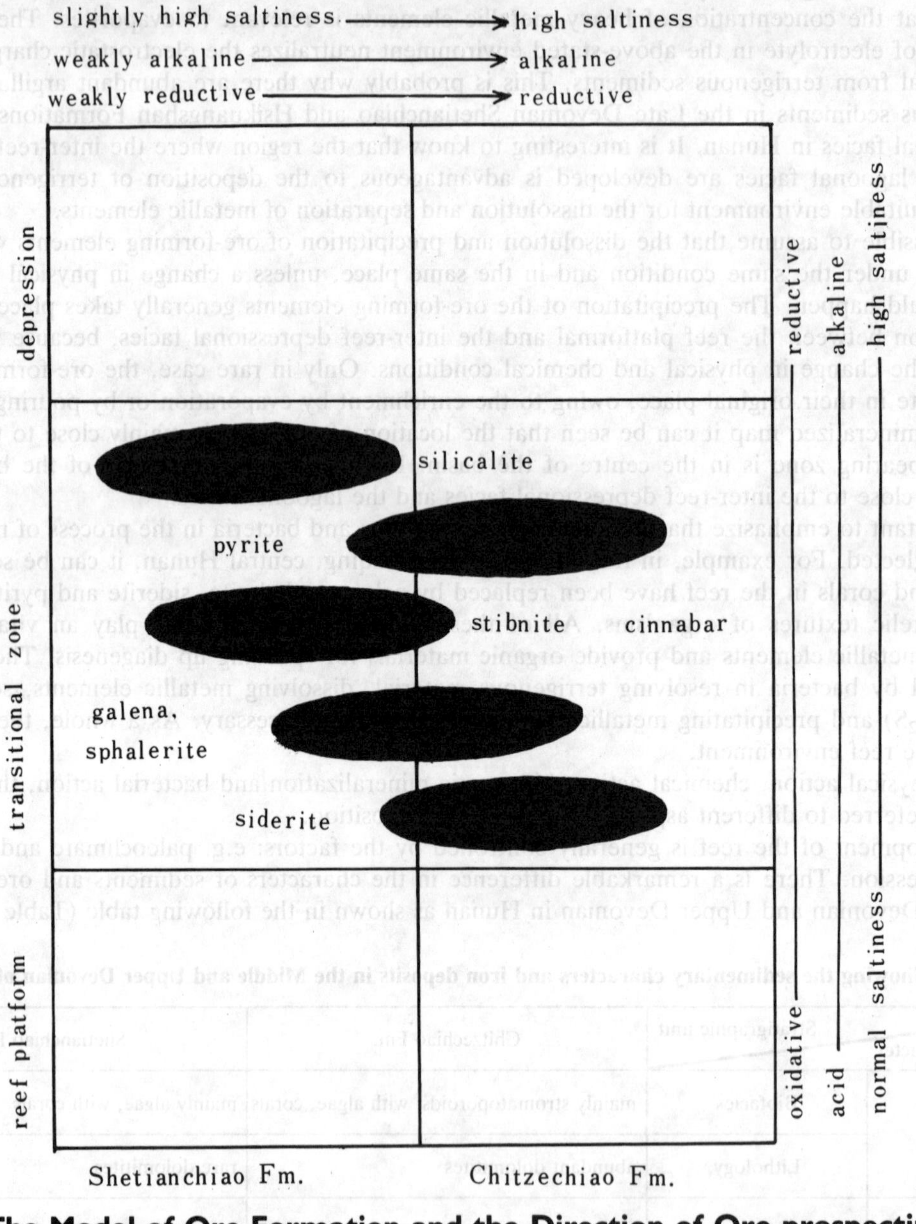

The Model of Ore Formation and the Direction of Ore-prospecting.

Devonian is the first ore-bearing series in the Late Palaeozoic mineralization cycle in Hunan. The Devonian in Hunan was deposited after a long period of break on the terrain close to the denuded old land under the favorable climate with a sufficient amount of source material for mineralization. It was deposited in a long-stablized sea basin with reef developing and under a favorable environmental condition for the differentiation of ore-forming materials.

As the Devonian of Hunan can be divided into the littoral facies and the shallow marine facies, correspondingly, it can be subdivided into the iron ore deposit of the Ningxiang type and the galena-sphalerite-siderite ore deposit association respectively. They are distinctive but inter-connected with each other. As a role, the source material for these ore deposits came from the denuded old land. The ore formation of these ore deposits includes the ore-generating process and the ore-forming process.

The ore-generating process for these ore deposits is attributed to the category of diagenesis, denoting the separation of metallic elements from their carrier. The ore-forming process mainly represents the processes of precipitation and enrichment of metallic elements. As for the iron ore deposit of the Ningxiang type, it is the process mainly of machanism; while as to the siderite and its sister ore deposits, it is the processes mainly of chemical action and biogenic agent. Owing to the effect of different mechanisms on ore formation, the controlling factors and the results are quite different, too. The process of ore formation can be summarized in Table 5.

Table 5. Showing the process of ore formation.

Terrigenous material	
↓	↓
deposited in inter-reef depression or lagoon	littoral deposits
↓	↓
hydrolization in surficial bittern under reductive condition	hydrolization caused by acid and reductive void water
↓	↓
condensation of solution in bottom water	oolites or colloid formed on the surface of bottom matters
↓	↓
precipitated allochthonously (migration of bottom solution) or autochthonously (change of nature of water body)	differentiation by natural panning
↓	↓
post-deformation	contact metamorphism occasionally occurred

From the above table, it is clear that the process of ore formation is not a simple course to be completed within one cycle. Instead, it should take a long historical period of evolution including the process of persistent migration, decomposition, combination and variation. What is more important is to distinguish and hold two principal links: the ore-generating and the ore-forming, which can help explain the existing form and the distributional pattern of metallic elements in time and space.

In summing up what we have discussed, the model of ore formation in Devonian of Hunan is shown in Text-fig. 19.

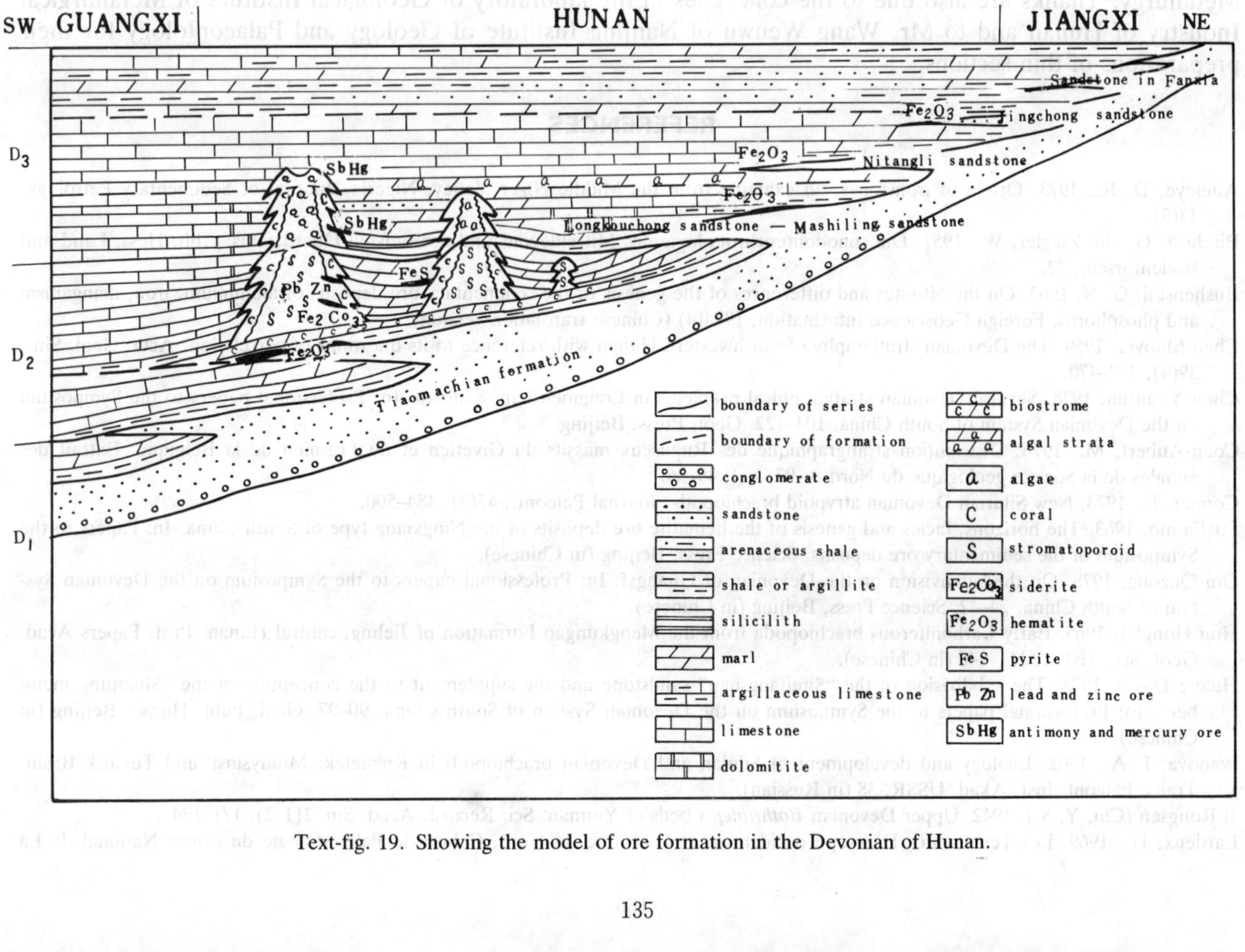

Text-fig. 19. Showing the model of ore formation in the Devonian of Hunan.

It is necessary to point out that the above text-figure only shows the general characters of sedimentation occurring in the peripheral part of the basin (exclusive of any detail of the measured sections), the lithological characters of marine transgression and regression, but does not represent the expanded or contracted scope of the sea basin.

The hematite and siderite are formed under the oxidative and reductive environmental conditions respectively. From the evolutionary history of the sedimentary iron deposits, the diversification of these environmental conditions became increasingly stronger along with the stronger diversification of the earth crust and the evolution of biosphere and atmosphere. In the hematitic facies the coexisting elements are rather simple; while in the siderite facies the coexisting elements are rather composite, especially when the independent siderite is formed, there coexist various coloured metallic sulfide ore deposits, the so-called sister ore deposit. This is the main trend of ore formation under the reductive environmental condition.

Since Hunan is well known for producing various coloured metallic ore deposits, it is called the "country of the coloured metallic ore deposits". A number of ore occurrences have been found in the Devonian carbonate sediments in Hunan. It is, therefore, worth mentioning the prospecting of the siderite ore deposit by tracing its sister ore deposits in Hunan.

The siderite and its sister ore deposits are controlled by lithofacies and occur in certain beds; they are referred to the stratabound deposit. Besides, as metallic elements are quite active, migrative and deformative, the tectonic controlling factor can not be neglected. This must be taken as a practical consideration for ore prospecting.

ACKNOWLEDGEMENTS

The authors are grateful to the colleagues of our institute who joined with us in the field work or helped us for fossil identification; we also wish to express our sincere thanks to the institutions and organizations concerned in Hunan for the help and assistance in the field work, in particular, to Xianghualing Mining Company, Xikuangshan Mining Company, Lushui Mining Company, Nos. 240, 238, 416 Explorating Teams, Petroleum Exploration Team of Hunan, Geological Institute of Hunan, Geological Survey Team of Hunan, Institute of Metallurgical Industry of Hunan and Central-South China Institute of Mining and Metallurgy. Thanks are also due to the colleagues in the laboratory of Geological Institute of Metallurgical Industry of Hunan and to Mr. Wang Wenwu of Nanjing Institute of Geology and Palaeontology for their preparations of thin sections.

REFERENCES

Adeleye, D. R., 1973. Origin of ironstones, an example from the Middle Niger Valley, Nigeria. Journal of Sedimentary Petrology, 43(3).

Bischoff, G. and Ziegler, W., 1957. Die conodontenchronologie des Mitteldevons und des tiefsten Oberdevons. Abh. Hess. Landsamt Bodenforsch., 22.

Bushenskii, G. N. 1963. On the affinites and differences of the genesis of the sedimentary ore deposits of aluminium, iron, manganism and phosphorus. Foreign Geoscience Information, 1974(4) (Chinese translation version).

Chen Maoyiu, 1959. The Devonian stratigraphy of northwestern Hunan with reference to its questions of correlation. Acta. Geol. Sin., 39(4), 457–470.

Chen Yuanren, 1978. Several Devonian stratigraphical problems in Longmenshan, Sichuan. In: Professional papers to the Symposium on the Devonian System of South China. 104–122. Geol, Press, Beijing.

Coen-Aubert, M., 1976. Distribution stratigraphique des Ruguseux massifs du Givetien et du Frasnien de la Belgique. Extrait des annales de la Societe geologique du Nord. t. 97, s. 4, 49–56.

Copper, P., 1973. New Silurian Devonian atrypoid brachiopods. Journal Paleont., 47(3), 484–500.

Fu Jia-mo, 1963. The horizons, facies and genesis of the hematite ore deposits of the Ningxiang type of South China. In: Papers to the Symposium of the sedimentary ore deposits, Science Press, Beijing (in Chinese).

Gui Quzong, 1978. On the subdivision on the Devonian of Guangxi. In: Professional papers to the Symposium on the Devonian System of South China. 24–42. Science Press, Beijing (in Chinese).

Hou Hongfei, 1965. Early Carboniferous brachiopoda from the Mengkungao Formation of Jieling, central Hunan. Prof. Papers Acad. Geol. Sci., (B), 1, 111–146 (in Chinese).

Huang Daxin, 1978. The subdivision of the "Shuifangshan" sandstone and the supplement to the conception of the "Shaotung member". In: Professional papers to the Symposium on the Devonian System of South China. 90–97. Geol. Publ. House. Beijing (in Chinese).

Ivanova, I. A., 1962. Ecology and development of Siluian and Devonian brachiopods in Kuznetzk, Minuysinsk and Tuvinsk Basin. Trans. Paleont. Inst. Akad. USSR. 38 (in Russian).

Ji Rongsen (Chi, Y. S.), 1942. Upper Devonian *Bothriolepis* beds of Yunnan. Sci. Record, Acad. Sin. 1(1–2). 171–174.

Lardeux, H., 1969. Les Tentaculites D'europe occidentale et D'afrique du Nord. Cahier de Paleontologie du centre National de La

Recherche Scientifique. Paris.
Liao Shifan, 1954. Outline of the regional geology of Chaling, Yong-Xin, Lianhua in the border area between Hunan and Jiangxi. Acta Geol. Sin, 34(1), 65–72 (in Chiness).
——— 1964. Some discussions on the paper entitled "Late Devonian stratigraphy, in the border area between Hunan and Jiangxi", by Liu Ya-guong, Geological Review, 22(5) (in Chinese).
Liu Guochang, 1942. Several sections of Devonian of Liling, Hunan. Geological Review, 7(4–5), 123–130 (in Chinese).
Liu Yaguong, 1963. Late Devonian stratigraphy in the border area between Hunan and Jiangxi, Geological Review, 21(2), 53–58 (in Chinese).
Ljaschenko, G. P., 1959. Devonian Coniconchia of Central and eastern part of Russian Platform. National technique Press, Moscow (in Russian).
McLean, R. A. and Pedder, A. E. H., 1984. Frasnian rugose corals of western Canada. Palaeontographica Abt. A, 185, 1-38.
Moore, R. C. (ed.), 1965. Treatise on invertebrate palaeontology, Pt. H, Brachiopoda. Geol. Soc. Amer. and Univ. Kansas Press.
Paeckelmann, W., 1942. Beiträge zur kenntnis devonischenr spiriferen. Abh. Reichsamte Bodenf. Berlin Neue Folge, 197. 1–188.
Pan Jiang, 1958. On the geological age of the Tiaomachien series. Acta Geol. Sin., 38(1), 135–144.
Pedder, A. E. H., 1963. *Alaiophyllum mackenziense* sp. nov. a Devonian tetracoral from Canada. Palaeontology, 6(1), 132–135.
———, 1973. Description and biostratigraphical significance of the Devonian coral genera *Alaiophyllum* and *Grypophyllum* in western Canada. Bull. Geol. Surv. Canada, 222, 93–127.
Rozkowska, M., 1969. Famennian Tetracoralloid and Heterocoralloid Fauna from the Holy Cross Mountains (Poland). Acta Palaeontologica Polonica, 14(1).
Sartenaer, P., 1956. Signification et importance du genre *Cyrtiopsis* dan les deposits Famennien inferieurs. Bull. Inst. Royal Science Natur. Belgique, 32(40), 1–11.
—— 1971. Redescription of the brachiopod genus *Yunnanella* Grabau, 1923 (Rhynchonellida). Smithsonian Contribution to Paleobiology, 3. 203–208
Shen Qiming, 1982. Conodont zonation of the Upper Devonian and opinions of the strata division of Hsianhua Mountain of Linwu, Hunan. Hunan Geology, 1(1), 32–55. (in Chinese).
Sidyjachenko, A. I., 1962. Famennian spiriferids and stratigraphy in central and south-eastern Karatau. Akad. Nauk SSSR, Siberian section, Inst. Geol. Geophy. Moscow (in Russian). 1–152
Si Xingjian (Sze, H. C.), 1937. On some *Psilophyton-like* remains from the Tiaomachien series in central Hunan. Bull. Geol. Soc. China, 17(2), 245–254.
Strahov, N. M. 1953. Diagenesis of sediments and its significance to sedimentary ore-forming. Professional papers on sedimentation, 1957(1) (Chinese translation version).
Sun Yunzhu, 1935. On the occurrence of the *Manticoceras* fauna in central Hunan. Bull. Geol. Soc. China, 14(2), 249–254.
———, 1958. The Upper Devonian coral fauna of Hunan. Palaeont. Sin., (B), 8, 1–28.
Sweet, W. C. and Bergstrom, S. M. (ed.), 1970-1971. Symposium on conodont biostratigraphy. Mem. Geol. Soc. America, 127.
Thiede, D. S. and Cameron, E. N., 1978. Concentration of heavy metals in the EIK point evaporite sequence, Saskatchewan. Economic Geology, 73(3).
Tian Jijun (Tian, C. C.,) 1938a. Devonian Brachiopoda of Hunan. Palaeont. Sin., (B), 4, 1–192.
—————, 1938b. Devonian of China. Geological Review, 3(4), 355–409. (in Chinese)
Troizkaya, T. D., 1970. Some Famennian bryozoans of Central Kazachstan (Salisuy-Tenizskii Watershed). In: Astrova, G. G. and Chudinova, I. I. (ed.), New species of Paleozoic bryozoans and corals. 40–46. Izd. Nauka, Moscow (in Russian).
Tsien, H. H., 1977. The sequence and distribution of Frasnian Rugose coral faunas in Belgium. Mem. BRGM, 9, 203–220.
Wang Chenyuan and Ziegler, W., 1982. On the Devonian-Carboniferous boundary in South China based on conodonts. Geologica et Palaeontologica, 16, 151–162.
Wang Xiaoqing (Wang, H. C,), Tian Jijun (Tien, C. C.) and Xu Yuandao (Hsu, Y. T.), 1933. The geology of Changsha, Hsiangtan, Hengshan, Hengyang, Hsianghsiang and shaoyang districts, central Hunan. Bull. Geol. Surv. Hunan, 15, Geol. 2, 1–112.
Wang Yu and Yu Changming, 1962. Devonian system of China. Science Press, Beijing.
—, — and Wu Qi, 1974. Advances in the Devonian biostratigraphy of South China. Mem. Nanjing Inst. Geol. Paleont., 6, 1–45 (in Chinese).
Wang Yuelun (Wang, Y. L.), Liu Zuyi (Liu, T. Y.) and Cheng Yuqi (Cheng, Y. C.), 1938. Geology of the iron ore deposits of Ninghsiang. Bull. Geol. Surv. China, 32, 1–32.
Wilson, J. L., 1975. Carbonate facies in geologic history. Springer-Verlag, Berlin.
Wu Wangshi, Zhao Jiaming and Jiang Shuigen, 1981. Corals from the Shaodong Formation (Etroeungt) of South China. Acta. Palaeont. Sin., 20(1), 1–14.
Xu HanKui, 1979. Upper Devonian *Yunnanella-Yunnanellina* (Brachiopoda) fauna from Hunan. Acta Strat. Sin., 3(2), 123–126 (in Chinese).
Yang Jingzhi and Mu Enzhi, 1953. Devonian stratigraphy of western Hupeh. Acta Palaeont. Sin., 1(2), 58–66.
Ye Lianjun, 1963. The terrestrial weathering products imbibition theory–new concept on genesis of sedimentary deposits. Sci, Geol. Sin., 2, 67–87.
Yue Senxun (Yoh Sen-shing) and Hou Hongfei, 1962. On the Devono-Carboniferous boundary and the existence of "Etroeungian" in South China. Acta Scientiarum Naturalium Universitalis Pekinensis, 8(3), 285–293.
Yu Changming, Liao Weihua and Deng Zhanqui, 1978. The sequence and distribution of Devonian coral assemblages. Acta strat. sin., 3(2), 96–102 (in Chinese).
——— Xu Hangkui, Pung Ji, Xiao Šongtao and Liu Zuhan, 1983. Devonian of Hunan with a bearing on the ore deposits. Papers submitted to the Symposium on the siderite oredeposits, 262–298, Science Press, Beijing (in Chinese).
Zhao Ruxuan, Zhang Caifan, Xia Zhifen, Zuo Zhibi and Liu Zuhang, 1978. Devonian system of Hunan Province. In: Professional Papers to the Symposium on the Devonian System of south China. 68–69. Geol. Publ. House, Beijing (in Chinese).

Ziegler, W. (ed.), 1973. Catalogue of conodonts. E. Schweizerbartsche Verlagsbuchhandlung. 1–4.
Ziegler, W., Klapper, G. and Johnson, J. G., 1976. Redefinition and subdivision of the Varcus-Zone (Conodonts, Middle-? Upper Devonian) in Europe and North Americ. Geologica et Palaeontologica, 10. 109–140.

PALAEONTOLOGIA CATHAYANA Volume 5, 1990, p.139–170

THE FIRST RADIATION OF SHELLED MOLLUSCS

Yu Wen

(Nanjing Institute of Geology and Palaeontology, Academia Sinica)

Abstract

In this paper are discussed 61 species and 48 genera, belonging to Polyplacophora, Merismoconchia, Monoplacophora, Gastropoda, Bivalvia and Rostroconchia, with emphasis laid on their first radiation. All the species and genera were discovered from the Meishucun Stage in the Yangtze Region of China and thus are called the Yangtze Micromolluscan Fauna. There is a probability of a close evolutionary relationship among the Polyplacophora, Merismoconchia and Monoplacophora. In the Cambrian, gastropods underwent two developmental stages. Morphologically, they changed from simple to complicated in shell. The Bivalvia first appeared in the Earliest Cambrian. As headless molluscs, they are also called "Acephala" and can not be regarded as advanced molluscs.

Excepting Cephalopoda, Scaphopoda and Aplacophora (the first two made their first appearance in the late Late Cambrian and the late Middle Ordovician respectively, and the last has no substantial fossils ever known), the species and genera belonging to Polyplacophora, Merismoconchia, Monoplacophora, Gastropoda, Bivalvia and Rostroconchia were discovered from the Earliest Cambrian Meishucun Stage, forming the Yangtze Micromolluscan Fauna. The appearance of the Yangtze Micromolluscan Fauna indicates another major event in the evolutionary history of invertebrates following the Ediacaran Faunas. It also represents the first stage of the diversification in the history of the Mollusca, and it can be taken as an important evidence for division and correlation of the International Precambrian-Cambrian boundary.

Since the finding in 1974 of Helcionellids from the Early Cambrian Maidiping Formation in Emei of Sichuan, there have been 12 scholars who successively described small shelly fossils of the Meishucun Stage as molluscs. By the year of 1985, about 31 papers on the Meishucun molluscan faunas have been published, with descriptions of 178 species and 85 genera altogether. After a review of these fossils, the writer thinks that 61 species and 48 genera may be recognized as valid (Table 1).

In the micromolluscan faunas, 10 genera of polyplacophora have so far been included. The polyplacophorans began to diversify at the beginning of the Cambrian. The merismoconchs occurring in the Meishucun Stage of China bear a single, pseudometameric and bilaterally symmetrical shell and a peculiar pattern of muscle scars, but few of them have so far been found. Also the monoplacophorans diversified for the first time in the Meishucun Stage, 14 genera including 16 species have been herein found, occupying about 3 / 4 of the forms in the contemporaneous sediments of the world. Among these 16 species, *Yangtzeconus priscus* Yu (pl. 6, figs. 3–6), *Huangshandongoconus pileus* Yu (pl. 6, figs. 1–2), *Eosoconus primarius* Yu (pl. 6, figs. 7–8), *Archaeotremaria polytremata* Yu (pl.7, figs. 1–3), *Truncatoconus yichangensis* Yu (pl. 6, figs. 9) and *Cambroconus minimus* Yu (pl. 6, figs. 10–12) show different types of muscle scars on their internal moulds. Many primitive gastropods also made their first appearance in the Meishucun Stage. They include *Cambrospira* and *Maclurites* with a hyperstrophic shell, *Archaeospira* with a sinistral shell and *Hubeispira* and *Tianzhushanospira* with a dextral shell. Since the finding of *Yangtzespira* with posterior canal, much attention and interest has been aroused among our colleagues both at home and abroad, for this very interesting gastropod provides important evidence for the study on the origin, evolution and classification of Class Gastropoda. *Yangtzedonta* is one of the primitive forms of Class Bivalvia and is also one of the impor-

Table 1. Geographical and stratigraphical distribution of the Yangtze Micromolluscan Fauna in the Yangtze Region of China.

Species \ Region / Localities	E. Yunnan												W. Sichuan						S.Shaaxi N.Sichuan		W. Hubei						
	Jinning			Chengjiang	Anning	Xundian			Huize		Luquan	Yongshan	Leibo	Mabian	Ebian	Leshan	Emei	Ningqian	Nanjiang	Yichang						Yuanan
	Meishucun	Wangjiawan	Shangsuan	Dadiyakou	Baiden	Xianfeng	Baizai	Dawan	Dahai	Yulu	Chahechang	Xiaotan	Niuniuzhai	Laoheba	Laogongshan	Fandian	Gaoqiao	Yuanjiaping	Shatan Changtanhe	Tianzhushan	Huangshandong	Hujingtan	Shipai	Songlinpo	Jijiapo	Yanhechi
Tchangsichiton notabilus Yu																				●	●					
Sinuconus clypeus Yu																				●	●					
Runnegarochiton modestus Yu	●					●	●																			
Meishucunchiton vulgarus Yu	●					●																				
Postestephaconus tentes Jiang	●		●			●	●																			
Stoliconus vomeres Jiang	●		●			●	●													●	●					
Yangtzechiton elongatus Yu	●					●	●																			
Luvanhaochiton spinus Yu	●																									
Gotlandochiton ? minimus Yu	●					●																				
Yunnanopleura biformis Yu						●	●																			
Yangtzemerisma rarum Yu																					●					
Yangtzemerisma ? cancellatum Yu																					●					
Minymerisma yunnanense Yu							●																			
Merismoconcha multisegmentata Yu																	●									
Yangtzeconus priscus Yu	●																●			●	●	●				●
Huangshandongoconus pileus Yu																					●					
Spatuloconus rudis Yu											●									●	●					
Actinoconus pyriformis Yu																						●				
Eosoconus primarius Yu																							●			
Latirostratus amplaperatus Yu		●																				●				
Truncatoconus yichangensis Yu																				●						
Obtusoconus paucicostatus Yu																				●	●	●		●	●	●
O. rostriptutus (Qian)										●										●	●	●				●
O. honorabilus (Qian, Chen et Y. Y. Chen)																				●	●					
Yochelcionella ? delicata Yu																	●				●					
Pollicina striata (Y.Y.Chen et Zhang)																					●			●	●	
Archaeotremaria polytremata Yu																	●									
Granoconus trematus Yu																						●				
Cambroconus minimus Yu																							●			
Bemella simplex Yu	●					●	●					●		●	●	●			●	●	●					
Bemella ? mirabilis Yu																				●	●					
Helcionella tianzhushanensis Yu																				●						
Helcionella ? liantuoensis Yu																				●	●	●	●			
Latouchella sanxiaensis Yu	●																			●						
Latouchella lauta Yu																				●	●					
Latouchella maidipingensis (Yu)	●	●	●		●	●	●		●								●	●		●	●					
Latouchella cf. *memorabilis* Missarzhevsky																				●						
Igorella hamata Yu	●				●	●	●													●						
Purella elegans Yu																						●				
Purella tianzhushanensis Yu												●								●						
Stenotheca sp.						●	●		●																	
Archaeospira ornata Yu	●		●			●	●										●			●	●					
Archaeospira imbricata Yu						●	●														●					
Archaeospira regularis (Jiang)	●				●	●	●						●													
Archaeospira sp.	●					●																				
Anabarella ? emeiensis Yu.																	●									
Latouchella cf. *karobkovi* (Vostokova)	●		●			●	●										●									
Cambrospira sinensis Yu																					●					
Maclurites hubeiensis Yu																					●					
Aldanella ? sp.																					●					
Hubeispira nitida Yu																					●					
Tianzhushanospira unicarinata Yu																					●					
Yangtzespira exima Yu				●			●																			
Heraultipegma yunnanense He et Yang						●			●											●	●					
Yangtzedonta primitiva Yu							●																			
Lepidites emeishanensis Zhong	●	●	●		●	●			●					●			●				●					
Aegides seperbes Jiang	●	●	●		●		●																			
Eohalobia diandongensis Jiang						●	●																			
Xiadongoconus luminosus Yu						●															●					
Protoconus crestatus Yu	●																						●			
Cassidina pristinis Jiang	●	●				●																				

tant members in the Yangtze Micromolluscan Fauna.

Of the Yangtze Micromolluscan Fauna, Polyplacophorans, monoplacophorans and gastropods are considered to have lived a benthic creeping life within the subtidal and intertidal zones. From the state of preservation, most specimens are preserved incompletely; some are fragments, some overlap with each other. Moerover, they are uneven in distribution, sometimes tangled at one place, sometimes sporadic at the other. The Meishucun Stage in Southwest China is composed mainly of phosphate and carbonate sediments, indicating that they were deposited in the shallow sea or in the turbulent flow of shallow sea or in the littoral areas.

From the elements of 6 classes of shelled molluscs in the Meishucun Stage of the Yangtze Region, we may conclude that in the Earliest Cambrian the Yangtze Sea is the birthplace of the Mollusca as well as the center from where their evolutionary radiation began (text-fig. 6).

The hyolithids are excluded from the Yangtze Micromolluscan Fauna. In view of their morphological features in the shells, many scholars considered them not to belong to the Mollusca, but to another group of fossils (Runnegar, Pojeta, Morris, Taylor, Taylor and McClung, 1975; Runnegar, 1980, 1983; Pojeta, 1980, 1985).

CLASS POLYPLACOPHORA DE BLAINVILLE, 1816

According to the recorded materials, fossil Polyplacophora made its first appearance in the Late Cambrian (Walcott, 1885; Bergenhayn, 1960; Runnegar, Pojeta, Taylor and Collins, 1979). In 1984, while making study of the Earliest Cambrian molluscs from the Yangtze Region, the writer discovered the fossil polyplacophorans and erected two genera and species: *Yangtzechiton elongatus* Yu and *Luyanhaochiton spinus* Yu (Yu, 1984a, b, 1987). In recent years, as a result of intense study on the Precambrian-Cambrian boundary, some new fossils have been found, mainly in the Meishucun Stage, including *Tchangsichiton notabilus* Yu (pl. 1, figs. 1–2), *Runnegarochiton modestus* Yu (pl. 1, figs. 3–6), *Meishucunchiton valgarus* Yu (pl. 1, fig. 8), *Stoliconus vomeres* Jiang, *Postestephaconus tentes* Jiang (pl. 1, figs. 9–11), *Sinuconus clypeus* Yu (pl. 1, fig. 7), *Yunnanopleura biformis* Yu (pl. 4, figs. 1–7) and *Gotlandochiton*? *minimus* Yu (pl. 2, figs. 9–10). Among them, *Tchangsichiton*, found from the Huangshandong Member of Tongying Formation in western Hubei, is characterized by having the intermediate valves nearly as long as wide and wedged or shield in shape, the anterior margin broadly embayed with sides strongly serrate, the tegmentum not divided into distinct central and lateral areas, the surface marked with transverse costae, the ventral surface shallow to depressed without apical hole, the tail valve longer than wide, with elongate mucro terminal. As compared with the other genera of this class, both extinct and extant hitherto known, it may represent a primitive form of Cambrian polyplacophorans.

Runnegarochiton is a peculiar genus of polyplacophorans from the Zhongyicun Member in Yunnan. At first glance, this genus is closely related to *Priscochiton* Dall, 1882 from the Middle Ordovician in Quebec of Canada, but the former possesses recurvated plates on the ventral surface, which makes it different from all other genera of Polyplacophora. The recurvated plates may be one of main features in the primitive polyplacophorans. But their function needs to be further studied (pl. 1, figs. 4–5).

Yangtzechiton, as one of polyplacophorans, is characterized by having a shell of complete coverage. Here it should be mentioned that in *Yangtzechiton elongatus* Yu, the ventral surface of the head valve is not smooth. This species has a nephroid depression near the anterior margin, which the writer calls the "nephroid depression". The anterior margin of this depression is similar to the inserting teeth present in the living polyplacophorans, in which it inserts into the girdle. Behind the nephroid depression are W-shaped, curved striae with a longitudinal furrow in the central part. On the sides of furrow the striae are raised and oblique toward the lateral margin (pl. 2, fig. 2). These morphological characters have not seen in other fossil species, nor in the extant species of the Polyplacophora. The present species bears the intermediate valve composed of a tegmentum and an articulamentum. Although the shell surface are poorly preserved and damaged during the preparation, the dividing line between tegmentum and articulamentum can still be seen. Based on the presence or absence of the articulamentum, many scholars divided the Polyplacophora into two subclasses, namely Paleoloricata and Neoloricata (Bergenhayn, 1955, 1960; Smith, 1960), and thus assigned the genera and species of the Lower Paleozoic to subclass Paleoloricata. However, some scholars consider that in fossil Polyplacophora it is not prectical to distinguish Paleoloricata from Neoloricata simply based on the presence or absence of the sutural laminae and insertion plates, because most parts of the fossil have already been silicified or partly replicated (Runnegar, Pojeta, Taylor and Collins, 1979). In the present specimens of *Yangtzechiton elongatus* Yu (pl. 2, figs. 1, 3) the intermediate valves possess, in addition to tegmentum, a well-developed articulamentum, which is overlapped by the preceding valve. For this reason the writer agrees with Runnegar *et al.* that the Polyplacophora should not be classified simply in the presence or absence of the articulamentum.

In the holotype of *Yangtzechiton elongatus*, there are many spicules on the dorsal and ventral sides of the intermediate valves, which are very rarely seen in the Lower Paleozoic polyplacophorans. They are different in shape and size, sparse in branches, and can be distinguished into three types: 1) needle-like spicules, 2) semicircular spicules and 3) bunch-like spicules (pl. 3, figs. 1-5).

At the first glance, *Yangtzechiton* is somewhat similar to *Paracarinachites* Qian et Jiang, 1980, from the contemporaneous deposits in eastern Yunnan. But by comparison, there are striking differences between them: In *Yangtzechiton* the shell is composed of 3–5 valves of complete coverage. The dorsal part of the intermediate valves has an elliptical hole, the head valve has a nephroid depression on the anterior margin of the ventral surface; whereas in *Paracarinachites* the sclerite is multi-layered, elongated and arcuated-univalve, the lower side is distinctly divided into two parts: adapicle curved part and posterior part. Adapical curved part with a subcircular dish which bears an opening in its subcentre; posterior part widely concave. Outer surface of the sclerite bears 5–6 nodular projections in the adapical curved part and 10–14 alternately arranged spines in the posterior part. (text-fig. 1). By these obvious differences the writer considers that *Paracarinachites* and its related form *Carinachites* Qian (Qian, 1984) belong to another group of shelly microfossils, not to the Polyplacophora, their systematic classification awaits a further study. *Yangtzechiton*

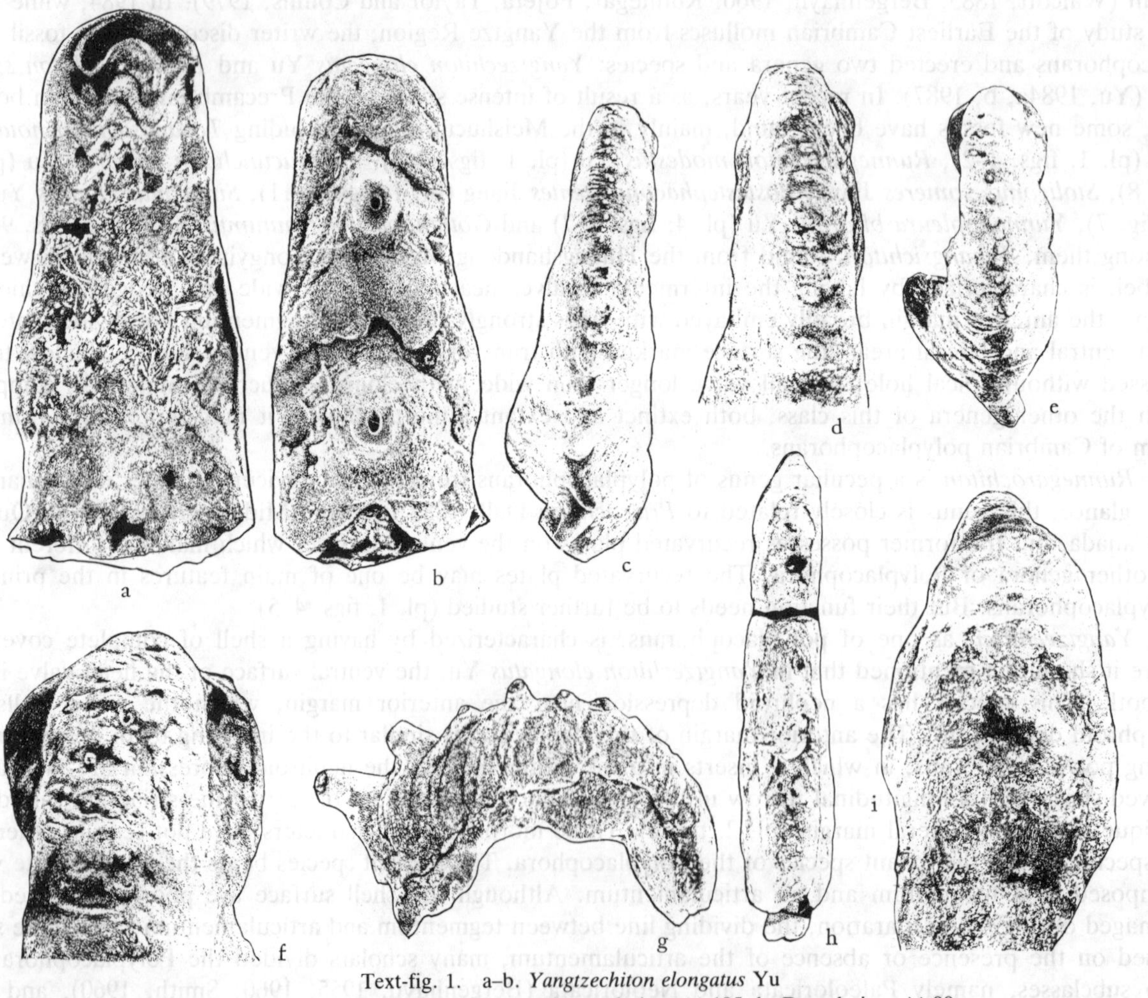

Text-fig. 1. a–b. *Yangtzechiton elongatus* Yu
a. Ventral view, × 105. b. Dorsal view, × 90.
c–i. *Paracarinachites sinensis* Qian et Jiang
c-d. Dorsal views, × 50 and × 75;
e-f. Dorsal views, ×70 and × 190;
g. Lateral view, ×80;
h-i. Ventral views, ×27 and × 90.

is similar to *Glyptochiton* Koninck, 1883 (Lower Carboniferous, Europe) in having a broad sinus on the posterior margin of the head valve and a concave hole on the dorsal surface of the intermediate valve. There is a probable evolutionary connection between the two genera, despite their very different geological ranges.

Luyanhaochiton, one of polyplacophorans from the Meishucun Stage of the Yangtze Region, is analogous to the post-Paleozoic ployplacophorans in the shape of the shell. But noticeably , in this genus the anterior part of the intermediate valve has a hollow spine obliquing backwards, a feature unknown in any other fossil or living genus of the Polyplacophora. Between the anterior margin of the central part of intermediate valve and the preceding valve, there is a space, semicircular in shape, with a hollow spine in the middle part of the anterior margin. The spine is stout and broadly rounded in the basal part, with its posterior part joining the central part of the valve and its anterior part extending toward and inserting into the preceding valve. The projecting portion of the spine overlaps the same valve (pl. 2, figs. 6–8) and occupies the whole semicircular space. Distally, the hollow spine gradually tapers and obliques backward. Taken as a whole, the spine possibly connects the anterior valve with the posterior one.

CLASS MERISMOCONCHIA YU, 1983

The Merismoconchia, a newly erected class of Mollusca, only has a small number of genera and species discovered so far. At present, there is little knowledge about its geological and geographical distributions. But studies on the Earliest Cambrian specimens led us to know something about the evolutionary relationship between this class and other classes of Mollusca. For example, *Yangtzemerisma rarum* Yu (pl. 5, figs. 1–6) and *Merismoconcha multisegmentata* Yu (pl. 5, figs. 11–14) both bear 8 pseudometameres. They are similar to polyplacophorans in the dorsal part of shell. The difference lies in that, in polyplacophorans, each valve bears two pairs of symmetrically arranged muscle scars on it, whereas in merismoconchs, muscle scars are located at the second and body segments respectively. More important difference is that in Merismoconchia the ventral part of shell has an elongate-oval aperture and a rostrate segment which protrudes out of the anterior part of the apertural margin. But it is the elongate-oval aperture and the protruding ros-

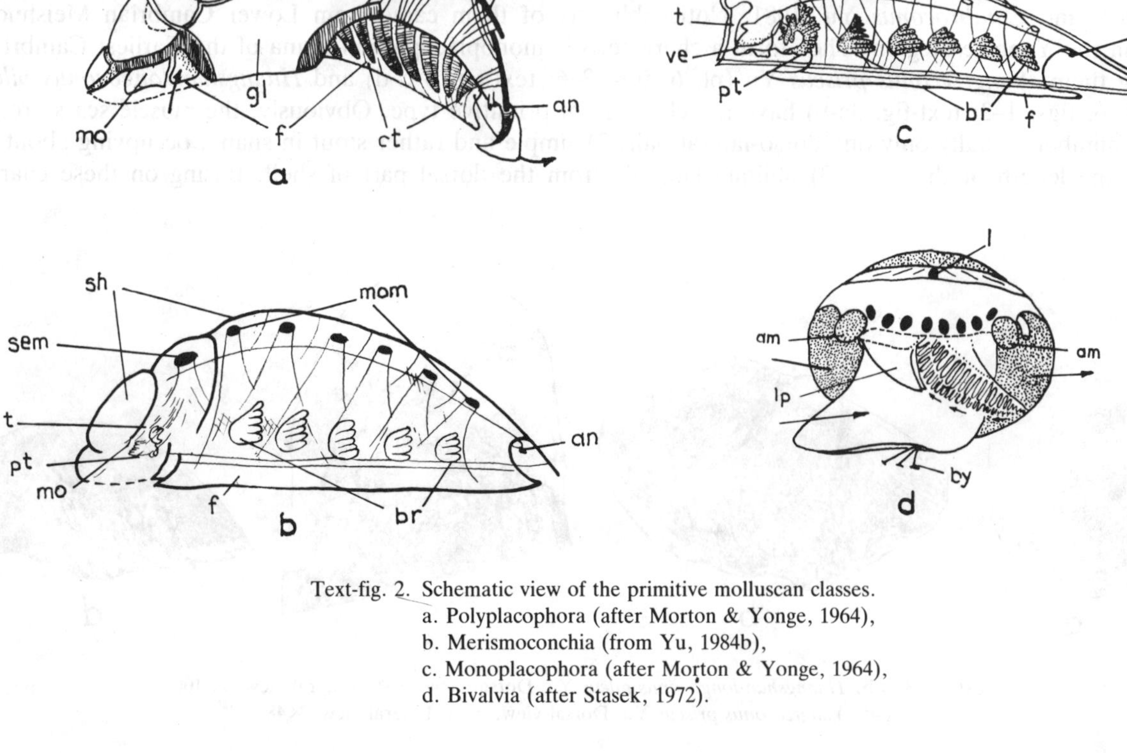

Text-fig. 2. Schematic view of the primitive molluscan classes.
a. Polyplacophora (after Morton & Yonge, 1964),
b. Merismoconchia (from Yu, 1984b),
c. Monoplacophora (after Morton & Yonge, 1964),
d. Bivalvia (after Stasek, 1972).

trate segment that show the merismoconchs to be analogous to the monoplacophorans. There are, however, some differences: 1) In merismoconchs, the shell is obviously divided into 3 parts of unequal lengths; while, in monoplacophorans, the shell is a complete univalve. 2) In merismoconchs, the muscle scars are located on the second and body segments of the shell respectively, while, in monoplacophorans, pairs of muscle scars are symmetrically arranged on the shell.

What has been mentioned above indicates that polyplacophorans, merismoconchs and monoplacophorans are similar in some respects but differ in some others. Their common characters are:1) they all belong to the molluscs with an antero-posterior axis of growth, and 2) their gills, kidneys, muscles and reproductive glands are pseudometamerically arranged. In comparison, polyplacophorans have the pseudometameres of the interior organs which are clearly shown in 8 separate valves, and merismoconchs bear them on the dorsal part of the shell, but in monoplacophorans, the pseudometameres are not visible on the shell. Therefore among the 3 groups of molluscs there has been a close evolutionary relationship and they probably evolved from the same branch in lineage, that is to say the 8 separate valves of polyplacophorans gradually developed to the rostrate, the second and body segments of merismoconchs, and then merged themselves in the cap-shaped or cyrtoconic univalve of monoplacophorans. Besides, the separately paired muscle scars of the oldest polyplacophorans gradually developed to the second segmental muscle scars and locomotive muscle scars of merismoconchs and then joined together as seen in some monoplacophorans (text-fig. 2)

CLASS MONOPLACOPHORA WENZ IN KNIGHT, 1952

Monoplacophora, as a class of the Mollusca, first appeared in the Earliest Cambrian. In the Yangtze Micromolluscan Fauna, the monoplacophorans are characterized by having a small shell. These small-shelled fossils are widely distributed over the world. From 1974 up to now, about 12 Chinese scholars have described these fossils, with 60 genera and 123 species erected. After making a review of these monoplacophorans, the writer believes that only 16 species and 14 genera belong to the Earliest Cambrian monoplacophorans. Of others, some are not monoplacophorans, some are hardly identified, or some should be rejected as synonyms.

According to the current information, the Earliest Cambrian is the stage in which the monoplacophorans made their first diversification. In the stage, there occurred such genera as *Spatuloconus* and *Eosoconus* of the Tryblidioidea and *Truncatoconus, Obtusoconus* of the Archinacelloidea, which are the forerunners of post Early Cambrian forms. From the previous literatures on the Early Cambrian Monoplacophora, there are at least 7 genera preserved with muscle scars, namely, *Aktugaria* Missarzhevsky, 1976, *Yangtzeconus* Yu, 1979, *Huangshandongoconus* Yu, 1979, *Eosoconus* Yu, 1979, *Archaeotremaria* Yu, 1979, *Truncatoconus* Yu, 1979 and *Cambroconus* Yu, 1981. Noticeably, six of them came from Lower Cambrian Meishucun Stage in the Yangtze Region. They form a characteristic monoplacophoran fauna of the Earliest Cambrian. Among them, *Yangtzeconus priscus* Yu (pl. 6, figs. 3–6; text-fig. 3 c–d) and *Huangshandongoconus pileus* Yu (pl. 6, figs. 1–2; text-fig. 3a-b) have muscle scars of primitive type. Obviously, the muscle scars are: 1) few in number, usually only one dorso-lateral pair, 2) simple and rather stout in shape, occupying about 1 / 4–1 / 5 the length of the shell, 3) oblique laterally from the dorsal part of shell. Basing on these charac-

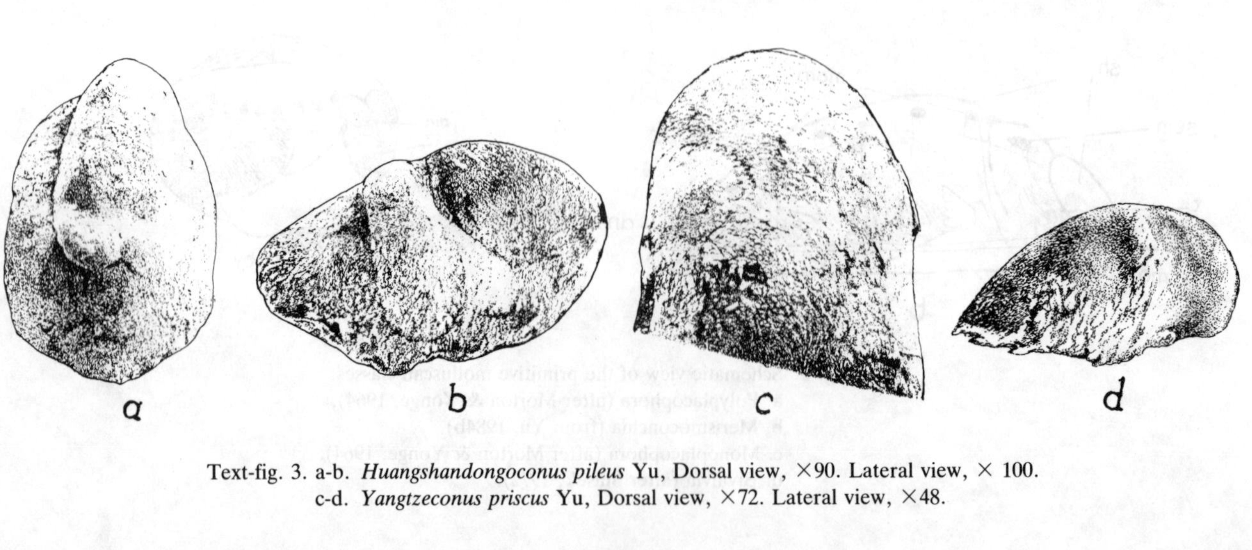

Text-fig. 3. a-b. *Huangshandongoconus pileus* Yu, Dorsal view, ×90. Lateral view, × 100.
c-d. *Yangtzeconus priscus* Yu, Dorsal view, ×72. Lateral view, ×48.

teristics, the writer has named this type of muscle scars Protomya (Yu, 1979). Moreover, in the oldest monoplacophorans, we can see the development and variation of muscle scars. For instance, the muscle scars increase in number or join together, and change from simple to complicated or to specialized ones. The genus *Archaeotremaria* from the Meishucun Stage is characterized by having a cap-shaped shell, four pairs of muscle scars in the central part of the dorsal side, a row of small tremata on the tubular projections, which, except the last one remaining open and serving as exhalant outlet, become filled up during the growth. In the Early Paleozoic molluscan faunas, there were few genera and species with tubular projections in the dorsal part of the shell. They include the Early Cambrian *Rozanoviella* Missarzhevsky, 1981, *Granoconus* Yu, 1979, and *Yochelcionella* Runnegar & Pojeta, 1974 and the Ordovician- Silurian *Tremanotus* Hall, 1865, *Salpingostoma* Roemer, 1876 and *Boitremus* Horny, 1962, etc. Whether there is an evolutionary relationship between these forms is a very complicated problem difficult to be solved at the present stage of knowledge.

The genus *Eosoconus* has a cyrtoconic shell, which makes it differ from all other genera of the Monoplacophora. Moreover, it bears 6 pairs of muscle scars and a pair of semicircular depression near the apex, which may be a pair of septa, an interesting feature for the primitive monoplacophorans. The function of them awaits our further study (pl. 6, figs. 7–8).

Truncatoconus, as a genus of the Monoplacophora, is characterized by having a cap-shaped shell, and by the presence of a pair of incompletely ring-shaped anterior muscle scars and a pair of small subcircular posterior ones. Evidently the genus is a primitive form in the Order Archinacelloidea (pl. 6, fig. 9).

CLASS GASTROPODA CUVIER, 1797

The gastropods made their first appearance in the earliest Cambrian. From then on, they developed rather rapidly. The ancestral forms of many important gastropods are characterized by having a small shell, fewer whorls, a large aperture and simple ornamentation on the shell. They include hyperstrophic, sinistral and dextral forms. In the Late Cambrian, the gastropods increased both in genera and in species, and were characterized by having a large-sized shell and complicated ornamentation on the shell. At the beginning of Ordovician, they were especially flourishing. For over 400 million years, they have lived constantly and can be adaptable to different ecological environments either in the sea, the brackish water, the fresh water or on the land. Furthermore, they are so many in genera and species that they rank the second largest class in the animal kingdom, only next to Class Insecta.

According to the present knowledge, about 20 genera of the Cambrian gastropods have been already reported over the world, more than half of the genera and species were found in Asia, mainly in China. From the Cambrian gastropods ever discovered in China, the writer has an impression that they underwent at least two developmental stages. In the first stage, or, exactly, in the earliest Cambrian, there are *Cambrospira sinensis* Yu (pl. 8, figs. 16–17) and *Maclurites hubeiensis* Yu (pl. 8, fig. 15), both of which are hyperstrophic; *Archaeospira ornata* Yu (pl. 8, figs. 4–11) which is sinistral, *Hubeispira nitida* Yu (pl. 8, figs. 1–3) and *Tianzhushanospira unicarinata* Yu (pl. 8, fig. 12) which are dextral. The fauna was found mainly in the phosphate rocks or phosphorus dolomites of the Meishucun Stage. Excepting *Maclurites* which has a rather long geological range, the others have a short geological range. Also they have rather simple morphological features, forming the main elements of the Yangtze Micromolluscan Fauna. In the second developmental stage, in the late Late Cambrian, the gastropod fauna is represented by *Matherellina walcotti* Kobayashi, "*Maclurites ulrichi* (Kobayashi) (Kobayashi, 1933), etc. This fauna was mainly found from the Fengshan Stage, the Wanwangou Limestome and the Qiushugou Member in North China. The gastropods are characterized by having a large shell. Besides, many of them have an extensive distribution.

In the interval between the two developmental stages appeared several important genera and species, including *Auriculaspira* Zhou et Xiao, 1984 and *Xinjispira* Yu et Rong, 1987 (text-fig. 4), which have been found from the Lower Cambrian Xinji Formation in the Southwestern part and the southern margin of North China Platform.

When establishing the genus *Yuwenia*, Runnegar made a point that this genus is like a mirror image of some living opisthobranchs (Runnegar, 1981). On such problems when the Subclass Opisthobranchia first appeared in geological history and how the fossil opisthobranchs are identified, many scholars have made discussion (d'Orbigny, 1842; Kinght *et al.* 1960; Robertson, 1963, 1973; Kollmann 1967; Kollmann and Yochelson, 1976; Sohl and Kollmann, 1985). Shortly, when classifying fossil opisthobranchs one should adhere the following criteria 1. the heterostrophic protoconch, 2. the downward depressed, involuted or

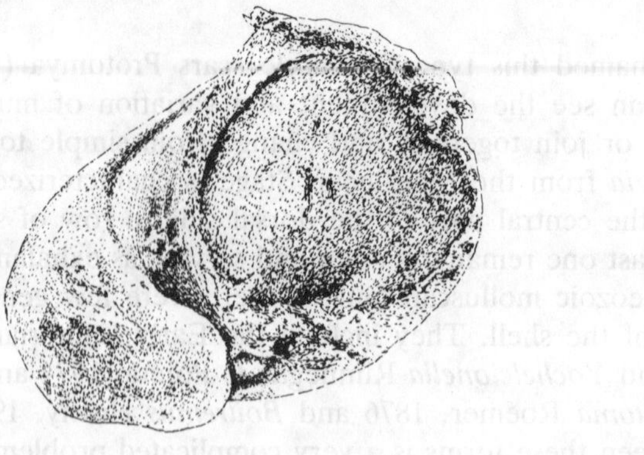

Text-fig. 4. *Xinjispira simplex* (Zhou et Xiao) Apertural view, ×65.

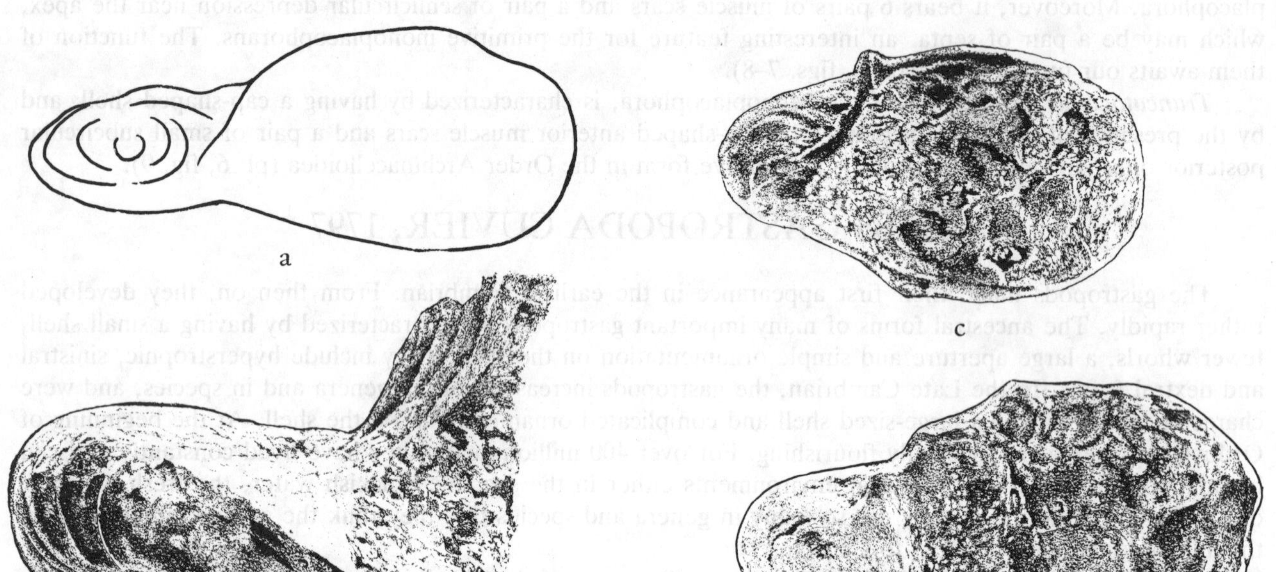

Text-fig. 5. *Yangtzespira exima* Yu
a. Cross-section, ×30; b. Apertural view, showing posterior canal, ×74;
c. Apertural view of a young form, ×55; d. Apertural view, ×70.

sometimes raised spire, 3. the narrow and long aperture, which is broad and concave anteriorly and has a sinus, sometimes, an exhalant siphonal canal near the suture, 4. the columella with spiral ridges or folds, 5. the surface smooth or sometimes with indistinct ornamentation, 6. the smaller and thicker shell and 7. the operculum degenerated or absent.

According to the above-mentioned criteria, the writer believes that the genus *Yangtzespira* may be a primitive form of Subclass Opisthobranchia. The reasons are: 1. The basal spire is concave and somewhat involute with one and half whorls in the exposed part as viewed from lower part of the shell. However, on studying the longitudinal sections of the shell, *Yangtzespira exima* Yu is composed of 3 whorls. The first half-whorl, apparently the protoconch, is very tiny and involute (text-fig. 5a). 2. The aperture is broad anteriorly, intersecting the columellar lip at a blunt angle, with the intersecting part overturned. The columellar lip is arcuate in appearance and composed of 4 thick ridges; its intero-lateral part appears smooth and is strongly overturned, but intersects the basal lip at an angle of 31°, with the intersecting part forming a posterior canal in the shape of an acute angle, which is probably the position of the anus. 3. the aperture

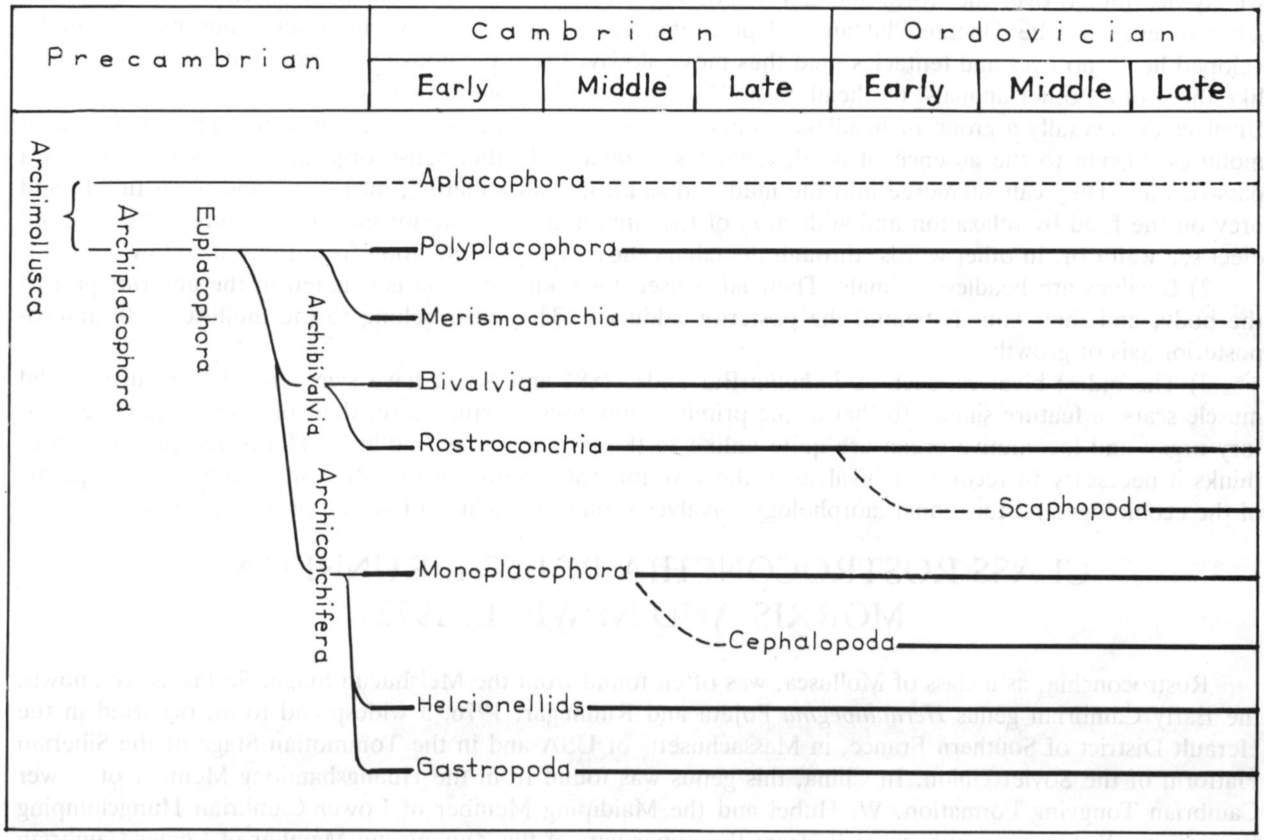

Text-fig. 6. Phylogeny of the phylum Mollusca.

is large, much wider than high, also elongatedly narrow in shape. 4. The inner lip wall of the last whorl is smooth, with thickened inductura (pl. 9, figs. 8–9), which are secreted from the surface of the outer mantle. Taking those characteristics into consideration, the genus *Yangtzespira* may belong to Subclass Opisthobranchia. Whether it is correct or not needs further vertification.

CLASS BIVALVIA LINNE 1758

It is generally considered that bivalves appeared later then univalved molluscs. However, in the past 20 years or so, new progress has been made in the research work on the Cambrian bivalves. The discovery of such genera as *Fordilla* Barrande, 1881. (Cambrian, Europe, North America, Asia), *Pojetaia* Jell, 1980 (Cambrian, Australia, Asia), *Tuarangia* MacKinnon, 1982 (Cambrian, New Zealand), particularly, the genus *Yangtzedonta* Yu, 1985 (pl. 10, figs. 1–3) from the Meishucun Stage of the Lower Cambrian have thrown light on the evolution of bivalves at least in the following points: 1) The first appearance of bivalves can be traced back to the Earliest Cambrian, 2) In primitive bivalves the hinge structure is simple, with a stout hinge tooth and a corresponding socket on the same valve, 3) They possess a row of pedal muscle scars (about 7–8 in number) and 4) The umbo is located in the anterior part, protruding over the hinge margin. All these characteristics mentioned above may probably explain why Subclass Pterimorphia appeared earlier than Subclass Palaeotaxodonta. Since bivalves appeared simultaneously with primitive polyplacophorans, monoplacophorans, gastropods and rostroconchs in the Earliest Cambrian, a question naturally arises whether bivalves are really advanced molluscs. The answer is "no".

The reasons why bivalves display primitive characters of molluscs are given below:

1) Bivalves as headless molluscs, have also been named "Acephala". The head, generally situated at the most anterior part of the body, is the most important part of the animal and has a mouth, nerve center and other sense organs. In contrast with a *Sepia*, the cephalopod has a developed body with head, eyes and other sense organs, moves quickly and therefore, can hunt for other small animals. Gastropods also have a well-developed body with head, sense organs and spiral shell. In comparison with bivalves, they are much widely distributed over the world and can be adaptable to different environments as seen in oceans, rivers, lakes or even on the Tibetan Plateau. Polyplacophorans are very primitive molluscs. They bear an undeveloped head, no eyes and tentacles, and thus move slowly. From the above statements, it can be seen that, like those of all other animals, the head of molluscs plays a vital role in their life.
Bivalves are actually a group of headless animals and by lack of the head, they may be a primitive form of molluscs. Owing to the absence of head, tentacles, radula and other sense organs, bivalves only live in a passive way. They can submerge into the mud and sand, live an adhering, burrowing and parasitic life and prey on the food by relaxation and adduction of the siphon at the posterior end of the body to take in and eject sea-water or, in other words, through the ciliary movement to draw food from the water flow.

2) Bivalves are headless animals. Their labia used for taking in food is situated in the anterior part of the body, and their anus is behind the posterior adductor. They also belong to the molluscs with antero-posterior axis of growth.

3) The oldest bivalves, such as *Babinka* Barrande, 1881 and *Synek* have symmetrically arranged pedal muscle scars, a feature similar to that of the primitive molluscs. Furthermore, their nervous system, respiratory organ and locomotive organ are quite unlike to those of advanced molluscs. Hence the present writer thinks it necessary to reconsider bivalves in the taxonomical position of the Mollusca. From the viewpoints of the ecobiology and functional morphology, bivalves should be included into the primitive molluscs.

CLASS ROSTROCONCHIA POJETA, RUNNEGAR, MORRIS AND NEWELL, 1972

Rostroconchia, as a class of Mollusca, was often found from the Meishucun Stage. So far as we known, the Early Cambrian genus *Heraultipegma* Pojeta and Runnegar, 1976, a widespread form, occurred in the Herault District of Southern France, in Massachusetts of USA and in the Tommotian Stage of the Siberian Platform of the Soviet Union. In China, this genus was found from the Huangshandong Member of Lower Cambrian Tongying Formation, W. Hubei and the Maidiping Member of Lower Cambrian Hungchunping Formation, W, Sichuan, and especially from the upper part of the Zhongyicun Member of Lower Cambrian Tongying Formation in Dahai of Huize county, E, Yunnan , where its specimens are so abundant as to be stratified. Morphologically, the Chinese specimen is different from *Heraultipegma varensalensa* (Cobbold), hence the name *Heraultipegma yunnanense* He et Yang (pl. 10, figs. 4–8).

ACKNOWLEDGMENTS

The writer wishes to express his sincere thanks to Dr. N. J. Morris of British Museum (Natural History), England for his careful reading of the manuscript. He is also grateful to Messrs. Cao Dehui, Mao Zhiqiang, Song Zhiyao and Xu Baorui for their technical assistance.

REFERENCES

Barrande, J., 1881. Systeme silurien du centre de la Boheme. Premiere Partie Recherches paleontologique, 6, Paris and Prague.

Bergenhayn, J. R. M., 1955. Die fossilen schwedischen Loricaten nebst einer vorlaufigen Revision des Systems der ganzen Klasse Loricata. Lunds Univ. Arssk., new ser. B, 51(8), 1–41.

———, 1960. Cambrian and Ordovician Loricates from North America. Jour. Paleont. 34(1), 168–178.

Berg-Madsen, V. and Peel, J. S., 1978. Middle Cambrian monoplacophorans from Bornholm and Australia, and the systematic position of the bellerophontiform molluscs. Lethaia, 11, 113–125.

Brasier, M. D., 1984. Microfossils and small shelly fossils from the Lower Cambrian Hyolithes Limestone at Nuneation, English Midlands. Geol. Mag., 121(3), 229–253.

Chen Meng'e, Chen Yiyuan and Zhang Shusen, 1981. The small fossil assemblages in the limestone of the uppermost part of Dengying Formation at Songlingpo, Yichang. Earth Science, 1, 32–41 (in Chinese).

Chen Ping, 1984. Discovery of Lower Cambrian small shelly fossils from Jijiapo, Yichang, West Hubei and its significance. Prof. Pap. Stratigr. Palaeont., 13, 49–64. Geol. Publ. House, Beijing (in Chinese).

Chen Yiyuan and Zhang Shusen, 1980. Earliest Cambrian small shelly fauna from Song Lin Pu of the Yangtze Gorge. Geol. Review, 26. 190–197 (in Chinese).

Cobbold, J. W., 1935. Lower Cambrian faunas from Herault, France. Ann. & Mag. Nat. Hist., ser, 10, 16(91), 25–48.

Haas, W., 1981. Evolution of Calcareous hard parts in primitive Molluscs. Malacologia, 21(1-2). 403–418.

He Tinggui, 1984. Discovery of *Lapworthella bella* Assemblage from Lower Cambrian Meishucun Stage in Niuniuzhai, Leibo County, Sichuan Province. Prof. Pap. Stratigr. Paleont., 13, 23–34. Geol. Publ. House, Beijing (in Chinese).

——— and Yang Xian-he, 1982. Lower Cambrian Meishucun Stage of the western Yuagtze stratigraphic region and its small shelly fossils. Bull. Chengdu Inst. Geol. M. R., 3, 69–95 (in Chinese).

———, Pei Fang and Fu Guang-hong, 1984. Some small shelly fossils from the Lower Cambrian Xinji Formation in Fangcheng County, Henan Province. Acta. Palaeont. Sinica, 23(3) 355–366.

Jell, P. A, 1980. Earliest Known pelecypod on earth-a new Early Cambrian genus from South Australia. Alcheringa, 4, 233–239.

Jiang Zhiwen, 1980. Monoplacophorans and gastropod fauna of the Meishucun Stage from the Meishucun section, Yunnan. Acta Geol. Sinica, 54(2), 112–123.

——— 1982. Small shelly fossils. In Luo et al: The Sinian-Cambrian boundary in Eastern Yunnan, China. Yunnan People's Press, Kunming (in Chinese).

Knight J. B., 1952. Primitive fossil gastropods and their bearing on gastropod classification. Smithson. Misc. Coll., 117 (13).

——— and Yochelson, E. L., 1960. Monoplacophora. In: Moore, R. C. (ed.), Treatise on invertebrate paleontology. Part I, Mollusca. Geol. Soc. Am. & Univ. Kansas Press.

Kobayashi, T., 1933. Upper Cambrian of the Wuhutsui Basin, Liaotung, with special reference to the limit of Chaumitian (or Upper Cambrian) of Eastern Asia and its subdivision. Japan. Jour. Geol. Geogr., 11(1–2), 55–155.

Kollmann, H. A., 1967. Die Gattung *Trochactaeon* in der ostalpinen Okerkreide; zur Phylogenie der Actaeonellidae. Naturhistoisches Museum Wien Annales, 71, 139–258.

——— and Yochelson, E. L., 1976. Survey of Paleozoic gastropods possibly belonging to the Subclass Opisthobranchia. ibid., 80, 207–220.

Lu Yanhao, 1979. Cambrian mineral deposits in China and the bioenvironmental contral hypothesis. Geol. publ. House, Beijing (in Chinese).

Luo Huilin, Jiang Zhiwen, Wu Xiche, Song Xueliang and Ouyang Lin, 1982. The Sinian-Cambrian Boundary in eastern Yunnan, China. People's Publishing House, Yunnan (in Chinese).

MacKinnon, D. L., 1982. *Tuarangia paparua* n. gen and n. sp., a late Middle Cambrian pelecypod from New Zealand. Jour. Paleont. 56, 589–598.

———, 1985. New Zealand late Middle Cambrian Molluscs and the origin of Rostroconchia and Bivalvia. Alcheringa, 9, 65–81.

Morton, J. E. and Yonge, C. M., 1964. Classification and structure of the mollusca. In: Wilbur, K. W. and Yonge, C. M. (eds.), Physiology of the Mollusca. 1, 1–58. Academic Press, New York.

Pojeta, J. Jr., Runnegar, B., Morris, N. J. and Newell, N. D., 1972. Rostroconchia: a new class of bivalved mollusks. Science, 177 (4045). 264–267.

Pojeta, J. Jr., 1980. Molluscan phylogeny. Tulane Studies in Geology and Paleontology, 16, 55–80.

Qian Yi, 1984. Several groups of bizarre fossils from the Earliest Cambrian in Eastern Yunnan. Bull. Nanjing Geol. Palaeont., Acad. Sinica, 6, 97–112.

Robertson, R. A. and Merrill, A. S., 1963. Abnormal dextral hyperstrophy of post-larval Heliacus (Gastropoda: Architectonicidae). The Veliger, 6(2), 76–79.

———, 1974. The biology of the Architectonicidae, gastropods combining prosobranch and opisthobranch traits. Malacologia, 14 (1-2), 215–220.

Rozanov, A. Yu, Missarzhevsky, V. V., et al., 1969. The Tommotian Stage and the Lower Cambrian Boundary problem. Tr. Geol. Inst. Acad. Nauk. SSSR. 206.

Runnegar, B., 1980. Hyolitha: Status of the phylum. Lethaia, 13, 21–25.

———, 1981. Muscle scars, shell form and torsion in Cambrian univalved mollusca. Lethaia, 14, 311–322.

———, 1983. Molluscan phylogeny revisited. Mem. Ass. Australas. Palaeontolos., 1, 121–144.

Runnegar, B. and Pojeta, J. Jr., 1974. Molluscan phylogeny: The Paleontological viewpoint. Science, 186, 311–317.

———, —————, Morris, N. J., Taylor, J. D., Taylor M. E. and McClung, G., 1975. Biology of the Hyolitha. Lethaia, 8, 181–191.

Runnegar, B. and Pojeta, J. Jr., Taylor, M. E. and Collins, D., 1979. New species of the Cambrian and Ordovician Chitons *Matthevia* and *Cheolodes* from Wisconsin and Queensland: evidence for early of history polyplacophoran molluscs. Jour. Paleont. 53, 1374–1394.

Salvini-Plawen, L. V., 1981. On the origin and evolution of the Molluscs. Atti. Dei Convegni Lincei, 49, 235–293.

Sohl, N. F. and Kollmann, H. A., 1985. Cretaceous Actaeonellid gastropods from the western hemisphere. U. S. Geol. Surv. Prof. Pap., 1304.

Smith, A. G. and Hoare, R. D., 1987. Paleozoic polyplacophora: A checklist and bibliography. Calif. Acad. Sci. Occas. Paper, 146. 1–71.

Walcott, C. D., 1885. Note on some Paleozoic Pteropods. Amer. Jour. Sci., 30, 17–21.

Yin Jichen, Ding Lianfang, He Tinggui, Li Shilin and Shen Lijuan, 1980. The Palaeontology and Sedimentary Environment of the Sinian Sysytem in Emei-Ganluo Area. Sichuan. Sichuan People's Press, Chengdu (in Chinese).

Yochelson, E. L., 1978. An alternative approach to the interpretation of the phylogeny of ancient mollusks. Malacologica, 17, 165–191.

Yu Wen, 1974. Cambrian gastropods. In "A handbook of the stratigraphy and Palaeontology of southwest China". 111–112. Science Press, Beijing (in Chinese).

Yu Wen, 1979. Earliest Cambrian monoplacophorans and gastropods from western Hubei with their biostratigraphical significance. Acta Palaeont. Sinica, 18, 233–270.

———, 1981. New Earliest Cambrian monoplacophorans and gastropods from W. Hubei and E. Yunnan. Acta Palaeont. Sinica, 20, 552–556.

———, 1984a. Early Cambrian molluscan faunas of Meishucun Stage with special reference to Precambrian-Cam brian Boundary. In: Contributions to 27th International Geological Congress, 21–35. Science Press, Beijing.

———, 1984b. On Merismoconchids. Acta Palaeont. Sinica, 23, 423–446.

———, 1985. *Yangtzedonta* —A problematic bivalvia from the Meishucun Stage, China. Acta Micropalaeont, Sinica, 3, 401–408.

———, 1986. Lower Cambrian univalved molluscs from Kuruktag, Xinjiang. Acta Palaeont. Sinica. 25, 10–16.

———, 1987. Yangtze Micromolluscan Fauna in the Yangtze Region of China with notes on Precambrian-Cambrian Boundary (1), 19–344. Nanjing Univ. Press, Nanjing.

Yu Wen, Wang Huiji and Li Zishun, 1963. Fossil gastropods of China. Science Press, Beijing (in Chinese).

EXPLANATION OF PLATES

The illustrated specimens are housed in Nanjing Institute of Geology and Palaeontology, Academia Sinica.

Plate I

1–2. *Tchangsichiton notabilus* Yu
 1. Dorsal view of an intermediate valve, ×90. Holotype. Cat. no. 84140. Coll. no. TF-1.
 2. Dorsal view, × 85. Paratype. Cat. no. 84141. Coll. no. TF-1. Upper part of Huangshandong Member of Lower Cambrian Tongying Formation, Tianzhushan, Yichang, W. Hubei.

3–6. *Runnegarochiton medostus* Yu
 3. Dorsal view of an intermediate valve, ×45.
 4. Lateral view, showing the recurvatural plate on the ventral side, ×40.
 5. Ventral view, ×40. Holotype. Cat. no. 84161. Coll. no. KM-7.
 6. Dorsal view, ×42. Paratype. Cat. no. 84162. Coll. no. KM-7.
 Upper part of Zhongyicun Member of Lower Cambrian Tongying Formation, Meishucun, Jinning, E. Yunnan.

7. *Sinuconus clapeus* Yu
 Dorsal view of an intermediate valve, × 72. Holotype. Cat. no. 54408. Coll. no. HF-2.
 Upper part of Huangshandong Member of Lower Cambrian Tongying Formation, Huangshandong, Yichang, W. Hubei.

8. *Meishucunchiton vulgarus* Yu
 Dorsal view of an intermediate valve, ×80. Holotype. Cat. no. 84191. Coll. no. XD-2.
 Upper part of Zhongyicun Member of Lower Cambrian Tongying Formation, Xianfeng and Baizai of Xundian, E. Yunnan.

9–11. *Postestephaconus tenes* Jiang
 9. Lateral view of an intermediate valve, ×105.
 10. Dorsal view, ×90. Cat. no. 84209. Coll. no. KM-7.
 11. Dorsal view, ×90. Cat. no. 84206. Coll. no. XB-28.
 Upper part of Zhongyicun Member of Lower Cambrian Tongying Formation, Meishucun of Jinning, Baizai of Xundian, E. Yunnan.

Plate 1

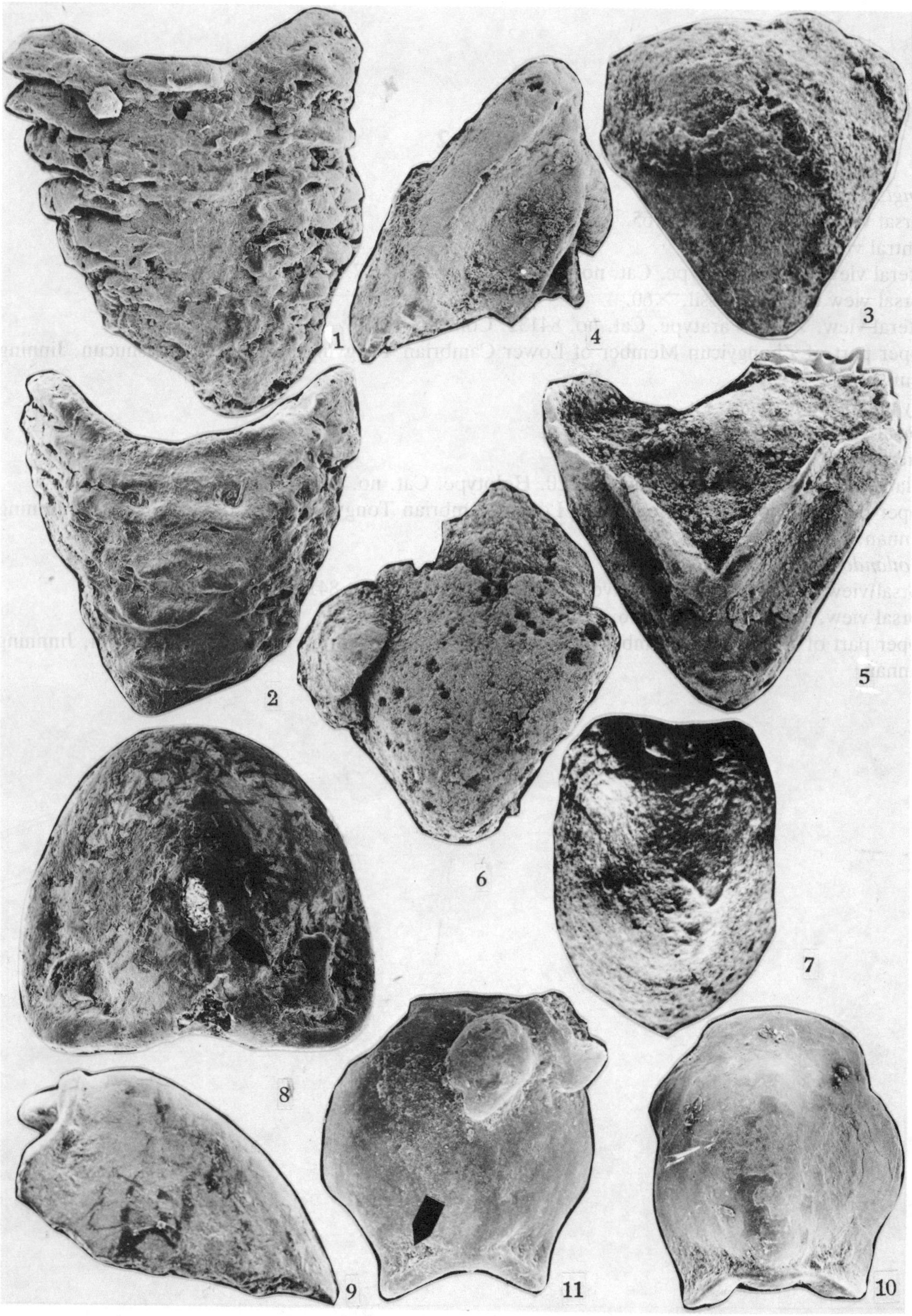

Plate 2

1–5. *Yangtzechiton elongatus* Yu
1. Dorsal view of shelly fossil, ×65.
2. Ventral view, ×70.
3. Lateral view, ×80. Holotype. Cat. no. 84131. Coll. no. KM-7.
4. Dorsal view of shelly fossil, ×60.
5. Lateral view, ×105. Paratype. Cat. no. 84132. Coll. no. KM-7.
   Upper part of Zhongyicun Member of Lower Cambrian Tongying Formation, Meishucun, Jinning, E. Yunnan.

6–8. *Luyanhaochiton spinus* Yu
6. Dorsal view of shelly fossil, ×80.
7. Lateral view, × 100.
8. Enlargement of intermediate valve. ×220. Holotype. Cat. no. 84135. Coll. no. KM-7.
   Upper part of Zhongyicun Member of Lower Cambrian Tongying Formation, Meishucun, Jinning, E. Yunnan.

9–10. *Gotlandochiton? minimus* Yu
9. Dorsal view of an intermediate valve, ×65. Holotype. Cat no. 84155. Coll. no KM-7.
10. Dorsal view, ×70. Cat. no. 84156. Coll. no.KM-7.
    Upper part of Zhongyicun Member of Lower Cambrian Tongying Formation, Meishucun, Jinnning, E. Yunnan.

Plate 2

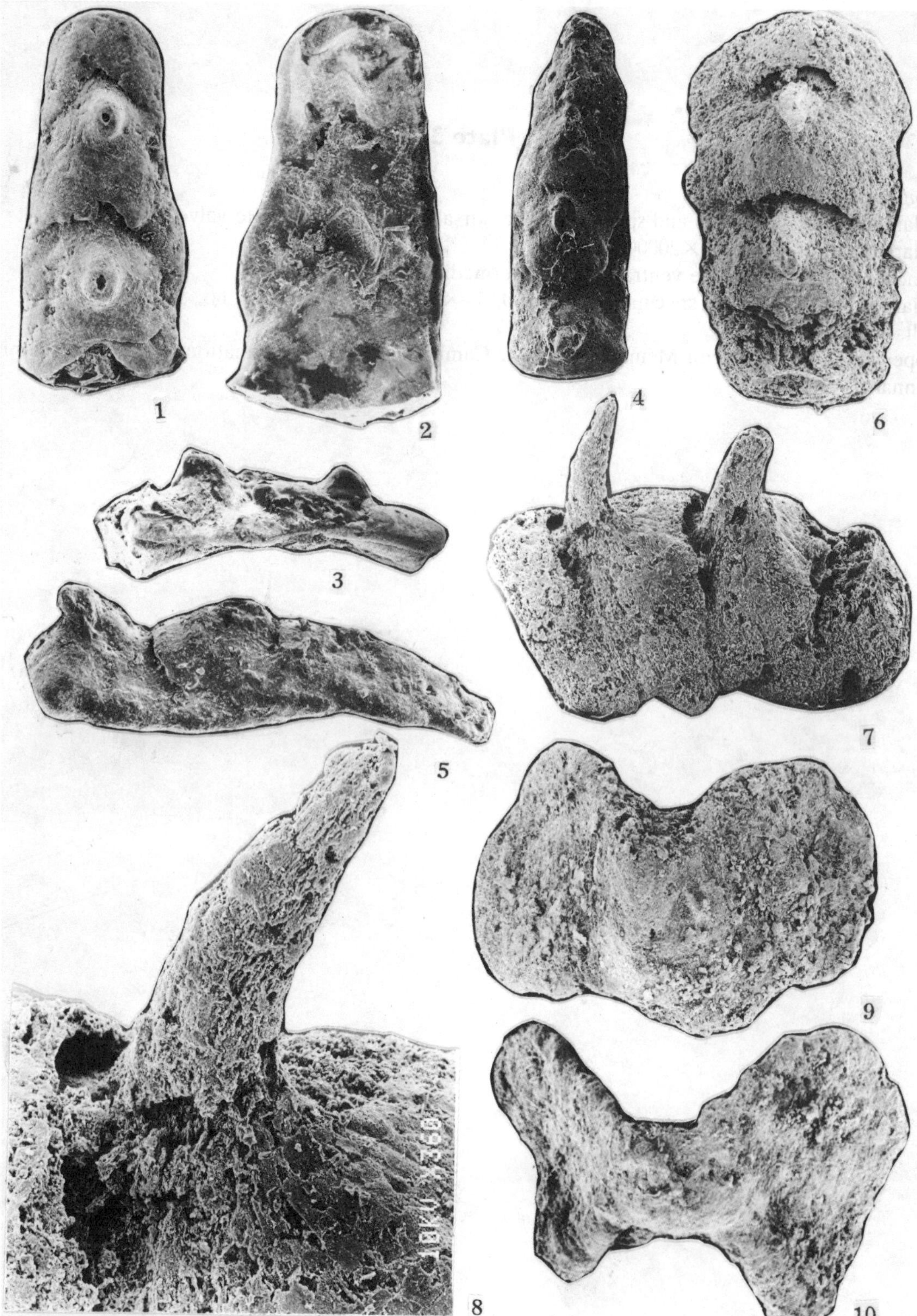

Plate 3

1–5. *Yangtzechiton elongatus* Yu

1. Enlargement of sculptures and spicules on the dorsal side of intermediate valve, ×1950.
2. Enlargement of spicules, ×20000.
3. Bunches of spicules on the ventral side of intermediate valve, ×240.

4–5. Enlargement of the same specimen. 4. ×1500, 5. ×5000. Holotype. Cat. no. 84131. Coll. no. KM-7.
Upper part of Zhongyicun Member of Lower Cambrian Tongying Formation, Meishucun, Jinning, E. Yunnan.

Plate 3

Plate 4

1–7. *Yunnanopleura biformis* Yu

1. Ventral view of an intermediate valve, ×30.
2. Dorsal view, showing the central and lateral areas, ×30.
3. Lateral view, ×30.
4. Enlargement of the central and lateral areas. Holotype. ×175. Cat. no. 84173. Coll. no. XB-52.
5. Ventral view of a head valve(?). ×40.
6. Dorsal view, ×45. Cat no. 84176. Coll. no. XB-52.
7. A cross-section of an intermediate valve, ×37. Cat. no. 84178. Coll. no. XB-52.

Upper part of Zhongyicun Member of Lower Cambrian Tongying Formation, Baizai, Xundian, E. Yunnan.

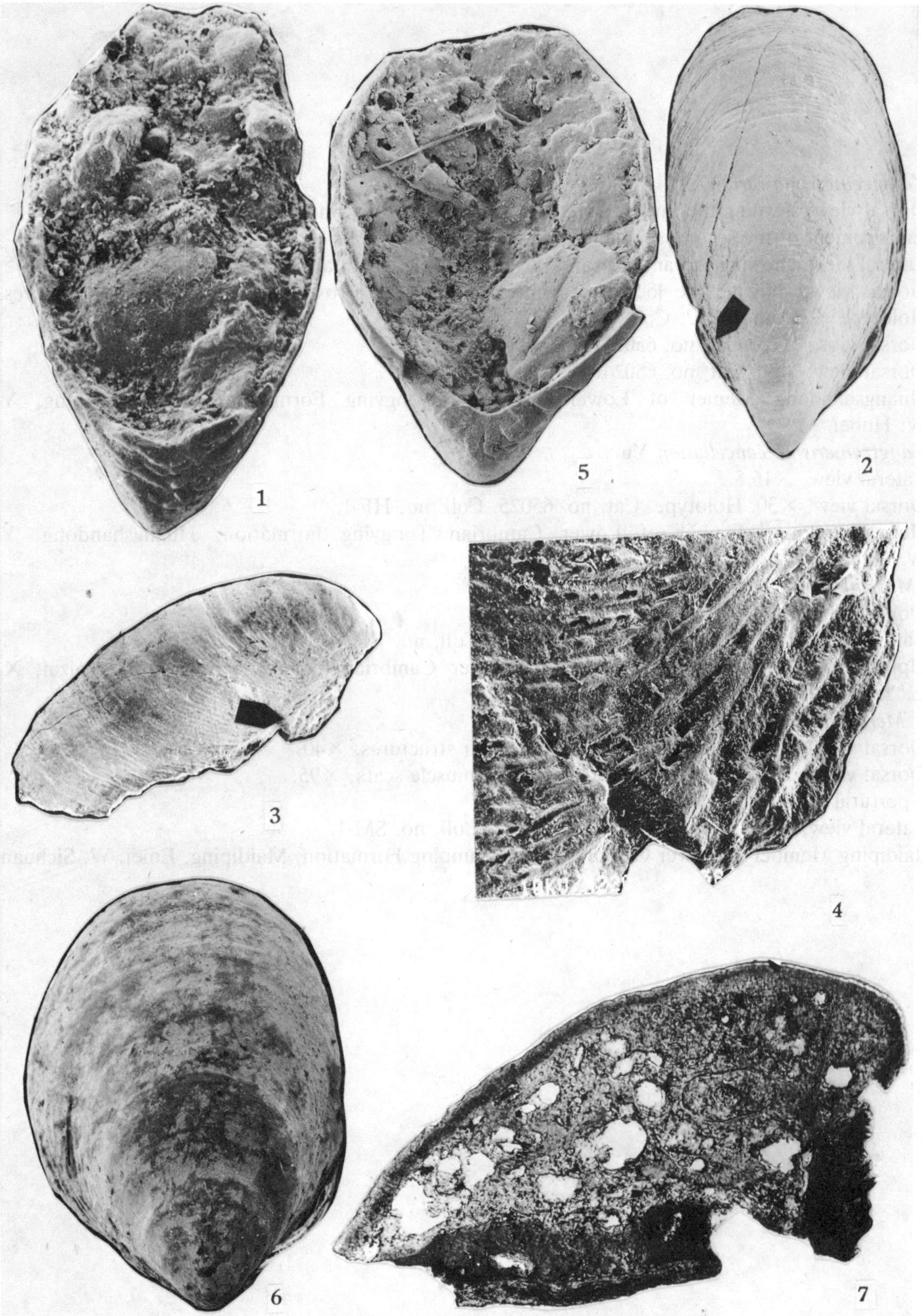

Plate 5

1–6. *Yangtzemerisma rarum* Yu
1. Dorsal view, shwoing the muscle scars (see arrow) and other internal structures, ×25.
2. Enlargement of the second segmental muscle scar, ×45
3. Lateral view, showing the anterior and posterior intersegmental furrows, ×22.5.
4. Dorsal view, showing the locomotive muscle scars (see arrow) and other internal structures, × 25. Holotype. Cat. no. 65022. Coll. no. HF-1.
5. Dorsal view, ×65. Cat. no. 65023. Coll. no. HF-1.
6. Dorsal view, ×30. Cat. no. 65024. Coll. no. HF-1.
   Huangshandong Memer of Lower Cambrian Tongying Formation, Huangshandong, Yichang, W. Hubei.

7–8. *Yangtzemerisma cancellatum* Yu
7. Lateral view, ×16.5.
8. Dorsal view, ×30. Holotype. Cat. no. 65025. Coll. no. HF-1.
   Huangshandong Member of Lower Cambrian Tongying Formation, Huangshandong, Yichang, W. Hubei.

9–10. *Minymerisma yunnanense* Yu
9. Dorsal view, ×60.
10. Lateral view, ×60. Holotype. Cat. no. 65026. Coll. no. XB-40.
   Upper part of Zhongyicun Member of Lower Cambrian Tongying Formation, Baizai, Xundian, E. Yunnan.

11–14. *Merismoconcha multisegmentata* Yu
11. Dorsal view, showing the muscle scars and other structures, ×40.
12. Dorsal view, enlargement of second segmental muscle scars, ×95.
13. Apertural view, ×37.5.
14. Lateral view, ×24. Holotype. Cat. no. 54464. Coll. no. SM-1.
   Maidiping Member of Lower Cambrian Hungchunping Formation, Maidiping, Emei, W. Sichuan.

Plate 5

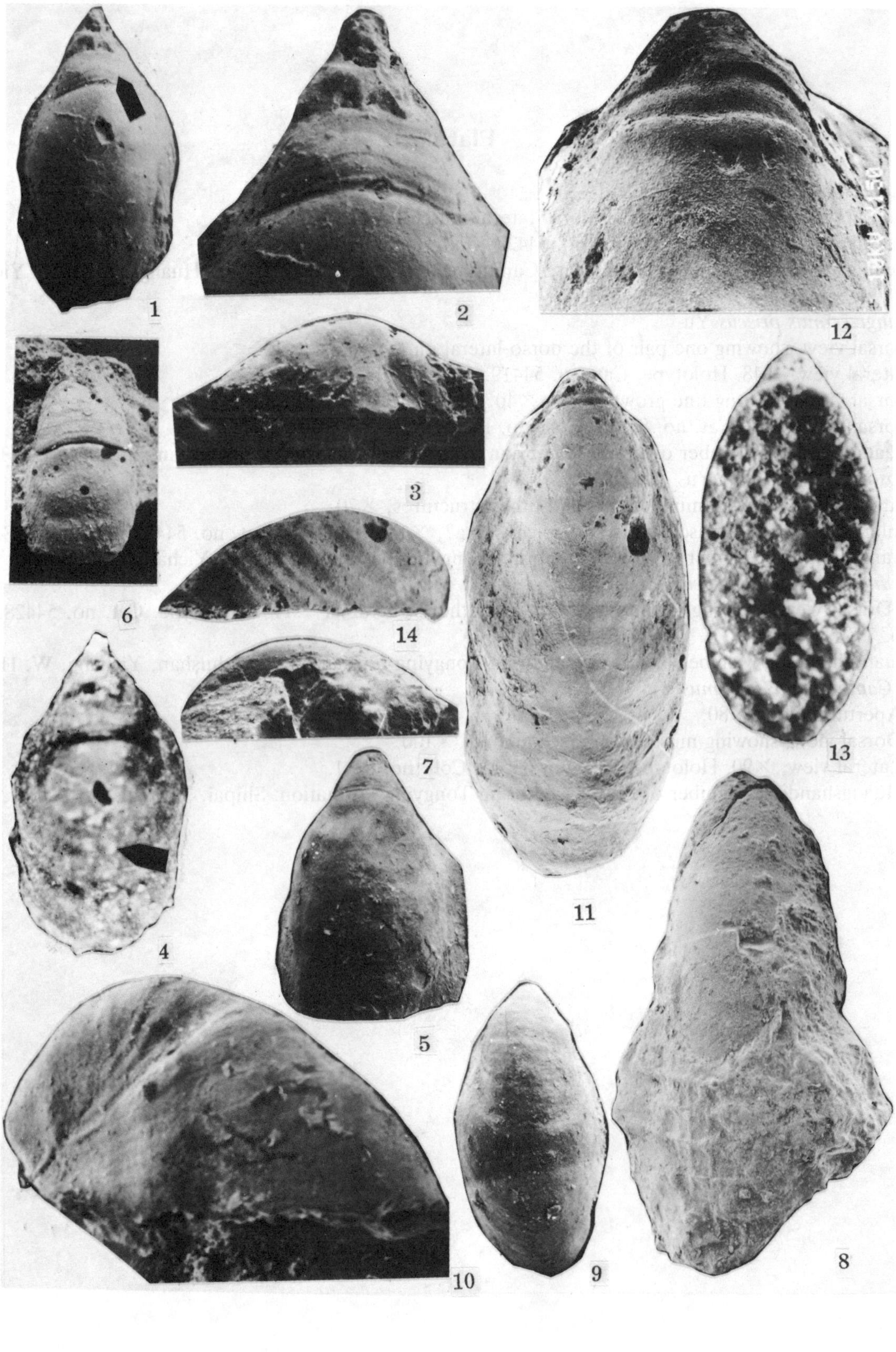

Plate 6

1–2. *Huangshandongoconus pileus* Yu

1. Lateral view, showing one pair of dorso-lateral muscle scars, ×100.
2. Dorsal view, ×90. Holotype. Cat. no. 54427. Coll. no. HF-1.
   Huangshandong Member of Lower Cambrian Tongying Formation, Huangshandong, Yichang, W. Hubei.

3–6. *Yangtzeconus priscus* Yu

3. Dorsal view, showing one pair of the dorso-lateral muscle scars, ×72.
4. Lateral view, ×48. Holotype. Cat. no. 54419. Coll. no. TF-1.
5. Dorsal view, showing fine growth lines, ×40. Cat. no. 84274. Coll. no. HS-1.
6. Dorsal view, ×55. Cat. no. 84275. Coll. no. HF-1.
   Huangshandong Member of Lower Cambrian Tongying Formation, Tianzhushan, Yichang, W. Hubei.

7–8. *Eosoconus primarius* Yu

7. Lateral view, showing muscle scars and other structures, ×70.
8. Enlargement of the muscle scars and apical septa?, ×100. Holotyp. Cat. no. 54425. Coll. no. SF-3.
   Huangshandong Member of Lower Cambrian Tongying Formation, Shipai, Yichang, W. Hubei.

9. *Truncatoconus yichangensis* Yu
   Dorsal view, showing the muscle scars and other structures, ×115. Holotype. Cat. no. 54428. Coll. no. TF-1.
   Huangshandong Member of Lower Cambrian Tongying Formation, Tianzhushan, Yichang, W. Hubei.

10–12. *Cambroconus minimus* Yu

10. Apertural view, ×80.
11. Dorsal view, showing muscle scars (see arrow), ×100.
12. Lateral view, ×90. Holotype. Cat. no. 65015. Coll. no. SF-1.
    Huangshandong Member of Lower Cambrian Tongying Formation, Shipai, Yichang, W. Hubei.

Plate 6

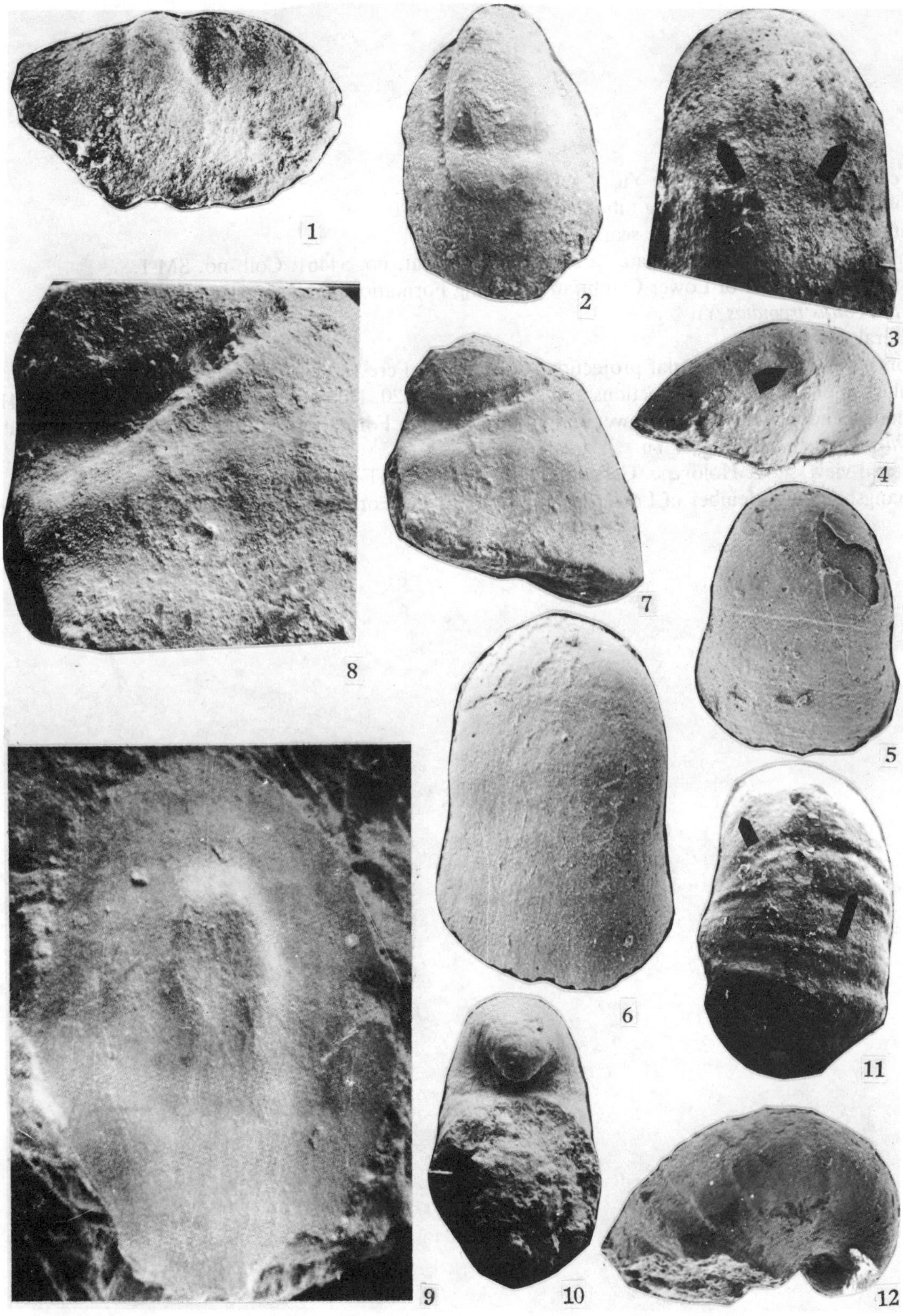

Plate 7

1–3. *Archaeotremaria polytremata* Yu
  1. Oblique dorsal view, showing tubular projections, ×120.
  2. Lateral view, showing muscle scars (see arrow), ×120.
  3. Enlargement of last two tremata, ×200. Paratype. Cat. no. 54461. Coll. no. SM-1.
  Maidiping Member of Lower Cambrian Tongying Formation, Gaoqiao, Emei, W. Sichuan.

4–6. *Granoconus trematus* Yu
  4. Lateral view, ×60.
  5. Dorsal view, showing tubular projections on the dorsal crest, ×60.
  6. Enlargement of tubular projections and sculptures, ×120. Holotype. Cat. no. 54455. Coll. no. HU-10.
  Huangshandong Member of Lower Cambrian Tongying Formation, Hujingtan, Yichang, W. Hubei.

7. *Obtusoconus paucicostatus* Yu
  Lateral view, ×72. Holotype. Cat. no. 54417. Coll. no. TF-1.
  Huangshandong Member of Lower Cambrian Tongying Formation, Tianzhushan, Yichang, W. Hubei.

Plate 7

Plate 8

1–3. *Hubeispira nitida* Yu
1. Apertural view, ×29.
2. Basal view, ×24.
3. Apical view, ×20. Holotype. Cat. no. 65022. Coll. no. TF-2.
Huangshandong Member of Lower Cambrian Tongying Formation, Shipai, Yichang, W. Hubei.

4–11. *Archaeospira ornata* Yu
4–5. Apertural and dorsal views, ×40.
6. Basal view, ×40. Holotype Cat. no. 54471. Coll. no. SF-2.
Huangshandong Member of Lower Cambrian Tongying Formation, Shipai, Yichang, W. Hubei.
7. Apertural view, ×30.
8. Apical view, ×35.
9. Dorsal view, ×40. Cat. no. 84359. Coll. no. XB-18.
10. Apertural view, ×13.
11. Enlargement of the parietal region, ×25. Cat no. 84360. Coll. no. XB-40.
Upper part of Zhongyicun Member of Lower Cambrian Tongying Formation, Baizai, Xundian, E. Yunnan.

12. *Tianzhushanospira unicarinate* Yu
Apertural view, ×36. Holotype Cat. no. 54459. Coll. no. TF-1-3.
Huangshandong Member of Lower Cambrian Tongying Formation, Tianzhushan, Yichang, W. Hubei.

13–14. *Aldanella* sp.
13. Apertural view, ×10.
14. Dorsal view, ×10. Cat. no. 84414. Coll. no. TF1-1.
Huangshandong Member of Lower Cambrian Tongying Formation, Tianzhushan, Yichang, W. Hubei.

15. *Maclurites hubeiensis* Yu
A cross-section, showing the widely concave umbilicus and acute umbilical edge, ×4. Cat. no. 54457. Coll. no. TF1-1.
Huangshandong Member of Lower Cambrian Tongying Formation, Tianzhushan, Yichang, W. Hubei.

16–17. *Cambrospira sinensis* Yu
16. Apertural view, ×10.
17. Dorsal view, ×10. Holotype Cat no. 54456. Coll. no. TF-1.
Huangshandong Member of Lower Cambrain Tongying Formation, Tianzhushan, Yichang, W. Hubei.

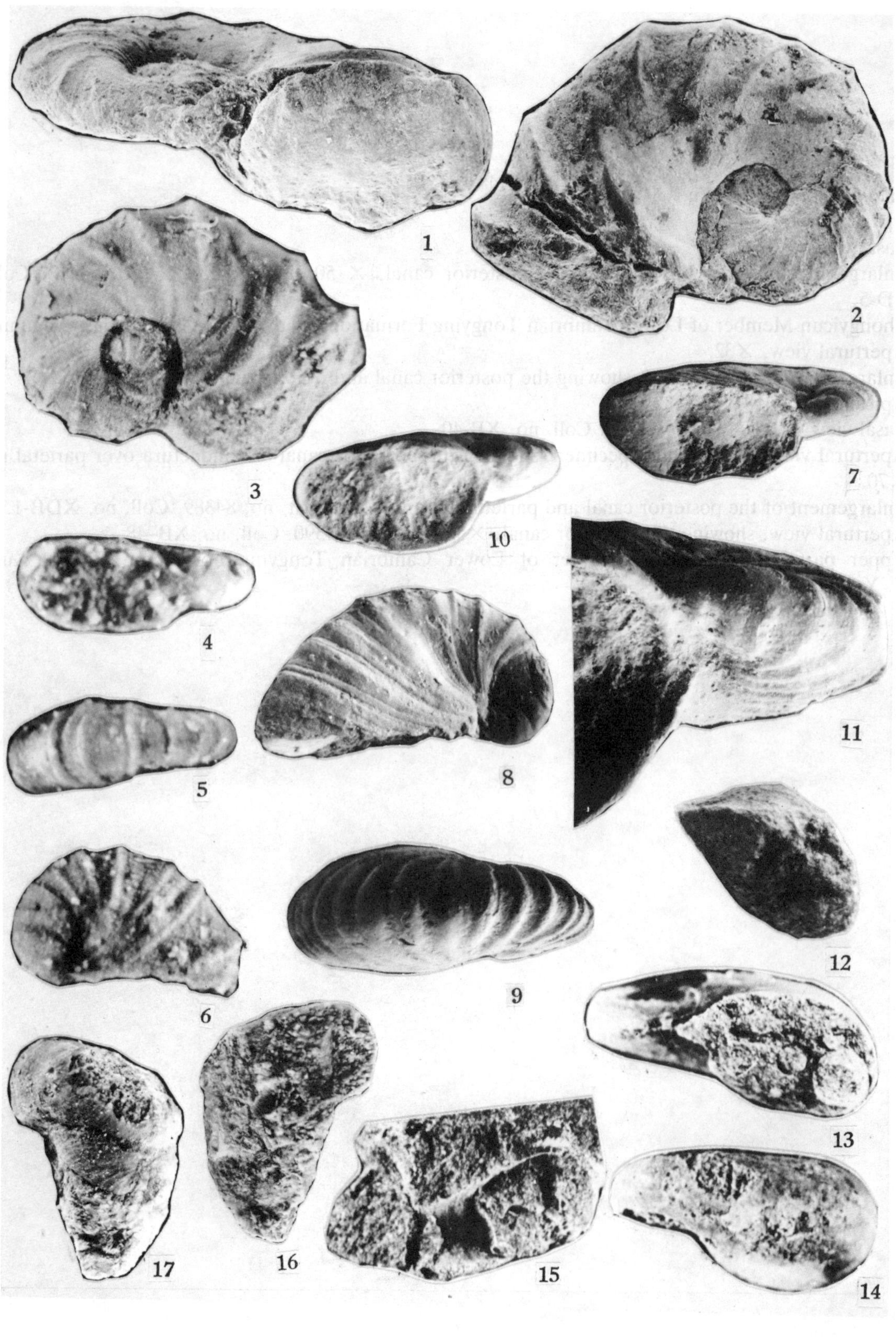
1
2
3
4
5
6
7
8
9
10
11
12
13
14
15
16
17

Plate 9

1–10. *Yangtzespria exima* Yu

1. Apertural view, ×15.
2. Basal view, ×15.
3. Enlargement of inner lip, showing the posterior canal, × 50. Holotype. Cat. no. 54462. Coll. no. YD-5.
   Zhongyicun Member of Lower Cambrian Tongying Formation, Dadiyakou, Chengjiang, E. Yunnan.
4. Apertural view, ×32.
5. Enlargement of the inner lip, showing the posterior canal and other structures, ×74.
6. Apical view, ×18.
7. Basal view, ×23. Cat. no. 84150. Coll. no. XB-40.
8. Apertural view of a juvenile specimen, showing the posterior canal and inductura over parietal region, ×70.
9. Enlargement of the posterior canal and parietal region, ×140. Cat. no. 84389. Coll. no. XDB-1.
10. Apertural view, showing the posterior canal, ×55. Cat. no. 84390. Coll. no. XB–48.
    Upper part of Zhongyicun Member of Lower Cambrian Tongying Formation, Baizai, Xundian, E. Yunnan.

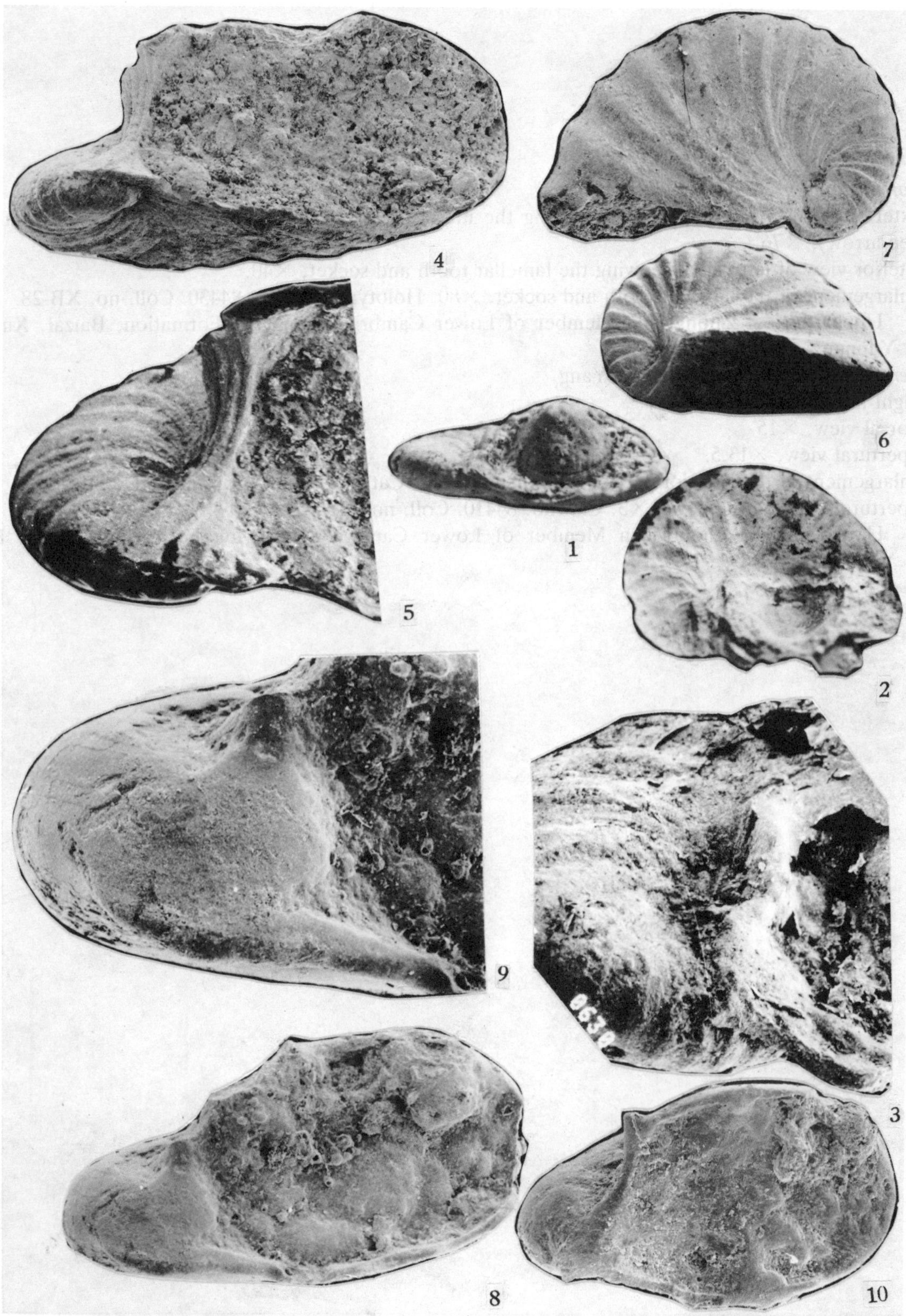
4
7
6
5
1
2
9
3
8
10

Plate 10

1–3. *Yangtzedonta primitiva* Yu

1. Exterior view of internal mould, showing the muscle scars of lower part of posterior umbonal ridge (see arrow), ×76.
2. Interior view of left valve, showing the lamellar tooth and socket, ×40.
3. Enlargement of the lamellar tooth and socket, ×80. Holotype. Cat no. 84430. Coll. no. XB-28.

Upper part of Zhongyicun Member of Lower Cambrian Tongying Formation, Baizai, Xundian, E. Yunnan.

4–8. *Heraultipegma yunnanense* He et Yang

4. Right lateral view, ×16.5.
5. Dorsal view, ×15.
6. Apertural view, ×13.5.
7. Enlargement of the anterior part of the dorsum, ×85. Cat. no. 84416. Coll. no. HD4-2.
8. Apertural and ventral views, ×3. Cat. no. 84410. Coll. no. HD4-2.

Upper part of Zhongyicun Member of Lower Cambrian Tongying Formation, Dahai, Huize, E. Yunnan.

Plate 10

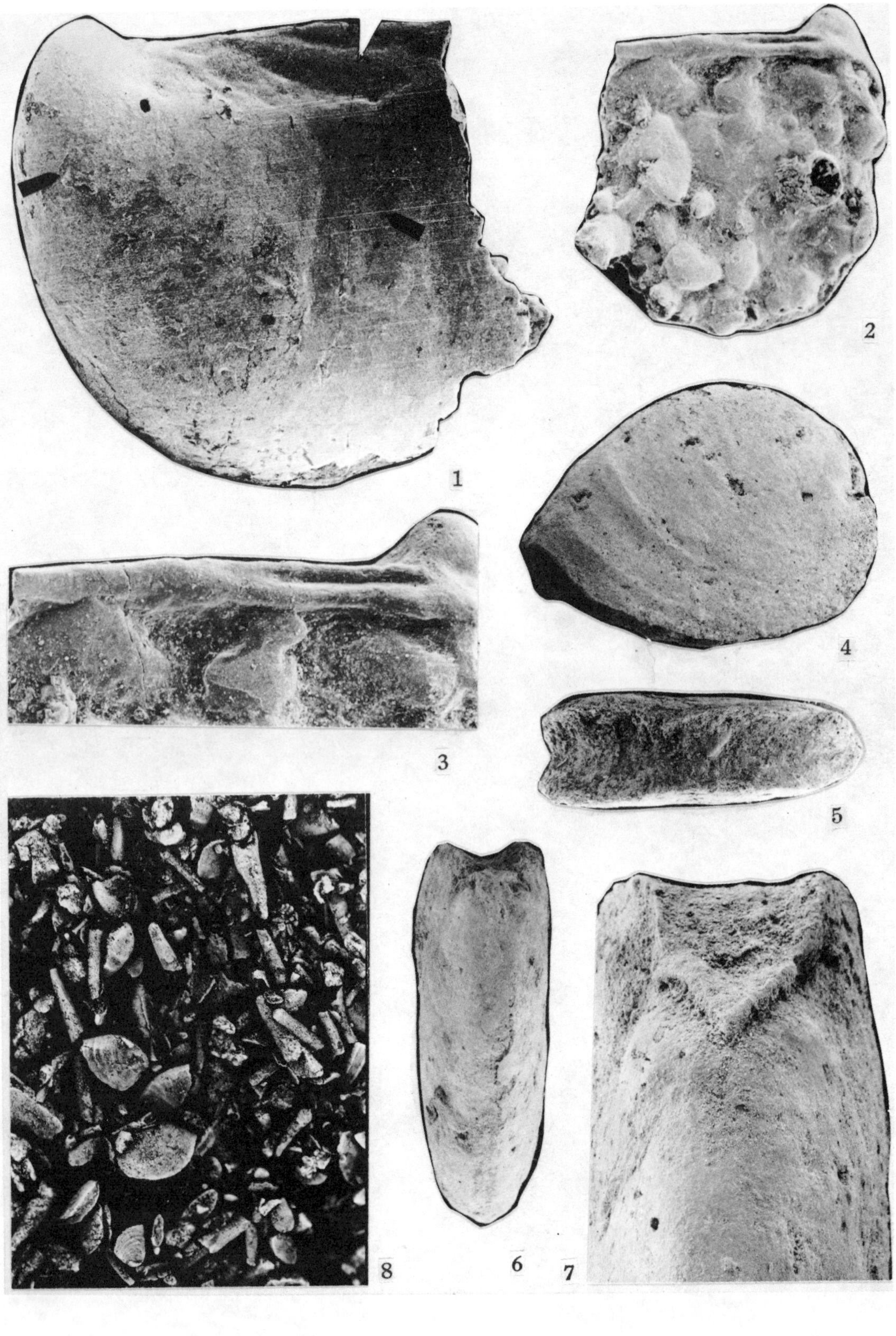

PALAEONTOLOGIA CATHAYANA Volume 5, 1990, p.171–180

NOTES ON *ARCHIKAINELLA* AND *PARAACIDASPIS* (TRILOBITA)

Zhang Wentang

(*Nanjing Institute of Geology and Palaeontology, Academia Sinica*)

Abstract

The type material of Chinese Upper Cambrian eoacidaspidid trilobites is here reviewed. *Belovia, Eoacidaspis, Paraacidaspis* and *Archikainella* are assigned to the family Eoacidaspididae. *Paraacidaspis* is a valid genus. *Metaacidaspis* Qiu, 1983, is a junior synonym of *Archikainella* Liu, 1982. *Archikainella acuta* Liu, *Metaacidaspis beigongensis* Qiu, *Paraacidaspis triangularis* Luo and *Paraacidaspis* sp. are junior synonyms of *Archikainella vomerinus* Liu, 1982. Judging from the geographic distribution of four genera of this family, they are good indicators for the typical Late Cambrian slope biofacies.

Thanks to Bruton's work on the "Cambrian origins of the Odontopleurid trilobites" (Bruton, 1983), it is possible to make a comparative study on the Chinese Late Cambrian eoacidaspidid trilobites with Russian genera of the same family. Bruton's paper was published in 1983, but he did not mention any published Chinese literature concerning these trilobites, for example, the work of Egorova et al. (1963), Lu et al. (1965), Yang (1978) and Liu (1982). One year later Luo (1983) and Qiu et al. (1983) also described Upper Cambrian eoacidaspidids from western Yunnan as well as from southern Anhui.

Bruton (1983, p. 879) considered that "Poletaeva listed *Paraacidaspis* nom. nud. (1957, pp. 162-163) and later diagnosed the genus (1960, p. 68, pl. 3, fig.1) and figured the holotype of the type species *P. hunanica* Jegorova in Poletaeva, 1960, nom. nud., from the Upper Cambrian Changshanian Stage, Hunan Province, South China. Since the type species was not formally described, the genus *Paraacidaspis* is invalid. The study on the cranidium of *P. hunanica* nom. nud. and a second species, *P. sibirica* Poletaeva, 1960 (p. 70, pl. 3, fig. 2) from the Upper Cambrian of western Siberia, leaves little doubt that *Paraacidaspis* is a subjective synonym of *Eoacidaspis*." So far as I know, the type species of *Paraacidaspis*, i. e. *P. hunanica* Jegorova, has been formally described and figured (Jegorova et al., 1963) and later repeatedly illustrated by Lu et al. (1965). Yang (1978) collected and described some new material of this species from Changshanian Stage of the same locality in western Hunan. As a result of a comparative study on *Eoacidaspis* and *Paraacidaspis*, the latter is quite different from the former. Therefore, *Paraacidaspis* should be considered as a valid genus.

With *A. vomerinus* as the type species, Liu (1982) erected an eoacidaspidid genus *Archikainella* from the uppermost Cambrian Fengshanian Stage in western Hunan. One year later, based on *M. beigongensis* from the Tangcun Formation (Fengshanian) of Beigong, southern Anhui, Qiu (Qiu et al., 1983) established a new generic name *Metaacidaspis*. Judging from the published descriptions and illustrations by Qiu and Liu, *Metaacidaspis* should be a junior synonym of *Archikainella* Liu, 1982. Luo's *Paraacidaspis triangularis* Luo (1983, pl. 6, fig. 13) from the uppermost Cambrian Baoshan Formation (Fengshanian) in western Yunnan and Apollonov's *Paraacidaspis* sp. (Apollonov et al., 1984, pl. 6, figs. 13, 14) from the uppermost Upper Cambrian (*Lophosaukia* Zone) in Malyi Karatau Range, southern Kazakhstan, should be referred to *Archikainella.*

Belovia Poletaeva, 1956, *Acidaspidina* Lazarenko, 1960, and *Eoacidaspis* Poletaeva, 1956 are assigned to the family Eoacidaspididae Poletaeva, 1956. Originally, *Belovia* was assigned by Poletaeva (1956) to Odontopleuridae, but later (Poletaeva, 1956, 1960) it was placed in Eoacidaspididae together with the younger *Eoacidaspis* and *Paraacidaspis* (Bruton, 1983). *Belovia, Eoacidaspis, Paraacidaspis* and *Archikainella* have in common the waisted glabella, sinuous dorsal furrows, the diverging anterior facial suture, and the

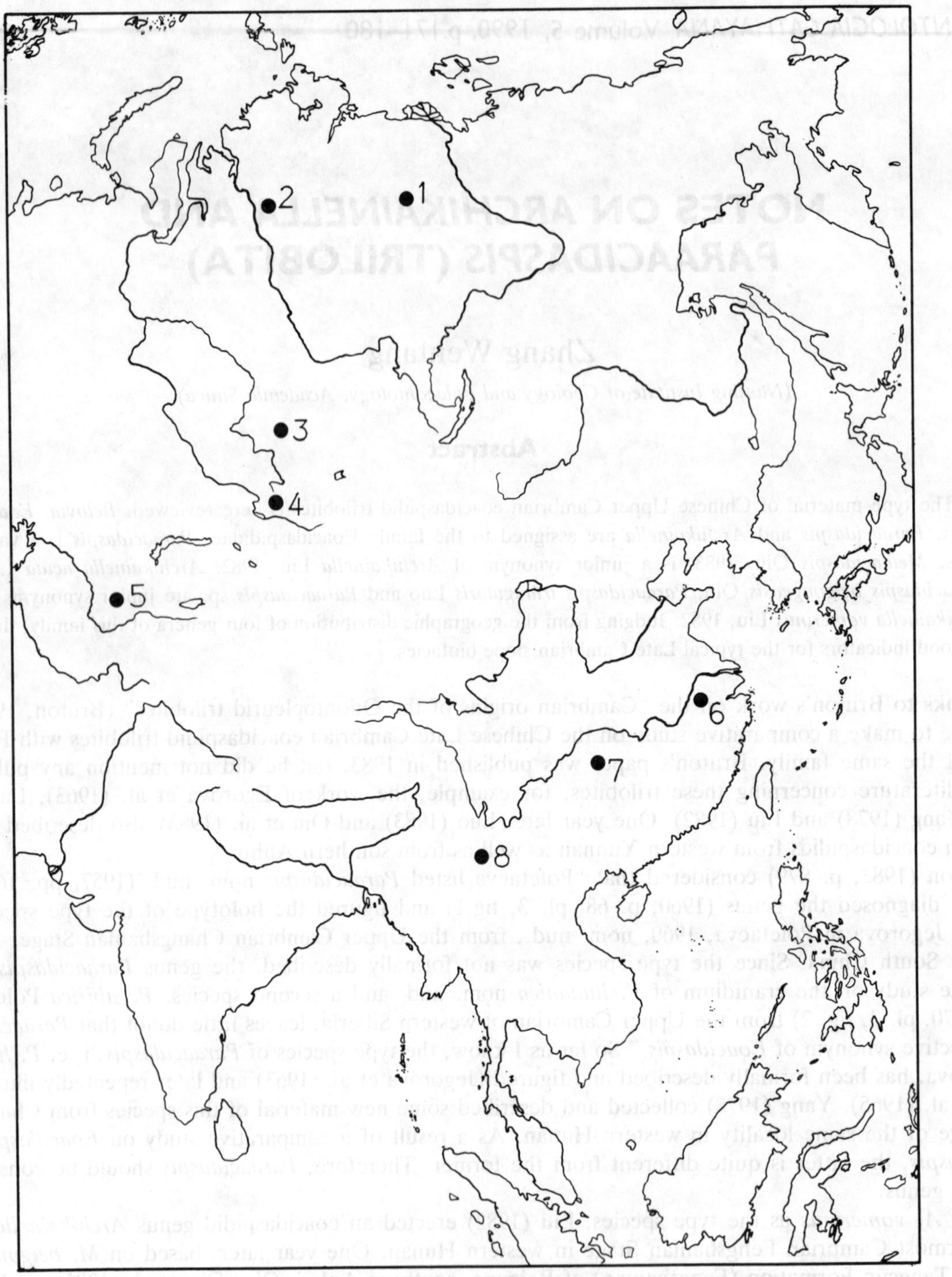

Text-figure 1. Map showing the distribution of the family Eoacidaspididae.

1. *Belovia clava* Tchernysheva in Poletava, 1956;
2. *Eoacidaspis amplicauda* Lazarenko, 1968; *Belovia aliquantula* Rozova, 1964; *Belovia laeta* Rozova, 1964; *Belovia cyclica* Rozova, 1964;
3. *Eoacidaspis salairica* Poletaeva, 1956; *Belovia salairica* Poletaeva, 1960; *Belovia radugini* Poletaeva, 1977;
4. *Belovia media* Poletaeva, 1977;
5. *Paraacidaspis* sp. (= *Archikainella vomerinus* Liu, 1982, Apollonov, et al., 1984. pl. 6, figs. 13, 14);
6. *Archikainella vomerinus* Liu, 1982 (= *Metaacidaspis beigongensis* Qiu in Qiu et al., 1983, pl. 88, fig. 9; Fengshanian Stage, southern Anhui; *Paraacidaspis hunanica* Jegorova (Qiu in Qiu et al., 1983, pl. 88, fig. 8; Changshanian Stage, southern Anhui);
7. *Paraacidaspis hunanica* Jegorova (Changshanian Stage, Fenghuang, western Hunan); *Paraacidaspis latilimbus* Yang (Baojing, western Hunan, Changshanian Stage); *Archikainella vomerinus* Liu, 1982 (= *Archikainella acuta* Liu, 1982);
8. *Archikainella vomerinus* Liu, 1982 (= *Paraacidaspis triangularis* Luo, 1983) (Fengshanian Stage, western Yunnan).

prominent rim along the anterior border. These four genera mentioned above should be accommodated under the family Eoacidaspididae. Moreover, *Acidaspidina* is different from these four genera in the shape of glabella, glabellar furrows, anterior border and dorsal furrows. It seems, therefore, that *Acidaspidina* should not belong to this family.

Acidaspidella Rozova, 1963 and *Acidaspides* Lermontova, 1951 are assigned to Odontopleuridae Burmeister, 1843. *Acidaspides* is a direct ancestor of the post-Cambrian Odontopleurinae. The origin of Eoacidaspididae is unknown and more skeletal parts are needed before comparison with earlier stocks is possible (Bruton, 1983). I agree with Bruton's conclusion.

Belovia occurred in the late Middle and Upper Cambrian of Olenek River, northern Siberia; Altai of northeastern Kazakhstan, and ca. 300 km southeast of Novosibirsk. *Eoacidaspis* was found from Upper Cambrian of Salair Range, western Siberia and lower Yenyse River. *Archikainella* occurred in the uppermost Cambrian (Fengshanian) of western Hunan, southern Anhui and Malyi Karatau Range of southern Kazakhstan, and *Paraacidaspis* is a trilobite genus of early Changshanian Stage in western Hunan, southern Anhui and probably western Siberia. In recent years, many authors have explained the global distribution of major Cambrian trilobite faunas in terms of restricted and unrestricted access to open oceans (Palmer, 1973; Robison, 1975); of shelf and slope biofacies (Taylor & Cook, 1975); of Yangtzean (and North China or Huangho) fauna and Chiangnan (or Machari) fauna (Lu et al., 1982; Xiang et al., 1981; Kobayashi, 1966, 1967). In China, the Yangtzean (or North China) fauna is equivalent to the shelf biofacies including endemic polymeroid ptychoparioids, corynexochids, redlichiids, etc., while the Chiangnan (or Machari) fauna represents the slope biofacies and includes many genera and species of pandemic agnostids and open-shelf polymeroid trilobites, such as olenids, paradoxidids, asaphids, damesellids, corynexochids, protolenids, eodiscids and probably old representative of odontopleurids. In the mobile belt or geosyncline belt in China, trilobites are very rare, if they are present at all, and always show the characters of slope or Chiangnan biofacies, e.g. the trilobite fauna of the northern part of western Tienshan within both Chinese territory and Kazakhstan, and the trilobite fauna found in the Chilian Geosyncline. *Paraacidaspis* and *Archikainella* occurred in the Chiangnan belt. Representatives of *Archikainella* was found from western Yunnan, which probably represents the slope biofacies of northeastern marginal belt of the Gondwanaland. It is very interesting to note that Olenek River and Lower Yenyse River where *Belovia* and *Eoacidaspis* occur, fall within the north marginal belt of the Siberian Platform, and definitely represent the slope biofacies. With regard to Altai and Malyi Karatau Range, it may be referred to the slope biofacies, or belong to the mobile belts. From the fact mentioned above, it seems clear that the genera of the family Eoacidaspididae have the biostratigraphical importance and the biogeographical significance as well. These genera are indicative of the typical Late Cambrian slope biofacies.

Acknowledgements I wish to express my gratitude to Prof. Xiang Liwen and Drs. Liu Yiren, Luo Huilin and Zhang Quanzhong, for providing the photographs of the Chinese type eoacidaspidid trilobites.

FAMILY EOACIDASPIDIDAE POLETAEVA, 1957

This family comprises *Belovia, Eoacidaspis, Paraacidaspis* and *Archikainella.*

Occurrence: Middle to Upper Cambrian; Siberia, Kazakhstan and South China.

Archikainella Liu, 1982

1982 *Archikainella* Liu, p. 982
1983 *Metaacidaspis* Qiu in Qiu et al., p. 252–253

Type species: *Archikainella vomerinus* Liu, 1982, from the uppermost Cambrian, Qiangongping, Fenghuang, Hunan.

Diagnosis: Glabella triangular, waisted at the posterior 1/3 sagittal length; dorsal furrows sinuous and deep posteriorly; 4 pairs of lateral glabellar furrows; proximal ends of S_1 and S_2 connected by deep longitudinal furrow; S_3 deep and oblique; S_4 short; occipital furrow narrow; occipital ring broad (sag.), gradually narrowing distally; fixigena widest opposite to S_1 and dying out opposite to distal end of S_3; eye ridge oblique; eye lobe at the midlength of glabella; frontal area and anterior area of fixigena broad (sag., exsag); surface finely gran ulated; anterior facial suture long and strongly divergent; librigena, thorax and pygidium unknown.

Occurrence: Late Cambrian Fengshanian Stage, Hunan, western Yunnan and southern Anhui, China and Malyi Kratau Range, southern Kazakhstan.

Archikainella vomerinus Liu, 1982

(Pl. 1, figs. 1–4)

1982 *Archikainella vomerinus* Liu, pl. 234, fig. 5.
1982 *Archikainella acuta* Liu, pl. 234, fig. 4.
1983 *Metaacidaspis beigongensis* Qiu in Qiu et al., pl. 88, fig. 9.
1983 *Paraacidaspis triangularis* Luo, pl. 6, fig. 13.
1984 *Paraacidaspis* sp., Apollonov et al., pl. 6, figs. 13, 14.

Holotype: Fragmentary cranidium illustrated by Liu (1982, pl. 234, fig. 5, Cat. no. HT81107) from the Upper Cambrian of Qiangongping, Fenghuang County, Hunan.

Description: Glabella with fixigena more or less triangular, waisted at the posterior 1/3 sagittal length; dorsal furrows sinuous and deep posteriorly; four pairs of glabellar furrows; proximal ends of S_1 and S_2 connected by deep longitudinal furrow; S_3 and S_4 short, isolated and pit-like, not connected with dorsal furrows distally. Occipital furrow narrow, deep laterally, shallow and bending forward medially. Occipital ring broad (sag., exsag), gradually narrowing distally. Fixigena convex, widest opposite to S_1 pair and dying out opposite to distal ends of S_3. Palpebral ridge oblique; palpebral lobe small, located at the midlength of glabella; surface smooth or with fine granules; frontal area and anterior area of fixigena broad (sag., exsag.); preglabellar field depressed, triangular; paradoublure line clear, oblique and near the palpebral ridges. Surface with two different sizes of granules. Border furrow distinct; border narrow (sag., exsag.) and convex; anterior margin rounded forward; anterior section of facial suture strongly divergent, posterior section of facial sutures, librigena, thorax and pygidium unknown.

Comparison: This genus is very similar to *Paraacidaspis* in the broad frontal area and anterior area of fixigena, in the prominent paradoublure lines on both sides of cranidium, in the waisted glabella and the shape of fixigena; but differs from *Paraacidapis* in having triangular glabella with fixigena, more divergent anterior section of facial suture and different glabellar furrows and glabellar lobes.

So far as I am aware, five species have been described and should be synonymized with the type species of this genus. The fragmentary cranidium of *A. acuta* Liu has the same cranidial characteristics of *A. vomerinus*. *Metaacidaspis beigongensis* Qiu, which is represented by a well preserved cranidium, should be a junior synonym of *A. vormerinus* because there is no distinguishable feature for separating these two species. However, *Paraacidaspis triangularis* Luo has an elongate triangular shape of glabella with fixigena, it is evident that this cranidium is preserved in shale and slightly compressed. Two fragmentary cranidia figured by Apollonov and Chugaeva (Apollonov et al., 1984) as *Paraacidaspis* sp., should undoubtedly be referred to *Archikainella vomerinus* Liu because the glabellar furrows, triangular glabella with fixigena and broader (sag., exsag.) granulated frontal area and anterior area of fixigena, are the same as in *A. vomerinus*.

Formation and locality: Uppermost Cambrian, Qiangongping, Fenghuang County, Hunan; Paoshan Formation (Fengshanian), western Yunnan; Tangcun Formation (Fengshanian), southern Anhui; Uppermost Cambrian, Malyi Kratau Range, Kazakhstan.

Paraacidaspis Poletaeva, 1963

Type species: *Paraacidaspis hunanica* Jegorova in Jegorova et al., 1963.

Diagnosis: Glabella waisted at the posterior 1/3 sagittal length, dorsal furrows sinuous and deep posteriorly. S_1 short and horizontal at level of the inward bending of dorsal furrows, proximally connecting S_2; S_3 and S_4 short and faint. Frontal lobe broadly rounded. Occipital ring broad (sag.), gradually narrowing distally. Fixigena widest opposite to S_1, and dying out opposite to the outer end of S_3. Eye ridge oblique and very short; eye lobe at the midlength of glabella. Frontal area broad (sag., exsag.), with fine striations and granules; paradoublural line distinct. Border furrow shallow, border narrow and convex. Librigena and thorax unknown. Fragmentary pygidium with a long and stout axial lobe.

Occurrence: Late Cambrian Changshanian Stage, Hunan and southern Anhui, China.

Paraacidaspis hunanica Jegorova, 1963

(Pl. 1, figs. 5–9)

1960 *Paraacidaspis hunanica*, Poletaeva, pl. 3, fig. 1.
1963 *Paraacidaspis hunanica*, Jegorova, pl. 12, figs. 6–8.
1965 *Paraacidaspis hunanica*, Lu et al., pl.. 130, figs. 15, 16.

1978 *Paraacidaspis hunanica*, Yang, pl. 13, figs. 13, 14. text-fig. 9a-b.
1982 *Paraacidaspis hunanica*, Liu, pl. 239, fig. 8.
1983 *Paraacidaspis hunanica*, Qiu in Qiu et al., pl. 88, fig. 8.

Holotype: Jegorova et al., 1963, pl. 12, fig. 6, cranidium from the Late Cambrian Changshanian Stage of Tingziguan, Chatian, Fenghuang County, Hunan. (Field No. 424; Cat. No. F312).

In 1963, Jegorova (in Jegorova et al., 1963, pp. 52–54, pl. 12, figs. 6–8) gave a full description of the genus *Paraacidaspis* Poletaeva, 1956, and the type species *Paraacidaspis hunanica* Jegorova, as well as the diagnosis of the family Eoacidaspididae Poletaeva, 1956. It seems necessary to translate the original diagnosis of the genus *Paraacidaspis* by Jegorova from Chinese into English as follows:

"Genus *Paraacidaspis* Poletaeva, 1956; type species: *Paraacidaspis hunanica* Jegorova (sp. nov.), middle part of the Upper Cambrian, Fenghuang, Hunan.

Diagnosis: Cranidium stout, more or less rectangular, with rounded anterior margin. Glabella large and convex, composed of the main portion of the cranidium, with 4–5 pairs of lateral glabellar furrows, by which the glabella is divided into 5–6 pairs of faint glabellar lobes; one pair of glabellar lobes circular or elliptical in outline. Frontal lobe of glabella broadly rounded. Fixigena narrow (tr.). Eye lobe small, near the posterior border. Occipital ring prominent, narrower (exsag.) laterally and broader (sag.) in the middle, with a small median tubercle. Occipital furrow broad (exsag., sag.) and deep laterally, slightly arched forward and shallow in the middle. Preglabellar field broad (exsag., sag.) and depressed, border narrow and roundly arched forward. Border furrow very shallow. Anterior section of facial suture long and divergent from the palpebral lobes, then gradually turning toward the frontal margin. Cranidium with fine granules and fine transverse striations.

Remarks: *Paraacidaspis* is a characteristic genus with biological significance. This genus was erected by O. K. Poletaeva in 1956 (Poletaeva, 1956, p. 162), but no description was given at that time. At present, as the generosity of Poletaeva, she placed her manuscript of the diagnosis and description of this genus and the photographs of *Paraacidaspis sibirica* Poletaeva at our disposal. The diagnostic characters of the family Eoacidapididae in Poletaeva's manuscript entitled "New material of the primitive representatives of the superfamily Odontopleuridae Prantel and Pribyl from the Cambrian of Asia" are as follows: The main characteristics of the family Eoacidaspididae are the broadly rounded anterior margin of the cranidium, broad (sag., exsag.) and depressed preglabellar field, well developed glabellar furrows, medium-sized glabellar lobes and well rounded and convex fixigena which is near the palpebral lobes. In comparison with all the materials at hand, we discussed with Poletaeva and designated *Paraacidaspis hunanica* Jegorova as the type species of the genus *Paraacidaspis*. It reveals that the Chinese species has the standard characteristics of this genus and more Chinese specimens have been obtained.

Paraacidaspis may be compared with genera of the superfamily Odontopleuridae, especially with *Eoacidaspis* Poletaeva (Poletaeva, 1956, p. 175) from the Upper Cambrian of Salair, Siberia. Poletaeva has pointed out that *Paraacidaspis* not only resembles *Eoacidaspis* but also keeps phylogenetic relationship within the two genera. However, when compared with these two genera, it is easy to find the differences, for instance, *Paraacidaspis* has broad (sag., exsag.) preglabellar field, and 4–5 pairs of glabellar furrows and 5–6 pairs of glabellar lobes, of which one pair is rounded in outline; while *Eoacidaspis* has 3–4 pairs of glabellar furrows and 4–5 pairs of glabellar lobes, of which one pair is lenticular in outline. Moreover, *Eoacidaspis* is also different from *Paraacidaspis* in the ridgelike anterior border and in the absence of preglabellar field. All those differences mentioned above enable us to establish the new genus *Paraacdiaspis*.

Occurrence: Upper Cambrian; Hunan of China, Siberia.

The translation of the original description of this species *Paraacidaspis hunanica* Jegorova (sp. nov.) (Jegorova et al., 1963, pl. 12, figs. 6-8) is as follows:

Holotype: Cranidium figured in pl. 12, fig. 6 from the middle part of the Upper Cambrian, Tingziguan Village, Fenghuang County, Hunan Province.

Diagnosis: Cranidium subrectangular, with broadly rounded anterior margin and slightly convex. Glabella large, with 5 pairs of glabellar furrows and 6 pairs of glabellar lobes. Fixigena about 1/4 the width of glabella. Palpebral lobe narrow and small. Preglabellar field broad (sag., exsag.) and depressed, with granules and transverse striations.

Material: Three well preserved specimens of cranidium.

Description: Cranidium large, rectangular, with broadly rounded anterior margin. Glabella rectangular and slightly elongate; frontal lobe of glabella convex, well rounded and highly convex along the sagittal line; posterior part of glabella more or less flat; maximum width of glabella located between two palpebral lobes, slightly contracted behind the maximum width of glabella, and slightly expanded near the base; with 5 pairs of glabellar furrows and 6 pairs of glabellar lobes; the posterior two pairs of glabellar lobes long and extending backward; 3rd pairs of glabellar lobes circular in outline, the remaining three pairs short and separated with a broader space in the middle, and without regular shapes. The clearness of glabellar furrows are variable; the anterior two pairs slightly and shallowly impressed and short with the distal ends not reaching the dorsal furrows; the middle two pairs connecting distally with the dorsal furrows; and the posterior two pairs not only connecting distally with the dorsal furrow, but also joining themselves proximally; a pair of small circular glabellar lobes enclosed within these furrows including the dorsal furrows. The posterior pair of glabellar furrows long and oblique backward. In spite of these glabellar furrows mentioned above, a small pit occurred on the glabellar front of a specimen. It is uncertain whether it represents the sixth pair of glabellar furrows. Dorsal furrows shallow in the anterior part, sinuous and deep posteriorly. Occipital furrow arched forward medially, and deep and straight laterially; occipital ring convex, with a prominent median node. Fixigena extremely narrow anteriorly, more or less confluent with the glabella, and suddenly expanded as a large convex elliptical node near the palpebral lobe. Palpebral lobe small, near the posterior border; palpebral ridge absent. Border very broad (exsag., sag.) and depressed. Anterior margin forward rounded, with small reticulated ornament formed by fine striations; border furrow obsoure. Anterior section of facial suture divergent, posterior section not preserved. Surface with fine granules.

Measurement of cranidium (in mm)

Length	17.0	25.5
Width between palpebral lobes		19.0
Length of glabella	10.5	16.0
Basal glabellar width	8.0	14.0
Frontal width of glabella	6.0	12.0
Length (sag.) of occipital furrow		4.0
Width of fixigena	2.0	3.5
Width of border	4.5	6.0

Discussion and comparison: So far only two species, *Paraacidaspis sibirica* Poletaeva and *P. hunanica* Jegorova, have been referred to this genus. The characteristic cranidium of this species may be summarized as follows:

1. Segmentation of glabella is clear;
2. With 5 pairs of glabellar furrows and 6 pairs of glabellar lobes;
3. Bended and uneven dorsal furrows;
4. Narrow and oblate fixigena;
5. Small palpebral lobes near the posterior margin;
6. Border broad (exsag., sag.) and depressed, and
7. With granules and fine striations.

Judging from the outline of cranidium, segmentation of glabella, peculiar glabellar lobes, relatively narrow fixigena and other features, this species can be compared with *Paraacidaspis sibirica*, but the latter differs from this species in the obscure glabellar furrows and lobes, in the shallow preglabellar field, and in having node on the fixigena. Surface ornaments of *P. sibirica* is unknown. From the morphologic characteristics mentioned above, it enables us to erect this new species, on which the genus *Paraacidaspis* Poletaeva is based.

Formation and locality: This species occurs in association with *Prochuangia* in the deep grey crystalline limestone from Tingziguan Village, Chatian, Fenghuang County, western Hunan.

Yang (1978) described and illustrated a new species of *Paraacidaspis, P. latilimbatus*, a fragmentary cranidium of *P.* sp., a cranidium of *P. hunanica* Jegorova, and an associated fragmentary pygidium being only doubtfully assignable to this species. The axial lobe of this fragmentary pygidium is long and stout, tapering gradually backward; two pairs of axial furrows and pleural furrows are visible. Yang's specimens show the well developed anterior cranidial border, long (sag., exsag.) preglabellar field, and well developed paradoublure line. This genus falls within the *Chuangia-Prochuangia* zone of the Changshanian Stage of Hunan.

Eoacidaspis and *Belovia* have the same type of cranidium and glabella, but the preglabellar field is very short (sag., exsag.) when compared with that of *Paraacidaspis hunanica*. Paradoublure line is the characteristic cranidial feature of *Paraacidaspis* but this feature is absent in both *Belovia* and *Eoacidaspis*. Judging from the illustration given by Poletaeva (1960, pl. 3, fig. 2), *Paraacidaspis sibirica* Poletaeva is too fragmentary to say anything definite. As mentioned above, only two species, *P. hunanica* Jegorova, 1963, and *P.*

latilimbatus Yang, 1978, are valid. *P. latilimbatus* Yang is based on only one cranidium which shows a slightly shorter glabella, the remaining features of cranidium are the same as those of *hunanica*. The publication of the genus *Paraacidaspis* Poletaeva should be in the year of 1963.

PS. Recently a monograph on "Classification and phylogeny of the trilobite order Lichida" has been published (Thomas, A. T. and Holloway, D. J., 1988. Phil. Trans. R. Soc. Land., B321, 179–262). Thomas and Holloway considered that the family Lichakephalidae Tripp, 1957, has priority because it was published five months earlier than Eoacidaspididae Poletaeva, 1957. Genera included within Lichakephalidae are *Lichakephalus* Sdzuy, 1955; *Lichakephalina* Antcygin in Varganov, 1973; *Acidaspidella* Rozova, 1963; *Eoacidaspis* Poletaeva, 1956; ?*Acidaspidina* Lazarenko, 1960; and ?*Brutonia* Thomas and Holloway, 1988. They followed Bruton (1983) in assigning *Acidaspides* Lermontova, 1951 to the family Odontopleuridae. The joined authors also considered that *Belovia* Poletaeva, 1956 may be a subjective synonym of *Eoacidaspis*. Judging from the holotype and paratype specimens of *E. salairica* Poletaeva (Poletaeva, 1956; Bruton, 1983) and the holotype specimen of *Belovia calva* Tchernysheva, 1956 (Poletaeva, 1956), *Eoacidaspis* and *Belovia* should be two valid genera. It seems suitable that the Ordovician Lichakephalidae Tripp, 1957 and the Cambrian Eoacidaspididae Poletaeva, 1957 may be considered at present as two taxonomic units of family or subfamily rank because more skeletal parts of both families are needed.

REFERENCES

Apollonov, M.K., Chugaeva, M.N. & Dubinina, S.V., 1984. Trilobites and conodonts from the Batyrbay section (uppermost Cambrian-Lower Ordovician) in Malyi Karatau Range. 1–16, Nauka Kazakh. S.S.R. (in Russian).

Bruton, D.L., 1983. Cambrian origins of the Odontopleurid trilobites. Paleontology, 26(4), 875–885.

Chugaeva, M.N. & Apollonov, M.K., 1982. The Cambrian-Ordovician boundary in the Batyrbaisai section, Malyi karatau Range, Kazakhstan, U.S.S.R. In: Bassett, M.G. and Dean, W.T. (eds.), The Cambrian-Ordovician boundary: section, fossil distribution and correlations. Nat. Mus. Wales, Geol. Ser. 3, Cardiff.

Jegorova, L.I., Xiang Liwen, Li Sanji, Nan Runsan and Guo Chenming, 1963. Cambrian trilobites fauna from Guizhou and western Hunan. Special Publication of Chinese Geological Academy, Ser.B, Stratigraphy and Paleontology, 3(1), 1–88 (in Chinese).

Kobayashi, T., 1966. The Cambro-Ordovician formations and faunas of South Korea, Pt. 10, Stratigraphy of Chosen Group of South Korea. Jour. Fac. Sci. Univ. Tokyo, sec. 2, 16(1), 1–84.

Kobayashi, T., 1967. The Cambro-Ordovician formations and faunas of South Korea, Pt. 10, Stratigraphy of the Chosen Group in Korea. Sect. C. The Cambrian of eastern Asia and other part of the continent. Jour. Fac. Sci. Univ. Tokyo, Sect. 2, 16(3), 381–535.

Liu Yiren, 1982. Trilobita. In: Paleontological Atlas of Hunan. Geol. Publ. House, Beijing (in Chinese).

Lu Yanhao, Zhang Wentang, Zhu Zhaoling, Qian Yiyuan and Xiang Liwen, 1965. Trilobites of China. Science Press, Beijing (in Chinese).

Lu Yanhao, Zhu Zhaoling, Qian Yiyuan, Lin Huanlin and Yuan Jinliang, 1982. Correlation chart of Cambrian in China with explanatory text. In: N.I.J.P. (ed.), Stratigraphic correlation chart in China with explanatory text. 38-54. Science Press, Beijing (in Chinese).

Luo Juilin, 1983. New finds of trilobites from late Cambrian in western Yunnan. In: Contr. Geol. Qinghai-Xizang (Tibet) plateau, 11, 1–24, Geol. Publ. House, Beijing (in Chinese).

Luo Huilin, 1984. Subdivision and correlation of the Cambrian System in southeastern Yunnan. Acta Geol. Sin., 58(2), 96–104.

Luo Huilin, 1985. Subdivision and correlation of the Cambrian System in western Yunnan. Geology of Yunnan, 1, 69–83 (in Chinese).

Qiu Hongan, Lu Yanhao, Zhu Zhaoling, Bi Dechang, Lin Tienri, Zhou Zhiyi, Zhang Quanzong, Qian Yiyuan, Ju Tienyin, Han Nairen and Wei Xiuzhe, 1983. Trilobita. In: Palaeontological Atlas of East China. 1, 28–254, Geol. Publ. House, Beijing (in Chinese).

Palmer, A.R., 1973. Cambrian trilobites. In: Hallam, A. (ed.), Atlas of Paleobiology, 3-11. Elsevier Scientific Publ. Co., Amsterdam.

Poletaeva, O.K., 1956. Trilobita. In: Material in Palaeontology, new families and genera. Trudy Vses. Nauchissled. Geol. Inst. (VSEGEI). Moscow, 12, 173–177.

Poletaeva, O.K., 1960. Material on the palaeontology and stratigraphy of western Siberia. Trudy SNIIGGIMS Leningrad, 8, 67–71 (in Russian).

Repina, L.N., 1981. Trilobite biostratigraphy of the Lower Cambrian. In: Short papers for the 2nd International Symposium on the Cambrian System. U.S. Geol. Surv. Open file report 81–743.

Robison, R.A., 1975. Middle Cambrian trilobite biostratigraphy of the Great Basin. Brigham Young Univ. Geol. Studies. 23(2), 93–110.

Rozova, A.V., 1968. Biostratigraphy and trilobites of the Upper Cambrian and Lower Ordovician of the northwest Siberian Platform. Otdel. Inst. Geol. Geofiz. Sibir. Moscow, 36, 1–195.

Rozova, A.V., 1984. Biostratigraphic zoning and trilobites of the Upper Cambrian and Lower Ordovician of the northwest Siberian Platform. Transactions of the Inst. Geol. & Geophysics, Acad. Sci. U.S.S.R., Siberian Branch, Vol.36, Amerind Publishing Co., Ltd., New Delhi.

Taylor, M.E. and Cook, H.E., 1975. Continental shelf and slope facies in the Upper Cambrian and Lower Ordovician of Nevada.

Brigham Young Univ. Geol. Studies, 23(2), 181–214.

Tchernysheva, N.E., 1960. New early Palaeozoic trilobites of eastern Siberia and Kazakhstan. In: Markovsk, B.P. (Ed.), New species of prehistoric plants and vertebrates of the U.S.S.R. Pt. 2. Trudy Vses. Nauch-issled. Geol. Inst. (VSEGEI).Moscow, 254–255 (in Russian).

Xiang Liwen, An Taixiang, Guo Zhenming, Li Sanji, Nan Runsan, Qian Yi, Sheng Xinfu, Yang Jialu, Yuan Kexin, Zhang Sengui, and Zhou Guoqiang, 1981. The Cambrian System of China. Stratigraphy of China. 4. Geol. Publ. House, Beijing (in Chinese).

EXPLANATION OF PLATES

Figs. 1–4. *Archikainella vomerinus* Liu

1. Cranidium, × 4, Cat. No. HITO455 (Holotype of *Metaacidaspis beigongensis* Qiu, In Qiu et al., 1983, pl.88, fig. 9). Upper Cambrian Tangcun Formation (Fengshanian), Beigong, Jingxian, southern Anhui.
2. Fragmentary cranidium, ×1.5, Cat. No.HT81106 (Holotype of *Archikainella acuta* Liu, In Liu, 1982, pl. 234, fig.4).
3. Fragmentary cranidium, × 2, Cat. No. HT81107 (Holotype of *Archikainella vomerinus* Liu. In Liu, 1982, pl. 234, fig. 5). Upper Cambrian (Fengshanian), Qiangongping, Fenghuang County, western Hunan.
4. Compressed cranidium, ×2, Field No. A-413-1, Cat. No. YF2041 (Holotype of *Paraacidaspis triangularis* Luo. In Luo, 1983, pl. 6, fig. 13).
   Upper Cambrian Baoshan Formation (Fengshanian), Liuxui, Shidian, western Yunnan.

Figs. 5–9. *Paraacidaspis hunanica* Jegorova

5. Cranidium, ×4, Cat. No. HITO451 (Qiu et al., 1983, pl. 88, fig. 8).
   Middle part of the Upper Cambrian (Changshanian), Beigong, Jingxian, southern Anhui.
6. Cranidium, × 2, Holotype, Field No. 424, Cat. No. F312, (Jegorova et al., 1963, pl. 12, fig. 6).

7–8. Cranidia, × 2 (Jegorova et al., 1963, pl. 12, figs. 7, 7a).

9. Cranidium, × 2 (Jegorova et al., 1963, pl. 12, fig. 8). Middle part of the Upper Cambrian (Changshanian), Tingziguan, Chatian, Fenghuang County, western Hunan.

Plate 1

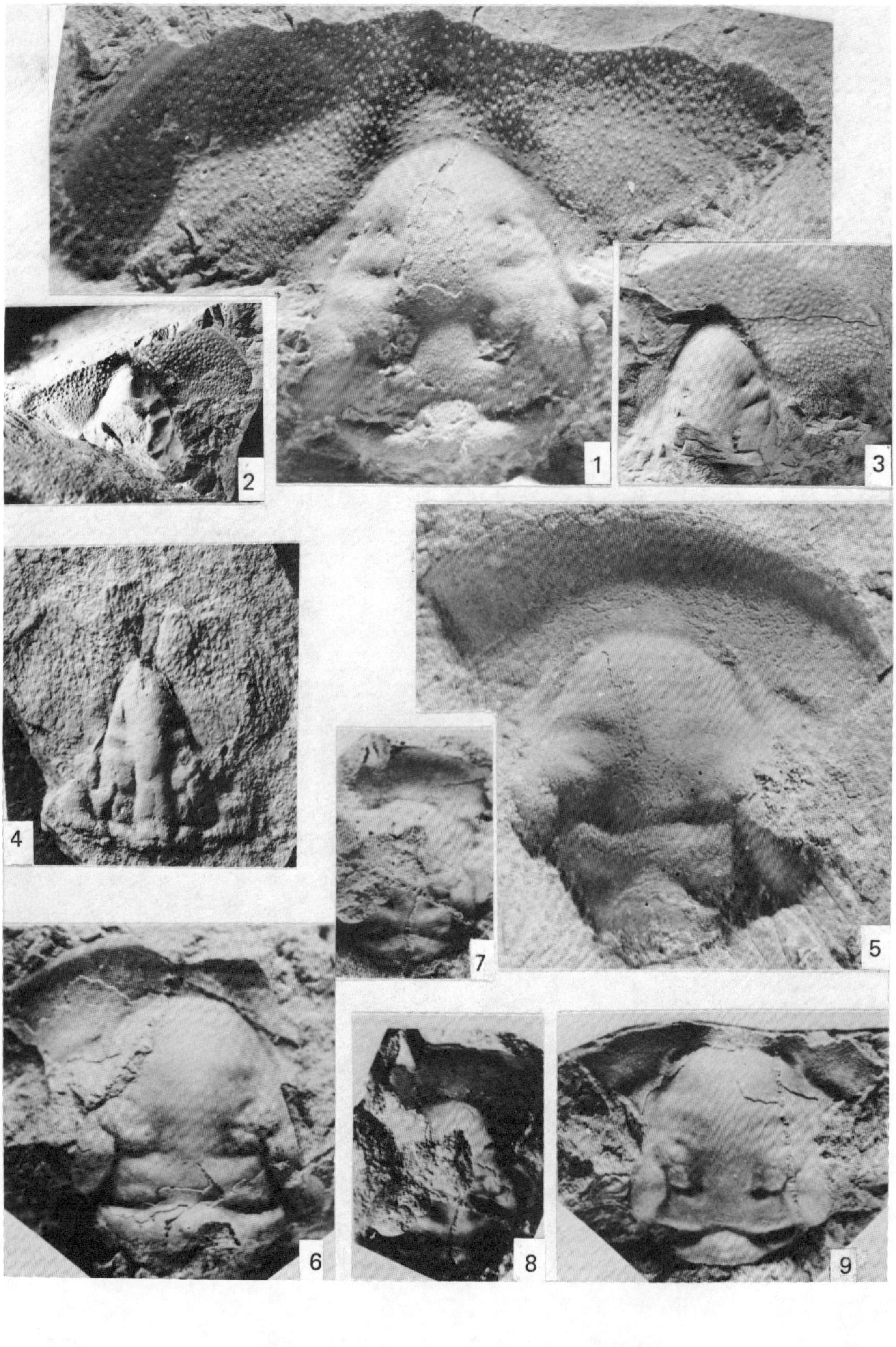

PALAEONTOLOGIA CATHAYANA Volume 5, 1990, p.181–204

PERMIAN SPORES AND POLLEN FROM KARAMAY REGION OF XINJIANG, CHINA

Zhang Lujin

(*Nanjing Institute of Geology and Palaeontology, Academia Sinica*)

Abstract

The present paper deals with palynological dating of the Baikouquan Formation in Karamay region, northwestern Junggar Basin, Xinjiang. The spores and pollen include 29 genera and 50 species, of which 5 species are new. The palynological assemblage is closely similar to that from the Upper Jijicao Formation of southern Junggar Basin. The evidence of palynology tends to support the Permian age (Lower Permian to Lower Upper Permian) for the Upper Jijicao Formation as well as the Baikouquan Formation.

INTRODUCTION

One of the most representative formations of Late Palaeozoic in Karamay region is the Urho Group. It is divided into two stratigraphical units. The lower part of Urho Group is named the Huangyangquan Formation. The upper and middle parts are called the Baikouquan Formation. Both the formations are composed mainly of conglomerates, intercalated with thin-layer mudstones. Most of the rocks are brown-red in colour. Only a small amount of fossils were collected from the Baikouquan Formation and lack sufficient evidences for the age determination. The list is as follows:

Fossil plant: *Cordaites* sp.

Spores and pollen: *Circella* sp., *Cycas* sp., *Ginkgo* sp., *Coniferales* sp., *Podocarpus* sp., *Protopinus* sp., *Spinasellea* sp., *Bennettitales* sp.

From 1979 on, the present writer went to northern Xinjiang five times for palynostratigraphical investigation. For this study, a great number of palynological samples from twenty five drillholes have been examined, and a conclusion has been drawn by the present writer for the first time and was being reported in the Karamay Academic Symposium in 1982.

The materials under study are obtained from the drill hole No. Xa-10, near the Xazgat village, about 110 km northeast of Karamay. The palynological materials from drill hole No. Xa-10 are the best of all in this region. The Baikouquan Formation is 1600m in thickness. It is composed mainly of conglomerates intercalated with brownish-red thin-layer mudstones in the upper part, suggesting the deposits under the oxidized environment. In the middle and lower parts, the formation consists mainly of conglomerates intercalated with dark and dark brown thin-layer mudstones and carbonaceous mudstones, coal-lines, pyrite, and plant fragments, showing a limited reduced environment. Spores and pollen were derived just from the limited reduced deposits of the formation. Overlying this formation is the Upper Cangfanggou Group. The underlying formation is invisible, because the drilling stopped at the base of the Baikouquan Formation.

CHARACTERS AND AGE OF THE PALYNOLOGICAL ASSEMBLAGE

Fourteen samples were taken from the drill hole section, four of them being productive (Nos. 2, 8, 10, 12). The palynological forms are given below: *Granulatisporites absonus* Foster, *Apiculatisporis* sp., *Apiculatisporis bulliensis* Helby et de Jersey, *Acanthotriletes filiformis* (Balme et Hennely) Tiwari, *Lophotriletes rec-*

tus Bharadwaj et Salujha, *Leiotriletes* sp., *Endosporites eminens Leschik*, *Endosporites sinensis* sp. nov., *Cycadopites linearis* sp. nov., *Cycadopites altilis* Zhang, *Cycadopites labrosus* Zhang, *Cycadopites* sp., *Crucisaccites xinjiangensis* Zhang, *Crucisaccites urumqiensis* Zhang, *Crucisaccites quadratus* sp. nov., *Crucisaccites* sp., *Tumoripollenites baculatus*? Bharadwaj, *Hamiapollenites humilis* Zhang, *Hamiapollenites gracilis* Zhang, *Hamiapollenites impolitus* Zhang, *Hamiapollenites* sp., *Protowelwitschiapollis exolescus* Zhang, *Protowelwitschiapollis orcula* Zhang, *Protowelwitschiapollis margelis* Zhang, *Protowelwitschiapollis grandis* Zhang, *Protowelwitschiaopollis elagetis* Zhang, *Protowelwitschiapollis* sp., *Protowelwitschiapollis*? sp., *Costapollenites*? sp., *Vittatina multistriata* (Balme et Hennelly) Anderson, *Striatopodocarpites* spp., *Vesicaspora obliqua* Singh, *Vesicaspora* sp., *Florinites mediapudens* (Loose) Potonie et Kremp, *Citrullesaccites badaowanensis*? Zhang, *Citrullesaccites* sp., *Stellapollenites*? sp., *Crustaesporites gracilis* Zhang, *Aganosaccites sinensis* Zhang, *Potonieisporites magnus* Lele et Karm, *Szeisaccites*? sp. and *Protohaploxypinus*? sp.

The palynological assemblage of the Baikouquan Formation is characterized by the abundance of striated disaccate pollen. The main genera are *Hamiapollenites, Vittatina, Striatopodocarpites, Protowelwitschiapollis* and *Sinosaccites*. A number of evolutive monosaccate pollen grains occupy the second place, such as *Virkkipollenites, Parasaccites, Crucisaccites, Corisaccites* and *Potonieisporites*. In the present assemblage, the Pteridophyta only plays a subsidiary role. The representative genera is *Endosporites* accompanied with *Apiculatisporites*, *Acanthotriletes*, *Granulatisporites, Lophotriletes* and *Leiotriletes*.

In the aspects of palynological assemblage, the Baikouquan Formation is strikingly comparable to the Upper Jijicao Formation in southern Junggar Basin, Xinjiang. The following species are common in both assemblages: *Granulatisporites absonus* Foster, *Acanthotriletes filiformis* (Balme et Hennely) Tiwari, *Leiotriletes* sp., *Endosporites sinensis* sp. nov., *Cycadopites labrosus* Zhang, *Crucisaccites xinjiangensis* Zhang, *Crucisaccites urumqiensis* Zhang, *Virkkipollenites* spp., *Hamiapollenites humilis* Zhang, *Hamiapollenites gracilis* Zhang, *Hamiapollenites impolitus* Zhang, *Protowelwitschiapollis exolescus Zhang*, *Protowelwitschiapollis*

Table 1. Comparison of the Permian stratigraphical sequences between southern and northwestern regions of Junggar Basin.

<table>
<tr><th></th><th colspan="2">Southern region</th><th colspan="2">Northwestern region</th></tr>
<tr><td>Triassic</td><td colspan="2">Upper Changfangou Group
487.3m</td><td colspan="2">Upper Changfangou Group
710m</td></tr>
<tr><td rowspan="7">Permian</td><td rowspan="2">Lower Chang-fangou Group</td><td>Wutonggou Formation
399.2m</td><td rowspan="7">Urho Group</td><td rowspan="6">Baikouquan Formation
419-1600m</td></tr>
<tr><td>Quanzhijie Formation
250.1m</td></tr>
<tr><td rowspan="4">Upper Jijicao Group</td><td>Hongyanchi Formation
733.7m</td></tr>
<tr><td>Lucaogou Formation
659.6m</td></tr>
<tr><td>Jingjingzhigou Formation
576.8m</td></tr>
<tr><td>Wulapo Formation
1301.8m</td></tr>
<tr><td colspan="2">Lower Jijicao Group
1872.2m</td><td>Huangyangquan Formation
749.5m</td></tr>
<tr><td colspan="3">Marine Carboniferus deposits</td><td colspan="2">Carboniferus</td></tr>
</table>

orcula Zhang, *Protowelwitschiapollis margelis* Zhang, *Protowelwitschiapollis grandis* Zhang, *Vittatina multistriata* (Balme et Hennely) Anderson, *Striatopodocarpites* spp., *Aganosaccites sinensis* Zhang, *Potonieisporites magnus* Lele et Karm. and *Sinosaccites* sp.

It is of interest to note that some Gondwana palynomorphs have been found in the northwestern region of Junggar Basin, such as *Virkkipolenites, Papasaccites, Crucisaccites, Hamiapollenites* and *Vittatina*. Moreover, the palynological forms from the Baikouquan Formation and the Upper Jijicao Formation bear a considerable resemblance to those of the Gondwana Permina formations, i. e. Karharbari Stage of the Girdih Coalfield (Lele et Maithy, 1964); Barakar Stage of the Koiba Coalfield (Tiwari, 1965), Raniganj Stage of the Raniganj Coalfield (Bharadwaj, 1962; Bharadwaj et Salujha, 1964), and also to those of the American Permian formations, i. e. the Lower Permian (Wolfcampian Series) of Texas (Tschudy et Kosake, 1966), the Wellington Formation of Oklahoma (Hedlund, 1965), the Flowerpot Shale Formation of Oklahoma (Wilson, 1962), and even to those of the Kungurian Formation of Aktubinsk region of Urals, USSR (Samoilovich, 1953).

From the above statement, the writer comes to the following conclusion:

1. The Baikouquan Formation in Karamay region, belonging to the Urho Group is very well comparable to the Upper Jijicao Formation in the southern Junggar Basin.
2. The Baikouquan Formation and Upper Jijicao Formation may be regarded as Permian in age, or may be range from the Lower Permian to the lower part of the Upper Permian.
3. According to the palynological characteristics, the writer is inclined to believe that the southern and the northwestern region in Junggar Basin of Xinjiang might be an important place for linking the floras between the Euramerica and the Gondwanaland during the Late Palaeozoic time.

DESCRIPTION OF MAIN FORMS OF SPORES AND POLLEN GRAINS

Pteridophyta-Incertae Sedies

Granulatisporites (Ibrahim) Potonie et Kremp 1954

Granulatisporites absonus Foster

(Pl. 1, fig. 1)

1962 cf. *Lophotriletes* sp. Bharadwaj, pl. 1, figs. 29-30.
1979 *Granulatisporites absonus* Foster, p. 31, pl. 2, figs. 1–9.
1983 *Granulatisporites absonus* Foster, Zhang, pl. 1, figs. 2–10.

Size range: 29–37 μm

Description: Trilete microspores, convexly subtriangular in equatorial outline. Trilete rays distinct, extending almost to the equator, accompanied by thin, narrow labra. Exine ca. 1.5 μm thick, comprehensively sculptured with small varying proportions of grana and coni.

Comparison: The specimens discovered from Baikouquan Formation and Upper Jijicao Formation of Xinjiang, China agree fairly well with the species *Granulatisporites absonus* described by Foster (1979) from Middle Permian of southeast Queensland.

Acanthotriletes (Naumova 1939 ex 1949) Potonie et Kremp1954

Acanthotriletes filiformis (Balme et Hennely) Tiwari

(Pl. 1, figs. 3–4, 8–9, 12)

1956 *Apiculatisporites filiformis* Balme et Hennely, pl. 2, fig. 22.
1965 *Acanthotriletes filiformis* (Balme et Hennely) Tiwari, P. 173, pl. 1, figs. 19, 20.
1983 *Acanthotriletes* sp. Zhang, pl. 1, figs. 11–12.

Size range: 40–50 μm.

Description: Microspores, subroundish to roundly subtriangular in equatorial outline. Trilete rays slender or indistinct, extending near to the equator. Exine covered with 2.6 μm long and 1–2 μm wide at base, sparse spines.

Comparison: In regard to the shape of spores and the ornamentation of exine, the specimens described here from Xinjiang, China are strikingly alike to the species *Acanthotriletes filiformis*.

Endosporites Wilson et Coe 1940

Endosporites eminens Leschik

(Pl. 8, figs. 1–5)

1959 *Endosporites eminens* Leschik, p. 61, pl. 2, fig. 18.

Total size range: 100–121 μm ×125–156 μm.

Size range of central body: 75–88 μm ×90–120 μm.

Description: Trilete microspores, large in size, roundish triangular to elliptical in equatorial outline. Central body approx. roundish to round triangular. Exine two-layered. Sexine sac-like, microreticulate, covering the whole central body except a large part of its proximal face. Trilete rays simple and gracile, reaching to the equator of the central body and sometimes slightly extending to the reticulate sexine on the proximal area.

Comparison: The present figured five specimens from Baikouquan Formation are identical in all respects with the *Endosporites eminens* described by Leschik from Karroo Sandstone (Lower Permian) of southwestern Africa. It is highly possible that there can be no doubt of their specific identity.

Endosporites sinensis sp. nov.

(Pl. 8, figs. 6–7)

1983 *Lycopodiacidites* sp. Zhang, pl. 1, figs. 17–19.

Total size range: 50–68 μm ×53–96 μm, occasionally 96×96 μm.

Size range of central body: 33–42 μm × 33–60 μm, occasionally 60×60 μm.

Description: Trilete microspores, elliptical or approx. roundish in equatorial outline of the spores as well as the central body. Exine two-layered . Sexine sac-like, microreticulate, covering the whole central body except a large part of its proximal surface. Trilete rays simple and slender, reaching to the equator of the central body and sometimes slightly to the reticulate sexine in the proximal area.

Remarks: By the small size and approx. roundish outline, the present species is distinguished from all other known species of the genus *Endosporites*.

Holotype: Pl. 8, fig. 6, No. Xa-10-2(4).

Spermatophyta

Gymnospermae

Cycadoales

Cycadaceae

Cycadopites (Wodehouse 1933) ex Wilson et Webster 1946

Cycadopites linearis sp. nov.

(Pl. 1, figs. 17–18, 24)

Size range: 15–27 × 32–38 μm.

Description: Monosulcate, approx. elliptical in equatorial outline. Ends of long axis of pollen more or less truncate. Distal sulcus as long as the axis, narrow at the center and wide at the ends. Exine of pollen ornamented with longitudinal lineation.

Remarks: By the longitudinal lineation, the present species is readily distinguished from all other forms of the genus *Cycadopites*.

Holotype: Pl. 1, fig. 18, No. Xa-10-12(14).

Gymnospermae (? Pteridospermae) - Incertae Sedis

Virkkipollenites Lele 1964

Virkkipollenites aff. *corius* Bose et kar

(Pl. 2, figs. 6–7, 9, 12–13)

Size range: 42–60 μm.

Description: Monosaccate pollen, subcircular-circular in equatorial outline. Central body lighter than saccus, generally intramicroreticulate, sometimes imperfectly intramicroreticulate or granulose. Trilete rays obscure in most specimens, when being ill-developed, extending to about 1/2 the length of radius. Proximal attachment of saccus to central body equatorial, distal attachment subeqatorial, attachment zone more or less indistinct. Saccus interreticulate, well developed, frilled in margin due to numerous, small radial folds.

Remarks: According to the aspects of the pollen and the characteristic radial folded structure of the saccus, the present specimens bear a striking resemblance to the African species *Virkkipollenites corius* described by Bose and Kar (1966, p. 79, pl. 21, figs. 1, 2) from the lowermost part of Lower Permian deposits of Walikale region, Congo. The only point of difference is that the Chinese specimens are smaller in size. The present writer is inclined to believe that the specific affinity between the specimens from Congo and northern Xinjiang, China is rather certain.

Virkkipollenites sinensis sp. nov.

(pl.2, figs.8, 10–11,14,16–17)

Size range: 47–63 μm.

Description: Monosaccate pollen, subcircular to slightly ovalish in equatorial outline. Central body tuberculate to granulate. Trilete rays visible or obscure, extending near to the equator of the central body. Saccus attached proximally at the equator of central body, distally along a narrow subequatorial zone. Distal saccus attachment not associated with a body infold system. Saccus reticulate, always folded radiately in equatorial region.

Comparison: In regard to the aspects of pollen, the present specimens can be compared with the species described by Lele as *Virkkipollenites mehtae* (1963, p. 159–160, pl. 2, fig. 16) from the Talchir Beds of India. The one point of difference is that the ornamentation of the central body of our specimens is rather rougher than those of the specimen described by Lele.

Holotype: pl. 2, fig. 8, No. Xa-10-12 (22).

Parasaccites Bharadwaj et Tiwari 1964

Parasaccites bilateralis Tiwari

(Pl. 2, fig. 23)

1964 *Parasaccites bilateralis* Bharadwaj et Tiwari, pl. 2, fig. 12.
1965 *Parasaccites bilateralis* Tiwari, pl. 4, fig. 73.

Size: 67×96 μm.

Description: Monosaccate pollen, bilaterally oval in equatorial outline. Central body subcircular, faintly visible. Trilete rays weak. Zone of saccus attachment subequatorial on both faces of the central body. Succus finely interreticulate, with more or less radial arrangement of muri. In regard to the shape of the pollen and the para-condition of saccus attachment.

Comparison: The present specimen is identical perfectly well with *Parasaccites bilateralis* described and figured by Tiwari (1965) from the Barakar Stage of W. Bakaro Coalfield of India.

Parasaccites karamayensis sp. nov.

(Pl. 2, figs. 20–22)

1966 *Parasaccites* sp. Navale et Tiwari, pl. 1, figs. 9–11.

Size range: 24–83 μm.

Description: Monosaccate pollen, circular or subcircular in equatorial outline. Saccus attached subequatorially both on proximal as well as on distal surface of the central body, leaving subcircular, equal bladder-free area on both faces, i.e. para-condition of saccus attachment. Central body distinct, reticulate. Trilete rays simple and gracile, extending to about 2/3 the length of central body radius. Saccus reticulate, more or less folded radiately in equatorial region.

Comparison: The Present described species bears a considerable resemblance to the species described by Tiwari under the name of *Parasaccites bilateralis* from Barakar Stage of Lower Gondwana. The main difference is that the central body of present species is ornamented by rough reticulation and the trilete rays are visible; the exine of *Parasaccites bilateralis* is thin and the trilete rays are not visible. Holotype: Pl. 2, Fig.22, No. Xa-10-2(10).

Crucisaccites Lele et Maithy 1964

Crucisaccites xinjiangensis Zhang

(Pl. 1, figs. 28–29, 31)

1983 *Crucisaccites xinjiangensis* Zhang, p. 331, pl. 2, figs. 19, 20, 26.

Size renge of pollen: 48-62 μm.
Size range of central body: 30–40 μm

Description: Monosaccate, subcircular to circular in equatorial outline, with crucial sulcus. The sulcus with thickened lips. Sulcus interreticulate. Central body smooth to scabridulous. Several specimens of this species are discovered originally from Lucaogou Formation of southern Junggar Basin of Xinjiang, China (Zhang, 1983). The present material is more abundant than before. The supplementary paratype seems to be necessary.

Holotype: See Zhang 1983, Pl. 2, fig. 26.
Paratype: Present paper, Pl. 1, fig. 29. No. Xa-10-12(1).

Crucisaccites urumqiensis Zhang

(Pl. 1, figs. 34–37)

1983 *Crucisaccites urumqiensis* Zhang, p. 332, pl. 2, figs. 22, 23, 25.

Size range of pollen: 30–47 μm × 42–74 μm.
Size range of central body: 24–45 μm ×28–55 μm.

Description: Monosaccate to primeval disaccate, oval to fusiform in equatorial outline, with crucial sulcus. Sulcus without lips. Central body faintly traceable, scabridulous. Saccus intermicroreticulate. This species is distingushed from *Crucisaccites xinjiangensis* by its characteristic oval to fusiform outline and its simple crucial sulcus. Some specimens of this species are described originally by Zhang (1983) from Lucaogou Formation of southern Junggar Basin of Xinjiang, China. The specimens studied here are preserved better than those from southern Junggar Basin.

Holotype: See Zhang 1983, Pl. 2, fig. 22.
Paratype: Present paper, Pl. 1, fig. 35, No. Xa-10-12(11).

Crucisaccites quadratus sp. nov.

(Pl. 3, figs. 1–4, 6–7, 17)

1964 *Crucisaccites* sp. Lele et Maithy, pl. 1, fig. 7.
1983 Crucisaccites ? sp. Zhang, pl, 2, figs. 28, 29.

Size range: 47–70 μm × 72–87 μm.

Description: Monosaccate pollen, roundish quadrate in equatorial outline. Central body distinct to indistinct, subcircular to elliptical. Saccus encloses most of the central body, but leaves a quadrate area on the central part of the body. Central body alete, granulate to microreticulate. Saccus reticulate and always folded at the four parts of roundish angles.

Remarks: This species differs from all other known species in the quadrate outline and the folded parts of the angles.

Holotype: Pl. 3, fig. 3, No Xa-10-10(31).

ACKNOWLEDGEMENTS

The writer wishes to express her deep gratitude to Karamay Institute of Xinjiang Petroleum Administration and its chief engineers Song Hanliang, Shang Zhenping, Zhang Guojun Wang Aimin, Xia Minsheng and Zhu Baisheng for their guidance and help in the investigation of this region.

REFERENCES

Adreeva, E. M., Luber, A. A. and Sedova, M. A., 1960. Spore and pollen complexes of Middle and Upper Paleozoic of the USSR. Int. Geol. Cong, 21 Ses., Report of Soviet Geologists, Problem 6. State Publishing Office of Scientific and Technical Literature in Geology and Conservation of Mineral Wealth. Moscow, 189–195 (in Russian).

Balme, B. E., 1969. The Permian-Triassic boundary in Australia. Spec. Publ. Geol. Soc. Aust., 2, 99–112.

_______, 1970. Palynology of Permian and Triassic strata in the Salt Range and Surghar Range, West Pakistan. In: Kummel, B. and Teichert, C. (eds.), Stratigraphic boundary problems: Permian and Triassic of West Pakistan. Univ. Kansas, Spec. Publ., 4, 306–453.

_______ and Helb, R. J., 1973. Floral modifications at the Permian-Triassic boundary in Australia. In: Loban, A. and Hills, L. V. (eds.), The Permian and Triassic Systems and their mutual boundary. Can. Soc. Petrol. Geol. (Mem. 2), Calgary, Alberta, 433–443.

_______ and Playford, G., 1967. Late Permian plant microfossils from the Prince Charles Mountains, Antarctica. Rev. Micropaleont. 10(3), 179–192.

Bharadwaj, D. C., 1962. The miospore genera in the coals of Raniganj Stage (Upper Permian), India. Palaeobotanist, 9(1–2), 68–106.

_______ and Salujha, S. K., 1964. Sporological study of seam 8 in Raniganj Coalfield, Bihar (India) -Part 1. Description of sporae dispersae. Palaeobotanist, 12(2), 181–215.

_____________, 1965. Asporogical study of seam 7 (Jote Dhem Colliery) in the Raniganj Coalfield, Bihar (India). Palaeobotanists, 13(1), 30–41.

________ and Tiwari, R. S., 1964. The correlation of coalseams in Karba Coalfield, Lower Gondwanas, India. C. R. 5th Congr. Int. Strat. Geol. Carb., Paris (1963), 1131–1143.

_____________, 1964. On two monosaccate genera from Barakar Stage, India. Palaeobotanist, 12(2), 139–146.

_____________, 1966. Sporological correlation of coalseams in Bachra area of North Karanpura Coalfield, Bihar, India. Palaeobotanist, 15(1–2), 1–10.

Bose, M. N. and Kar, R. K., 1966. Palaeozoic sporae from Congo 1. Kindu-Kalima and Walikale regions. Ann. Mus. Ro. Africa Centr., Ser. 80, 53, 1–168.

Foster, C. B., 1974. Stratigraphy and palynology of the Permian at Waterloo Bay, Yorke Peninsula, South Australia. Trans. Roy. Soc. S. Aust., 98(1), 29–42.

_______, 1975. Permian plant microfossils from the Blair. Palaeontographica B, 154(5–6), 121–171.

________, 1979. Permian plant microfossils of the Blair Athol Coal Measure, Baralaba Coal Measures, and Basal Reman Formation of Queensland. Geol. Surv. Queensland Publ., 372, 1–154.

Grebe, H. and Schweizer, H. J., 1964. Die Sporae dispersae des niederrheinischen Zechsteins. Fortschr. Geol. Rheinld Westf., 12, 201–224.

Hart, G. F., 1960. Microfloral investigations of the Lower Coal Measures (K2); Ketewaka–Mchuchuma Coalfield, Tanganyika. Bull. Geol. Surv. Tanganyika, 30.

Hedlund, R. W., 1965. Palynological assemblage from the Permian Wellington Formation, Noble County, Oklahoma. Oklahoma Geol. Notes, 25(8), 236–241.

Helby, R., 1966. Sporologische Untersuchungen an der Karbon-Perm Grenze im Pfäler Bergland. Fortschr. Geol. Rheinld. Westf., 13, 645-704.

Helby, R. J., 1968. The Carboniferous-Permian boundary in eastern Australia: an interpretation on the basis of palynological information. Spec. Publ. Geol. Soc. Aust., 2, 69–72.

de Jersey, N. J., 1979. Palynology of the Permian-Triassic transition in the western Bowen Basin, Geol. Surv. Queenland Publ., 374, 1–61.

Kaiser, H., 1976. Die Permische mikro-flora der Cathaysia-schichten von Nordwest-Schansi, China. Palaeontolographica B, 159(4–6), 83–157.

Kemp, E. M., Balme, B. E., Helby, R. J., Kyle, R. A., Playfor, G. and Price, P. L., 1977. Carboniferous and Permian palynostratigraphy in Australia and Antarctica–a review. BMRJ. Aust. Geol. Geophys., 2, 177–208.

Lele, K. M., 1964. Studies in the Talchin Flora of India: 2. Resolution of the spore genus *Nuskoisporites* Pot. et Kl. Palaeobotanist, 12(2), 147–168.

_______ and Maithy, P. K., 1964. An unusual monosaccate spore from the Karharbari Stage, Giridih Coalfield, India. Palaeobotanist, 12(3), 307–312.

Leschik, G., 1956. Sporen aus dem Salzton des Zechsteins von Nouhof (bei Fulda). Palaeontographica B, 100, 122–142.

_______, 1959. Sporen aus den "Karrn-Sandsteinen" von Narronaub (Sudwest-Afrika). Senckenbergiana Lethaea, 40(1/2), 51–95.

Maheshwari, H. K., 1966. Studies in the *Glossopteris* flora of India-29. Miospore assemblage from the Lower Gondwana exposures along Bansloi River in Rajmahal Hill, Bihar. Palaeobotanist, 15(3), 258–280.

Maithy, P. K., 1965. Studies in the *Glossopteris* flora of India–27. Sporae dispersae from the Karharbari Beds in the Giridih coalfield, Bihar. Palaeobotanist, 13(3), 291–307.

Rigby, J. F., and Hekel, H., 1977. Palynology of the Permian sequence in the Spingsure Anticline, Central Queensland. Publ. Geol.

Surv. Queensland, 363, Paleont. Pap., 37.
Salujha, S. K., 1965. Miospore assemblage of Seam 9 of East Raniganj Coalfield (India). Palaeobotanist, 13(3), 227–238.
Samoilovich, S. R., 1953. Pollen and spores from the Permian deposits of the Cherdynsk and Aktyubinsk Pre-Urals. Trudy VNIGRI, 75, 5–92.
Segroves, K. L., 1969. Saccate plant microfossils from the Permian of Western Australia. Grana Palynol., 9(1–3), 17–227.
Tiwari, R. S., 1965. Miospore assemblage in some coals of Barakar Stage (Lower Gondwana) of India. Palaeobotanist, 13(2), 168–214.
Tschudy, R. H. and Kosanke, R. M., 1966. Early Permian vesiculate pollen from Texas, U. S. A. Palaeobotanist. 15(1–2), 59–71.
Venkatachala, B. S. and Kar, R. K., 1966. *Corisaccites* gen. nov., a new saccate pollen genus from the Premian of Salt Range, West Pakistan. Palaeobotanist, 15(1–2), 107–109.
Wilson, L. R., 1962. Permian plant microfossils from the Flowerpot Formation, Green county, Oklahoma. Circ. Okla. Geol. Surv., 49.
Zhang Lujin, 1983a. On the characters of Permian microflora in the Junggar Basin of Xinjiang. Palaeontologia Cathayana, 1, 327–365.
Zhang Lujin, 1983b. On the Age of Badaowan Formation in northern Xinjiang. Scientia Sinica(b), 26(7), 774–784.

EXPLANATION OF PLATES

All palynological samples are derived from the Baikouquan Formation in Karamay region of northwestern Junggar Basin of Xinjiang, China, and macerated by He Cuiling. All the figures are photographed by Fan Xiaoyi and the writer. Magnifications are × 600, unless otherwise stated. The original specimens are kept in the Nanjing Institute of Geology and Palaeontology, Academia Sinica.

Plate I

(Figs. 1–14, 16–38 ×600; Fig. 15 × ca.200)

1. *Granulatisporites absonus* Foster
2,6,7. *Apiculatisporis* sp.
3,4,8,9,12. *Acanthotriletes filiformis* (Balme et Hennely) Tiwari
5, 10–11. *Apiculatisporis bulliensis* Helby et de Jersey
13. *Lophotriletes rectus* Bharadwaj et Salujha.
14. *Alatisporites* ? sp.
15. *Endosporites* sp.
16. *Leiotriletes* sp.
17,18,24. *Cycadopites linearis* sp.nov.
19,20. *Cycadopites altilis* Zhang
21–23. *Cycadopites labrosus* Zhang
25–27. *Cycadopites*? spp.
28,29,31. *Crucisaccites xinjiangensis* Zhang
30,32,33. *Crucisaccites*? sp.
34–37. *Crucisaccites urumqiensis* Zhang
38. *Monosulcites* sp.

Plate 1

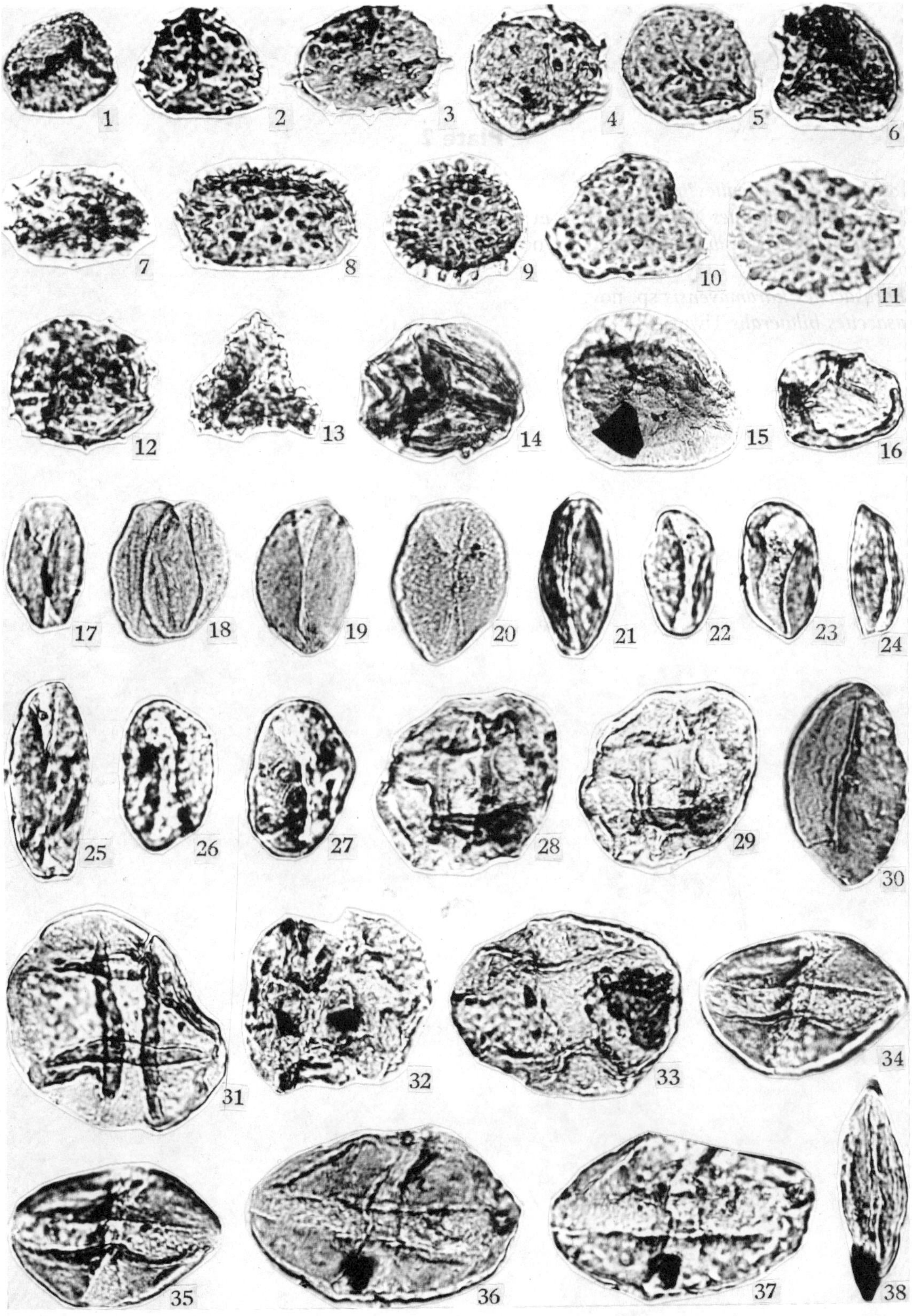
1
2
3
4
5
6
7
8
9
10
11
12
13
14
15
16
17
18
19
20
21
22
23
24
25
26
27
28
29
30
31
32
33
34
35
36
37
38

Plate 2

1–5,15,18,24. *Virkkipollenites*? spp.
6,7,9,12,13. *Virkkipollenites* aff. *corius* Bose et Kar.
8,10,11,14,16,17. *Virkkipollenites sinensis* sp. nov.
19. *Sinosaccites*? sp.
20–22. *Parasaccites karamayensis* sp. nov.
23. *Parasaccites bilateralis* Tiwari.

Plate 2

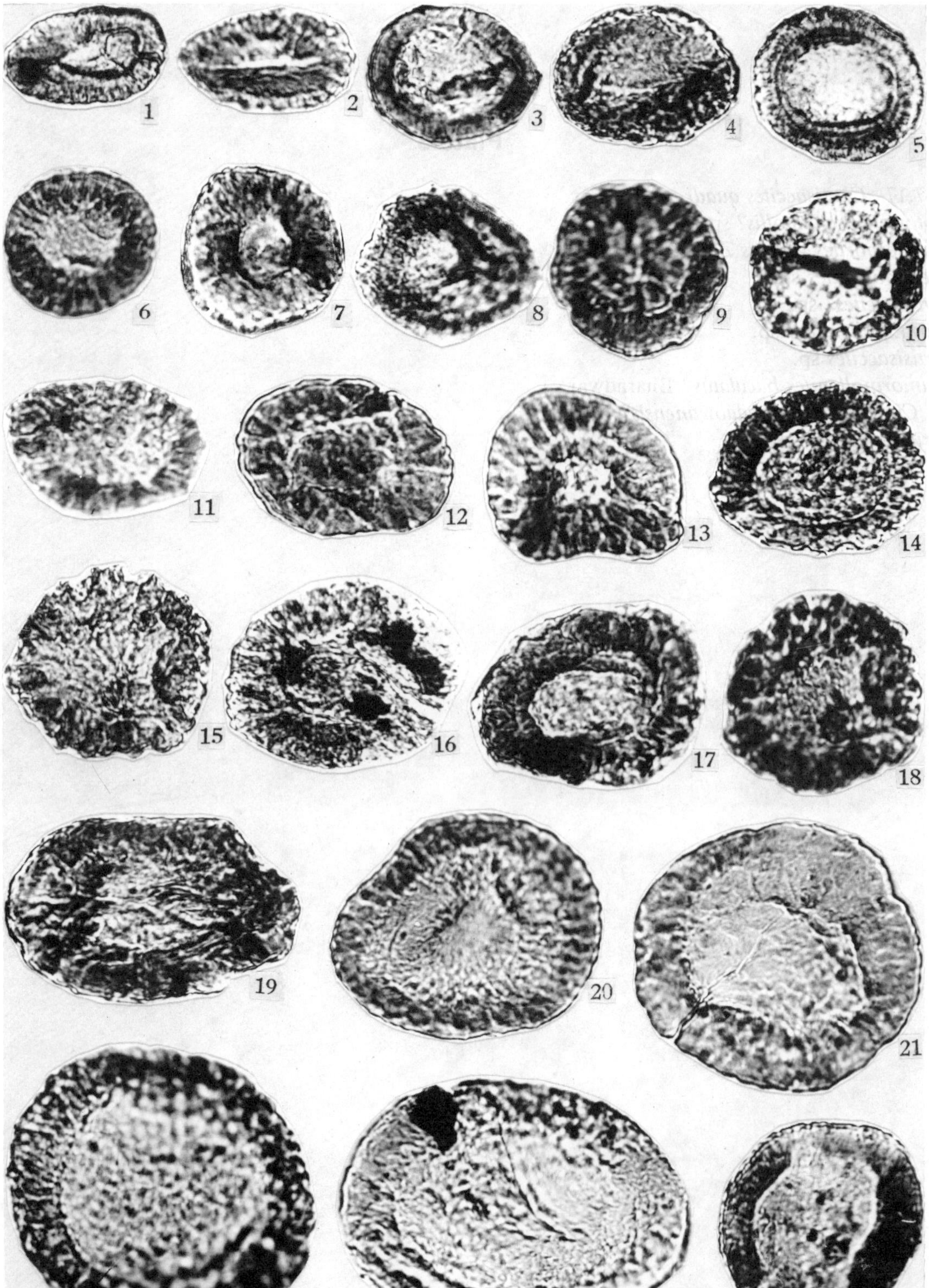

Plate 3

1–4,6–7,17. *Crusisaccites quadratus* sp. nov.
8. *Protowelwitschiapollis*? sp.
9. *Crustaesporites gracilis* Zhang
10. *Stellapollenites* sp.
11. *Sinosaccites*? sp.
12. *Virkkipollenites*? sp.
13. *Crusisaccites* sp.
14. *Tumoripollenites baculatus*? Bharadwaj
15,16. *Citrullesaccites badaowanensis*? Zhang
5. *Citrullesaccites*? sp.

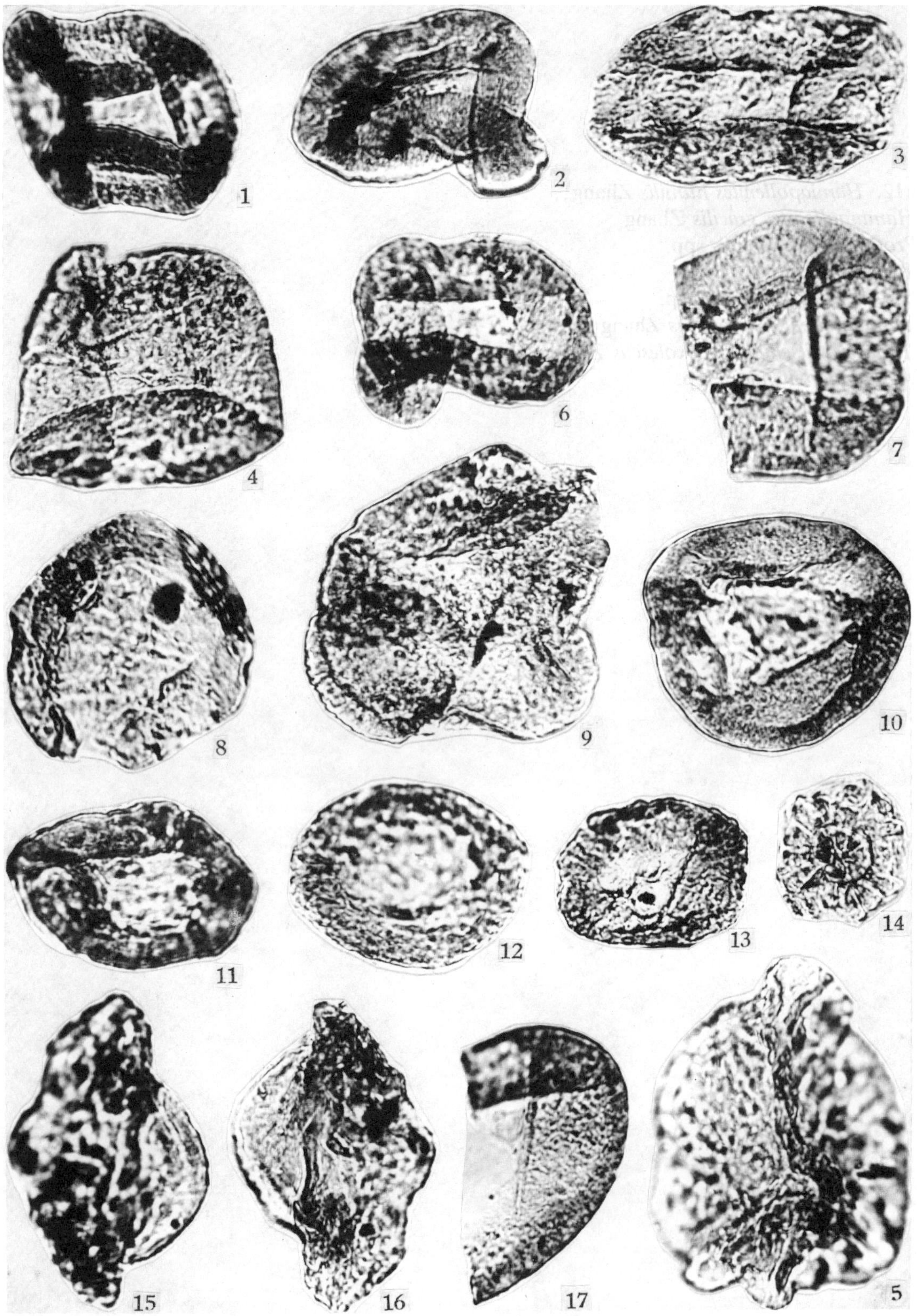
1
2
3
4
6
7
8
9
10
11
12
13
14
15
16
17
5

Plate 4

1–3,5,6,12. *Hamiapollenites humilis* Zhang
4,7,8. *Hamiapollenites gracilis* Zhang
9–11. *Protowelwitschiapollis* spp.
13,14,18. *Sinosaccites* sp.
15,22. *Protowelwitschiapollis*? sp.
16,17. *Hamiapollenites impolitus* Zhang
19,20, *Protowelwitschiapollis exolescus* Zhang
21,23,24. *Hamiapollenites*? spp.

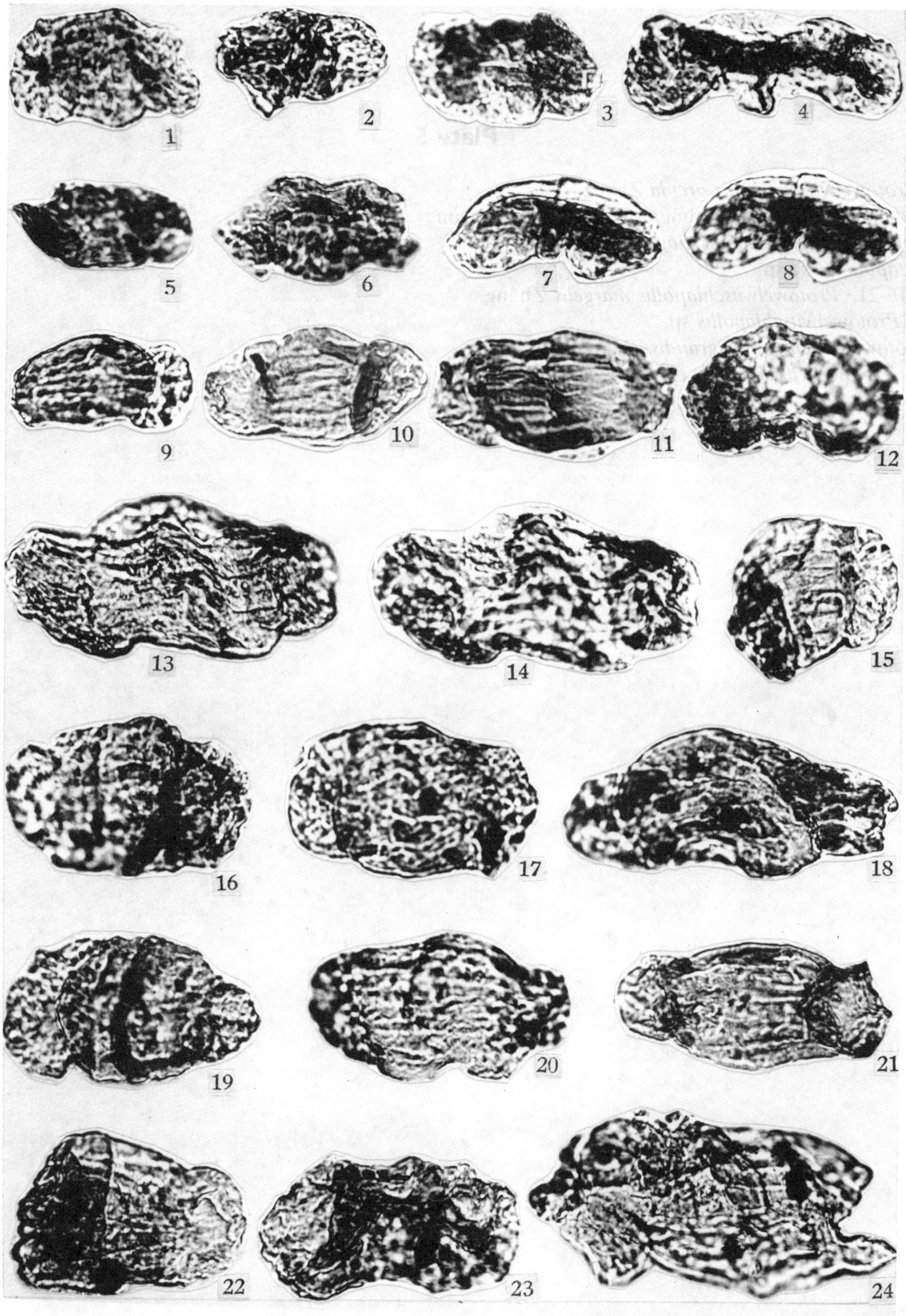

Plate 5

1–3. *Protowelwitschiapollis orcula* Zhang
4,5. *Vittatina multistriata* (Balme et Hennelly) Anderson
6,8,9. *Protowelwitschiapollis elagetis*? Zhang
7. *Costapollenites*? sp.
10–14,16-21. *Protowelwitschiapollis margelis* Zhang
15,22. *Protowelwitschiapollis* sp.
23. *Protowelwitschiapollis grandis* Zhang
24–28. *Protowelwitschiapollis exolescus* Zhang

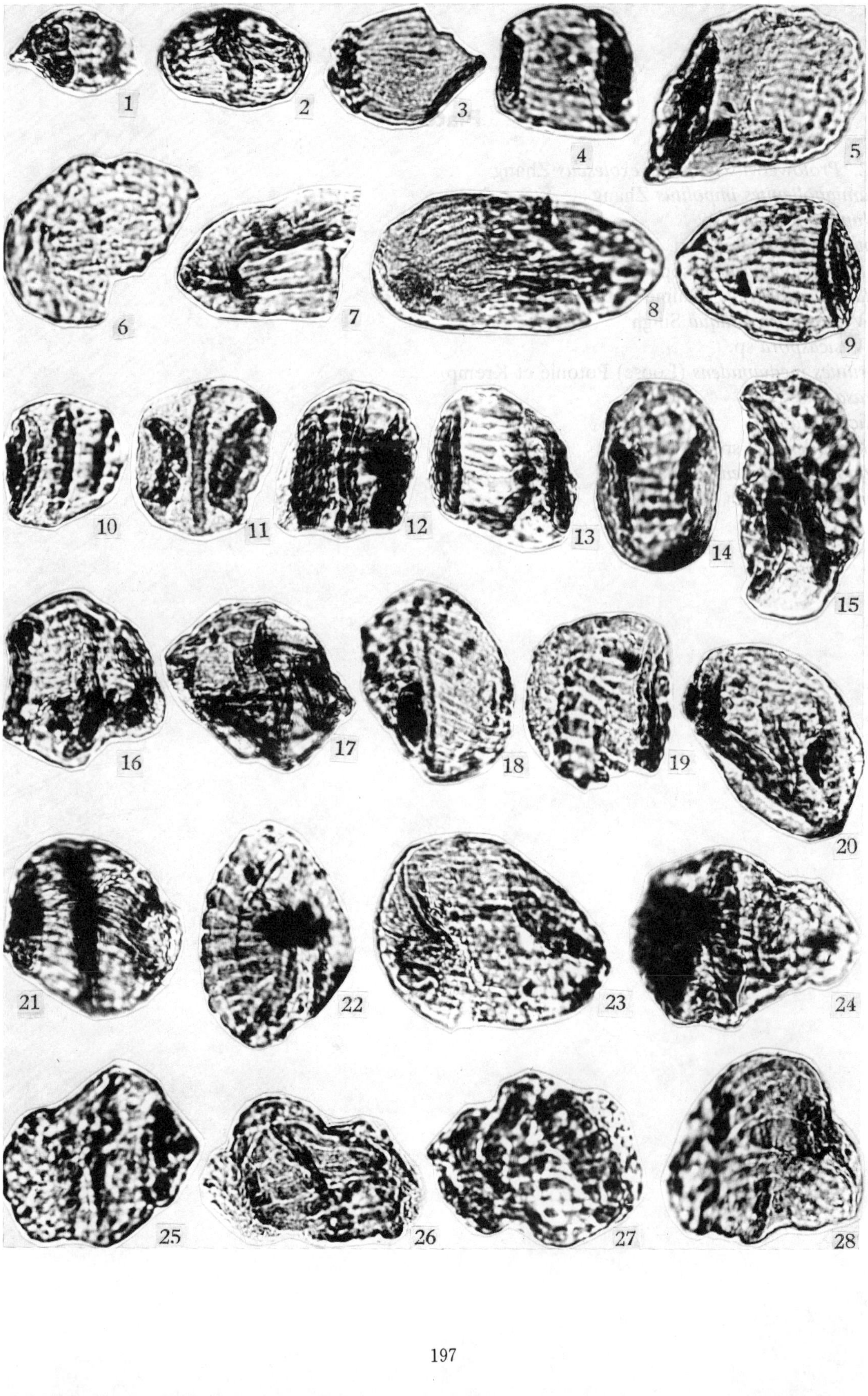

Plate 6

1–4,8,12. *Protowelwitschiapollis exolescus* Zhang
5–7. *Hamiapollenites impolitus* Zhang
9,10. *Hamiapollenites*? sp.
11. *Protowelwitschiapollis* sp.
13–16. *Striatopodocarpites* spp.
17. *Vittatina multistriata* (Balme et Hennelly) Anderson
18–21. *Vesicaspora obliqua* Singh
22–24. *Vesicaspora* sp.
25. *Florinites mediapudens* (Loose) Potonié et Kremp
26. *Parasaccites*? sp.
27. *Vesicaspora*? sp.
28. *Virkkipollenites*? sp.
29. *Aganosaccites sinensis* Zhang
30 *Aganosaccites*? sp.

Plate 6

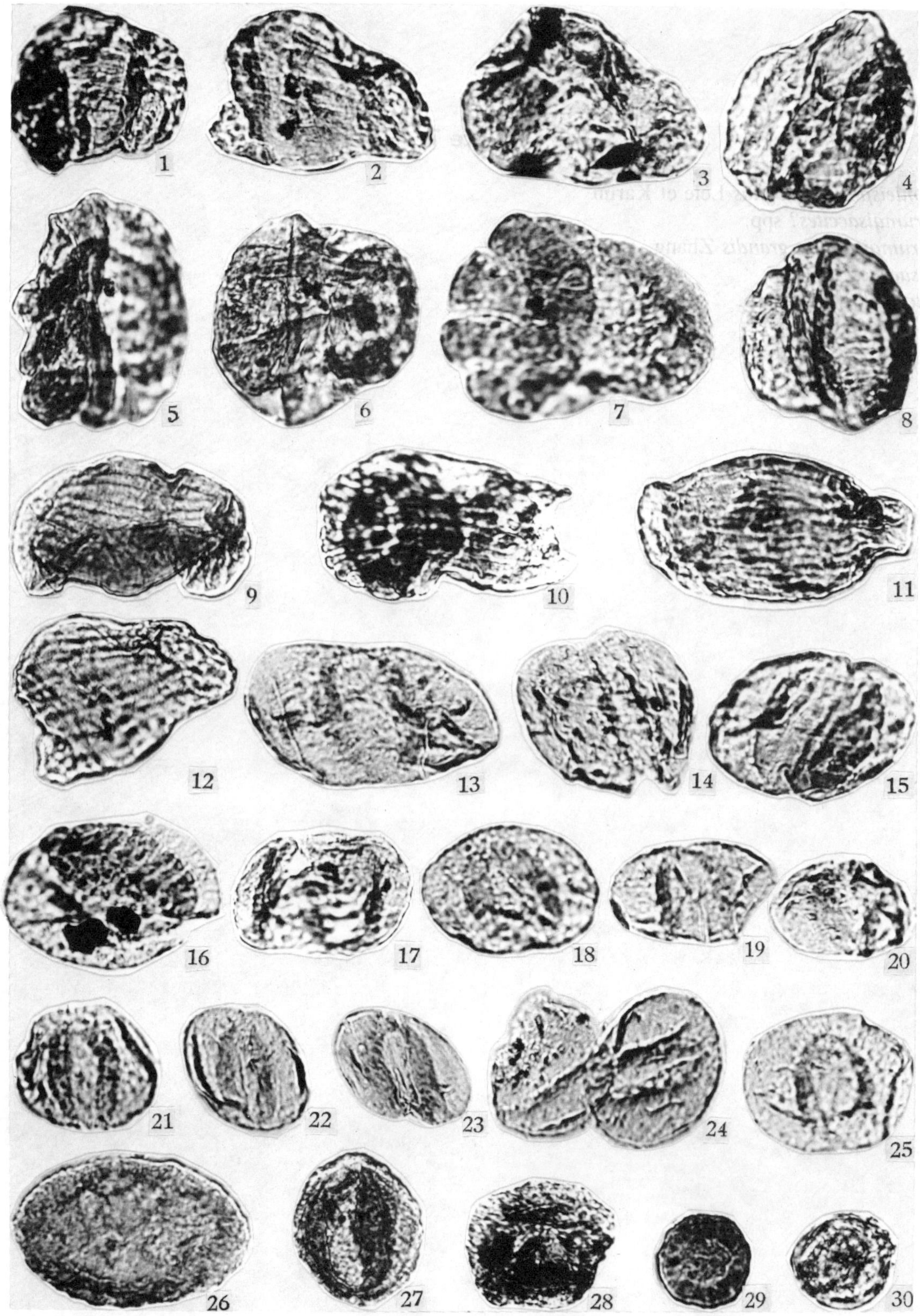

Plate 7

1. *Potonieisporites magnus* Lele et Karim
2,6. *Urumqisaccites*? spp.
3–5. *Urumqisaccites grandis* Zhang
7. *Szeisaccites*? sp.
8. *Protohaploxypinus*? sp.

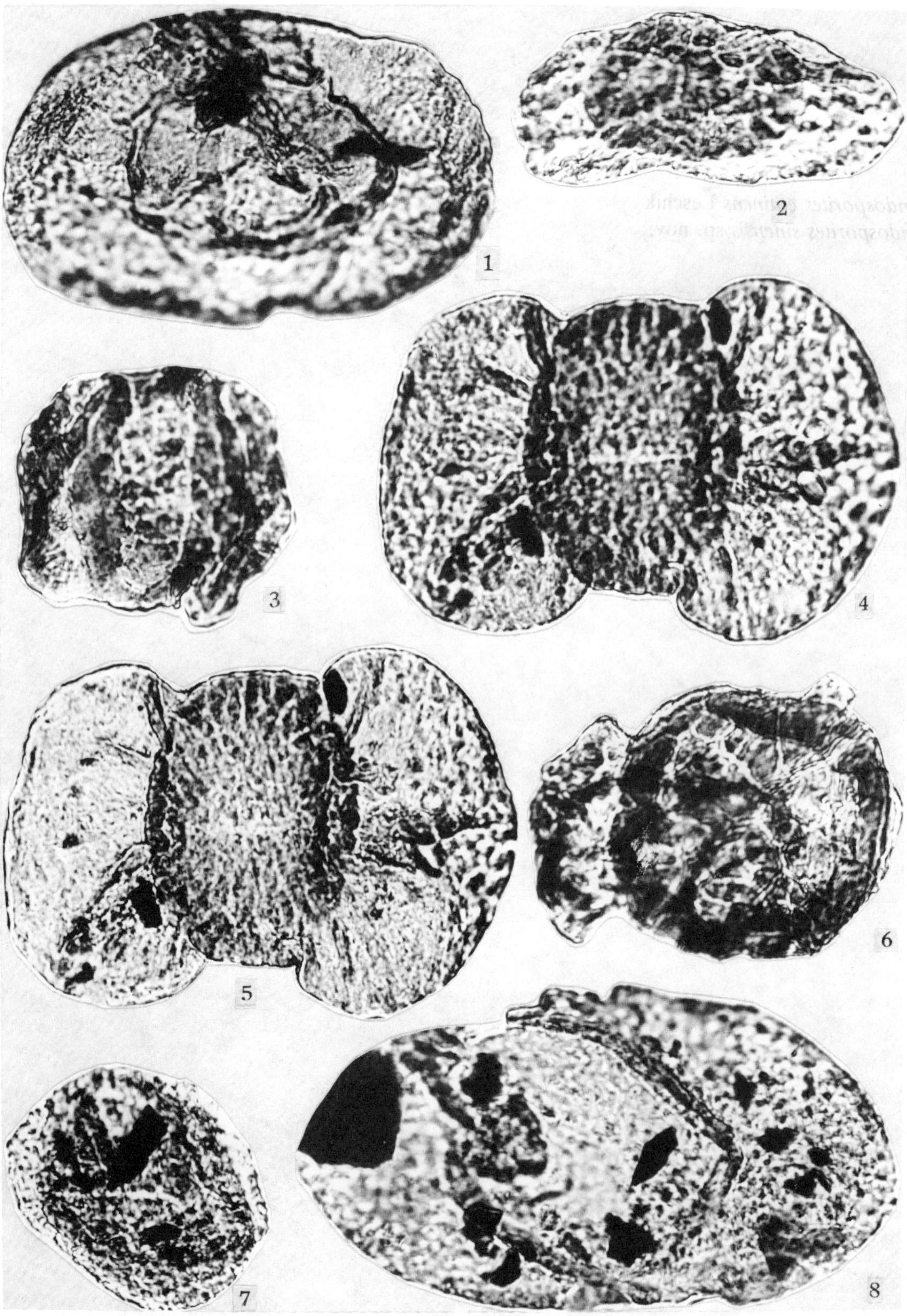
1
2
3
4
5
6
7
8

Plate 8

1–5. *Endosporites eminens* Leschik
6,7. *Endosporites sinensis* sp. nov.

Plate 8

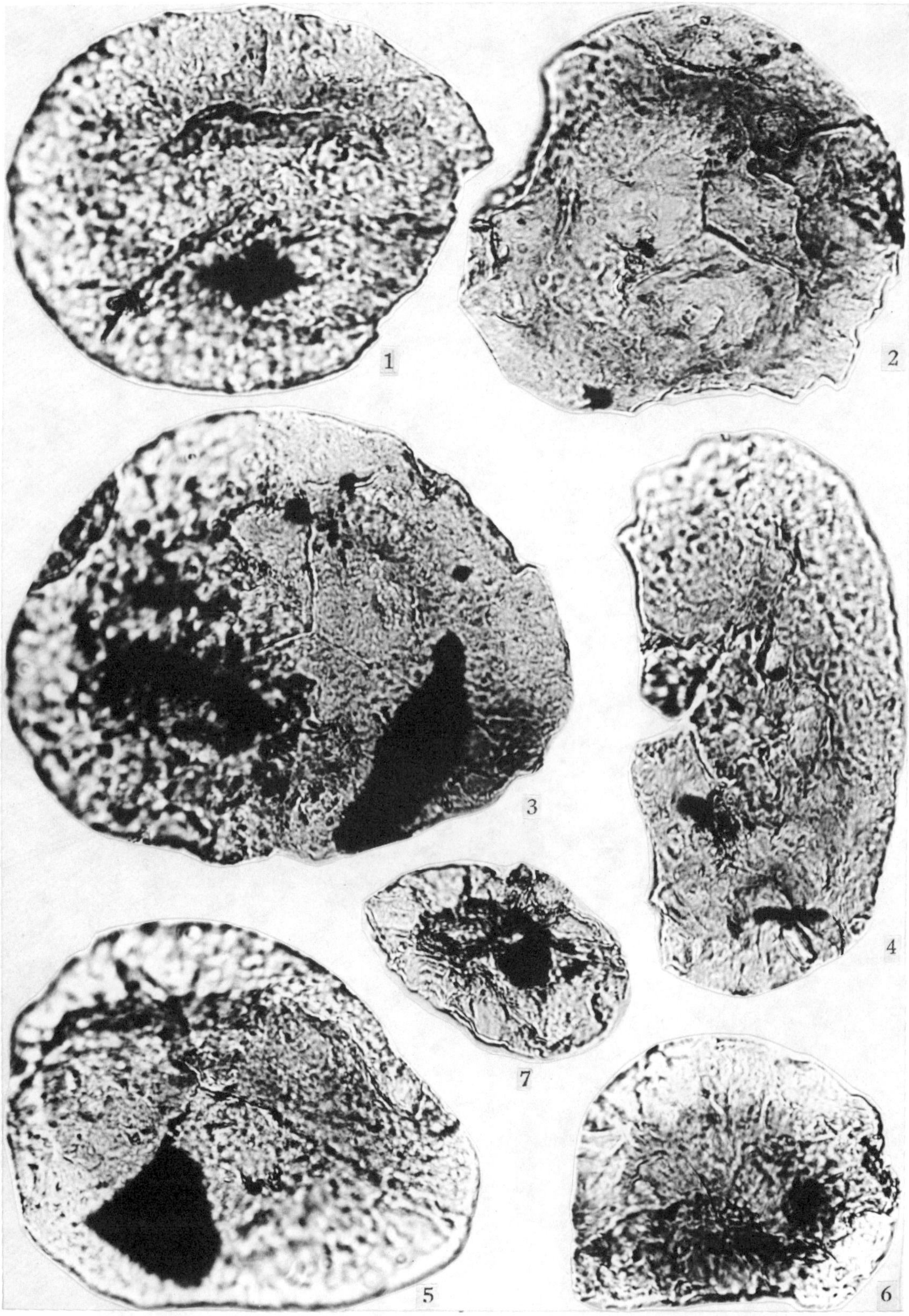

PALAEONTOLOGIA CATHAYANA Volume 5, 1990, p.205–214

CONODONT BIOFACIES AND PALEOECOLOGY OF THE CARBONIFEROUS OF CHINA

Wang Chengyuan

(Nanjing Institute of Geology and Paleontology, Academia Sinica)

David L. Clark

(Department of Geology and Geophysics, University of Wisconsin, Madison, WI, USA)

ABSTRACT

Serious study of Carboniferous conodonts in China has been in progress only during the last ten years and encouraging achievements, especially in conodont biostratigraphy, have been made. A complete Carboniferous conodont sequence has been recognized and a preliminary biofacies framework has been established. It is now possible to recognize at least three broadly defined conodont biofacies at positions along an environmental gradient that includes basinal, outer shelf and inner shelf facies. Basinal facies are limited in distribution to South China. Outer shelf facies occur over most of the area of the platform of South China and inner shelf facies are distributed over a limited area of South China as well as the vast area of the platform of North China.

During the Carboniferous, basinal and outer shelf conditions existed over most parts of the South China platform. The conodont genera *Siphonodella* and *Gnathodus* occur in the deeper water deposits and define a basinal biofacies. *Siphonodella sulcata* is more widely distributed than the related Late Devonian species *Siphonodella praesulcata.* The latter species defines a basinal biofacies. *Declinognathodus* and *Paragnathodus* are part of the outer shelf biofacies.

Most of the Late Carboniferous conodonts are more widely distributed. *Idiognathodus* and *Streptognathodus* occur both in outer shelf and inner shelf deposits. *Neogondolella bisselli,* considered to indicate Early Permian in North America, is limited to outer shelf and basinal biofacies.

More detailed work on Carboniferous conodonts in the vast platform of North China should result in the definition of even more specific biofacies positions along the marine environmental gradient.

CARBONIFEROUS CONODONT SEQUENCES OF CHINA

The first record of Carboniferous conodonts in China was made by Wang Chengyuan (1974), who described several Lower Carboniferous conodonts from the Tonchejiang Formation of Guangxi. The conodonts were found on the bedding surface of black shale of the Formation. Later, Wang Chengyuan and Wang Zhihao (1978) reported typical Lower Carboniferous conodonts including *Siphonodella duplicata* and *S. lobata* in the uppermost "Daihua Formation" which was once considered to be Late Devonian. They revised the definition of the Daihua Formation and proposed a new name (Wangyou Formation) for the lowermost Carboniferous part.

Little work had been done on Carboniferous conodonts before 1980, but a series of publications appeared in the early 1980's. In 1981, Zhao Songyin published the first report on the Carboniferous conodonts from the North China platform. He recognized two conodont assemblages from the Taiyuan Formation: the *Streptognathodus elegantulus-Idiognathodus* A. Z. and *S. elongatus* A. Z. During the past five years, more than ten papers on the Upper Carboniferous conodonts of the North China platform have been published by Wan Shilu, Ding hui, Zhao Songyin, Wang Zhihao, Li Yinlan, Zhang Wensheng and others. This work has clearly established the Carboniferous conodont biostratigraphy of the North China platform. The well-known sections of West Hill near Taiyuan City have been studied by a number of scientists. Of especial importance is the works of Prof. Wan Shilu and his colleagues on the Carboniferous conodonts in the North China platform.

The conodont sequence of the North China platform was summarized as follows:

Taiyuan Formation

Permian 6. *Sweetognathus whitei* zone

Carboniferous

5. *Streptognathodus elongatus-S. wabaunsensis-S. fuchengensis* A. Z.
4. *Streptognathodus elegantulus-Idiognathodus hebeiensis* A. Z.

Penchi Formation

3. *Neognathodus roundyi-Streptognathodus cancellosus* A. Z.
2. *Neognathodus bassleri-Idiognathodus shanxiensis* A. Z.
1. *Idiognathoides sinuatus-I. corrugatus* A. Z.

There are no marine Lower Carboniferous deposits in the vast area of the North China platform.

Conodont studies in Northwest China have been accomplished by Wang Zhihao and Wang Chengyuan (1983), Zhao Zhixin (1984), Shi Meiliang and Zhao Zhixin (1985). Zhao Zhixin (1985) first established a Carboniferous conodont sequence on the southwestern margin of the Tarim Basin in Xinjiang. The sequence is as follows (revised):

Permian

Keziliqiman Formation

Sweetognathus whitei Zone

Carboniferous

Tahaqi Formation

Streptognathodus gracilis Zone *Streptognathodus elegantulus-S. oppletus* A. Z.

Azigan Formation

Streptognathodus parvus-S. suberectus A. Z.

Kalawuyi Formation

Idiognathodus delicatus-Neognathodus bassleri-Idiognathoides corrugata A. Z.

Heshilafu Formation

Gnathodus bilineatus-G. homopunctatus A. Z.

New data show that the lowermost Carboniferous conodonts like *Siphonodella* are definitely present in Xinjiang. Besides, there are continuous sections across the Devonian-Carboniferous boundary in this region.

The conodont biostratigraphy in the Qilianshan area has been summarized as follows (Wang Zhihao and Wang Chengyuan, 1983; Shi Meiliang and Zhao Zhixin, 1985).

Jingyuan Formation

6. *Streptognathodus elongatus-S. wabausensis* A. Z.
5. *Idiognathoides delicatus-I. incurtus* A. Z.
4. *Idiognathoides corrugata-I. sulcatus* A. Z.
3. *Declinognathodus noduliferus - D. lateralis - Idiognathoides sinuatus* A. Z.
2. *Gnathodus bilineatus-G. commutatus* A. Z.
1. *Gnathodus girtyi - Neoprioniodus peracutus* A. Z.

A similar conodont sequence has also been recognized in the Ningxia area (Wang Chengyuan, 1987) . There are no marine deposits of lowermost Carboniferous in either the Qilianshan or the Ningxia areas

In South China, the most important research of Carboniferous conodonts during the last few years has been centered on the Devonian-Carboniferous boundary strata. The Muhua section, one of two candidates for the Devonian-Carboniferous boundary stratotype, has been studied in detail. A Lower Carboniferous conodont sequence in the deep water facies (Muhua and Nandan) has been established (Xiong Jianfei, 1983; Wang Chengyuan and Yin Baoan, 1984, Hou Hongfei et al., 1985). The Lower Carboniferous conodont sequence is as follows:

11. *Adetognathus lautus* Zone
10. *Gnathodus bilineatus* Morphotype *C-Adetognathus unicornis* A. Z.
9. *Gnathodus bilineatus* Morphotype *B-Paragnathodus commutatus* A. Z.
8. *Scaliognathus anchoralis* Zone
7. *Gnathodus simiglaber-G. taxus* A. Z.
6. *Siphonodella isosticha*-Upper *crenulata* Zone

5. Lower Siphonodella crenulata Zone
4. *Siphonodella sandbergi* Zone
3. Upper *Siphonodella duplicata* Zone
2. Lower *Siphonodella duplicata* Zone
1. *Siphonodella sulcata* Zone

In 1985, Ji Qiang revised the "standard" scheme of *Siphonodella* zonation but his new zonation needs additional work before it can be used in other areas.

In the vast area of South China, the Upper Carboniferous conodont sequences are very similar to those of the North China platform, but are more detailed. In this area it is clear that *Neogondolella bisselli* appears earlier than *Sweetognathus whitei*, and *Streptognathodus wabausensis* appears earlier than *S. elongatus*. The Upper Carboniferous conodont sequence in South China is as follows:

Permian
Sweetognathus whitei Zone
Carboniferous
Neogondolella bisselli Zone
Streptognathodus elongatus Zone
Streptognathodus wabausensis-S. elegantulus A. Z.
Streptognathodus elegantulus Zone
Streptognathodus parvus Zone
Streptognathodus suberectus - Neognathodus bassleri A. Z.
Idiognathoides corrugatus - I. sinuatus A. Z.
Declinognathodus noduliferus Zone

The upper boundary of the Carboniferous system in China is placed at a level different from those used in North America and Europe. Selection of an internationally accepted level has not been made. For now in China the first appearance of *Sweetognathus whitei* in the evolutionary lineage from *S. merrilli* to *S. whitei* is a convenient boundary between the Carboniferous and Permian. As defined, this boundary falls within the upper part of the Maping Formation or within the Sazhitang Formation.

CARBONIFEROUS PALEOGEOGRAPHY OF CHINA

The maps of Carboniferous paleogeography of China edited by Guang, Shicong et al. (1984) show clearly that an extensive regression took place in the Early Carboniferous as a response to Caledonian tectonics.

During the Early Carboniferous, the South China Sea was restricted to the present Zhujiang drainage area. This sea area was linked with the Lower Yangtze area. The sediments of the South China Sea can be roughly subdivided into three different lithofacies, an inner shelf facies, an outer shelf facies and a trough-platform facies (=basinal facies) (Fig.1)

The inner shelf facies covered the Lower Yangtze area and parts of Hunan and Jiangxi. The outer shelf included the platform-margin facies and platform facies. Salinity was normal and there was good circulation with the open sea. A large part of the South China Sea in Grangxi, Guangdong, Hunan, Guizhou and Yunnan was occupied by the outer shelf facies. There were at least six deep water basins in the South China Sea. These were located either in the open platform belt or between the open platform belt and the platform margin zone. Lithofacies of the trough-bearing platform units are not common and their extent was comparatively small. Fossils have been considered to be similar to those of the pelagic basin facies. No classical eugeosynclines and miogeosynclines existed in the South China Sea.

The North China platform was uplifted, and the West China Sea and Xinjiang-Mongolian Sea were mostly trough-basin facies. No conodont biofacies definitions are possible for this area yet. Our discussion is limited to the North and South China Platforms.

During the Late Carboniferous there was a significant transgression. In North China, Yinshan Land, Jiao-Liao-Huai-Yang Land and Qinling Island (Fig. 2) were the areas of the North China Sea which included dominantly an inner shelf facies. This sea was linked with the South China Sea in its southern part and the Xinjiang-Mongolian Sea in the northeastern and northwestern parts. The North China Sea was occupied by shallow water with terrestrial, lagoonal, deltaic, and fluvial sediment accumulating. Several coal beds are present. It is possible that in some small areas the Sea was relatively deeper. The entire interval could be subdivided into several different lithofacies, but there are no enough data. The entire region of the

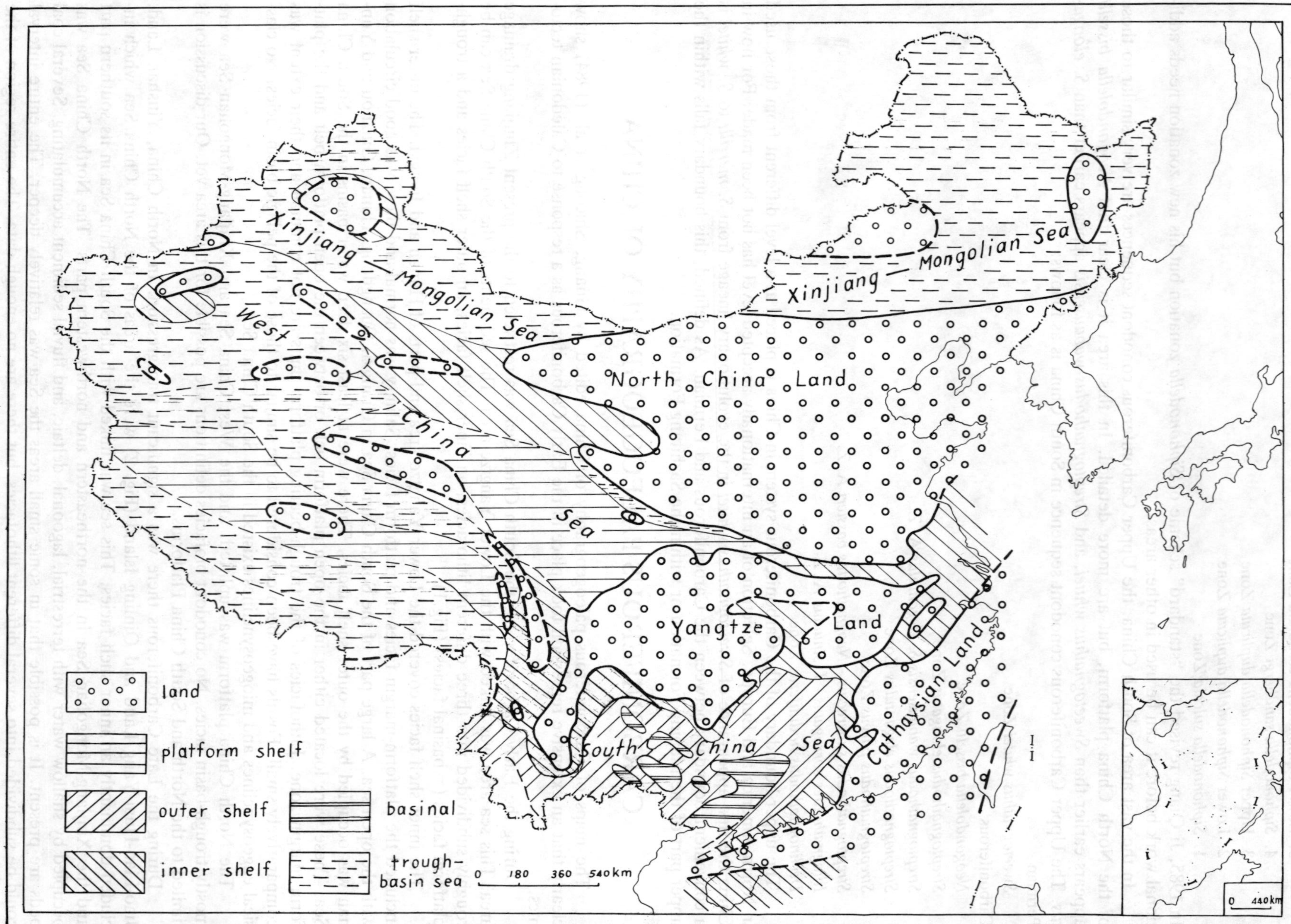

Text-fig. 1. Palaeogeographical map of Early Carboniferous of China (After Guang et al., 1984, revised).

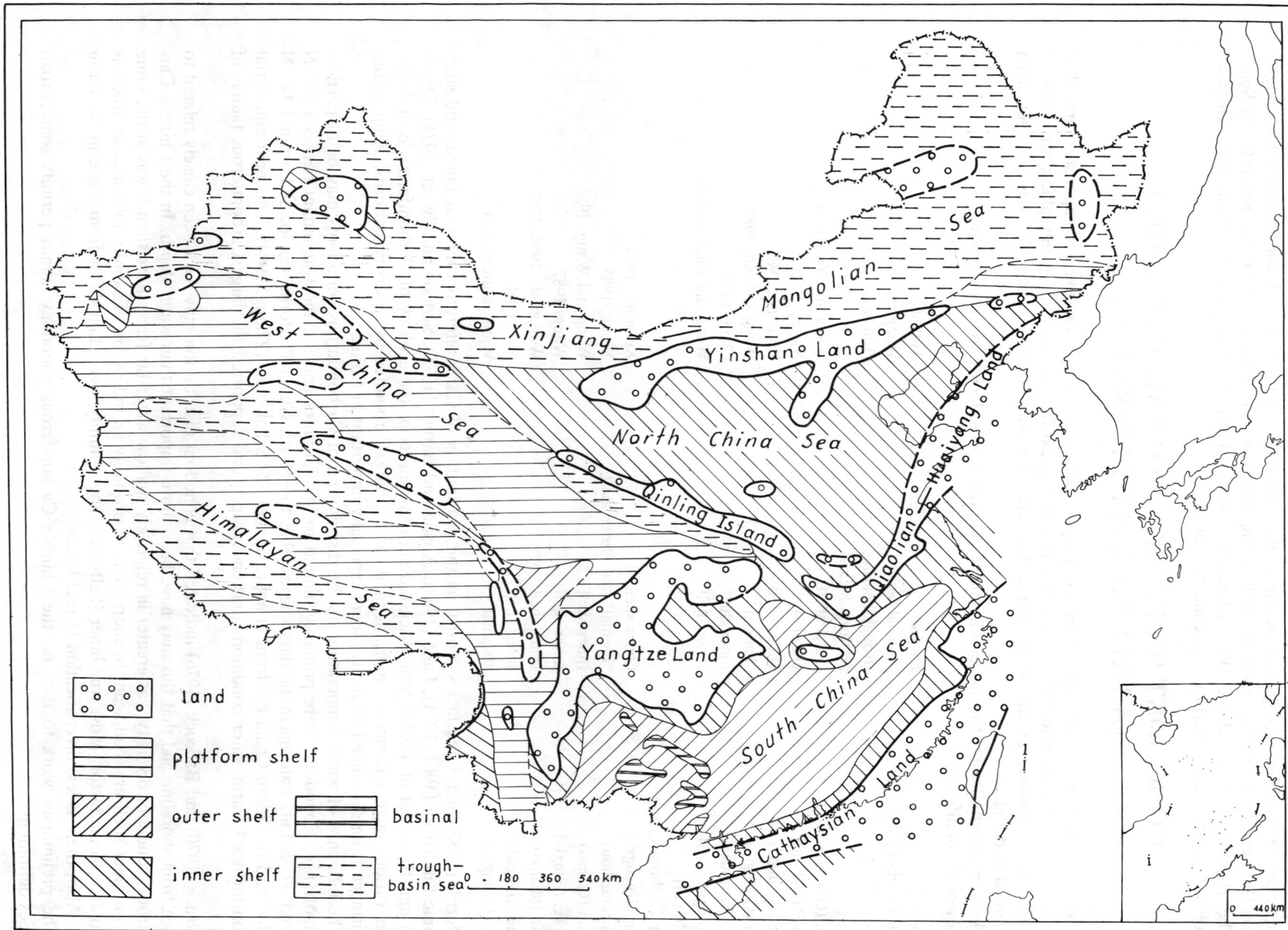

Text-fig. 2. Palaeogeographical map of Late Carboniferous of China (After Guang et al., 1984, revised).

North China Sea is considered to belong to an Inner Shelf Facies.

The South China Sea was flanked by the Cathaysian Land on the east and southeast and by Yangtze Land on the west. Several small basins (troughbearing platform facies) are present in the South China Sea.

The Himalayan Sea and the West China Sea occupied the western part of China. The Xinjiang-Mongolian Sea covered the north geosyncline area of China.

CONODONT BIOFACIES AND PALEOECOLOGY (A PRELIMINARY DEFINITION)

The Carboniferous conodont biofacies and paleoecology discussed in the present paper are limited to the North China Platform and the South China Platform (Text-fig. 3). Regional geographic distribution of Carboniferous conodonts in the West China Sea and the Xinjiang-Mongolian Sea has not yet been documented in detail.

The sections studied for this discussion and the representative facies are listed below:

Lower Carboniferous

Section	**Facies**	**Authors**
Muhua (Guizhou)	"basinal facies" (trough-bearing platform facies)	Wang and Yin, 1984 Hou *et al.*, 1985
Xiakou (Guangxi)	Outer shelf	Wang and Yin, 1985
Nashui (Guizhou)	Outer shelf	Xiong, 1985
Baping (Guangxi)	Outer shelf	Xu *et al.*, 1986
Limiao (Guangxi)	"basinal"	Wang and Xu, 1988
Binhai (Jiangsu)	Inner shelf	Chen, 1984
Monggongao (Part) (Hunan)	Inner shelf	Wang and Ziegler, 1981
Upper Carboniferous		
Nandan (Guangxi)	Outer shelf	Xu *et al.*, 1986
Nashui (Guizhou)	"basinal" and Outer shelf	Xiong, 1985
Jingyuan (Gansu)	Inner shelf	Wang and Wang, 1983
Zhongwei (Ninxia)	Inner shelf	Wang, 1987
Taiyuan (Shanxi)	Inner shelf	Wan *et al.*, 1984, 1985
Nanjing (Jiangsu)	Inner shelf	Yin, 1984
Debao (Guangxi)	"basinal"	Wang, unpublished

Merrill (1962) first noted facies control on the distribution of Carboniferous conodont faunas and since that time Lindström (1964; 1974), Lane (1968), Austin and Barnes (1975), Seddon and Sweet (1971), Druce (1970) and Barnes et al. (1973) have discussed various aspects of the ecology of Carboniferous conodonts. Of the various factors deemed important in understanding Carboniferous diversity patterns, temperature and distance from shoreline (=salinity) seem especially important in China.

The influence of water temperature on the distribution of Carboniferous conodonts may be of particular importance. Conodonts were primarily limited to low latitudes and are most abundant between 30° N Lat. and 30° S Lat. Paleomagnetic data suggest that the North China Platform was located around 14° N Paleolat. and the South China Platform at approximately 2° S Paleolat. (Zhang, 1984). This near equatorial position included warm water conditions that encouraged and supported the large Carboniferous faunas of China.

Druce (1970) and Babcock (1976) indicated that conodont distribution may have been closely related to distance from shoreline and that this may have been more important than water depth. In the Chinese Carboniferous, conodont diversity was greater in relatively shallow water far from shore than it was at the same depth on inner shelf areas adjacent to shore. This is probably an indication of salinity tolerance as much as any other ecologic factor. Most true inner shelf areas have salinity levels that differ from those in the basin edges. Also, there is less diversification in reef facies.

The preliminary work to date on the Chinese Carboniferous conodont faunas permits some broad biofacies definitions.

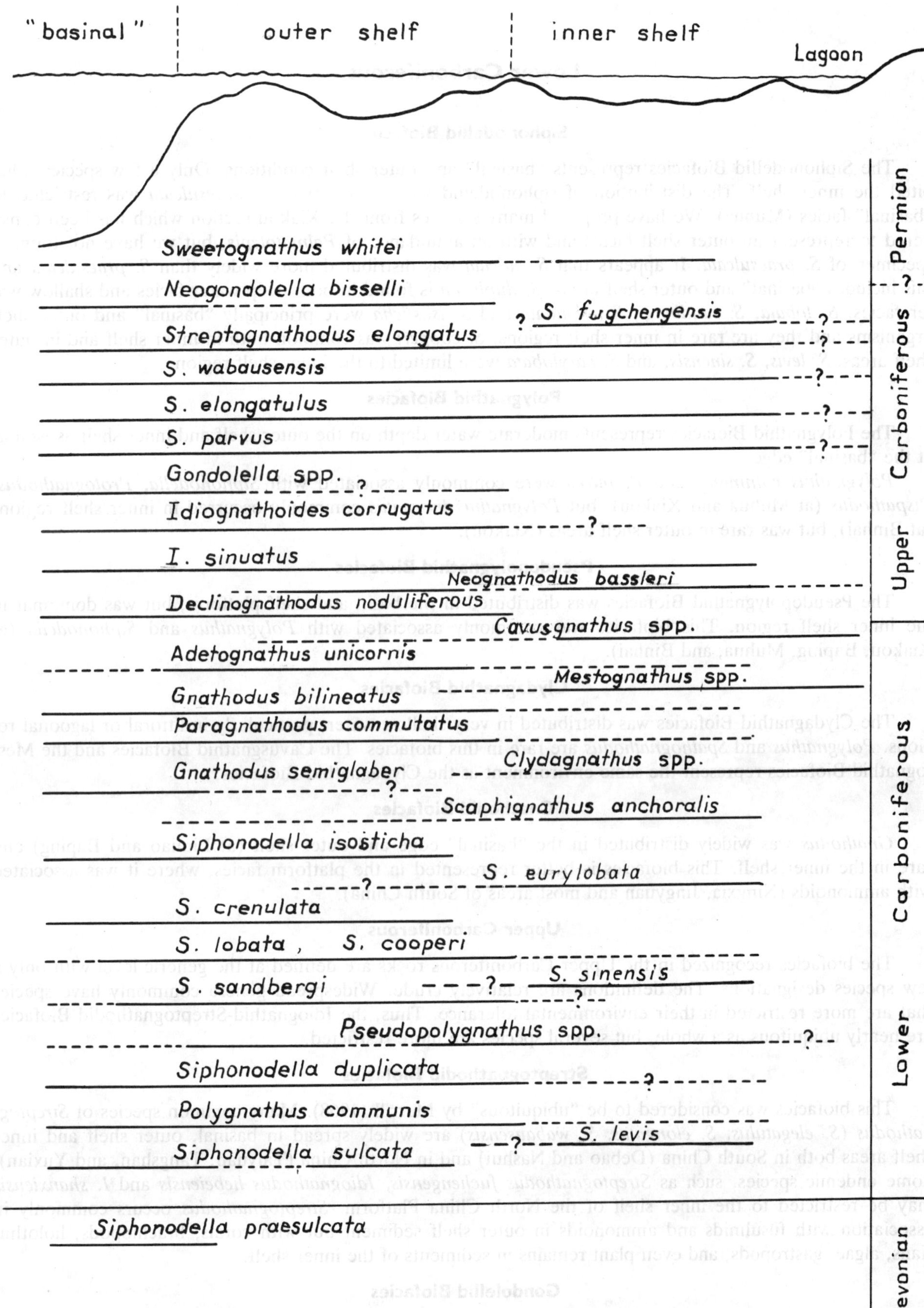

Text-fig. 3. Carboniferous conodont distribution in China through time and biofacies.

Lower Carboniferous

Siphonodellid Biofacies

The Siphonodellid Biofacies represents "basinal" and outer shelf conditions. Only a few species inhabited the inner shelf. The distribution of siphonodellid species is varied. *S. praesulcata* was restricted to "basinal" facies (Muhua). We have prepared many samples from the Xiakou section which has been considered to represent an outer shelf facies and with an abundance of *Palmatolepis*, but we have not found a specimen of *S. praesulcata*. It appears that *S. sulcata* was distributed more widely than *S. praesulcata* and this includes "basinal" and outer shelf areas. *S. duplicata* is found in both deep water facies and shallow water facies. *S. lobata*, *S. sandbergi, S. crenulata* and *S. isosticha* were principally "basinal" and outer shelf organisms and they are rare in inner shelf regions. *S. cooperi* was common both in inner shelf and in outer shelf areas. *S. levis, S. sinensis*, and *S. eurylobata* were limited to the inner shelf region.

Polygnathid Biofacies

The Polygnathid Biofacies represents moderate water depth on the outer shelf and inner shelf as well as at the "basinal" edge.

Polygnathus communis and *P. purus* were commonly associated with *Siphonodella, Protognathodus, Bispathodus* (at Muhua and Xiakou), but *Polygnathus inornatus* commonly occurred in inner shelf regions (at Binhai), but was rare in outer shelf areas (Xiakou).

Pseudopolygnathid Biofacies

The Pseudopolygnathid Biofacies was distributed in the basin and shelf platform, but was dominnat in the inner shelf region. This biofacies was commonly associated with *Polygnathus* and *Siphonodella* (at Xiakou, Baping, Muhua, and Binhai).

Clydagnathid Biofacies

The Clydagnathid Biofacies was distributed in very shallow water, particularly in littoral or lagoonal regions. *Polygnathus* and *Spathognathodus* are rare in this biofacies. The Cavusgnathid Biofacies and the Mestognathid Biofacies represent the same environment as the Clydagnathid Biofacies.

Gnathodid Biofacies

Gnathodus was widely distributed in the "basinal" edge and outer shelf (at Limiao and Baping) and rare in the inner shelf. This biofacies is better represented in the platform facies, where it was associated with ammonoids (Ningxia, Jingyuan and most areas of South China).

Upper Carboniferous

The biofacies recognized in the Upper Carboniferous rocks are defined at the generic level with only a few species designations. The definitions are relatively crude. Widespread genera commonly have species that are more restricted in their environmental tolerance. Thus, the Idiognathid-Streptognathodid Biofacies are nearly ubiquitous as a whole, but several species are more restricted.

Streptognathodid Biofacies

This biofacies was considered to be "ubiquitous" by Merrill (1968). Many common species of *Streptognathodus* (*S. elegatulus, S. elongatus, S. wabausensis*) are widely spread in basinal, outer shelf and inner shelf areas both in South China (Debao and Nashui) and in North China (Taiyuan, Tangshan, and Yuxian). Some endemic species, such as *Streptognathodus fuchengensis, Idiognathodus hebeiensis* and *I. shanxiensis* may be restricted to the inner shelf of the North China Platform. *Streptognathodus* occurs commonly in association with fusulinids and ammonoids in outer shelf sediment but with corals, brachiopods, holothurians, algae, gastropods, and even plant remains in sediments of the inner shelf.

Gondolellid Biofacies

This biofacies represents outer shelf and basinal facies and occurs at several localities in South China (e.g. Nashui, Baping, Limiao and Debao). More than one thousand conodont samples were prepared from

North China platform deposits but none contained *Gondolella*. Only a few specimens of *Gondolella* were found in South China and in the North geosynclinal region (Xinjiang). Merrill and von Bitter (1985) suggested that the controlling environmental factor for species of *Gondolella* may be low PH. Low PH environments commonly existed in close proximity to more open, high-energy water conditions.

Neogondolellid Biofacies

This biofacies is distributed as widely as the *Gondolella* Biofacies. The earliest species of *Neogondolella* in China is *N. bisselli*. Apparently, it lived only in the basinal and outer shelf regions (e.g. Nashui and Debao), but it is unknown in the North China Platform.

It is possible that the *S. elongatus* Zone of the North China Platform may be equivalent to the sum of the *Streptognathodus elongatus* and *Neogondolella bisselli* Zones of the South China Platform. No *Neogondolella* species appears in the North China Platform prior to the first occurrence of *N. bisselli* together with *Sweetognathus whitei*, and this represents a much younger time interval.

REFERENCES

Austin, R. L., 1976. Evidence from Great Britain and Lreland concerning west European Dinantian conodont paleoecology. Geol. Assoc. Canada, spec. Paper, 15, 201–217.

Austin, R. L., and Barnes, C. R., 1975. The biostratigraphic limitations of conodonts with particular reference to the base of the Carboniferous. In: Contribution a L'etude biostratigraphique et micropaleontologique du Dinantien en Europe Occidental. Bull. Soc. Belge. Geol. Paleont. Hydrol., 82, 351–374.

Babcock. L. C., 1976. Conodont paleoecology of the Lamar Limestone (Permian), Delaware Basin, West Texas. In: Barnes, C. R. (ed.), Conodont paleoecology. Geol. Assoc. Canada Spec. Paper, 15, 279–294.

Barnes, C. R., Rexroad, C. B., and Miller, J. F., 1973. Lower Paleozoic conodont provincialism. In: Rhodes, F. H. T. (ed.), Conodont paleozoology. Geol. Soc. America, spec. Paper, 141, 157–190.

Carey, S. P., 1984. Conodont biofacies of the Triassic of northwestern Nevada. In: Clark, D. L. (ed.), Conodont biofacies and provincialism. Geol. Soc. America, spec. Paper, 196, 295–306.

Clark, D. L., and Carr, T. R., 1984. Conodont biofacies and biostratigraphic schemes in western North America: A model. In: Clark, D. L. (ed.), Conodont biofacies and provincialism Geol. Soc. America, Spec. Paper, 196, 1–9.

Ding Hui, Wan Shilu and Liang Xiangyuan, 1983. Preliminary studies of the conodont biostratigraphy of the Taiyuan Formation, Late Carboniferous in Hebei, Henan Provinces. Shanxi Mining college learned Journal, 3(1), 8–18 (in Chinese).

Ding Hui and Zhao Qingsheng, 1985. Conodont from the Dongdayao Limestone of Taiyuan Formation in the western Mountain of Taiyuan and its stratigraphical significance, Shanxi Mining college Learned Journal 3(1), 8–18 (in Chinese).

Druce, E. C., 1973. Upper Paleozoic and Triassic conodont distribution and the recognition of biofacies. In: Rhodes, F. H. T. (ed.), Conodont paleozoology. Geol. Soc. America, Spec. Paper, 141, 191–237.

Guang Shicong and others, 1984. The changes of land and sea, marine sedimentary facies and hydrocarbons in China, Sci. Press, Beijing.

Hou Hongfei, Ji Qiang, Wu Xianghoe, Xiong Jianfei, Wang Shitao, Gao Lianda, Sheng Huibin, Wei Jiarong and Susan Turner, 1985. Muhua sections of Devonian-Carboniferous boundary beds.Geol. Publishing House, Beijing.

Ji Qiang, 1985. Study on the phylogeny, Taxonomy, zonation and biofacies of *Siphonodella* (conodonta), Bull, Inst. Geol. Chinese Acad. Geol. Sci. 11, 51–78.

Lindström, M., 1964. Conodonts. Elsevier, Amsterdam.

Merrill, G. K., 1962. Facies relationship in Pennsylvanian conodont faunas. Texas Journal of Science, 14.

Merrill, G. K., 1973. Pennsylvanian conodont paleoecology. In Rhodes, F. H. T. (ed.), Conodont paleozoology. Geol. Soc. America, Spec. Paper, 141, 239–274.

Merrill, G. K., and von Bitter, P. H., 1984. Facies and frequencies among Pennsylvanian conodonts: Apparatuses and abundances, In: Clark, D. L. (ed.), Conodont biofacies and provincialism. Geol. Soc. America, Spec. Paper, 196, 251–261.

Seddon, G., and Sweet, W. C., 1971. An ecologic model for conodonts. J. Paleont., 45, 869–880.

Shi Meiliang and Zhao Zhixing, 1985. Carboniferous conodont sequence in North Qilianshan area. Kexue Tongbao (Science Bulletin), 1985 (15), 1179–1182 (in Chinese).

Wan Shilu, Ding Hui and Zhao Songyin, 1983. Middle and Late Carboniferous conodont biostratigraphy of North China. Journal of China Coal Society, 1983(2), 62–72 (in Chinese).

Wang Chengyuan, 1974. Carboniferous conodonts from Guangxi and Permian conodonts from Sichuan. In: Handbook of the stratigraphy and paleontology in Southwest, China, 383–384. Sci. Press. Beijing (in Chinese).

Wang Chengyuan, 1987. Conodonts from the Tsingyuan Formation of Zhongwei, Ningxia. In: Namurian strata and fossils of Ningxia, China. Nanjing University Press, Nanjing.

Wang Chengyuan and Wang Zhihuo, 1978. Upper Devonian and Lower Carboniferous conodonts from southern Guizhou. Mem. Nanjing Inst. Geol. Palaeont. 11, 56–96 (in Chinese).

Wang Chengyuan and Yin Baoan, 1984. Conodont zonation of Early Lower Carboniferous and Devonian-Carboniferous boundary in pelagic facies, South China. Acta Palaeont. Sinica, 23(2), 224–242.

Wang Chengyuan and Yin Baoan, 1985. An important Devonian-Carboniferous boundary stratotype in neritic facies of Yishan County, Guangxi. Acta Micropalaeont. Sinica, 2(1), 28–51.

Wang Zhihuo and Li Yinlan, 1984. Discovery of Late Carboniferous conodonts from Shanxi Province and its significance. Acta

Palaeontol, Sinica, 23(2), 196–205.

Wang Zhihuo and Wang Chengyuan, 1983. Carboniferous conodonts form the Jingyuan Formation of Jingyuan, Gansu. Acta Palaeont. Sinica, 22(4), 438-448.

Wang Zhihuo and Zhang Wensheng, 1985. Discovery of conodonts from the Upper part of the Taiyuan Formation in Yuxian, Henan. Journal of Stratigraphy, 9(3), 228-230. (in Chinese).

Xiong Jianfei, 1983. On the Devonian-Carboniferous boundary and correlation in south China. Oil & Geology, 4(4), 337-356 (in Chinese).

Xiong Jianfei and Zhai Zhiqiang, 1980. Carboniferous conodont and fusulinid biostratigraphy in black area (Wangmo Ruya, Luodian Nashui), Guizhou Province. Geology of Guizhou, 3(5), 269 — 287 (in Chinese).

Xin Zhong-e, Wang Yunhui, Chen Huacheng, Zhang Ying and Bi Zhongqi, 1986. Age of the Laohudong dolomite in Nanjing area. Journal of Stratigraphy, 10(3), 216-220 (in Chinese).

Zhang Zhengkun, 1984. Sino-Korean block and Yangtze block as part of the Pacifica continent in the Late Paleozoic. Bull. Chinese Acad. Geol. Sci., 9, 45-53.

Zhao Songyin, 1981. Late Carboniferous conodonts from the Qinshui Basin in Shanxi Province, Bull. Tianjin Inst. Geol. Min. Res., 4, 97-101 (in Chinese).

Zhao Songyin and Wan Shilu, 1983. Some conodonts from the Upper Carboniferous Kaiping Formation in Tangshan area and their significance. Bull. Tianjin Inst. Geol. Min. Res. 8, 95-104. (in Chinese).

Zhao Zhixing, Han Jianxiu and Wang Zhengji, 1984. The Carboniferous strata and its fauna from southwestern margin of Tarim Basin in Xinjiang. Geol. Publ. House. Beijing (in Chinese).

PALAEONTOLOGIA CATHAYANA Volume 5, 1990, p.215–224

MIDDLE JURASSIC GASTROPODS FROM XIZANG (TIBET), CHINA

Wang Huiji

(Shanghai Museum of Natural History)

Yang Shengqiu

(Shanghai Meishan Metallurgical Corporation)

Abstract

This paper describes the Middle Jurassic gastropods from the Marl member of the Daheishan Formation in the Chuanba Coal Mine of Gerze, Xizang (Tibet). 9 species of gastropods occurred in there. Among them 3 species are new, the other 6 species are all known to appear in the Middle Jurassic Bathonian. Accordingly, the fossil-bearing beds are of Middle Jurassic age. Besides, 4 species of gastropods from there belong to the Cerithiacea, 1 species to the Littorinaces. Their extant species are living in the littoral area or shallow sea. Therefore, the fossil-bearing beds probably belong to littoral or shallow sea deposits.

Although the Jurassic gastropods from Xizang were recorded by Yin Jixiang (1962), Yang Shengqiu, Wang Huiji (1985), Yu Wen and Xia Jinbao (1985), the Middle Jurassic ones from Gerze Xian, Xizang (Tibet) are so far described for the first time. These specimens were collected by Yang Shengqiu from the marl member of the Daheishan Formation in the Chuanba Coal Mine of Gerze Xian. Here the stratigraphic sequence is as follows (in descending order):

Cretaceous Xiaqu Formation : Limestone

———— Unconformity ————

Late Jurassic-Early Cretaceous:

Chuanba Formation

Upper part: Grey, medium-bedded, crystalline limestone, containing brachiopods 200m

Lower part: Grey-black silty mudstone intercalated with yellow-grey medium or thin-bedded sandstone and two exploitable coal beds (each bed about 1.4 m. thick) and some uneven coal beds (35-50 cm in thickness), yielding plant fossils 400m

———— Covered (20–30m) ————

Middle Jurassic Daheishan Formation

Marl member: Yellowish grey medium to thin-bedded marl, yielding abundant bivalves and gastropods 200m

Limestone member: Black-grey, grey medium to thick-bedded crystalline limestone with silty calcite veins 700m

The coal-bearing Chuanba Formation yields abundant plant fossils of *Cladophlebis* (*Klukia*) *browniana,* suggesting Late Jurassic or early Early Cretaceous in age. This formation is unconformably overlain by the Xiaqu Formation and probably conformably rests on the Daheishan Formation. Here the strata are hard to be determinated by the faulting, but the study of gastropods from the Daheishan Formation has been made. On these specimens its geologic age has been based. In the marl member of the Daheishan Formation, there are 9 species of gastropods, namely *Pseudomelania* (*Rhabdoconcha*) *aspasia* (d'Orbigny), *Oonia luciensis* (Cossmann), *Rigauxia fluens* (Piette), *Trypanostylus* (*T.*) *elegantulus* sp. nov., *Purpurina* (*Pseudalaria*) *adsutura* sp. nov., *Procerithium* (*Cosmocerithium*) *leckenbyi* (Morris and Lycett), *Rhabdocolups trilicus* sp. nov., *Rhabdocolpus strangulatus* (d'Archiae), *Cryptaulax* (*Xystrella*) *echinatum* (von Buch). Among them, 3 are considered as new. The stratigraphic range of other 6 species is shown in Fig. 1.

Species / Strata	Bajocian (Inferior Oolite)	Bathonian (Great Oolite)	Callovian
Pseudomelania (*Rhabdoconcha*) *aspasia* (d'Orbigny)			
Oonia luciensis (Cossmann)			
Procerithium (*Cosmocerithium*) *leckenbyi* (Morris and Lycett)			
Rigauxia fluens (Piette)			
Cryptaulax (*Xystrella*) *echinatum*(von Buch)			
Rhabdocolpus strangulatus (d'Archiae)			

Text-fig. 1. Stratigraphic range of 6 gastropod species from the Daheishan Formation.

The above six species of gastropods are known to occur in the Bathonian, but only 3 species are confined to it. Hence the Daheishan Formation most probably belongs to the Middle Jurassic Bathonian age. These gastropods ranged in a short geological age and can thus be used for the correlation of biostratigraphic units.

Besides, in the Daheishan Formation 4 species belong to the Cerithiacea and 1 species belongs to the Littorinacea. They occupy 55.5% of the total species. The recent species of Littorinacea and Cerithiacea are generally living in the shallow sea or the littoral area. It is presumed that these fossil gastropods have lived in the similar environments.

Pseudomelania (*Rhabdoconcha*) *aspasia* (d'Orbigny) has been discovered from the Callovian of Tanganyika and the Bathonian of France; *Oonia luchiensis* (Cossmann) from the Great Oolite of Switzerland; *Rigauia fluens* (Piette) from the Bathonian of France; *Procerithium* (*Cosmocerithium*) *leckenbyi* (Morris and Lycett) from the Great Oolite of England; *Rhabdocolpus strangulatus* (d'Archiae) from the Bathonian of France and from the Bathonian to the Callovian of Tunisia; *Cryptaulax* (*Xystrella*) *echinatum* (von Buch) from the Dogger of Kleby, Poland. Since these fossils are mostly collected from the Western Europe, it has been proved that during the Middle Jurassic the Tethyan Realm extended to Xizang (Tibet) from the west to the east, sharing the similar genera and species.

DESCRIPTION OF SPECIES

Subulitacea

Pseudomelaniidae

Pseudomelania (Rhabdoconcha) aspasia (d'Orbigny)

(Pl. 1, figs. 1–2)

1965 *Pseudomelaxia aspasia* (d'Orbigny), Cox, Bull. Brit. Mus. (Nat. Hist.) Geol. Suppl. 1, p. 147, pl. 24, fig.10.

Shell moderate, turrical-conical; whorls regularly increasing; periphery located below the middle part of whorl; sutures deep and slightly oblique; Last whorl regularly increasing; periphery bluntly angular; base straight-oblique; aperture ovate, rhombic; parietal lip oblique, columellar lip concave; inner lip curving in the middle.

Oonia luciensis (Cossmann)

(Pl. 1, figs. 3–6)

1885 *Odontostonia luciensis* Cossmann, Mem. Soc. Geol. France, ser. 3, Tome 3, p. 220, pl. 4, fig.33.
1888 *Odontostonia* (?) *luciensis* Cossmann, Greppin. Abh. Schweiz Paläont. Ges., vol. 15, p. 55, pl. 2, fig. 17.

Shell small, conical-ovate; spire slightly convex, conical, with about 3-4 whorls, which are slightly convex and regularly increase on the spire; last whorl rapidly ventricose; aperture oval; posterior end acute; parietal lip oblique; columellar lip slightly ocncave. Anomphalous.

Loxonematacea

Zygoperuidae

***Rigauxia fluens* (Piette)**

(Pl. 1, figs. 7–8)

1857 *Turritella fluens* Piette, Bull. Soc. Geol. France, ser. 2, Tom 14, p. 560, pl. 7, fig. 16.

Shell small, awl-like; suture slightly oblique; prominent circle near adapical suture; surface concave, ornamented with growth lines, obscure spiral lines and 16 opisthocyrt crescent slender ribs, which become thickened at base; periphery bluntly angular; aperture probably ovate. Anomphalous.

Coelostylinidae

***Trypanostylus* (*T.*) *elegantulus* sp. nov.**

(Pl. 1, figs. 9–10)

Shell small, turrical, 12mm in height, 2mm in width, with at least 6 whorls; whorls slightly convex, regularly increasing; suture deep and slightly oblique; protoconch ornamented with axial ribs; the last 4th whorl with 11 strong and slightly oblique axial ribs; interspace of ribs wide; last whorl rapidly increasing; axial ribs vanished; periphery rounded; base straight-oblique; aperture rhombic; parietal lip oblique; columellar lip straight. Perforate.

Remarks: This new species is similar to *Trypanostylus* (*T.*) *funiculi gerus* (Piette) from the Bathonian stage in Aisne of France (1857, Bull. Soc. Geol. France, ser. 2, Tom 14, p. 552, pl. 8, fig. 17) in the protoconch with axial ribs, but differs from the latter in the small-sized shell and in the axial ribs, which gradually weaken and disappear in the last whorl.

Littorinacea

Purpurinidae

***Purpurina (Pseudalaria) adsutura* sp. nov.**

(Pl. 1, figs. 11–12)

Shell small, conical, 12.6mm in height, 9.3 mm in width; wholes regularly increasing; surface with stronger spiral carinae crossed by growth lines to form granular pattern; penultimate whorl with 4 spiral carinae; a strong one at the adapical suture; two at the middle whorl, of which the one at the adapical part is the strongest and the other moderately strong; the last one at the adapical suture line partly enveloped by the last whorl; last whorl rapidly increasing with 6 spiral carinae; from the suture downward, the second carina being the strongest, the first, third, fourth ones coming next, and the fifth and sixth ones rather slender; aperture broken, inner lip probably arched.

Remarks: This new species closely resembles *Pseudalaria ornata abbas* from the Inferior Oolite Formation of England (1892, Palaeont. Soc. Monogr. vol. 46, p. 280, pl. 21, figs. 16–18) in the spiral carinae, but differs from the latter in having one carina of the adapical suture on the whorl surface close to the suture.

Cerithiacea

Procerithiidae

***Procerithium (Cosmocerithium) leckenbyi* (Morris and Lycett)**

(Pl. 2, figs. 1–3)

1985 *Chemnitzia leckenbyi* Morris and Lycett, Palaeont. Soc. Monogr. vol. 1, p. 50, pl. 7, fig. 4.

Shell small, turrical; whorl regularly increasing; suture prominent; adapical surface of whorl slightly convex; spire terraced; surface of whorl flat and straight, with slightly oblique and arched axial and spiral ornamentations crossed with each other to form granular pattèrn; periphery bluntly angular; aperture rounded-rhombic; inner lip bluntly angular. Anomphalous.

Rhabdocolpus trilicius sp. nov.

(Pl. 2, figs. 7–12)

Shell turrical; whorls rapidly increasing, suture slightly oblique; whorl ornamented with 14-15 strong and slightly oblique axial ribs, upper part (adapical) of each axial rib slightly concave, other part (abapical) strong; 3 spiral lines at the adapical suture, others at the abapical suture obscure; periphery angular; base slightly oblique; aperture nearly oblique and rhombic; parietal lip oblique; columellar lip abaxially oblique; inner lip angular at middle; outer lip straight. Anomphalous.

Remarks: The present species is similar to *Procerithium (Rhabdocolpus) mandawaense* Cox 1965 from the Callovian stage in Tanganyika (Bull. Brit. Mus. (Nat. Hist.) Geol. Suppl. 1, p. 156, pl. 27, figs. 9–12) in the whorls which are flattened in the lateral surface with spiral and axial ornamentations, but differs from the latter in the 3 spiral lines at the adapical suture, in the obscure spiral lines at the abapical suture and in the strong axial ribs near the abapical suture.

Rhabdocolpus strangulatus (d'Archiae)

(Pl. 2, figs. 5–6)

1850 *Cerithium strangulatum* Archiae, Morris and Lycett, Palaeont. Soc. Monogr. vol. 1, p. 81, pl. 9, fig. 18.
1969 Exelissa strangulata (d'Archiae), Cox, Annales de Paleontologie vol. 55, p. 255, pl. 1, fig. 10

Shell small, high-conical, surface nearly flat; each whorl ornamented with 7 strong axial ribs; 14 slender spiral lines present between the interspace of axial ribs at the penultimate whorl, 20 at the last whorl.

The shell is ill in preservation, the upper part of aperture assumes an angular shape with thickened collumellar lip. Based on the aperture, it is tentatively referred to *Rhabdocolpus.*

Cryptaulax (Xystrella) echinatum (von Buch)

(Pl. 2, fig. 4)

1974 *Cryptaulax (Xystrella) echinatum* (von Buch) Grunderl, Z. Geol. Wiss. Berlin, Bd. 2, Heft. 7, S. 842, Taf. 2, Bild. 5-8, Abb. 11.

Shell small, high-conical, whorl regularly increasing, ornamented with strong axial ribs and 2 spiral lines; one of the spiral lines forming a shoulder, the other lying near the abapical suture; suture deep, peripheral carina strong on the last whorl; basal spiral ones thickened.

REFERENCES

Arkell, W. J., 1931. The upper Great Oolite Bradiford and Forest Marble of South Oxfordshire and succession of Gastropod faunae in the Great Oolite. Quart. Jour. Geol. Soc., London, 87.

Base, E., Karrenberg H., Lehman J.P., Alloiteau J. and Lefrance J. P., 1954. Fossiles du Jurassiques superieur et des (Gres de Nubie) de la region de Sana (Yemen). Bull. Soc. Geol. Fr., ser. 6, tome 4.

Brosamlen, R., 1909. Beitrag zur Kenntnisder Gastropoden des Schwabischen Jura. Palaeontographica, Bd. 56.

Cossmann, M., 1885. Contribution a l'etude de la faune de L'Etage Bathenien en France (Gastropodes). Mem. Soc. Geol. France, ser. 3, tome 3.

_______, 1899. Note sur les gastropodes du gisement Bathonien de Saint-Gaultier (Indre.). Bull. Soc. Geol. Fr., ser. 3, tome 27.

_______, 1913. Contribution a la paleontologie francaise des terrains jurassiques. III Cerithiacea et Loxonematacea. (suite et fin). Mem. Soc. Geol. Fr., Paleont., 46.

_______, 1925. Description des do siles. In: Douville (H.) Le Callovien dans le massif du Moghara. Bull. Soc. Geol. Fr., ser. 4, tome 25.

Cox, L. R., 1965, Jurassic Bivalvia and Gastropoda from Tanganyika and Kenya. Bull. Brit. Mus. (Nat. Hist.), Geol. Suppl., 1.

_______, 1969. Gastropods Jurassique du Sud-Est Tunisien. Annal. Palent., Paris, 55.

_______ and Arkell. W. J., 1948-1950. A survey of the Mollusca of the British Great Oolite series. Palaeontogr. Soc. Monogr., 102 and 103.

Delpey, G., 1940. Les gastropodes Mesozoiques de la Region Libanaise. Notes et Mem. Haut. Commissariat R. F. en Syrie et au Liban, tome 3.

Greppin, E., 1888. Description des fossiles de la Grande Oolite des Environs de Bale. Abh. Schweiz Paläont. Ges., 15.

Grundel, J., 1974. Bemerkungen zur Fassung der Gattunges *Procerithium* Cossmann 1902 und *Cryptaulax* Tate 1869 (Gastropoda, Cerithiacea) in Jura. Z. Geol. Wiss. Berlin, 2(6).

_______, 1974. Gastropoden aus dem Dogger 2, Procerithiidae, ibid., 2(7).

Hudleston, W. H., 1887-1896. A Monograph of the Inferior Oolite gastropod, Palaeont. Soc. Monogr., 46.

Morris, J. and Lycett, J., 1851-1855. A monograph of the Mollusca from the Great Oolite. ibid, 1–2.

Piette, E., 1857. Description des Cerithium enfouis dans les depots Bathoniens de l'Aisne et des Ardennes. Bull. Soc. Geol. Fr., ser. 2, tome 14.

Sohl, N. F., 1965. Marine Jurassic gastropods, central and southern Utah. Geol. Surv. U.S. Prof. Pap., 503-D.

Walther, J., 1951. Jurassische mikrofossilien insbesondere gastropoden, am Sudrand des Hils (Ergebnisse einer Kartierung nach Microfaunen). Paläont. Z. Band 25, Heft 1–2.

Wang Yigang and Sun Dongli, 1984. The Triassic and Jurassic palaeogeography and evolution of the Qinghai-Xizang (Tibet) Plateau. Can. J. Earth Sci., 22.

Yang Shengqiu and Wang Huiji, 1985. Jurassic-Cretaceous *Nerinea* from Xizang. Acta Palaeont. Sinica, 24(4).

Yin Jixiang, 1962. On the occurrence of *Nerinea* in north Tibet. ibid, 10(4).

Yu Wen, 1975. The gastropod fossils from the Mount Jolmo Lungma Region. In: Report of Scientific Expedition in the Mount Jolmo Lungma Region (1966–1968). Palaeontology, Fasc. 1, Science Press, Beijing (in Chinese).

Yu Wen and Xia Jinbao, 1985. Late Jurassic-Early Cretaceous Nerineids from Northern Xizang. Acta Paleont. Sinica, 24(5).

EXPLANATION OF PLATES

Plate I

Figs. 1–2. *Pseudomelania* (*Rhabdoconcha*) *aspasia* (d'Orbigny) × 1.3, Cat. No. 1369, Field No. 71-Gai-×26.

Figs. 3–6. *Oonia luciensis* (Cossmann)

3–4. × 7.1, Cat. No. 1370, Field No. 71-Gai-×26.

5–6. × 7.1, Cat. No. 1371, Field No. 71-Gai-×26.

Figs. 7–8. *Rigauxia fluens* (Piette)

× 3.3, Cat. No. 1372, Field No. 71-Gai-×26.

Figs. 9–10. *Trypanostylus* (*T.*) *elegantulus* sp. nov. Holotype, × 5.5, Cat. No. 1373, Field No. 71-Gai-×26.

Figs. 11–12. Purpurina (*Pseudalaria*) *adsutura* sp. nov. Holotype, × 4.4, Cat. No. 1374. Field No. 71-Gai-×26.

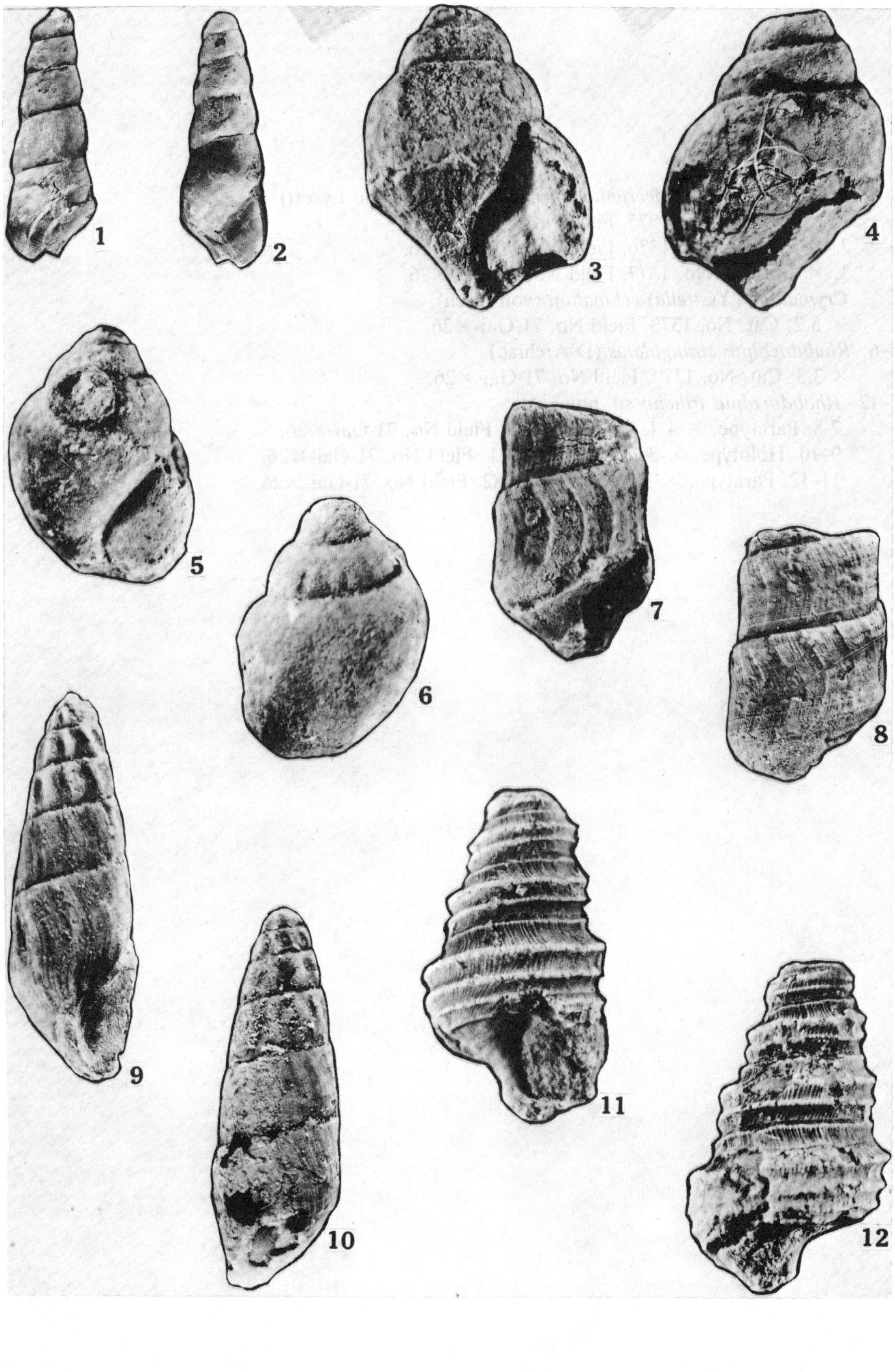

Plate 2

Figs. 1–3. *Procerithium* (*Cosmocerithium*) *leckenbyi* (Morris and Lycett)
1. × 7.5, Cat. No. 1375, Field No. 71-Gai-×26.
2. × 5.1, Cat. No. 1376, Field No. 71-Gai-×26.
3. × 7.5, Cat. No. 1377, Field No. 71-Gai-×26.

Fig. 4. *Cryptaulax* (*Xystrella*) *echinatum* (von Buch)
× 5.2, Cat. No. 1378, Field No. 71-Gai-×26.

Figs. 5–6. *Rhabdocolpus strangulatus* (D'Archiae)
× 3.3, Cat. No. 1379, Field No. 71-Gai-×26.

Figs. 7–12. *Rhabdocolpus trilicius* sp. nov.
7-8. Paratype, × 4.4, Cat. No. 1380, Field No. 71-Gai-×26.
9–10. Holotype, × 3.7, Cat. No. 1381, Field No. 71-Gai-×26.
11–12. Paratype, × 3.3, Cat. No. 1382, Field No. 71-Gai-×26.

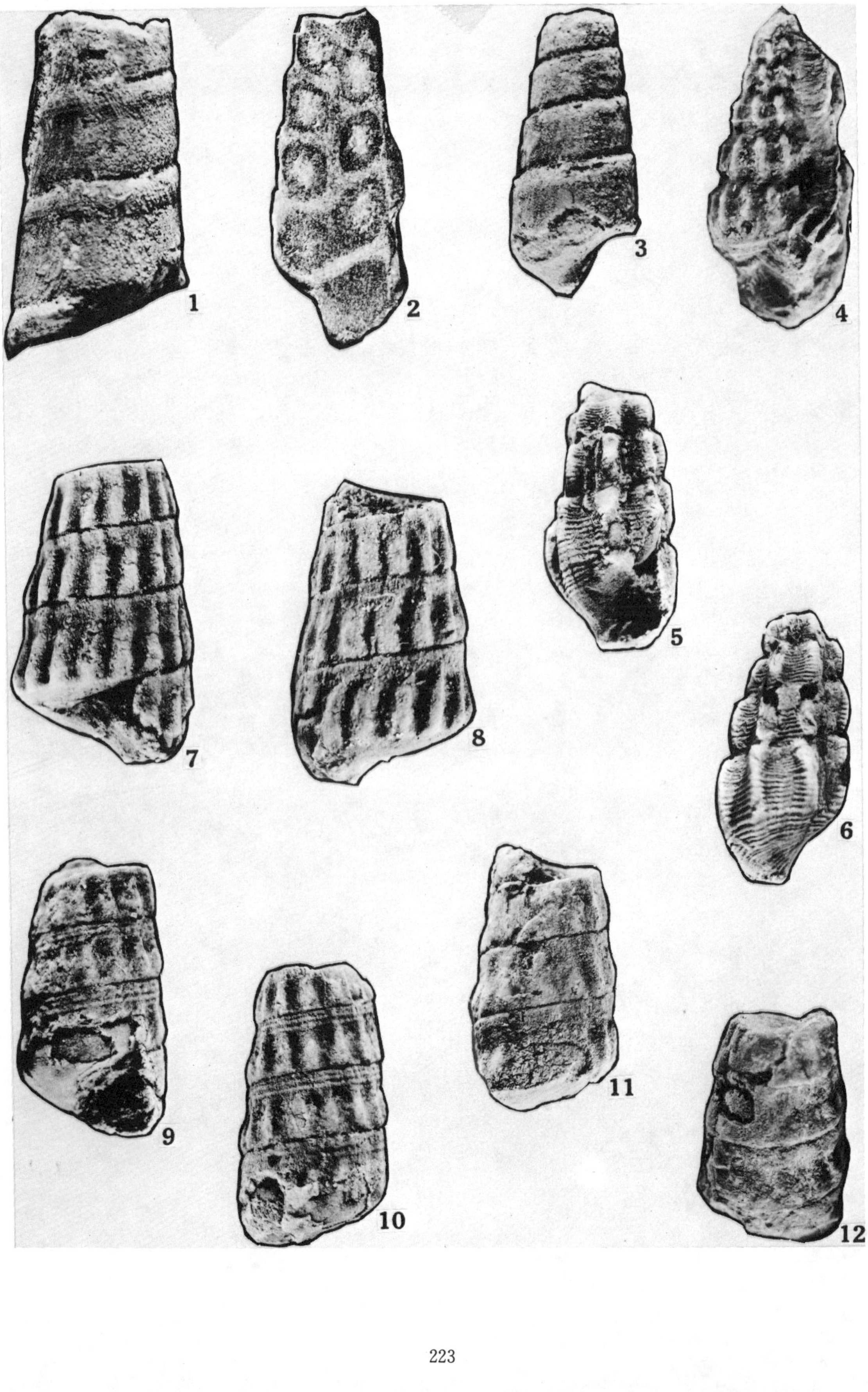
1
2
3
4
5
6
7
8
9
10
11
12

PALAEONTOLOGIA CATHAYANA Volume 5, 1990, p.225–262

JURASSIC BRACHIOPODS FROM WESTERN YUNNAN, CHINA

Sun Dongli

(Nanjing Institute of Geology and Palaeontology, Academia Sinica)

Abstract

19 named species and 1 indeterminate species belonging to the genera *Burmirhynchia, Moquellina, Holcothyris, Kutchithyris* and *Orthotoma* are here described, including 6 new species: *Moquellina semicostata, Holcothyris liuwanensis, H. yunnanensis, H. nongkanensis, Kutchithyris datianyakouensis* and *Kutchithyris longlingensis*. The described specimens were obtained from the lower and middle parts of the Liuwan Formation (Lower and Middle Jurassic) in western Yunnan, China. A systematic revision is made of the genera and species from this area described by Reed (1927), and the geographical and stratigraphical distribution of the brachiopod fauna is also discussed.

INTRODUCTION

A large number of Jurassic brachiopods were collected from Liuwan of Baoshan, Mengga and Nongkan of Luxi, Datianyakou of Longling, Wanding of Ruili, western Yunnan in the year of 1981 when an expedition to the Hengduan Mountain region organized by the Chinese Academy of Sciences was carried out (see Fig. 1).

The marine Jurassic rocks in western Yunnan were investigated long since. As early as in 1908–1910, Brown (1916, 1936) made geological journeys along the Nujiang River valley in western Yunnan and collected Jurassic fossils. But it was Reed (1927) who first gave a detailed description of the Jurassic brachiopods of western Yunnan. In his paper he indicated a close similarity of the brachiopod fauna of the Liuwan beds with that of the Namyau beds of Burma. But when discussing the age of the Liuwan beds, he claimed that a Kimmerridgian age might be ascribed to this fauna, since the evidence of other fossils did not point to a Bathonian age as deduced by Buckman from the brachiopods of the Namyau beds.

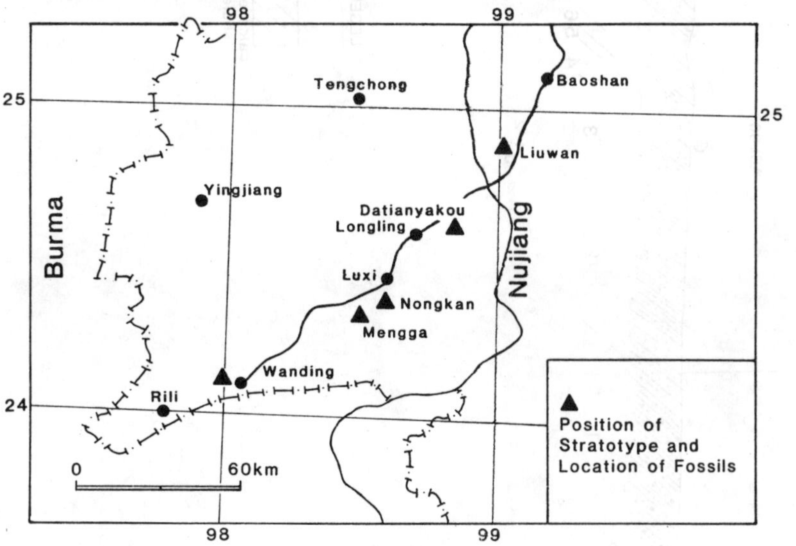

Text-fig. 1. Location of brachiopod-bearing beds in western Yunnan.

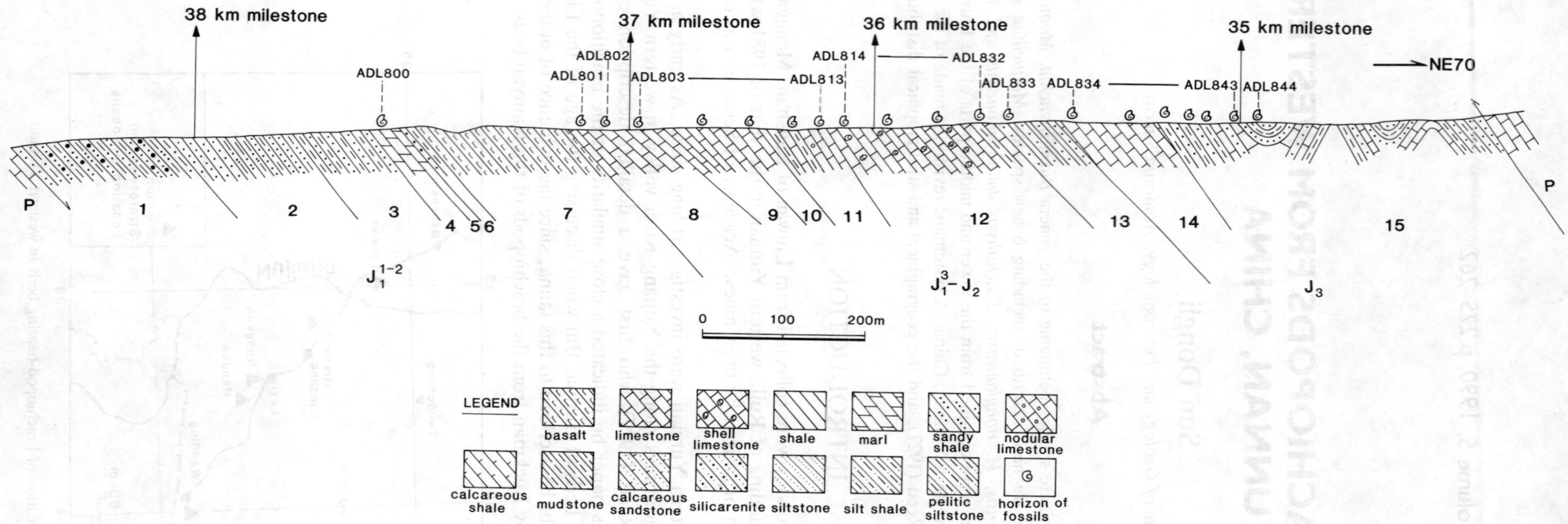

Text-fig. 2. The Jurassic section in Liuwan, Baoshan county, western Yunnan.

According to the members of the Yunnan Geological Bureau (1965), the marine Jurassic rocks in Liuwan of Baoshan were divided initially into six main units (in ascending order): 1. purplish red clastic rocks; 2. interbedding of purplish marl and shale; 3. volcanic rocks; 4. dark-grey thick-bedded limestone; 5. interbedding of mottled mudstone and marl; 6. grey medium-bedded limestone. Later, Yin Jixiang and Fang Zhongjing (1973) divided the Jurassic rocks in western Yunnan into four formations: the Nongkan Formation (J_3), the Longhai Formation (J_2^2), the Liuwan Formation (J_2^2) and the Mengga Formation (J_2^1).

In 1981, Chen Peiji, Chen Jinhua, Cao Meizhen, Pen Huazhang, He Guoxiong and the author restudied the stratotype of the Mengga Formation, the Liuwan Formation, the Longhai Formation and the Nongkan Formation in this area, and put forward the following suggestions for a revision of the previous Jurassic division: 1. the Jurassic section in Liuwan is the best in western Yunnan and can be considered as the standard one for the Jurassic division and correlation in the area; 2. the Liuwan Formation may be roughly equal to the Middle Jurassic in age, with its lower part perhaps including some Lower Jurassic deposits, the Liuwan beds named by Reed for the beds with the brachiopod *Holcothyris angulata-Burmirhynchia trilobata* assemblage virtually correspond to the middle part of the Liuwan Formation and its age is therefore Bathonian; 3. the middle and upper parts of the original Mengga Formation correspond to the lower part of the Liuwan Formation, the present concept of the Mengga Formation only includes the lower part of the original Mengga Formation in Mengga of Luxi; 4. the Longhai Formation erected by Yin and Fang (1973) should be considered as a facies change of the Liuwan Formation and, therefore, the term of Longhai Formation should be discarded. (see Fig. 2 and Table 1.)

Table 1. Marine Jurassic division in western Yunnan.

SERIES	THIS PAPER: LIUWAN BAOSHAN COUNTY	Bed	YIN AND FANG (1973): MENGGA LUXI COUNTY		NONGKAN LUXI COUNTY	DATIANYAKOU LONGLING COUNTY	WANDING RUILI COUNTY
J_3	NONGKAN FORMATION	15			NONGKAN FORMATION		
J_2	LIUWAN FORMATION	14	"LONGHAI FORMATION"		"LIUWAN FORMATION"	"LONGHAI FORMATION"	"LONGHAI FORMATION"
		13					
		12	"LIUWAN FORMATION"			"LIUWAN FORMATION"	"LIUWAN FORMATION"
		11	"MENGGA FORMATION"	MIDDLE AND UPPER MEMBER			
J_1		10					
		9					
		8					
	MENGGA FORMATION	7		LOWER MEMBER	MENGGA FORMATION	MENGGA FORMATION	MENGGA FORMATION
		3-6					
		2					
		1					

Lithologically, the stratigraphical sequence shown in Fig. 2 is made up of four units (in ascending order): 1. lower red clastic rocks (Beds 1–6); 2. volcanic rocks (Bed 7); 3. limestone with marl and mudstone in the middle part (Beds 8–14); 4. upper red clastic rocks (Bed 15). It suggests a cycle: regression-transgression-regression. Such sedimentary characters are very common in the Jurassic rocks of western Yunnan and at other places of West China.

GENERAL CHARACTERS OF THE BRACHIOPOD FAUNA

The brachiopod fauna occurred together with bivalves in the interbedding of grey calcareous shale and argillaceous limestone of the middle part of the Liuwan Formation (e. g. Bed 12 in the Liuwan section). The fauna seems to represent those living on the sublittoral sea floor and consists mainly of rhynchonellids and terebratulids, including 4 genera and 19 species, among which 6 species are new. The range of the species in the middle part of the Liuwan Formation shows that *Burmirhynchia trilobata* Ching, Sun et Ye, *B. adjudicata* Reed, *B. conjurata* Reed, *Holcothyris angulata* Buckman mainly occur in the lower levels; *Holcothyris nongkanensis* sp. nov., *H. yunnanensis* sp. nov., *Kutchithyris datainyakouensis* sp. nov., *K. longlingensis* sp. nov. usually appear in the high levels; and the rest are almost found from all the brachiopods-bearing beds. Therefore the author considers the fauna as a unitary assemblage biozone and does not think that it should be subdivided into more biozones.

Although the major species of the fauna are endemic, it still shows a close similarity with those of the Namyau beds of Burma (Buckman, 1918) and the Yansheping Group of northern Tibet and southern Qinghai (Ching, Sun and Ye, 1979; Sun, 1981). For example, *Holcothyris angulata* Buckman has been found in the above-mentioned area. *Burmirhynchia trilobata* Ching, Sun et Ye, *B. cuneata* Ching, Sun et Ye, *B. flabilis* Ching, Sun et Ye, *B. luchiangensis* Reed and *Kutchithyris olivaeformis* are also very common elements of the Yansheping Group.

The fauna characterized by *Burmirhynchia* and *Holcothyris* is known to be wide-spread in Asia (Afghanistan, Pamir, Southwest China, Burma, Laos and Japan). Some species of *Burmirhynchia* even occur commonly in Europe and Africa (Buckman, 1918; Laurin, 1984; Dubar, 1967; Muir-Wood, 1935, 1937). *Holcothyris* has been reported from the Bathonian stage in Premeyzel of France (Almeras, 1971). Recently, Almeras and Gupta (1986) indicated a close affinity of *Cererithyris, Holcothyris* and *Kutchithyris*. They also mentioned the Ovcharenko's view on the synonym of *Holcothyris* with *Kutchithyris* based on the material from southeastern Pamirs. The fact mentioned above shows the connection of brachiopod faunas between the western and eastern parts of the north shore of Tethys and even between the northern and southern shores of the Tethys. The *Burmirhynchia trilobata-Holcothyris angulata* assemblage of the middle part of the Liuwan Formation is here considered as Bathonian in age, because the genera *Burmirhynchia* and *Holcothyris* mainly appear in the Bathonian stage on the northern shore of the Tethys (Sun, 1980). The Late Lias foraminifer *Lobyrinthina* assemblage appears in Beds 8–9 of the Liuwan section. The Callovian brachiopod *Kutchithyris dongqoensis-Thurmannella rotunta* assemblage from the upper part of the Yunsheping Group in northern Tibet and southern Qinghai and the *Kuchithyris ingluvisa-Kutchirhynchia kutchensis* assemblage from the Lower Callovian stage in Kutch of India have not been found in western Yunnan. Besides, it should be mentioned that a single brachiopod found from the lower part of the Liuwan Formation (Bed 9 in the Liuwan section) has been here identified as *Orthotoma* sp. This species appears to be similar to *O. spinati* Rou. from the Middle Lias of Germany.

SYSTEMATIC DESCRIPTION

Order Rhynchonellida Kuhn, 1949

Superfamily Rhynchonellacea Gray, 1848

Family Rhynchonellidae Gray, 1848

Genus *Burmirhynchia* Buckman, 1918

1918 *Burmirhynchia* Buckman, pp. 49–50.
1965 *Burmirhynchia*, Ager, p. 612.
1972 *Burmirhynchia*, Ager, Childs & Pearson, p. 199.
1979 *Burmirhynchia*, Ching, Sun & Ye, pp. 158–159.
1984 *Burmirhynchia*, Laurin, pp. 387–388.

Type species: *Burmirhynchia gutta* Buckman

Diagnosis: Shell medium in size, globose in outline; fold low and sulcus shallow; anterior commissure uniplicate; linguiform extension sub-square or broadly arched; shell surface ornamented with many rounded or subangular costae; beak massive, strongly incurved, without beak-ridges; foramen large, elongate, hypothyrid and bordered with two narrow deltidial plates.

Pedicle collar absent; dental plates slightly divergent; teeth stout; median septum strong; septalium present; crura base trigonal; crura radulifer, thin and incurved ventrally at the end, forming channels or erect plates.

Remarks: The diagnosis mentioned above is synthesized according to Buckman's definition and others' views (Ager, 1965, 1972; Ching, Sun & Ye, 1979; Laurin, 1984). In young individuals of *Burmirhynchia*, the shell is usually elongate and depressed; anterior commissure flat or slightly arched; beak suberect; septalium wide; crura extending horizontally, radulifer; pedicle collar present. During the growth, the shell becomes transversely wide. In the adult stage, the shell is trilobate in outline and globose in profile; anterior margin arched or sub-square; septalium very narrow and short; crura incurved ventrally and channel-shaped at the end; pedicle collar absent.

Buckman (1918) founded his genus *Burmirhynchia* on the type species *B. gutta* from the upper limestone of the Namyau beds, Northern Shan States, Burma. The genus was world-wide in the Middle Jurassic, especially in the Bathonian age. Buckman originally named more than 52 species of the genus among them 40 species were found from the Namyau beds. These species were mostly erected on their trivial external differences. It should be pointed out that he laid undue emphasis on the use of muscle scars for classifacation. In fact, the Burmese species figured by Buckman were not well preserved from umbonal view. By using the method of the shell burning, the beak had been more or less destroyed. He further attributed the 40 species to 8 series. The specimens of the *Burmirhynchia obesa* series and the *B. transversalis* series are very few, and the others were collected mainly from the Linglawng, Se-Eng and Naksy. In my opinion, such series might roughly correspond to one species erected on a population as type. Most species included in each series, are only variants of a single group.

Kallirhynchia Buckman is very similar to *Burmirhynchia* and differs from the latter in having a straight beak, conjunct deltidial plates, a flat fold with prominent slopes and truncated anterior margin. Muir-Wood (1934) defined a type of crura as calcarifer in a more thorough study of the type species *Kallirhynchia yaxlayensis* (Davidson). But Laurin (1984) considered the crura of *K. yaxlayensis* as radulifer according to British and French materials. He thought 10 species of *Burmirhynchia* described by Buckman from England and *Kallirhynchia concinna* (Sowerby) to be synonyms. He also attributed "*Rhynchonella*" *quadricristata* Rullier and "*R*". *decorata* Schlotheim with stout angular ribs and strongly incurved beaks to *Burmirhynchia*. Many species of *Burmirhynchia* from Burma figured by Buckman (1918), which possibly have Straight beaks, might belong to *Kallirhynchia*.

Stratigraphical range and geographical distribution: Bathonian-Callovian; Europe, Asia and Africa.

Burmirhynchia praestans Reed

(Pl. 1, Figs. 17–20; text-fig. 3)

1927 *Phynchonella (Burmirhynchia) praestans* Reed, pp. 256–258, pl. 20, figs. 12–17a.

Description (*based on topotype specimens*): Shell medium in size, rounded subtrigonal or subcircular in outline, much inflated, subspherical in profile; length nearly equal to width; umbonal angle about 90°; beak pointed, incurved; beak-ridge subangular; planarea small; pedicle opening large, suboval and hypothyrid; deltidial plates disconjunct; 22–28 coarse subangular costae arising at umbo, 5–7 lying on the fold or sinus; fold and sulcus nearly obsolete, appearing on the anterior part of shell; tongue rounded or subquadrate, suddenly arched down ward, with a rounded or subtruncated tip; anterior commissure uniplicate.

Pedicle collar absent; dental plates subdivergent; delthyrial cavity subquadrate; umbonal cavities semicircular; denticula and accessory sockets developed; teeth deeply inserted, terminally expanded; hinge plates inseparable from inner socket-ridges with ventral convexity, anteriorly becoming subhorizontal and straight; septalium short, nerrow; median septum stout and extending to the half way of the shell length; crura radulifer and folded longitudinally, forming darsally facing channel.

Remarks: *Burmirhynchia praestens* is the most common species in the Liuwan Formation, western Yunnan. The original description of muscle marks is as follows: interior of pedicle valve with large sub-pyriform muscle scar extending about half the shell length; interior of brachial valve with inner muscle scars forming a hatchet-shaped area widening rapidly in front, having cancave outer edges and the anterior end suddenly truncated and straight; outer pair of scars oval, faintly outlined; genital areas coarsely marked without definite boundaries. The author doesn't think that muscle scars are very important for classification and the muscle scars were not shown by burning the topotype specimens.

Dimension of specimens:

Cat. No.	Length	Width	Thickness	Umbonal angle	Number of costae	Number of costae on sulcus / in fold
93799	26.3mm	25.0mm	22.2mm	90°	27–29	6 / 7
93800	23.9mm	23.7mm	19.6mm	89°	26–28	6 / 7
93801	21.8mm	23.0mm	16.8mm	90°	23–25	4 / 5
93802	22.8mm	21.5mm	18.2mm	96°	22–24	5 / 6
93803	23.4mm	22.8mm	19.8mm	92°	22–24	5 / 6
93804	20.5mm	20.4mm	18.3mm	95°	22–24	5 / 6
93805	20.0mm	20.6mm	17.5mm	90°	22–24	5 / 6

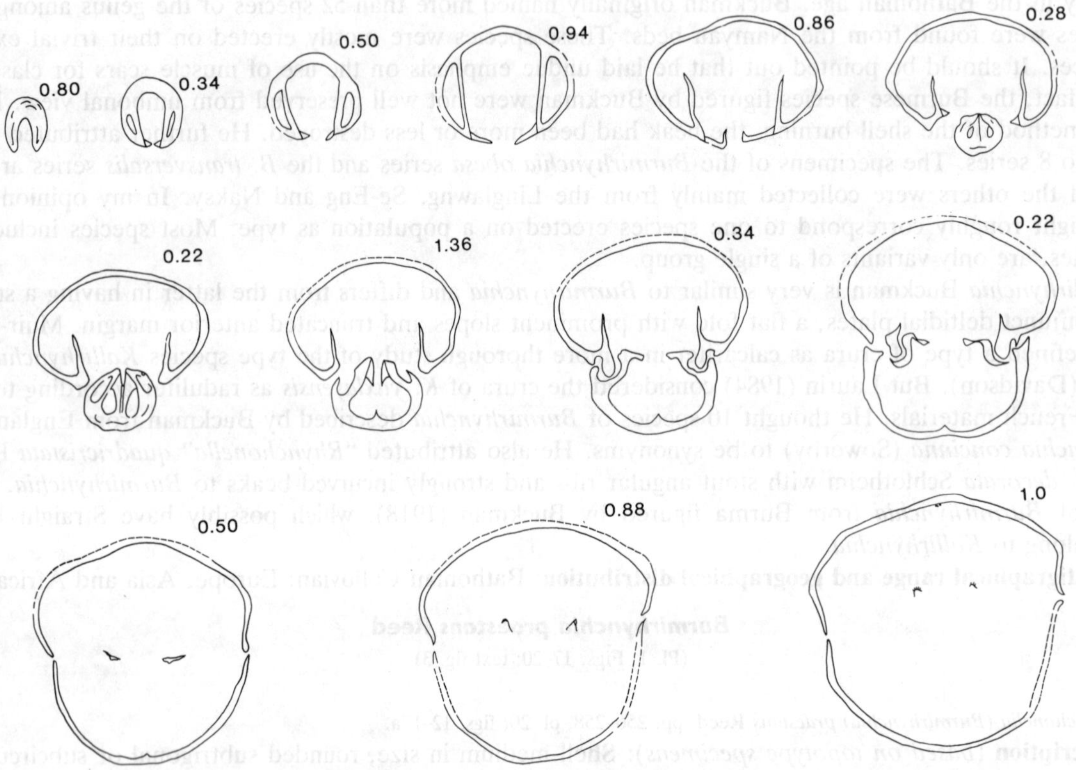

Text-fig. 3. Serial transverse sections of *Burmirhynchia praestans* Reed, × 1.7; Cat. No. 93803 (Topotype).

The present species is very similar to *B. costata* Buckman from the Namyau beds, Burma, (Buckman, 1918) but the latter is distinguished by its subtrigonal outline, fewer costae and small umbonal angle (about 80°). This species also resembles *Somalirhynchia africana* Weir (Muir-Wood, 1935) from the Upper Jurassic, Somaliland in the outline, but the latter is distinguished by its large size, stout subangular ribs and conjunct deltidial plates.

Locality and horizon: Liuwan, Baoshan county and Datianyakou, Longling county; the Middle Jurassic Liuwan Formation.

Burmirhynchia trilobata Ching, Sun et Ye

(Pl. 1, Figs. 13–16; text-fig. 4)

1979 *Burmirhynchia trilobata* Ching, Sun et Ye, pp. 162–163, pl. 46, figs. 29–38.
1981 *Burmirhynchia trilobata*, Sun, pp. 201, pl. 4, figs. 13–16.

Dimension of specimens:

Cat, No.	Length	Width	Thickness	Umbonal angle	Number of costae	Number of costae on fold / in sulcus
93806	22.9mm	25.8mm	17.1mm	110°	28–30	9/8
93807	19.8mm	20.9mm	14.1mm	105°	28–30	8/7

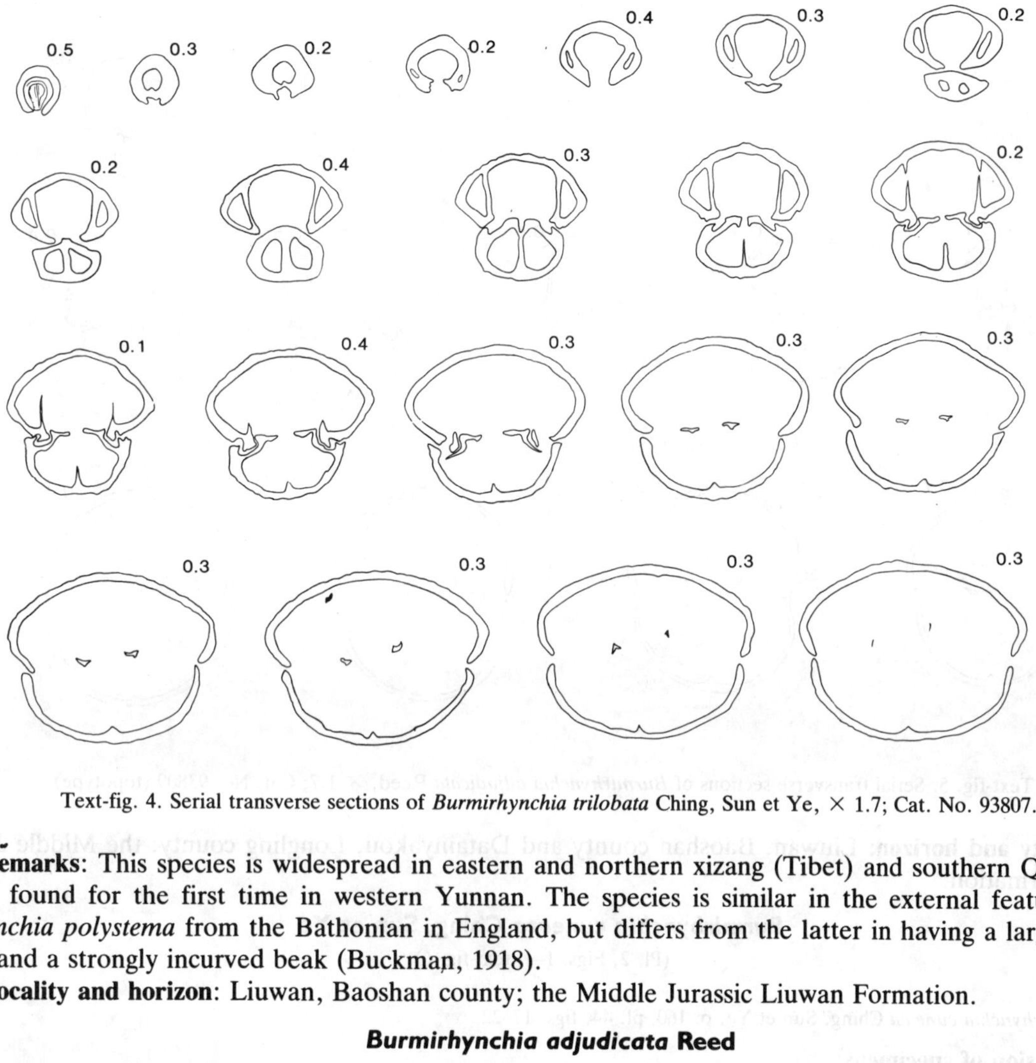

Text-fig. 4. Serial transverse sections of *Burmirhynchia trilobata* Ching, Sun et Ye, × 1.7; Cat. No. 93807.

Remarks: This species is widespread in eastern and northern xizang (Tibet) and southern Qinghai. But it was found for the first time in western Yunnan. The species is similar in the external feature to *Burmirhynchia polystema* from the Bathonian in England, but differs from the latter in having a large umbonal angle and a strongly incurved beak (Buckman, 1918).

Locality and horizon: Liuwan, Baoshan county; the Middle Jurassic Liuwan Formation.

Burmirhynchia adjudicata Reed

(Pl. 1, Figs. 5–8; text-fig. 5)

1927 *Rhynchonella* (*Burmirhynchia*) *praestans* var. *adjudicata* Reed, pp. 259–260, pl. 20, figs. 18, 18a-b, 19.

Dimension of specimens:

Cat. No.	Length	Width	Thickness	Umbonal angle	Number of costae	Number of costae on fold / in sulcus
93810	19.1mm	13.6mm	15.3mm	59°	22	5 / 4
93808	24.2mm	18.2mm	20.8mm	60°	26	7 / 6
93809	20.6mm	18.5mm	16.5mm	60°	24	7 / 6

Remarks: The main diagnosis of the species is its elongated oval to subpyriform outline and the marvellous thickness of shell. According to the sections of topotype, the internal structure is as follows: pedicle collar absent; dental plates parallel; teeth strong and inserted deeply into sockets with terminal expansions; umbonal cavity sub-square; denticula developed; hingeplates flat and merged with inner socket ridges; median septum stout; septalium short; crura bases trigonal; crura radulifer.

Burmirhynchia namtuensis Buckman found from the Namyau beds, Burma (Buckman, 1918) and the Yansheping Group, Tibet and Qinghai, China (Ching, Sun and Ye, 1979) is very similar to the present species in the outline. But it differs from the latter, in having fewer costae (about 16–18). Some specimens of *B. hopkinsi* (Davidson) from England and France (Laurin, 1984, pl. 7, fig. 8; Buckman, 1918, pl. 16, fig. 12) also resemble this species. But they have a strongly incurved beak.

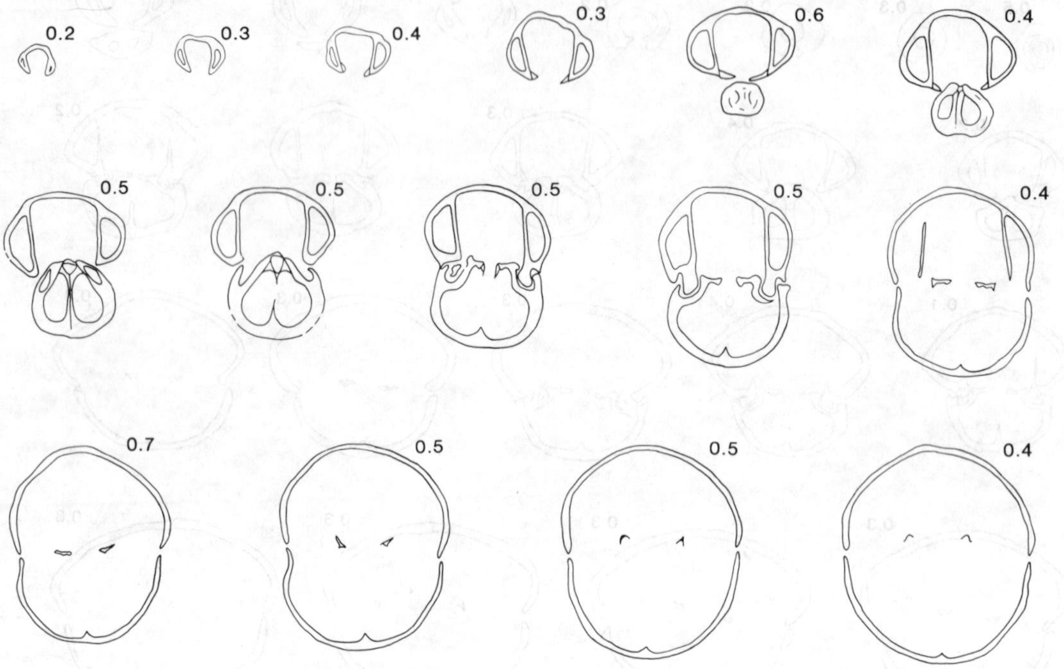

Text-fig. 5. Serial transverse sections of *Burmirhynchia adjudicata* Reed, × 1.7; Cat. No. 93809 (topotype).

Locality and horizon: Liuwan, Baoshan county and Datainyakou, Longling county; the Middle Jurassic Liuwan Formation.

Burmirhynchia cuneata Ching, Sun et Ye

(Pl. 2, Figs. 1–4; text-fig. 6)

1979 *Burmairhynchia cuneata* Ching, Sun et Ye, p. 160, pl. 44, figs. 17–22.

Dimension of specimens:

Cat. No.	Length	Width	Thickness	Umbonal angle	Number of costae	Number of costae on fold / in sulcus
93811	15.3mm	14.4mm	12.9mm	94°	18	5/4
93812	17.5mm	18.1mm	11.8mm	100°	16	5/4
93813	15.7mm	13.3mm	13.7mm	84°	16	5/4
93814	13.5mm	13.0mm	8.9mm	90°	16	5/4
93815	15.3mm	15.2mm	11.5mm	97°	16	5/4
93816	17.1mm	17.2mm	12.0mm	98°	17	6/5
93817	13.7mm	14.2mm	9.7mm	90°	16	6/5

Remarks: This species is widespread in western Yunnan, eastern and northern Xizang (Tibet) and southern Qinghai. It is similar to *B. luchiangensis* Reed in the number of costae but differs from the latter in its subangular costae arising at the umbo and in its greater angle. *B. parva* Buckman from the Namyau

Beds, Lenglawng, Northern Shan States, is distinguished from *B. cuneata* by its small umbonal angle and elongate outline.

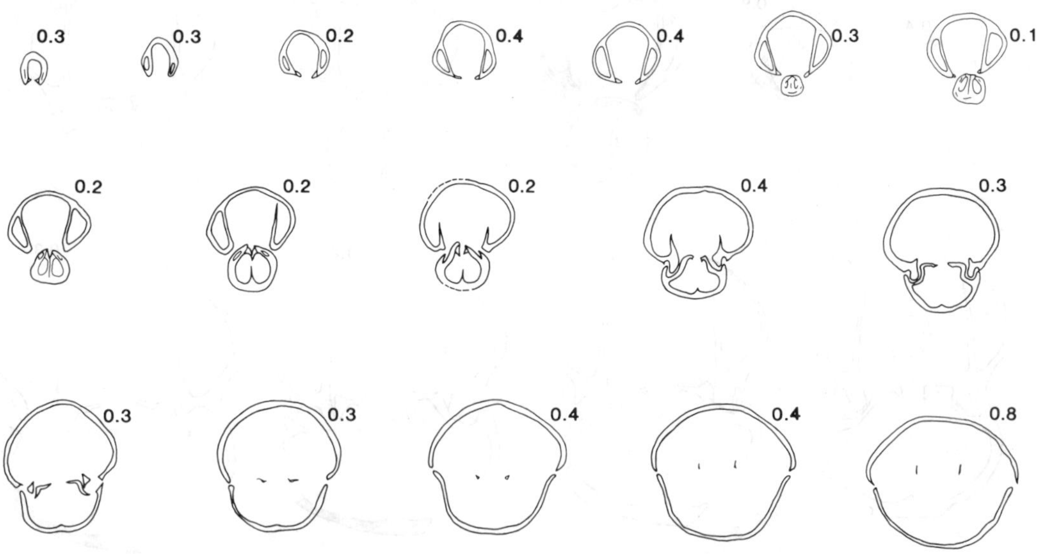

Text-fig. 6. Serial transverse sections of *Burmirhynchia cuneata* Ching, Sun et Ye, × 1.7; Cat. No. 93817.

Locality and horizon: Liuwan, Baoshan county and Datainyakou, Longling county; the Middle Jurassic Liuwan Formation.

Burmirhynchia luchiangensis Reed

(Pl. 2, Figs. 13–16; text-fig. 7)

1927 *Rhynchonella* (*Burmirhynchia*) *praestans* var. *luchiangensis* Reed, p. 261, pl. 19, figs. 21, 22, 22a.
1979 *Burmirhynchia luchiangensis*, Ching, Sun and Ye, p. 164, pl. 44, figs. 1–11.

Dimension of specimens:

Cat. No.	Length	Width	Thickness	Umbonal angle	Number of costae	Number of costae on fold / in sulcus
93818	18.0mm	16.9mm	15.9mm	98°	14–16	4/3
93819	18.2mm	20.8mm	15.5mm	100°	14–16	4/3
93820	18.7mm	19.9mm	14.2mm	92°	16–18	4/3
93821	18.8mm	16.7mm	15.1mm	89°	14–16	4/3

Remarks: The original description of the species does not conform completely with the characters shown by the figures of the type specimens. Actually the umbonal angle should not be 75° but 90°. A smooth area on the posterior part of the shell was not mentioned in the original description. The internal structure of this species, according to the study of the topotype specimens, is as follows: pedicle collar absent; dental plates thin, long, subparallel and extending to plane of articulation; delthyrial cavity sub-square; teeth strong, crenulated and inserted with a terminal expansion; denticula developed; sockets wide, crenulated; both inner and external socket ridges obvious; hinge-plates flattened and slightly convex ventrally; median septum high; septalium short and narrow; crura radulifer.

The species is similar in the outline to *Rutorhynchia quadplicata* Sun (Sun, 1981) from the Upper Jurassic Jieshan Formation in western xizang (Tibet). But the latter is distinguished by its small foramen, symphytium, separated hinge-plates and canalifer crura.

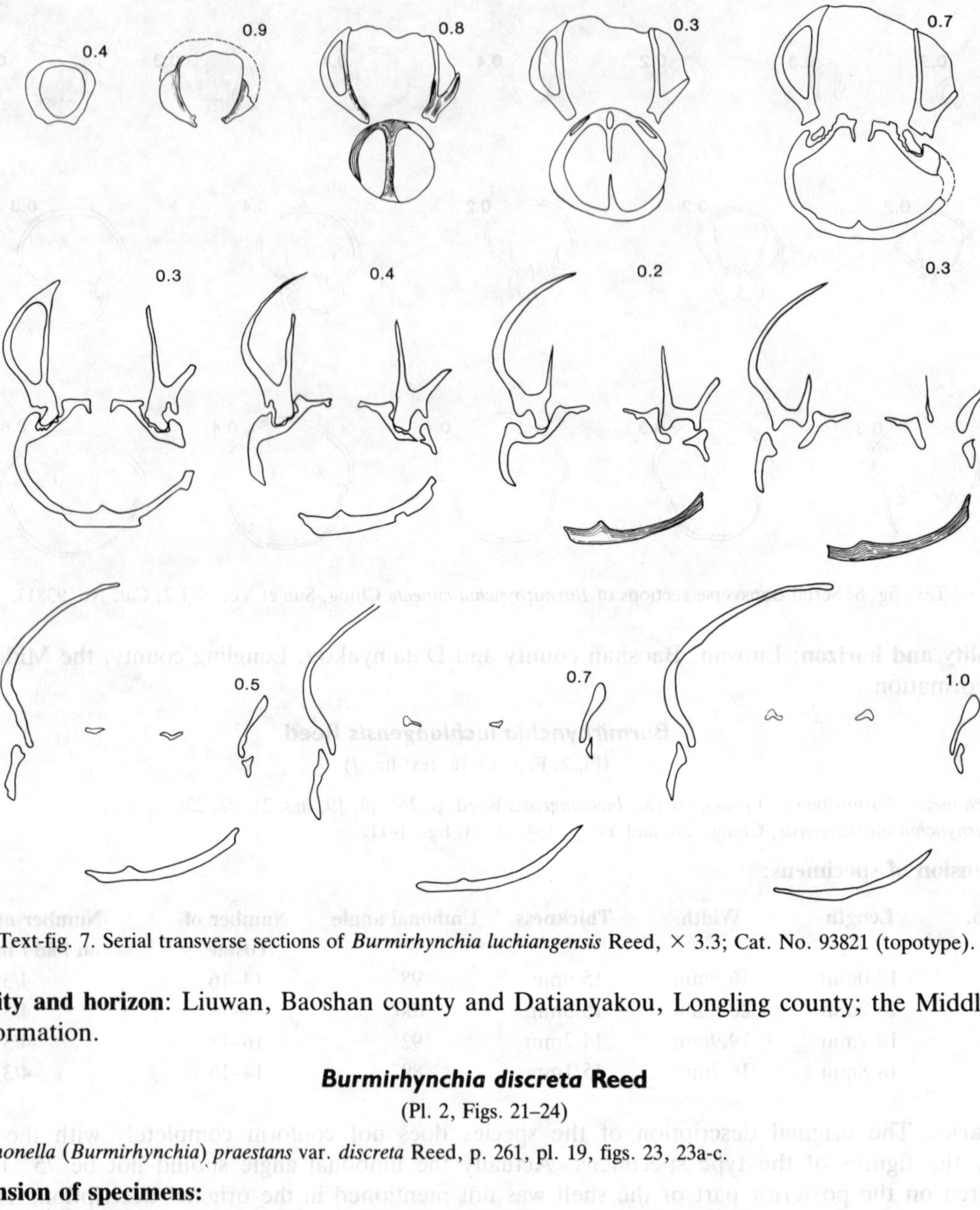

Text-fig. 7. Serial transverse sections of *Burmirhynchia luchiangensis* Reed, × 3.3; Cat. No. 93821 (topotype).

Locality and horizon: Liuwan, Baoshan county and Datianyakou, Longling county; the Middle Jurassic Liuwan Formation.

Burmirhynchia discreta Reed

(Pl. 2, Figs. 21–24)

1927 *Rhynchonella* (*Burmirhynchia*) *praestans* var. *discreta* Reed, p. 261, pl. 19, figs. 23, 23a-c.

Dimension of specimens:

Cat. No.	Length	Width	Thickness	Umbonal angle	Number of costae	Number of costae on fold / in sulcus
93822	20.5mm	20.5mm	14.0mm	95°	22	7/6
93823	18.5mm	20.3mm	14.0mm	95°	22	7/6

Remarks: Reed (1927) indicated in his description that this species was different from *B. luchiangensis* in the size of the middle and the lateral ribs, particularly in the brachial valve in which the 4 median ones are broader and stronger than the 9–10 on each side. As a matter of fact, the specimen illustrated by him was incomplete. The anterior margin is broken. In my opinion the species differs from *B. luchiangensis* in having more costae (about 22–24), in its low and wide fold with 6–7 costae and in its broadly arched linguiform extension. *B. flabilis* Ching, Sun et Ye is very similar to this species, but differs from the latter in its small umbonal angle (about 70°) and subangular costae arising at umbo.

Locality and horizon: Datainyakou, Longling county; the Middle Jurassic Liuwan Formation.

Burmirhynchia tenuiplicata Reed

(Pl. 1, Figs. 1–4, 9–12; text-fig. 8)

1927 *Rhynchonella* (*Brumirhychia*) *praestans* var. *tenuiplicata* Reed, p. 260, pl. 20, figs. 23–26.

Dimension of specimens:

Cat. No.	Length	Width	Thickness	Umbonal angle	Number of costae	Number of costse on fold / in sulcus
93824	25.9mm	23.0mm	17.2mm	98°	40	11/10
93825	15.5mm	18.0mm	10.8mm	96°	30	10/9
93826	18.3mm	16.6mm	11.8mm	83°	32	10/9
93827	17.4mm	17.4mm	12.7mm	90°	32	10/9
93828	18.5mm	16.5mm	9.8mm	85°	32	10/9

Remarks: The species is characterized by the shell with 30–40 fine rounded costae on the surface. It is very similar to *B. shanensis* Buckman (Buckman, 1918) and differs from the latter in having a larger umbonal angle and more costae on the fold.

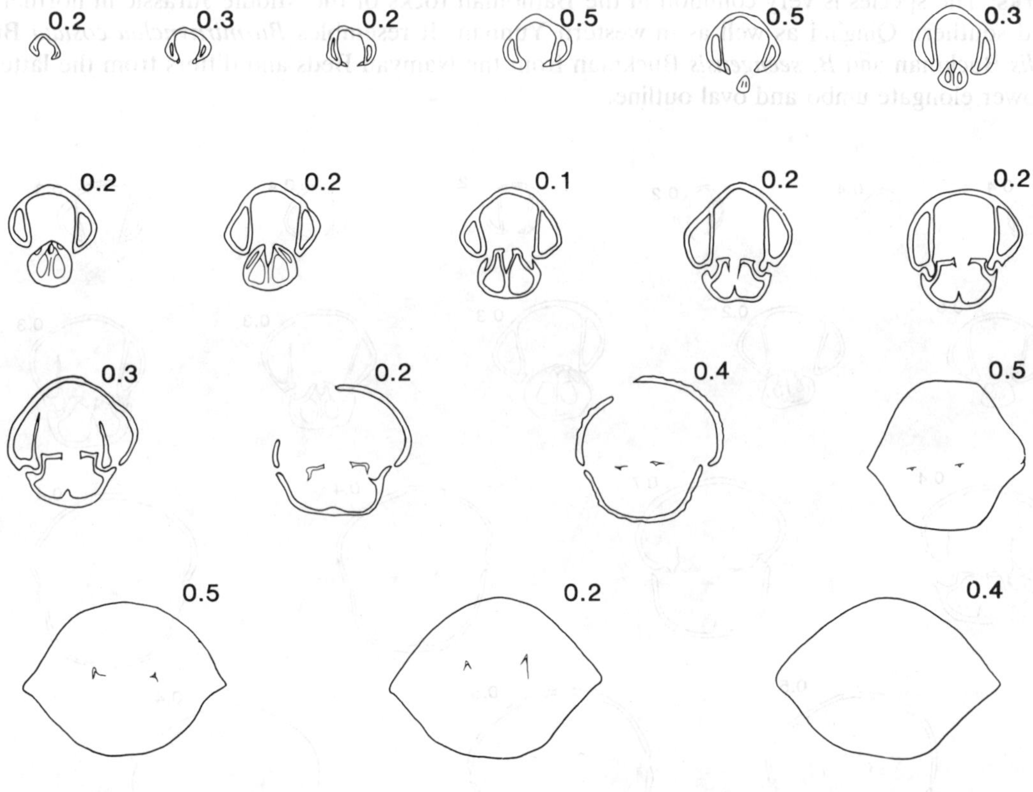

Text-fig. 8. Serial transverse section of *Burmirhynchia tenuiplicata* Reed, × 2.6; Cat. no. 93828 (topotype).

Locality and horizon: Liuwan, Baoshan county; the Middle Jurassic Liuwan Formation.

Burmirhynchia flabilis Ching, Sun et Ye

(Pl. 2, Figs. 25–28; text-fig. 9)

1979 *Burmirhynchia flabilis* Ching, Sun et Ye, p. 160, pl. 43, figs. 25–39.
1981 *Burmirhynchia flabilis*, Sun, p. 201, pl. 6, figs. 25–44.

Description: Shell medium in size, oval in outline (but elongated in juvenile), nearly equally biconvex in profile; dorsal a little more convex than ventral valve; fold and sinus distinctly marked anteriorly; sulcus shallow and fold flat; tongue short gently arched; widest in front of the middle; maximum thickness at middle; anterior commissure sligtly uniplicate; shell surface ornamented with 24–26 rounded costae arising near umbo; averagely 8 costae in a width of 10 mm in the anterior of shell; generally 7 costae on fold; beak small, elevated and slightly incurved; beak ridges round; umbonal angle about 90°; pedicle opening elliptical, hypothyrid; deltidial plates narrow, disjunct.

Pedicle collar narrow in juvenile; dental plates slightly divergent; umbonal cavities semicircular; teeth long and rounded; denticula developed; hinge-plates thick and flat, merged with inner socket ridges; septum stout; septalium short, deep and narrow; crura bases trigonal; crura radulifer, with the terminal part bending ventrally to form erect plate.

Dimension of specimes:

Cat. No.	Length	Width	Thickness	Umbonal angle	Number of costae	Number of costae on fold / in sulcus
93829	23.5mm	20.5mm	18.1mm	90°	26	7/6
93830	22.6mm	20.0mm	16.0mm	89°	24	7/6
93831	17.2mm	16.7mm	11.1mm	90°	26	7/6
93832	20.8mm	20.9mm	12.4mm	93°	24	6/5
93833	21.7mm	21.7mm	16.5mm	96°	24	6/5
93834	22.0mm	19.6mm	13.5mm	90°	24	6/5
93835	20.9mm	17.5mm	14.3mm	90°	24	6/5

Remarks: The species is very common in the Bathonian rocks of the Middle Jurassic in northern xizang (Tibet) and southern Qinghai as well as in western Yunnan. It resembles *Burmirhynchia costata* Buckman, *B. subovalis* Buckman and *B. seengensis* Buckman from the Namyau Beds and differs from the latter mainly in its narrower elongate umbo and oval outline.

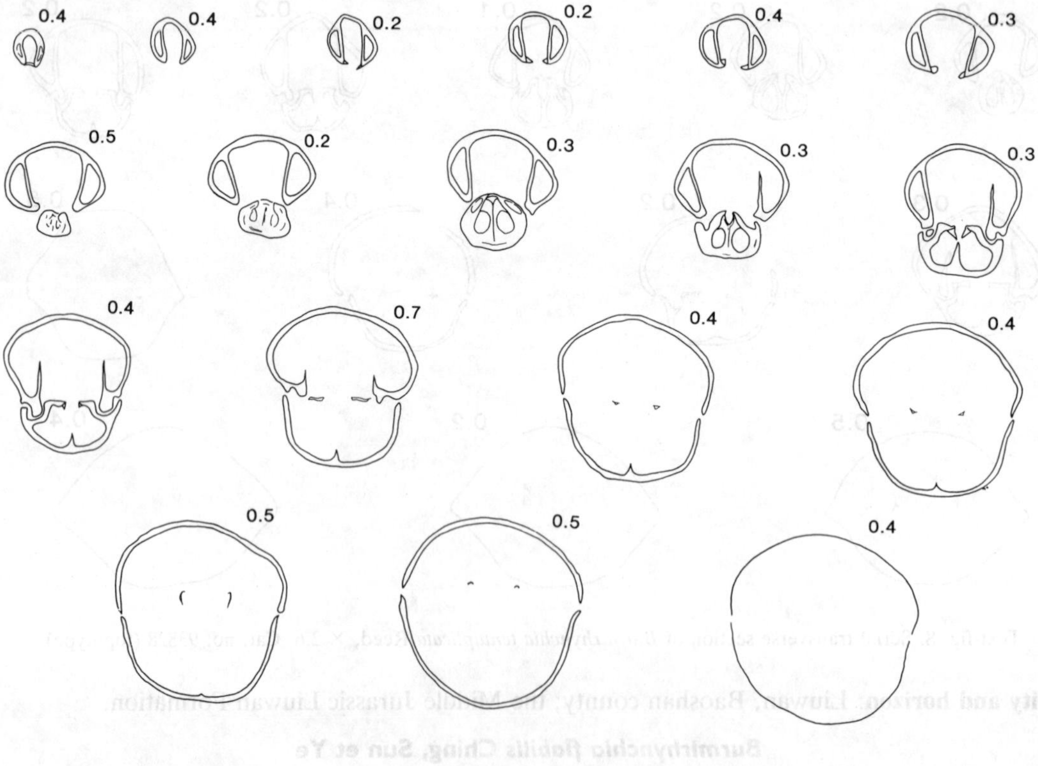

Text-fig. 9. Serial transverse sections of *Burmirhynchia flabilis* Ching, Sun et Ye, × 1.7; Cat. No. 93835.

Locality and hroizon: Liuwan Baoshan county and Datainyakou, Longling county; the Middle Jurassic Liuwan Formation.

Burmirhynchia conjurata **Reed**

(Pl. 2, Figs. 9–12; text-fig. 10)

1927 *Rhynchonella* (*Burmirhynchia*) *praestans* var. *conjurata* Reed, pp. 258–259, pl. 20, figs. 20–22.

Dimension of specimens:

Cat. No.	Length	Width	Thickness	Umbonal angle	Number of costae	Number of costae on fold / in sulcus
93836	15.6mm	15.5mm	13.0mm	100°	29	9/8
93837	16.2mm	15.6mm	14.6mm	80°	27	7/6
93838	15.0mm	13.5mm	11.1mm	90°	27	7/6
93839	15.2mm	13.4mm	10.2mm	92°	27	7/6

Remarks: The species is a common form in the Liuwan Formation. It resembles *Burmirhynchia praestans* in the external features and differs from the latter in its smaller size, trigonal-shaped shell, finer and more costae, elevated fold, better marked sulcus, truncated anterior margin and subsquare tongue. The internal structure of the species, as shown in the serial transverse sections (text-fig. 11), coincides with the general character of *Burmirhynchia*.

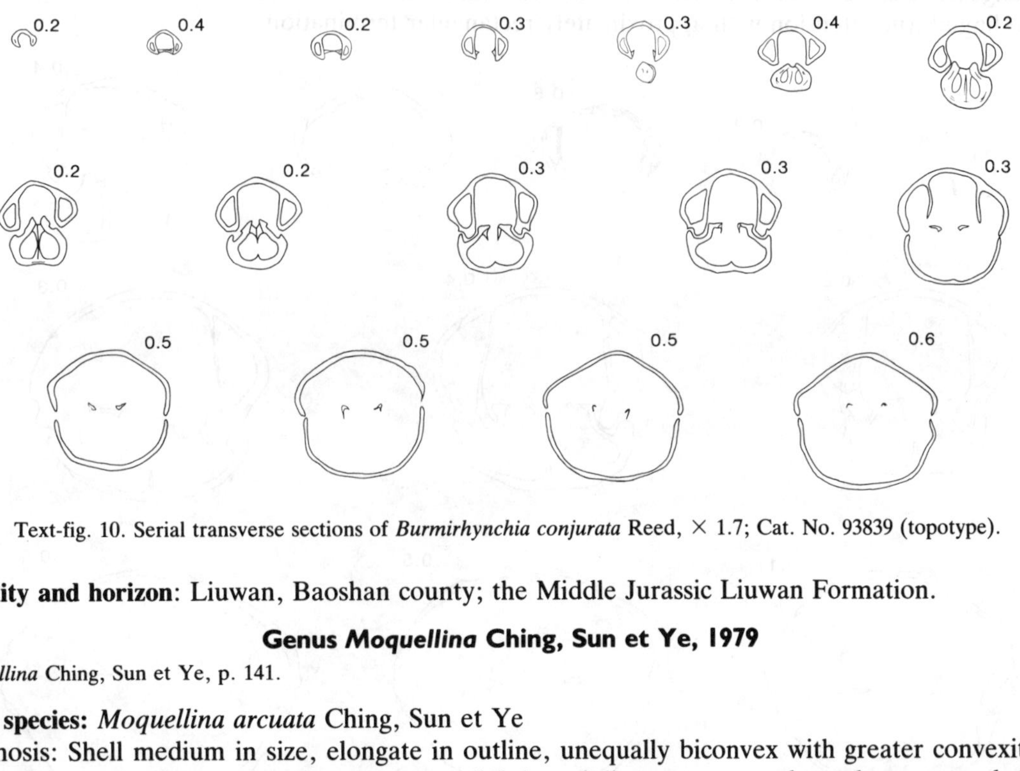

Text-fig. 10. Serial transverse sections of *Burmirhynchia conjurata* Reed, × 1.7; Cat. No. 93839 (topotype).

Locality and horizon: Liuwan, Baoshan county; the Middle Jurassic Liuwan Formation.

Genus *Moquellina* Ching, Sun et Ye, 1979

1979 *Moquellina* Ching, Sun et Ye, p. 141.

Type species: *Moquellina arcuata* Ching, Sun et Ye

Diagnosis: Shell medium in size, elongate in outline, unequally biconvex with greater convexity of dorsal valve; anterior commissure rectimaginate or slightly uniplicate; costae subangular or rounded and developed on all shell surface or in the anterior part of the shell; beak erected or incurved; foramen hypothyrid; deltidial plates conjunct or disjunct; pedicle collar short; crura plates descending directly to the floor of dorsal valve and extending to the plane of articulation; crura radulifer.

Stratigraphical range and geographical distribution: Bathonian–Callovian; West China.

Moquellina semicostata **sp. nov.**

(Pl. 2, Figs. 5–8; text-fig. 11)

Description: Shell medium in size, elongate oval in outline, gently biconvex with greater convexity of dorsal valve in profile; anterior commissure slightly uniplicate or rectimarginate; umbonal angle about 90°; beak short, incurved and sharply pointed; beak ridges subangular; planarea small; foramen large, subcircular and hypothyrid; deltidial plates disjunct; maximum thickness at the posterior of the shell; greatest width at the anterior end of the shell; large smooth area lying posteriorly; 14–15 subrounded costae on the anterior half of the shell surface.

Pedicle collar short; dental lamellae long, subparallel; delthyrial cavity quadrate; teeth strong and denticula developed; hinge-plates separated, short, flattened and fused with inner socket ridges; crural plates directly descending from the dorsal side of hinge-plates to the brachial valve and extending to articulation; crural base trigonal; crura radulifer; median septum low, extending to half way of the shell length.

Dimension of specimens:

Cat. No.	Length	Width	Thickness	Umbonal angle	Number of costae	Number of costae on fold / in sulcus
93840	17.0mm	15.4mm	10.3mm	100°	14	5/4
93841	18.2mm	16.0mm	9.7mm	95°	15	5/4

Remarks: The new species is very similar in the external features to *Scalpellirhynchia scalpellum* (Quenstedt) described by Muir-Wood (1936) and Ager (1965) from the Lower Jurassic of Europe but differs from the latter in its large size and in the way that its crural plates descend to the floor of brachial valve. The type species *M. arcuata* is distinguished from the new species by its subangular costae arising near umbo, conjunct deltidial plates, small umbonal angle (about 75°) and uniplicate anterior commissure. *Burmirhynchia luchiangensis* also resembles the new species, but differs from the latter in its sub-globular shell, short sub-square linguiform extension with approximately rectangular termination.

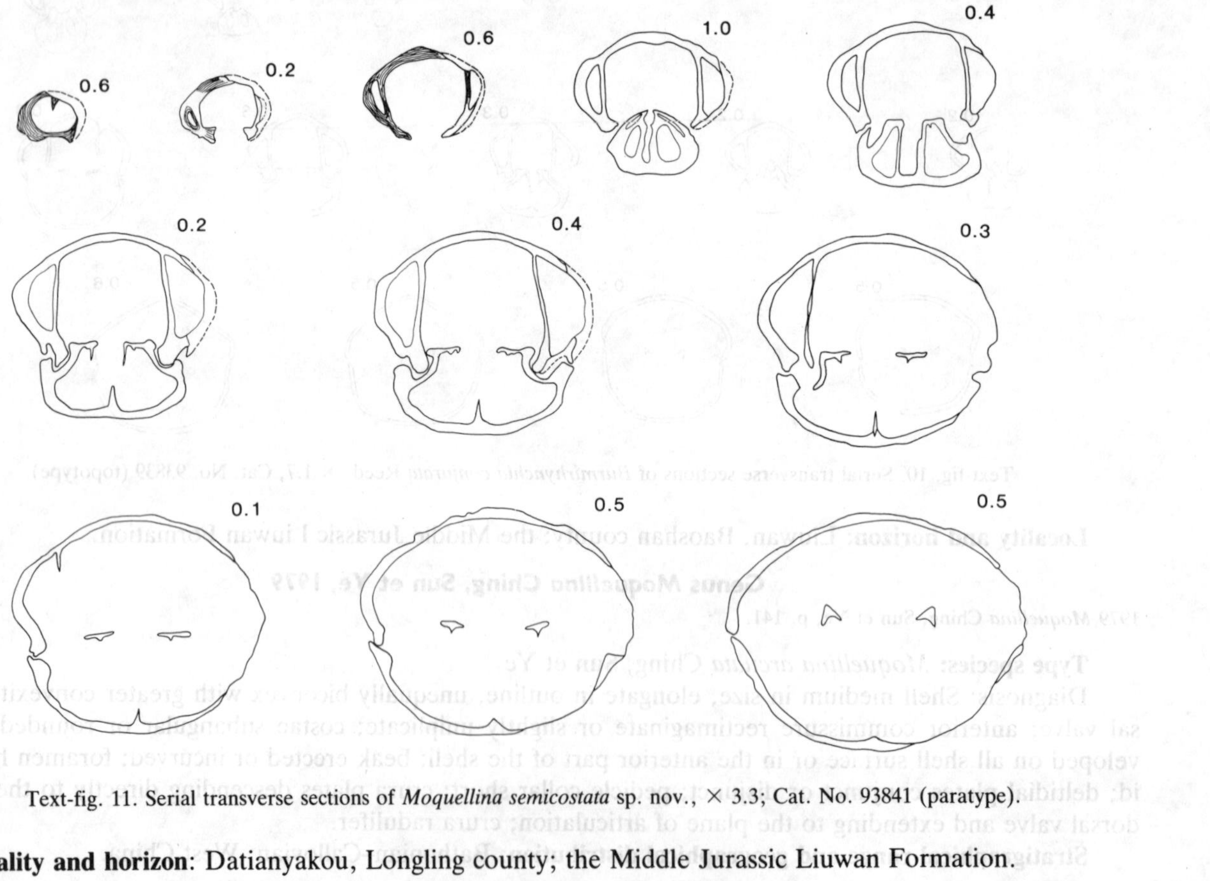

Text-fig. 11. Serial transverse sections of *Moquellina semicostata* sp. nov., × 3.3; Cat. No. 93841 (paratype).

Locality and horizon: Datianyakou, Longling county; the Middle Jurassic Liuwan Formation.

Order Terebratulida Waagen, 1883
Suborder Terebratulidina Waagen, 1883
Superfamily Terebratulacea Gray, 1840
Family Terebratulidae Gray, 1840
Genus *Holcothyris* Buckman, 1918

1918 *Holcothyris* Buckman, p. 125.
1965 *Holcothyris*, Miur-Wood, p. 780.
1971 *Holcothyris*, Almeras, p. 331.
1979 *Holcothyris*, Ching, Sun and Ye, p. 198.
1983 *Holcothyris*, Cooper, pp. 91–92.
1986 *Holcothyris*, Almeras and Gupta, pp. 419–429.

Type species: *H. angulata* Buckman

Emended diagnosis: Shell medium-large in size, roundly trigonal to subpentagonal in outline, moderately biconvex or planoconvex in profile; beak massive, short and incurved; foramen circular or labiate, permesothyrid-epithyrid; symphytium broad, usually concealed; dorsal sulcus developed and persistent from umbo to anterior; ventral fold obvious and carinate; anterior commissure sulcate to paraplicate-sulciplicate; shell surface smooth or ornamented with very fine capillate; pedicle collar developed; dental plate absent; cardinal process medianly depressed and bilobate; myophragm long, sometimes grooved posterriorly; hinge-plates in section slightly demarcated from inner socket ridges, slightly concave ventrally and tapering, clubbed anteriorly; adductor scars long, narrow, tapering posteriorly, widely divergent anteriorly; loop about half the length of the dorsal valve with a high-arched transverse band.

Remarks: The genus originally described by Buckman (1918) from the Namyau Beds, North Shan States, Burma is characterized by well-developed capillation, a dorsal furrow and a massive beak.

Euidothyris, Ptyctothyris and *Epithyris* from the Middle Jurassic bear a resemblance to *Holcothyris* in the external feature, but differ from the genus in having no dorsal sulcus and no capillation. *Avonothyris* Buckman, a common form in the Bathonian stage of Europe, is very similar to *Holcothyris* in its sometimes capillate shell and internal structure. It has a close affinity to *Holcothyris* but is distinguished from the latter by rather small and narrow beak, narrow subparallel adductor scars and moderately swollen dorsal umbo.

Buckman (1918) established the genus *Holcothyris* on the type species *H. angulata* from the Namyau Beds, Burma and believed that the genus embraced 22 species, which fell into 4 types and 7 series: 1. sulcate–the sulcate series; 2. trigonal–the trigonalis series; 3. biplicate – a. the angulata series, b. the pinguis series, c. the macra series and d. the flexa series; 4. dwarfs. According to the concept of population, each series might roughly correspond to one species. In fact the variation of individuals of *Holcothyris* is considerable. For example, in young individuals, length is smaller than width, anterior commissure sulcate, roundly trigonal in outline, dorsal beak straight, pedicle collar and cardinal process undeveloped, transverse band directly joined to crural processes; and in adult, shell subpentagonal in outline, anterior commissure paraplicate–sulciplicate, dorsal beak incurved, pedicle collar developed, cardinal process arched -plate-like and grooved, descending belt growing. But both in junior and adult, hinge-plate in section slightly concave ventrally, tapering and clubbed anteriorly, loop half the length of the dorsal valve and transverse band high arched.

Cooper (1983) proposed an etching methord to expose the loop. This is a reliable way to obtain correct measurements of the loop. I think that the method of serial sectioning is still useful for studying the loop and other internal structures of terebratulids, although it has some faults. Because most specimens can not be etched. Of course, one should pay much attentcon to orientation in the preparation of serial sections, so that he can establish the accurate position of the crural processes and obtain the full length of the loop.

The loop and other details of the Burma type specimen are not known. No serial sections are given in Muir-Wood (1965). Almeras (1971) gave a detailed account of the interior of French specimens identified as *Holcothyris angulata, H. depressa* (Sahni), *H. rotundata* (Sahni), *H. tragonalis* Buckman and had statistics for the loop. Ovcharenko (1967) claimed that *H. angulata* Buckman is a subspecies of *Kutchithyris acutiplicata* (Kitchin), because of similar exterior characters including the presence of radial capillae in both forms. Therefore, he proposed that the names *Kutchithyris* and *Holcothyris* are simultaneously proposed synonyms based on different subspecies of one species. In the author's opinion, *Holcothyris* and *Kutchithyris* are two independent genera having close affinity and apparent differences.

The loop statistics of some specimens from western Yunnan seem to be similar to that of the specimens from France, Pamir, India (see Table 2).

Stratigraphical range and geographical distribution: Upper Bajocian-Bathonian of Middle Jurassic; Asia, Europe and Africa.

Table 2. Loop statistics for *Holcothyris* and *Kutchithyris**

Proportion	*H. angulata* (93847) W. Yumman	*H. yunnanensis* (93850) W. Yunnan	*H. angulata* France Almeras, 1971	*K. ancile* (93864) W. Yunnan	*K. acutiplicata* Pamir Ovcharenko, 1967	*K. acutiplicata* Cutch Mitra & Ghosh, 1973
A	46°	52°	31°	45°	40°	/
a/ L1	0.42	0.40	0.42	0.51	0.50	0.36
b/ L1	0.58	0.60	0.58	0.49	0.50	0.64
c/ L1	0.21	0.21	0.25	0.30	0.25?	0.24
d/ L1	0.21	0.19	0.17	0.22	0.25?	0.12
e/ L1	0.32	0.30	0.24	0.22	0.19	0.18
f/ L1	0.26	0.30	0.34	0.26	0.31	0.46
g/ W1	0.66	0.67	1.10	0.68	/	/
h/ f	0.22	0.12	0.15	0.33	0.10	/
h/ L1	0.06	0.04	0.04	0.09	0.03	0.06
WD/ LD	1.00	0.88	0.93	0.88	/	/
W1/ L1	0.96	0.96	0.55	0.80	0.81	0.74
L1/ LD	0.45	0.53	0.48	0.49	/	/
W1/ WD	0.43	0.57	0.32	0.45	/	/
g/ WD	0.28	0.38	0.28	0.38	/	/

* Abbreviation of terminology for loop measurement: (After Cooper, 1983, 14–15)
A = loop angle; a = distance from cardinal process to tip of crural process (a focal point in the loop); b = distance from tip of crural process to end of terminal point; c = measure of outer hinge plate; d = measure of crus, from end of outer hinge plate to tip of crural process; e = distance from crural process to bridge (or apex) of transverse band; f = length of terminal points; g = width of hinge; h = measure of transverse band at its apex; L1 (a + b) = length of loop; W1 = width of loop; LD = length of the dorsal valve; WD = width of the dorsal valve.

Holcothyris liuwanensis sp. nov.

(Pl. 3, figs. 5–8; text-fig. 12)

Description: Shell small to medium in size, oval-rounded in outline, plano-convex to unequally biconvex in profile; beak massive, short, suberect and obliquely truncated; beak ridge rounded; planarea small; foramen large, circular and epithyrid; umbonal angle about 90°; symphytium very short; dorsal sulcus and ventral fold slightly developed and only appearing at anterior margin area; anterior commissure sulcate in junior and paraplicate in adult; leteral commissure straight; shell surface marked with capillae and fine concentric growth lines; punctum dense arranged concentrically or slightly undulated.

Pedicle collar short; teeth thick, plate-shaped with denticula; cardinal process low, bilobate; hinge-plates flat, wide, tapering, slightly concave ventrally, fused with inner socket ridges and clubbed anteriorly; adductor scars long, narrow and widely divergent; loop about half the length of the dorsal valve, with a high arched transverse band.

Dimension of specimens:

Cat. No.	Length	Width	Thickness	Umbonal angle	Anterior commissure	Lateral commissure
93842	24.4mm	22.5mm	13.6mm	95°	paraplicate	straight
93843 (holotype)	26.4mm	25.0mm	14.2mm	95°	paraplicate	straight
93844	27.3mm	23.0mm	16.3mm	90°	paraplicate	straight
93845	27.6mm	27.2mm	15.6mm	92°	paraplicate	straight
93846	28.7mm	26.3mm	15.6mm	90°	paraplicate	straight

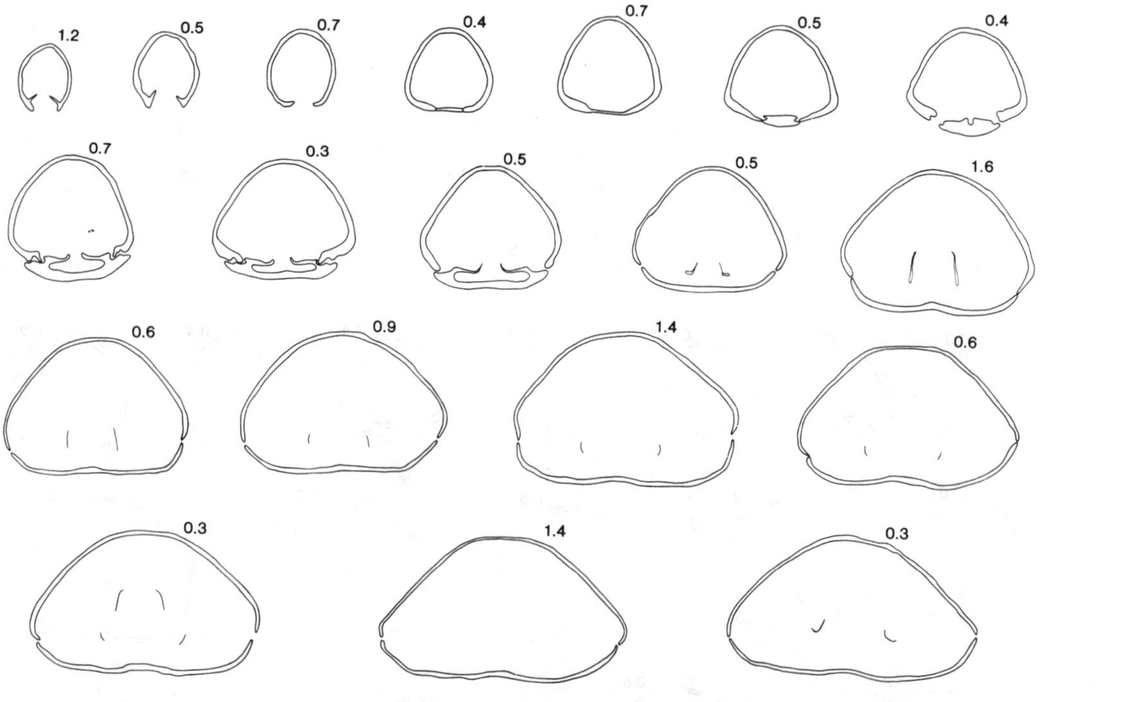

Text-fig. 12. Serial transverse sections of *Holcothyris liuwanensis* sp. nov., × 1.4; Cat. No. 93846(paratype).

Remarks: The species is very similar to species of the *Holcothyris triqonalis* series i. e. *H. trigonalis, H. rostrata, H. delta* and *H. acuminata* described by Buckman (1918, pl. 9, figs. 5–8) in the external features. However, the internal structures of the Buckman's species were not described. So their comparison can not be made in detail.

Loclity and horizon: Liuwan, Baoshan county and Datainyakou, Longling county; the Middle Jurassic Liuwan Formation.

Holcothyris angulata Buckman

(Pl. 5, figs. 5–8; text-fig. 13)

1918 *Holcothyris angulata* Buckman, pp. 182–193, pl. 10, figs. 1–4.
1918 *Holcothyris subangulata* Buckman, p. 193, pl. 10, fig. 5.
1971 *Holcothyris angulata*, Almeras, p. 331, pl. 58, figs. 7–1ſ.
1979 *Holcothyris tanggulaica* Ching, Sun et Ye, pl. 57. figs. 10–13.

Dimension of specimens:

Cat. No.	length	width	thickness	umbonal angle	anterior commissure	lateral commissure
93847	26.8mm	23.2mm	14.1mm	68°	paraplicate	slightly curved
93848	29.6mm	22.8mm	16.1mm	70°	paraplicate	slightly curved
93849	34.6mm	28.8mm	18.0mm	70°	paraplicate	slightly curved

Remarks: *Holcothyris angulata* is fairly common in the Namyau Beds, North Shan States, Burma. Unfortunately the internal structure of the species was not revealed in Buckman's description. Our specimens assigned to *Holcothyris angulata* coincide completely with the Burmese ones in the external characters, and the internal structure is now illustrated in serial transverse sections (text-fig. 13). Almeras (1971) described the detailed internal structure of the species from the lower Bathonian (zigzag zone), Premeyzel, Ain, France. As shown in Table. 2, the loop statistics indicate the similarity between the France specimen and western Yunnan one. It also seems to the author that the species is widespread in northern Tibet and southern Qinghai.

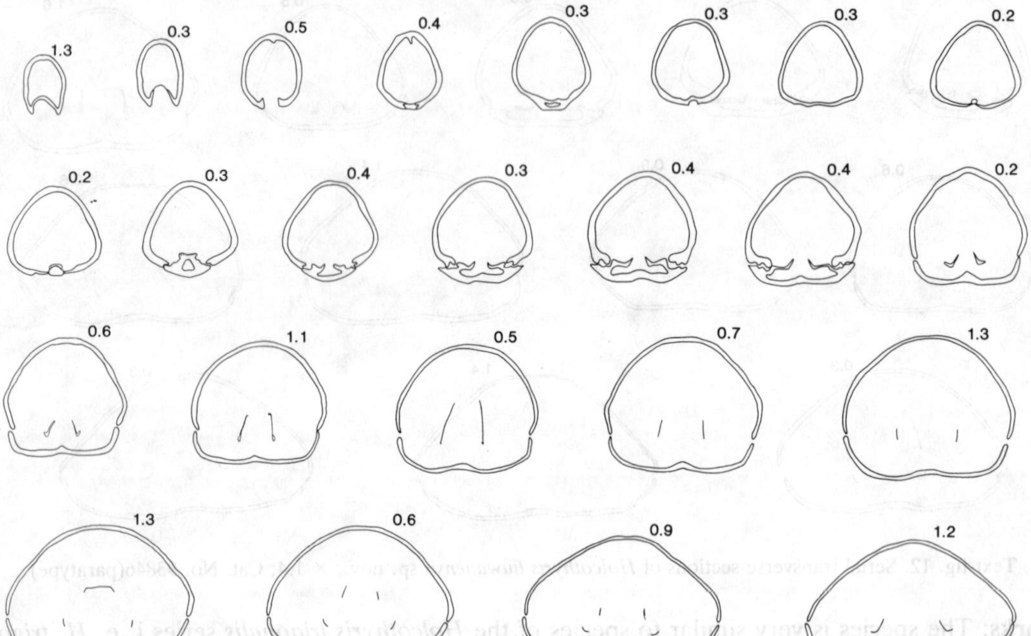

Text-fig. 13. Serial transverse sections of *Holcothyris angulata* Buckman, × 1.3; Cat. No. 93847.

Locality and horizon: Liuwan, Baoshan county; the Middle Jurassic Liuwan Formation.

Holcothyris yunnanensis sp. nov.

(Pl. 4, figs. 9–12; text-fig. 14)

Description: Shell large in size; elongately oval in outline, subequally biconvex; ventral valve a little more convex than dorsal valve; dorsal sulcus throughout the whole valve, shallow posteriorly and deeping anteriorly; lateral subangular plications developed on the anterior half; ventral valve gibbous with median fold bordered by rounded shallow sulcus on the anterior half; beak stout, strongly incurved and projecting over dorsal umbo; ventral umbo carinate-shaped; beak ridge rounded; planarea small; foramen large, elliptic and epithyrid; symphytium high, thick and concave; umbonal angle about 65°; shell surface ornameted with fine capillae and concentric growth lines; anterior commissure paraplicate and lateral commissure sigmoidal.

Pedicle collar absent; teeth short and thick with denticula; cardinal process bilobate with comb-shaped surface; hinge-plates flat, wide, fused with inner socket ridges, tapering, slightly concave ventrally and clubbed anteriorly; socket broad and shallow; crural process high and slightly incurved; loop about more than half the length of valve; descending branches thin, sort, vertical-plate-like; transverse band saddle-shaped; myophragm stout, starting from umbo; adductor scars depressed, long, narrow, tapering posteriorly, and slightly divergent anteriorly.

Dimension of specimens:

Cat. No.	length	width	thickness	Umbonal angle	anterior commissure	lateral commissure
93850	41.2mm	30.0mm	26.5mm	68°	paraplicate	sigmoidal
93851	44.5mm	33.4mm	24.0mm	65°	paraplicate	sigmoidal

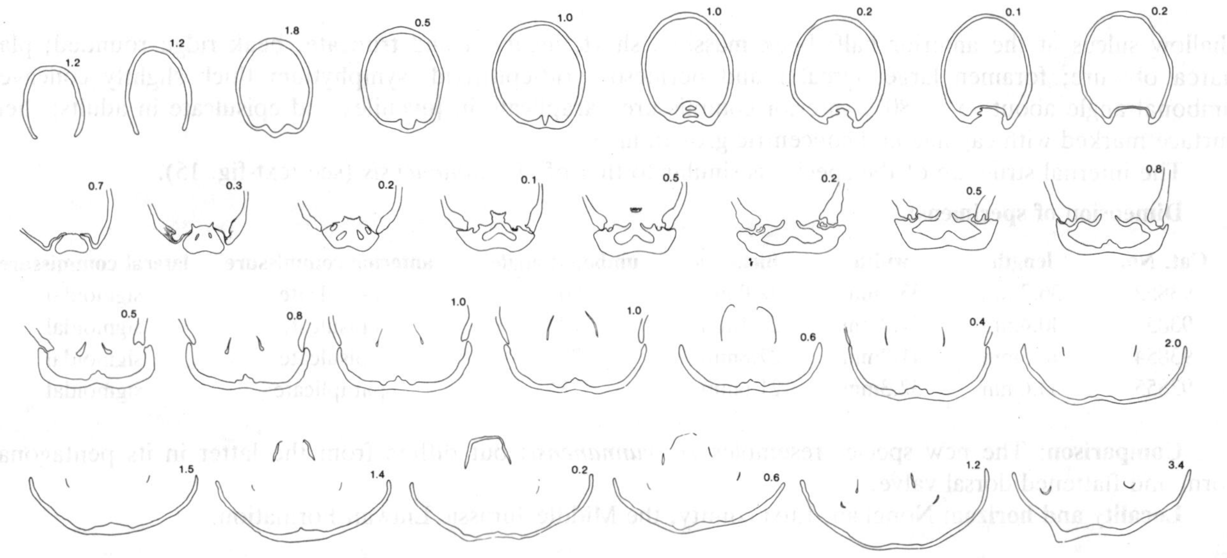

Text-fig. 14. Serial transverse sections of *Holcothyris yunnanensis* sp. nov. × 0.7; Cat No. 93850(paratype).

Comparison: The new species bears a resemblance to *Kutchithyris ingluviosa* (Kitchin) from Curth, India and differs from the latter by having dorsal sulcus throughout the whole valve. *Holcothyris pinguis* Buckman is also similar to the new species but differs from the latter in the smaller size and in having no dorsal sulcus throughout the whole valve.

Locality and horizon: Nongkan, Luxi county; the Middle Jurassic Liuwan Formation.

Holcothyris nongkanensis sp. nov.

(Pl. 5, figs. 9–12; text-fig. 15)

Description: Shell large in size; roundly pentagonal in outline, unequally biconvex to plano-convex in profile; maximum width at the anterior part of the shell; greatest thickness in the middle of the shell; dorsal sinus throughout, shallow posteriorly, wide and deep anteriorly; subangular lateral plications developed on the anterior half; ventral valve swollen, carinate near umbo; fold broad, rounded and borderd by rounded

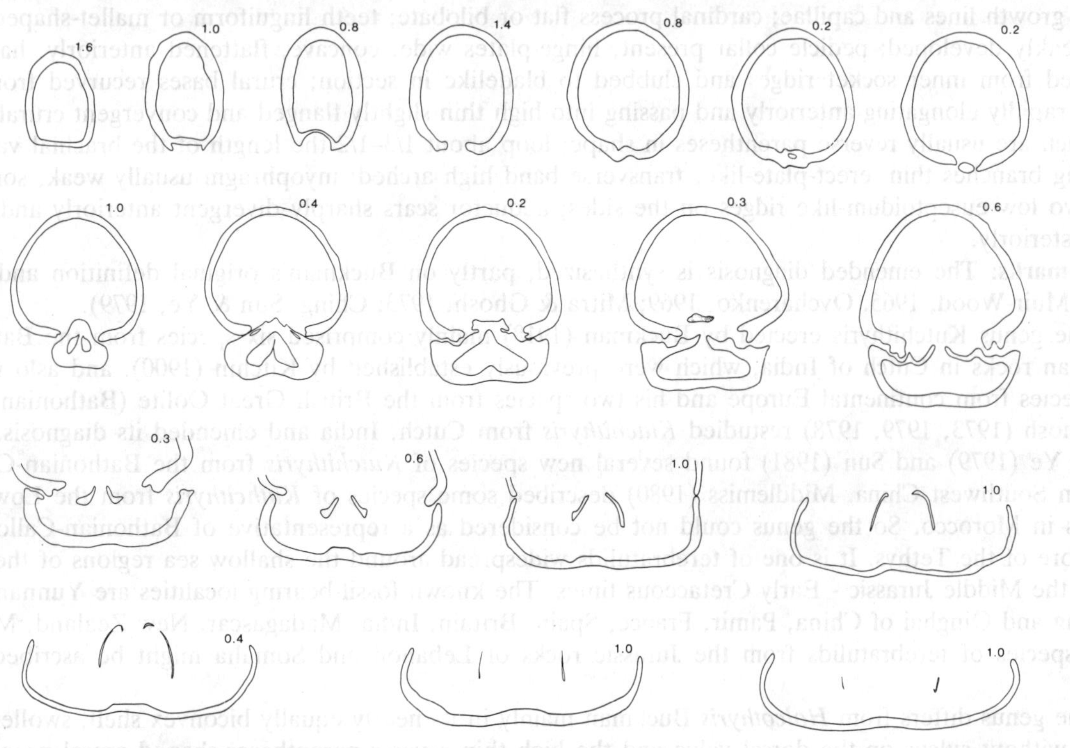

Text-fig. 15. Serial transverse sections of *Holcothyris nongkanensis* sp. nov., × 1.7; Cat. No. 93855(paratype).

shallow sulcus at the anterior half; beak massive, short, incurved and truncate; beak ridge rounded; planarea obscure; foramen large, circular and permesothyrid-epithyrid; symphytium thick slightly concave; umbonal angle about 75° – 80°; anterior commissure paraplicate in juveniles and episulcate in adults; shell surface marked with capillae and concentric growth lines.

The internal structure of the species is similar to that of *H. yunnanensis* (see text-fig. 15).

Dimension of specimens:

Cat. No.	length	width	thickness	umbonal angle	anterior commissure	lateral commissure
93852	36.2mm	33.7mm	22.0mm	80°	episulcate	sigmoidal
93853	40.4mm	34.7mm	26.1mm	78°	episulcate	sigmoidal
93854	40.5mm	41.2mm	27.8mm	75°	episulcate	sigmoidal
93855	33.6mm	32.3mm	21.6mm	75°	paraplicate	sigmoidal

Comparison: The new species resembles *H. yunnanensis* but differs from the latter in its pentagonal form and flattened dorsal valve.

Locality and horizon: Nongkan, Luxi county; the Middle Jurassic Liuwan Formation.

Genus *Kutchithyris* Buckman, 1918

1915 *Kutchithyris* Buckman, p. 78. (nomen nudum)
1918 *Kutchithyris* Buckman, p. 113.
1952 *Kutchithyris*, Roger, p. 124.
1965 *Kutchithyris*, Miur-Wood, Hp. 783.
1969 *Kutchithyris*, Ovcharenko, pp. 59–60.
1973 *Kutchithyris*, Mitra and Ghosh, pp. 185-188.
1979 *Kutchithyris*, Ching, Sun and Ye, p. 206.
1983 *Kutchithyris*, Cooper, pp. 96–97.

Type species: *Terebratula acutiplicata* Kitchin, 1900, p. 6, pl. 1, figs. 1–7.

Emended diagnosis: Shell medium to large in size, unequally biconvex in profile; development of anterior commissure from rectimarginate - uniplicate - sulciplicate, rarely to episulcate; some forms omitting the middle stage; lateral commissure extending posteriorly without any ventral or dorsal curvature, but anteriorly with ventral convexity; beak short, suberect to incurved; foramen large, circular and permesothyrid - epithyrid; symphytium usually concealed by beak; shell surface smooth and ornamented with numerous concentric growth lines and capillae; cardinal process flat or bilobate; teeth linguiform or mallet-shaped; denticula weakly developed; pedicle collar present; hinge-plates wide, concave, flattened anteriorly, hardly demarcated from inner socket ridges and clubbed to bladelike in section; crural bases recurved from hinge plates, rapidly elongating anteriorly and passing into high thin slightly flanged and convergent crural processes which are usually reverse parentheses in shape; loop about 1/3–1/2 the length of the brachial valve; descending branches thin, erect-plate-like; transverse band high arched; myophragm usually weak, sometimes with two low euseptoidum-like ridges on the sides; adductor scars sharply divergent anteriorly and thread-like posteriorly.

Remarks: The emended diagnosis is synthesized, partly on Buckman's original definition and others' views (Muir-Wood, 1965; Ovcharenko, 1969; Mitra & Ghosh, 1973; Ching, Sun & Ye, 1979).

The genus Kutchithyris erected by Buckman (1918) mainly comprised six species from the Bathonian-Callovian rocks in Cutch of India, which were previously established by Kitchin (1900), and aslo included two species from continental Europe and his two species from the British Great Oolite (Bathonian). Mitra and Ghosh (1973, 1979, 1978) restudied *Kutchithyris* from Cutch, India and emended its diagnosis. Ching, Sun & Ye (1979) and Sun (1981) found several new species of *Kutchithyris* from the Bathonian-Callovian rocks in Southwest China. Middlemiss (1980) described some species of *Kuthcithyris* from the Lower Cretaceous in Morocco. So the genus could not be considered as a representative of Bathonian-Callovian on one shore of the Tethys. It is one of terebratulids widespread around the shallow sea regions of the Tethys during the Middle Jurassic - Early Cretaceous times. The known fossil-bearing localities are Yunnan, Tibet, Xinjiang and Qinghai of China, Pamir, France, Spain, Britain, India, Madagascar, New Zealand, Morocco. Some species of terebratulids from the Jurassic rocks of Lebanon and Somalia might be ascribed to the genus.

The genus differs from *Holcothyris* Buckman mainly in its nearly equally biconvex shell, swollen dorsal umbo, without sulcus on the dorsal valve and the high thin reverse-parentheses-shaped crural process. *Sel-*

lithyris Middlemiss is very similar to the genus, but it is distinguished from the latter by its high erected beak, low crural process and apparent differentiation between hinge-plates and socket ridges.

The known species are as follows:

Bathonian stage: *Kutchithyris hypsogonia* (Kitchin), *K. acutiplicata* (Kitchin), *K. propingua* (Kitchin), *K. circumdata* (Deslongchamps), *K. fulia* Buckman, *K. egregia* Buckmam, *K. planiconvexa* (Kitchin), *K. jumarensis* (Kitchin), *K. tibetica* Ching, Sun et Ye, *K. olivaeformis* (Reed), *K. ancile* (Reed), *K. luchiangensis* (Reed), *K. pygmae* Ching, Sun et Ye, *K. anongensis* Sun, *K. mulaensis* Sun, *K. andoensis* Sun, *K. longlingensis* (sp. nov.), *K. datainyakouensis* (sp. nov.)

Callovian-Oxfordian stages: *K. dhosaensis* (Kitchin), *K. longplicata* (Kitchin), *K. aurata* (Kitchin), *K. ingluvisa* (Kitchin), *K. subcanaliculata* (Deslongchamps), *K. denggenensis* Sun, *K. zamdaensis* Sun, *K. pentagonalis* Sun.

Oxfordian-Valanginian stages: *K. subsella* (Leymerie).

Valanginian-Barremian: *K. kennedyi* Middlemiss, *K. brivesi* (Roch).

Stratigraphical range and geographical distribution: Bathonian-Barremian; Asia, Europe, Africa and New Zealand.

Kutchithyris olivaeformis (Reed)

(Pl. 3, Figs. 13–16; text-fig. 16)

1927 *Terebratula* (*Holcothyris*) *pinguis* var. *olivaeformis* Reed, pp. 264–265, pl. 19, figs. 18, 18a–c.
1979 *Kutchithyris olivaeformis* (Reed), Ching, Sun et Ye, p. 207, pl. 55, figs. 13–17.

Dimension of specimens:

Cat. No.	Length	Width	Thickness	Umbonal angle	Anterior commissure	Lateral commissure
93856	30.5mm	23.5mm	22.2mm	65°	sulciplicate	slightly sigmoidal
93857	28.6mm	23.0mm	18.2mm	62°	sulciplicate	sigmoidal
93858	27.7mm	22.6mm	19.9mm	65°	episulcate	sigmoidal
93859	32.4mm	24.2mm	21.2mm	60°	episulcate	sigmoidal

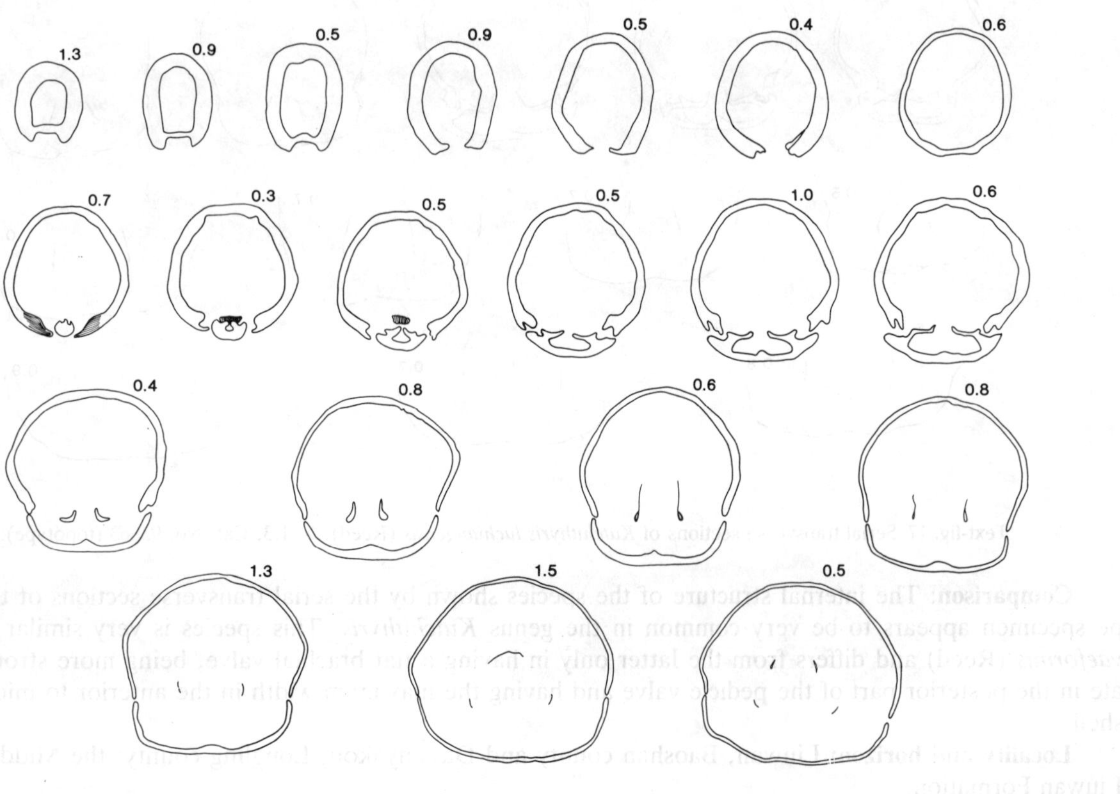

Text-fig. 16. Serial transverse sections of *Kutchithyris olivaeformis* (Reed), × 1.4; Cat. No. 93857 (topotype).

Comparison: The species is vary similar to *Holcothyris pinguis* Buckman and *H. curvata* Buckman from the Namyau Beds, Burma in the outline and the shell ornamentation, but its median sulcus on the brachial valve is limited to the anterior margin and does not extend throughout the shell as in *H. pinguis*. *Kutchithyris acutiplicata* of the Patcham Formation, Cutch, India resembles the species in the external features and differs from the latter in its pentagonal shell and more strong biplication.

Locality and horizon: Liuwan, Baoshan county, Datianyakou, Longling county and Nongkan, Luxi county; the Middle Jurassic Liuwan Formation.

Kutchithyris luchiangensis (Reed)

(Pl. 3, Figs. 1–4; text-fig. 17)

1927 *Terebratula* (*Holcothyris*) *pinguis* var. *luchiangensis* Reed, p. 264, pl. 19, figs. 19, 19a–c.

Dimension of specimens:

Cat. No.	Length	Width	Thickness	Umbonal angle	Anterior commissure	Lateral commissure
93860	22.4mm	20.3mm	13.9mm	78°	sulciplicate	straight
93861	28.2mm	23.2mm	17.2mm	75°	sulciplicate	slightly sigmoidal
93862	26.2mm	22.5mm	16.5mm	70°	sulciplicate	slightly sigmoidal
93863	24.3mm	21.3mm	14.9mm	74°	sulciplicate	straight

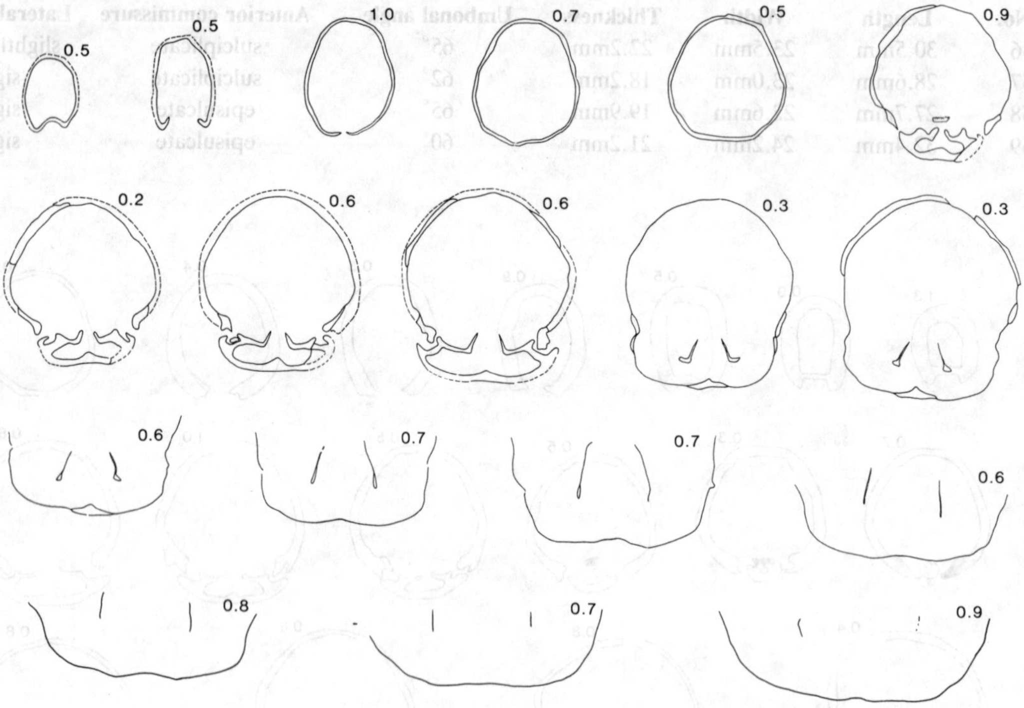

Text-fig. 17. Serial transverse sections of *Kutchithyris luchiangensis* (Reed), × 1.3; Cat. No. 93863 (topotype).

Comparison: The internal structure of the species shown by the serial transverse sections of the topotype specimen appears to be very common in the genus *Kutchithyris*. This species is very similar to *K. olivaeformis* (Reed) and differs from the latter only in having a flat brachial valve, being more strongly carinate in the posterior part of the pedicle valve and having the maximum width in the anterior to middle of the shell.

Locality and horizon: Liuwan, Baoshan county and Datianyakou, Longling county; the Middle Jurassic Liuwan Formation.

Kutchithyris ancile (Reed)

(Pl. 3, figs. 9–12; Pl. 5, figs. 1–4; text-fig. 18.)

1927 *Terebratula* (*Holcothyris*) *ancile* Reed, p. 263, pl. 19, figs. 17, 17a-b.
1979 *Kutchithyris lingularis* Ching, Sun et Ye, p. 207, pl. 55, figs. 9–12.

Dimension of specimens:

Cat. No.	Length	Width	Thickness	Umbonal angle	Anterior commissure	Lateral commissure
93864	30.2mm	22.3mm	18.3mm	65°	sulciplicate	slightly sigmoidal
93865	34.3mm	27.7mm	20.3mm	65°	sulciplicate	slightly sigmoidal
93866	25.0mm	19.7mm	13.8mm	67°	sulciplicate	nearly straight
93867	30.3mm	26.4mm	18.3mm	75°	sulciplicate	slightly sigmoidal
93868	36.2mm	29.1mm	22.7mm	70°	sulciplicate	sigmoidal

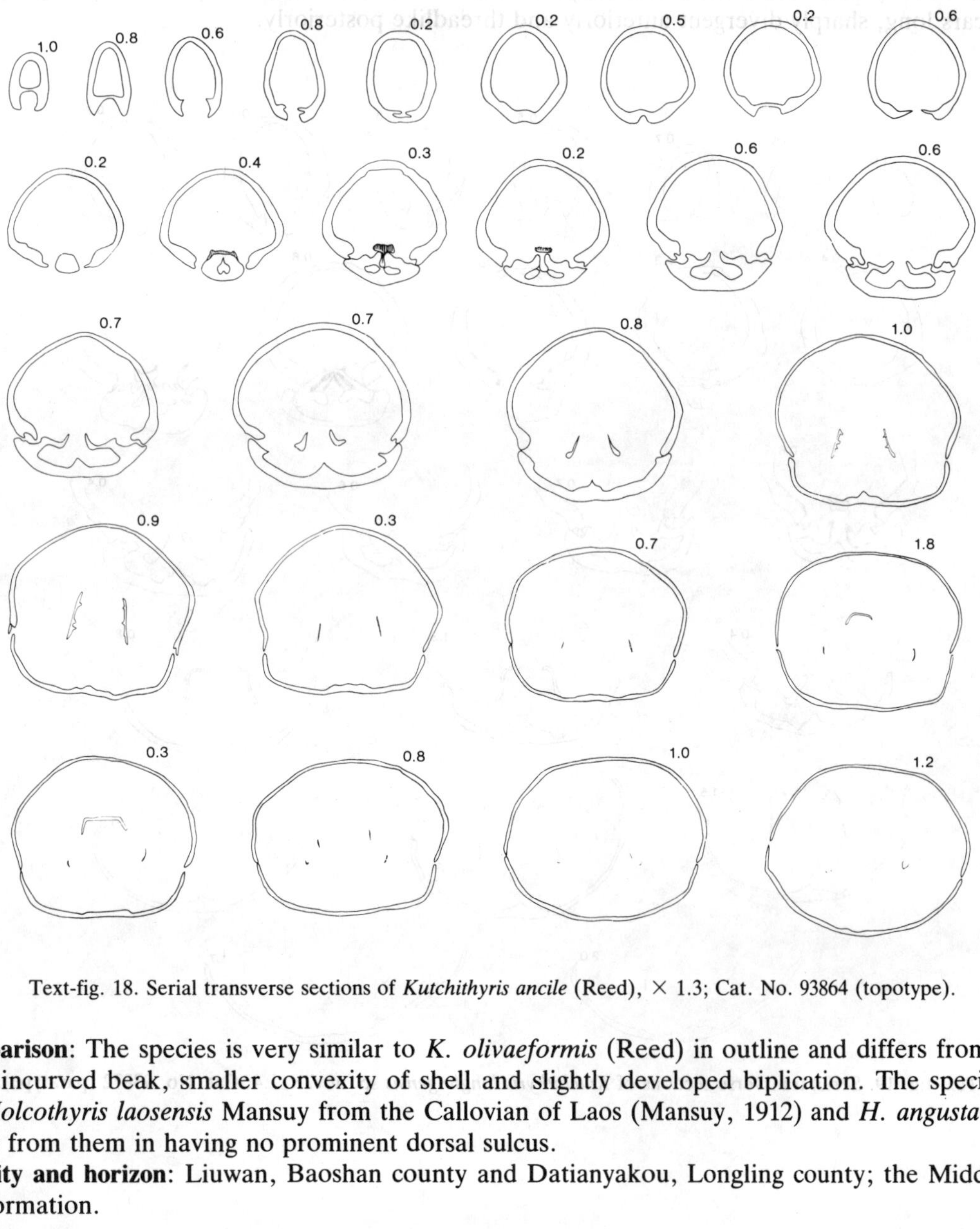

Text-fig. 18. Serial transverse sections of *Kutchithyris ancile* (Reed), × 1.3; Cat. No. 93864 (topotype).

Comparison: The species is very similar to *K. olivaeformis* (Reed) in outline and differs from the latter in its less incurved beak, smaller convexity of shell and slightly developed biplication. The species also resembles *Holcothyris laosensis* Mansuy from the Callovian of Laos (Mansuy, 1912) and *H. angusta* Buckman, but differs from them in having no prominent dorsal sulcus.

Locality and horizon: Liuwan, Baoshan county and Datianyakou, Longling county; the Middle Jurassic Liuwan Formation.

***Kutchithyris longlingensis* sp. nev.**

(Pl. 4, Figs. 1–4; text-fig. 19)

Description: Shell medium-sized, roundly trigonal in outline, unequally biconvex in profile; pedicle valve more convex than brachial valve; maximum width located at the anterior end of shell, greatest thickness at the middle of shell; anterior margin gently arched to truncate; anterior commissure sulciplicate; lateral commissure straight without any ventral or dorsal curvature; beak massive, suberect and truncate; beak ridges rounded; planarea small; foramen large, circular and epithyrid; delthyrium coverd by symphytium; umbonal angle about 70°; dorsal fold low and only appearing at 1/3 the valve length anteriorly in adults; a broad and shallow sulcus on the dorsal fold; border plicae low and rounded; ventral sulcus broad and shallow with a very short rounded median tongue; shell surface smooth and marked with dense concentric growth lines or lamellae and capillae, forming reticulated ornamentation.

Pedicle collar short; symphytium thick and slightly concave; teeth mallet-shaped; denticula present; caridnal process bilobate; hinge-plates strongly incurved, V-shaped in section and fused with inner socket ridges; sockets shallow; crural processes high, converging, with some spines on their external sides; descending branches of loop thin, like erect and short plate; transverse band saddle-shaped; myophragm prominent; adductor scars long, sharply divergent anteriorly and threadlike posteriorly.

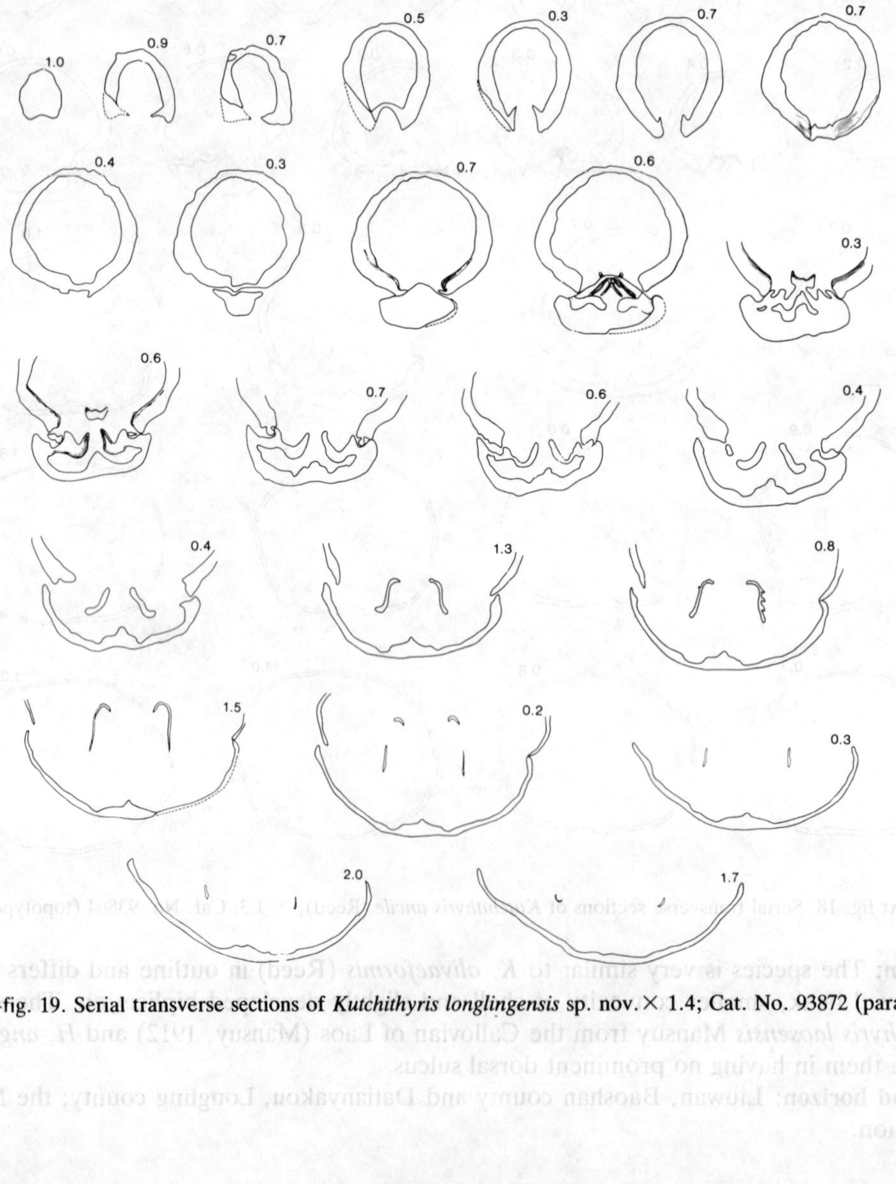

Text-fig. 19. Serial transverse sections of *Kutchithyris longlingensis* sp. nov. × 1.4; Cat. No. 93872 (paratype).

Dimension of specimens:

Cat. No.	Length	Width	Thickness	Umbonal angle	Anterior commissure	Lateral commissure
93869	24.9mm	24.3mm	15.5mm	73°	sulciplicate	straight
93870	30.7mm	29.6mm	19.0mm	75°	sulciplicate	slightly sigmoidal
93871	35.3mm	32.8mm	22.2mm	62°	sulciplicate	sigmoidal
93872	35.4mm	33.0mm	23.0mm	65°	sulciplicate	sigmoidal

Comparison: The new species is characterized by its trigonal form. It is easily differentiated from other species of the genus by this feature. *Holcothyris trigonalis* Buckman (Buckman, 1918, pl. 9, fig. 5) bears a resemblance to the new species in the external features but differs from the latter in its flattened dorsal valve with a prominent sulcus throughout the whole valve.

Locality and horizon: Liuwan, Baoshan county and Datianyakou, Longling county; the Middle Jurassic Liuwan Formation.

Kutchithyris datianyakouensis sp. nov.

(Pl. 4, Figs. 5–8; text-fig. 20)

Description: Shell medium in size; oval in outline, unequally biconvex in profile; pedicle valve more convex than brachial valve; convexity gentle; maximum width and greatest thickness both at the middle of shell; anterior commissure nearly rectimarginate or slightly sulciplicate; beak short, stout, suberect and truncate; beak ridges rounded; foramen big, circular to elliptical, epithyrid; symphytium short; fold and sulcus obliterated in adult forms, or very faint near anterior margin; shell surface ornamented with fine concentric growth lines and capillae.

Pedicle collar present; teeth linguiform; denticula slightly developed; cardinal process flattened; hinge-plates narrow, incurved, fused with inner socket ridges, tapering, and nearly V-shaped in section anteriorly; inner socket ridges low; sockets broad and shallow; crural processes high, converging; loop short; descending branches like erect and short plate; transverse band not known; myophragm low; adductor scars long, divergent anteriorly and threadlike posteriorly.

Dimension of specimens:

Cat. No.	Length	Width	Thickness	Umbonal angle	Anterior commissure	Lateral commissure
93873	38.2mm	35.3mm	20.2mm	85 °	near rectimarginate	stright
93874	38.0mm	38.5mm	20.8mm	88°	rectimarginate	stright

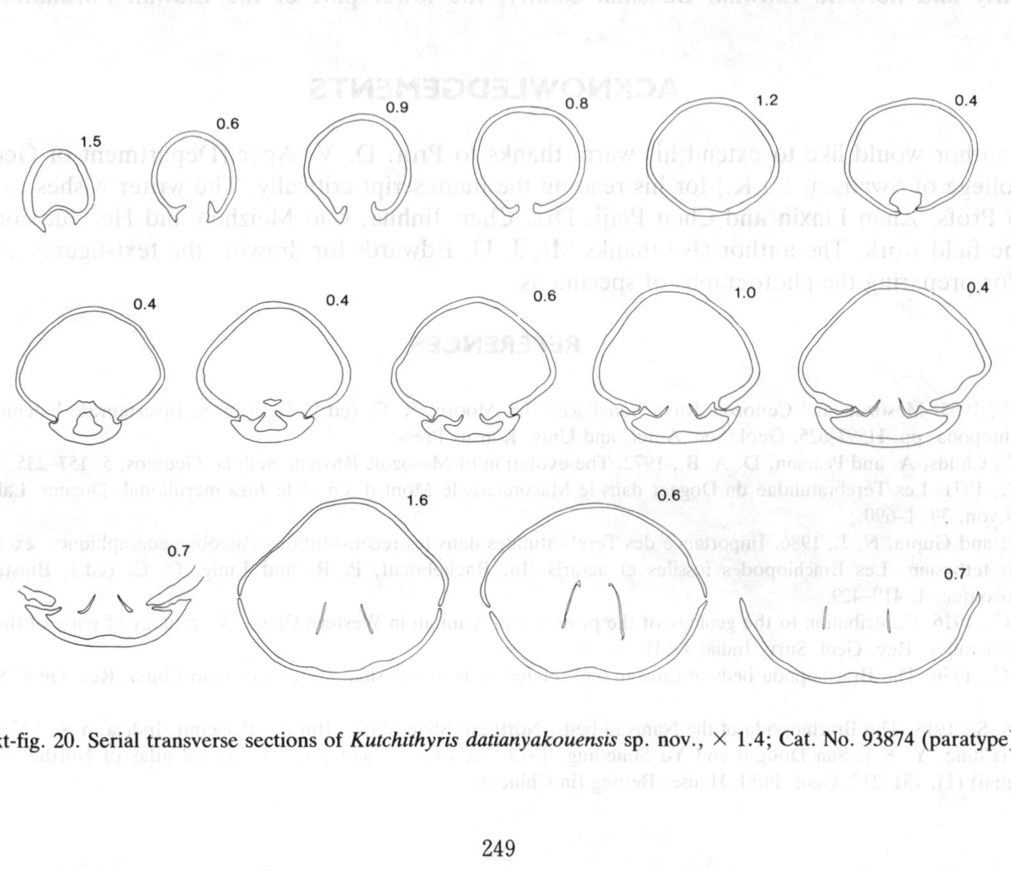

Text-fig. 20. Serial transverse sections of *Kutchithyris datianyakouensis* sp. nov., × 1.4; Cat. No. 93874 (paratype).

Remarks: It should be mentioned that the new species included in the genus *Kutchithyris* shows some external characters of the genus *Charltonithyris*, for example, the straight lateral commissure, the rectimaginate anterior commissure, the short and suberect beak and the gently biconvex shell. This implies that the species bears a relationship with *Charltonithyris*. But according to the obvious capillae ornamented on the shell, the new species is attributed to *Kutchithyris*. *Charltonithyris uptoni* Buckman from the Bajocian, middle Inferior Oolite, Gloucestershire, England is very similar to the new species in the external feature and differs from the latter in the roundly rhomboidal outline and in having no capillae. This species also differs from others of *Kutchithyris* in its oval form, gentle convexity and indistinctive fold and sulcus.

Locality and horizon: Datianyakou, Longling county; the Middle Jurassic Liuwan Formation.

Family Orthotomidae Muir-Wood, 1936
Genus *Orthotoma* Quenstedt, 1869

1869 *Orthotoma* Quenstedt, p. 315.
1876 *Orthoidea* Friren, p. l.
1918 *Orthotoma*, Buckman, p. 96.
1965 *Orthotoma*, Muir-Wood, Hp. 773.

Type species: *Terebratula heyseana* Quenstedt

Stratigraphical range and geographical distribution: Lias; Europe and Asia.

Orthotoma sp.

(Pl. 2, Figs. 17–20)

Description: Shell small (in figured specimen, length 6.8mm, width 6.1 mm, thickness 4.0 mm), subcircular in outline, equally biconvex in profile; beak small, suberect, hypothyrid; beak ridge angular; delthyrium bordered by jugate deltidial plates; fold and sulcus absent; anterior commissure rectimarginate; shell surface smooth.

Remarks: The specimen described above is only one of the brachiopods found from the lower part of the Liuwan Formation in western Yunnan so far. Its external features are roughly sinilar to those of the genus *Orthotoma*. But unfortunately, the internal structure of the specimen has not been known. It resembles *Orthotoma spinati* Rau from the Middle Lias, Germany in the external features and differs from the latter in its smaller size, smaller umbonal angle and short incurved cardinal margin.

Locality and horizon: Liuwan, Baoshan county; the lower part of the Liuwan Formation of Lower Jurassic.

ACKNOWLEDGEMENTS

The author would like to extend his warm thanks to Prof. D. V. Ager (Department of Geology, University College of Swansea. U. K.) for his reading the manuscript critically. The writer wishes to express his thanks to Profs. Zhan Linxin and Chen Peiji, Drs. Chen Jinhua, Cao Meizhen and He Guoxiong for their help in the field work. The author also thanks Mr. J. U. Edwards for drawing the text-figures and Mr. Zhu Chunliu for preparing the photographs of specimens.

REFERENCES

Ager, D. V., 1965. Mesozoic and Cenozoic Rhynchonellacea. In: Moore, R. C. (ed.), Treatise on Invertebrate Paleontology, pt. H, Brachiopoda, pp. H597–625. Geol. Soc. Amer. and Univ. Kansus Press.

Ager, D. V., Childs, A. and Pearson, D. A. B., 1972. The evolution of Mesozoic Rhynchonellida. Geobios, 5, 157–235.

Almeras, Y., 1971. Les Terebratulidae du Dogger dans le Maconnais, le Mont d' Or et le Jura meridional. Docum. Lab. Geol., Fac. Sci. Lyon, 39, 1–690.

Almeras, Y. and Gupta, N. J., 1986. Importance des Terebratulides dans les reconstitutions Paleobiogeographiques: exemple du Bathonien tethysien. Les Brachiopodes fossiles et acturls. In: Racheboeuf, P. R. and Emig, C. C. (ed.), Biostratigraphie du Paleozoique, 4, 419–429.

Brown, Z. C., 1916. Contribution to the geology of the provence of Yunnan in Western China. V. geology of parts of the Solween and Wekon valley. Rec. Geol. Surv. India, 47(4).

Brown, Z. C., 1936. The Brachiopoda beds of Liuwan and related beds in the Shan States and Indo-China. Rec. Geol. Surv. India, 71 (2).

Buckman, S. S., 1918. The Brachiopoda of the Namyau beds, Northern Shan States, Burma. Palaeont. Indica, n. s. 3 (2).

Jin Yugan (Ching, Y. K.), Sun Dong-li and Ye Song-ling, 1979. Mesozoic brachiopods. In: Fossil atlas of Northwest China (Fasc. Qinghai) (1), 131–217. Geol. Publ. House, Beijing (in Chinese).

Cooper, G. A., 1983. The Terebratulacea (Brachiopoda), Triassic to Recent: A study of the brachidia (Loops). Smithsonian Contributions to Paleobiology, 50, 1–455.

Dubar, G., 1967. Brachiopodes Jurassiques du Sahara Tunisien. Annales de Paleontologie Invertebrates, 53, 33–101.

Laurin, B., 1984. Les Rhynchonellides des Plates-formes du Jurassique moyen en Europe Occidentale. Cahiers de Paleontologie Section Invertebres. Editions du Centre National de la Recheche Scientifique (Paris).

Kitchin, F. L., 1900. Jurassic fauna of Cutch, pt. 1. The Brachiopoda. Palaeont. Indica, Ser. 9, 3 (1), 1–87.

Mansuy, H., 1912. Mission du Laos. I. Geologie des Environs de Luang-Prabang. Mem. Serv. Geol. Indo-chine, 1(4), 1–32.

Mitra, K. C. and Ghosh, D. N., 1973. Emended diagnosis of one terebratulid and two rhynchonellid genera of Buckman from Jurassic of Kutch Gujarat. Quart. Jl. Geol. Mining and Metallurgical Soc. India, 45 (4), 175–190.

Mitra, K. C., 1978. Jurassic terebratulidae from Jhura Dome, Kutch. Indian Journal of Earth Sciences, 5(2), 141–153.

Mitra, K. C., Bershan, S. and Bhattacharya, D., 1979. A study of Mesozoic stratigraphy of Kutch, Gujarat with special reference to rock - stratigraphy and biostratigraphy of Keera Dome. Bull. Ind. Geol. Assoc., 12 (2), 129–143.

Middlemiss, F. A., 1980. Lower Cretaceous Terebratulidae from South-Western Morocco and their biogeography. Palaeontology. 23 (3), 515–556.

Muir-Wood, H. M., 1934. On the internal structure of some Mesozoic brachiopoda. Royal Soc. London, Philos. Trans., Ser. B, 223, 511–567.

Muir-Wood, H. M., 1935. Jurassic Brachiopoda. In: Geology and palaeontology of British Somaliland, 75–147. London.

Muir-Wood, H. M., 1937. The Mesozoic Brachiopoda of the Attock district, Palaeont. Indica, n. s. 20 (6), 1–34.

Muir-Wood, H. M., 1965. Mesozoic to Recent Terebratulidina. In: Treatise on Invertebrate Palaeontology. (H) 2, 762–818. Univ. Kansas Press, Lawrence.

Ovcharenko, V. N., 1967. On the Taxanomic stutus of some generic names of Jurassic terebratulids. Doklady Akademii Nauk Tadzhikskoi SSR, 10 (2), 27–30 (in Russian).

Ovcharenko, V. N., 1969. Transitional forms and species differentiation of brachiopods. Paleontologicheskii Zhurnal, 3, 57–63 (in Russian).

Quenstedt, F. A., 1869. Brachiopoden. In: Petrefactenkunde Deutschlands. 2, 1–748. Tubingen and Leipzig.

Reed, F. R. C., 1924. Provisional list of Palaeozoic and Mesozoic fossils collected by Coggin Brown in Yunnan. Rec. Geol. Surv. India, 55 (4).

Reed, F. R. C., 1927. Palaeozoic and Mesozoic fossils from Yunnan. Palaeont. Indica, n. s. 10 (1).

Roger, J., 1952. Brachiopoda. In: Piveteau, J. (ed.), Traite de Paleontologie, 2, 3–160.

Sahni M. R., 1939. The Mesozoic brachiopoda of the Bannu district. Palaeont. Indica, n. s. 27 (1).

Sun Dongli, 1981. Mesosoic brachiopods of Xizang. In: Palaeontology of Xizang (3), 177–260. Science Press, Beijing (in Chinese).

Sun Dongli, 1982. Jurassic brachiopod fauna of China. Journal of Stratigraphy, 6 (1), 56–58 (in Chinese).

Yin Jixiang and Fang Zhongjing, 1973. Marine Jurassic in West Yunnan. Scientia Geologica Sinica, 3, 236–243 (in Chinese).

EXPLANATION OF PLATES

All the specimens are kept in Nanjing Institute of Geology and Palaeontology, Academia Sinica. All the figures are photographed at 1.5 times of natural size except those with special note.

Plate I

Figs. 1–4, 9–12. *Burmirhynchia tenuiplicata* Reed

1. Dorsal, 2. Ventral, 3. Lateral, 4. Anterior; Cat. No. 93827(Topotype).

9. Dorsal, 10. Ventral, 11. Lateral, 12. Anterior; Cat. No. 93824(Topotype).

Figs. 5–8. *Burmirhynchia adjudicata* Reed

5. Dorsal, 6. Ventral, 7. Lateral, 8. Anterior; Cat. No. 93808(Topotype)

Figs. 13–16. *Burmirhynchia trilobata* Ching, Sun et Ye

13. Dorsal, 14. Ventral, 15. Lateral, 16. Anterior; Cat. No. 93806.

Figs. 17–20. *Burmirhynchia praestans* Reed

17. Dorşal, 18. Ventral, 19. Lateral, 20. Anterior; Cat. No. 93799(Topotype).

Plate 1

Plate 2

Figs. 1–4. *Burmirhynchia cuneata* Ching, Sun et Ye
1. Dorsal, 2. Ventral, 3. Lateral, 4. Anterior; Cat. No. 93816.

Figs. 5–8. *Moguellina semicostata* sp. nov.
5. Dorsal, 6. Ventral, 7. Lateral, 8. Anterior; Cat. No. 93840(Holotype).

Figs. 9–12. *Burmirhynchia conjurata* Reed
9. Dorsal, 10. Ventral, 11. Lateral, 12. Anterior; Cat. No. 93836(Topotype).

Figs. 13–16. *Burmirhynchia luchiangensis* Reed
13. Dorsal, 14. Ventral, 15. Lateral, 16. Anterior; Cat. No. 93818(Topotype).

Figs. 17–20. *Orthotoma* sp.
17. Dorsal, 18. Ventral, 19. Lateral, 20. Anterior, × 3; Cat. No. 93875.

Figs. 21–24. *Burmirhynchia discreta* Reed
21. Dorsal, 22. Ventral, 23. Lateral, 24. Anterior; Cat. No. 93822(Topotype).

Figs. 25–28. *Burmirhynchia flabilis* Ching, Sun et Ye
25. Dorsal, 26. Ventral, 27. Lateral, 28. Anterior; Cat. No. 93832.

Plate 3

Figs. 1–4. *Kutchithyris luchiangensis* (Reed)
1. Dorsal, 2. Ventral, 3. Lateral, 4. Anterior; Cat. No. 93861(Topotype).

Figs. 5–8. *Holcothyris liuwanensis* sp. nov.
5. Dorsal, 6. Ventral, 7. Lateral, 8. Anterior; Cat. No. 93843(Holotype).

Figs. 9–12. *Kutchithyris ancile* (Reed)
9. Dorsal, 10. Ventral, 11. Lateral, 12. Anterior; Cat. No. 93867(Topotype).

Figs. 13–16. *Kutchithyris olivaeformis* (Reed)
13. Dorsal, 14. Ventral, 15. Lateral, 16. Anterior; Cat. No. 93859(Topotype).

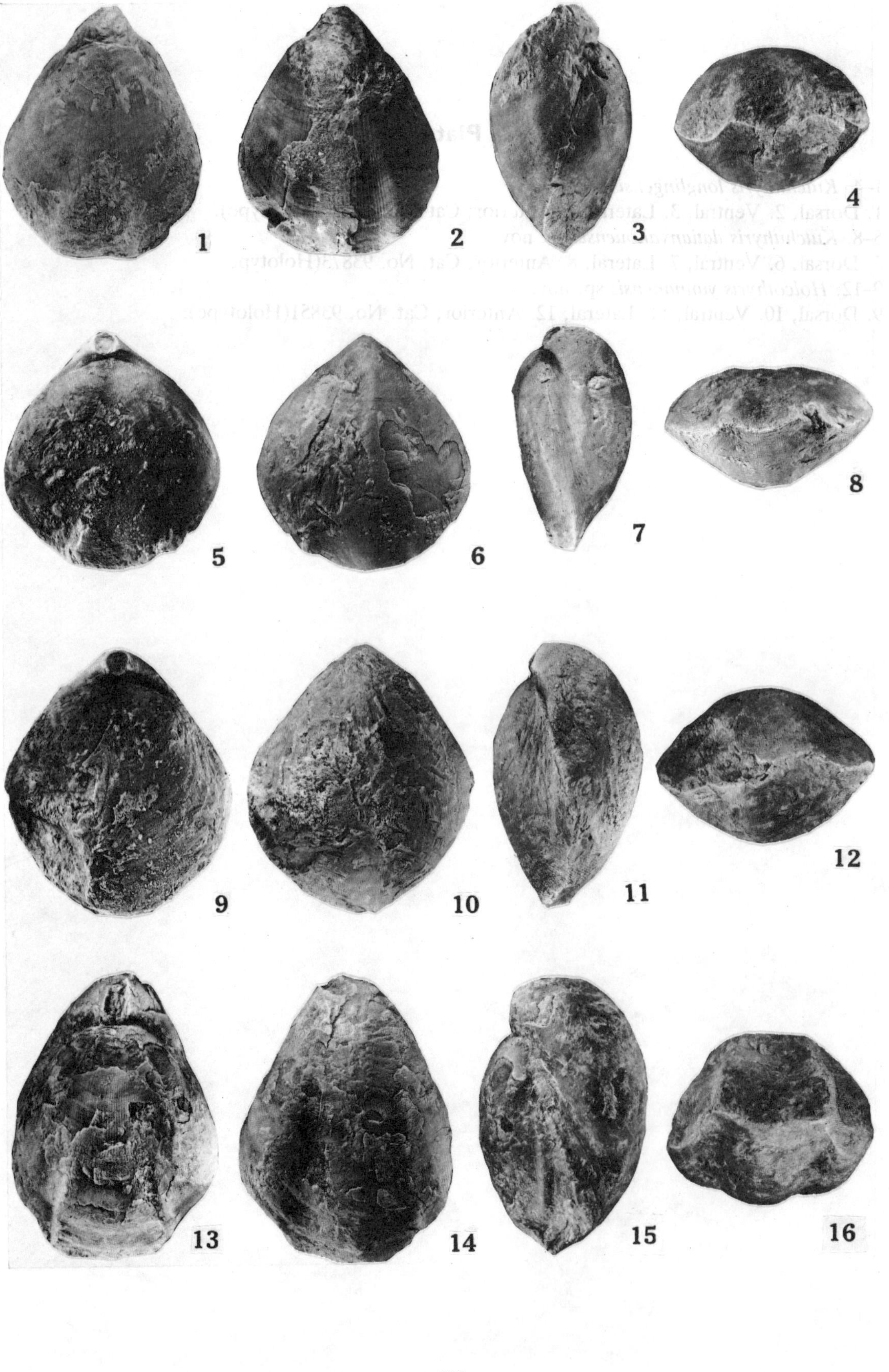
1
2
3
4
5
6
7
8
9
10
11
12
13
14
15
16

Plate 4

Figs. 1–4. *Kutchithyris longlingensis* sp. nov.
1. Dorsal, 2. Ventral, 3. Lateral, 4. Anterior; Cat. No. 93870(Holotype).

Figs. 5–8. *Kutchithyris datianyakouensis* sp. nov.
5. Dorsal, 6. Ventral, 7. Lateral, 8. Anterior; Cat. No. 93873(Holotype).

Figs. 9–12. *Holcothyris yunnanensis* sp. nov.
9. Dorsal, 10. Ventral, 11. Lateral, 12. Anterior; Cat. No. 93851(Holotype).

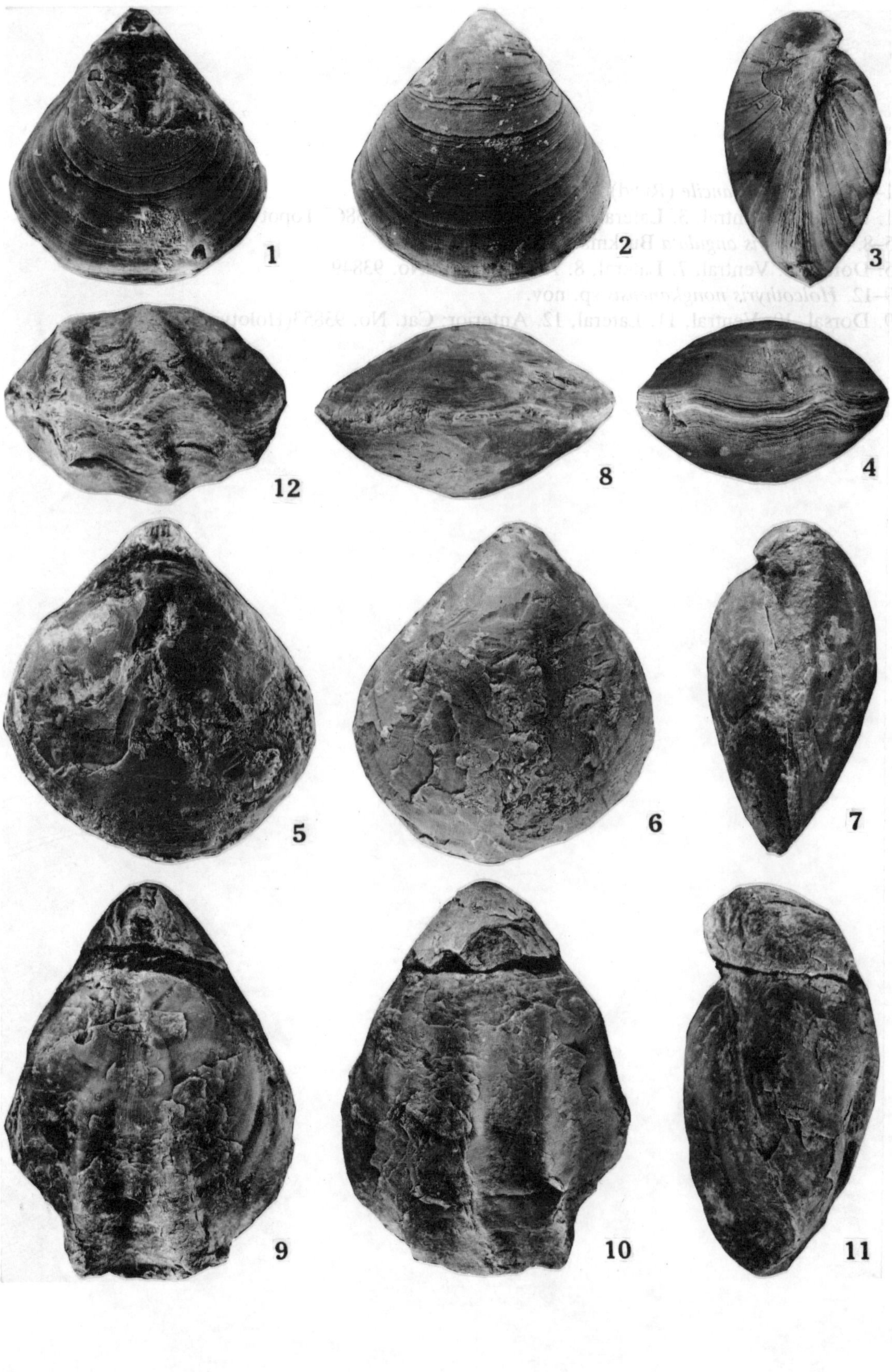
1
2
3
12
8
4
5
6
7
9
10
11

Plate 5

Figs. 1–4. *Kutchithyris ancile* (Reed)
1. Dorsal, 2. Ventral, 3. Lateral, 4. Anterior; Cat. No. 93865(Topotype).

Figs. 5–8. *Holcothyris angulata* Buckman
5. Dorsal, 6. Ventral, 7. Lateral, 8. Anterior; Cat No. 93849.

Figs. 9–12. *Holcothyris nongkanensis* sp. nov.
9. Dorsal, 10. Ventral, 11. Lateral, 12. Anterior; Cat. No. 93853(Holotype).

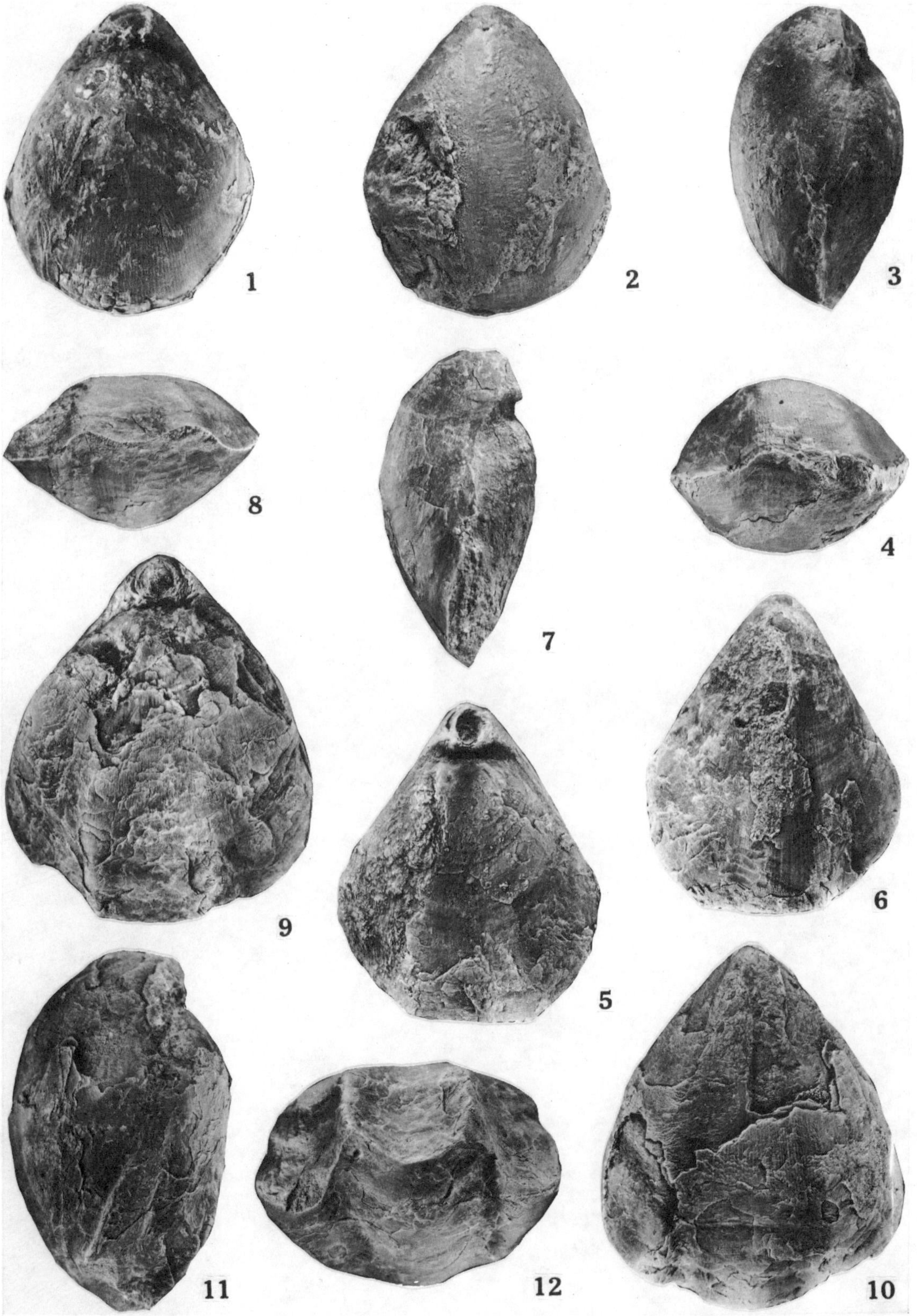
1
2
3
8
4
7
9
6
5
11
12
10

PALAEONTOLOGIA CATHAYANA Volume 5, 1990, p.263–266

SOME PROBLEMS ON THE GURYUL RAVINE SECTION OF KASHMIR AS PERMIAN-TRIASSIC BOUNDARY STRATOTYPE

Wang Chengyuan

(Nanjing Institute of Geology and Palaeontology, Academia Sinica)

The selection of a Global Boundary Stratotype Section and Point (GSSP) should be strictly treated on the basis of the requirements of International Commission on Stratigraphy (ICS). This is espicially necessary for the Permian-Triassic boundary which is not only an intersystem boundary but also the boundary between the Palaeozoic and the Mesozoic.

In the past years, the Working Group on the Permian-Triassic boundary recommended two sections as candidate of GSSP of the Permian-Triassic: The Changxing section in China and the Guryul Ravine section in Kashmir. Neither the Changxing nor the Kashmir section could completely satisfy the requirements of GSSP. Some insist that the oldest deposits of the Triassic in Kashmir are better than those in the Changxing area, while the youngest deposits of the Permian are much better in Changxing area than in Kashmir. Others argue that, for defining the boundary of the Permian-Triassic, the lower boundary of the Triassic is more important than the upper boundary of the Permian. The ammonoids from the Khunamuh Formation are well-preserved, whereas those from the Changxing area are not in a good state of preservation. But there are still problems on the Lower Triassic strata of the Guryul Ravine. Among them at least, the following should be noted:

1. Kapoor, H. M. and Nakazawa, K. (1981), in their authoritative monograph entitled "The Upper Permian and Lower Triassic Faunas of Kashmir", described the Guryul Ravine section as follows (p. 17):

"A detailed sedimentological analysis was carried out for the Guryul Ravine section, especially from Member C of the Zewan Formation to Member I of the Khunamuh Formation". "It was first clarified in the previous paper(Nakazawa et al., 1975) that the alternation of limestone and shale in the Khunamuh Formation is mostly caused by turbidity current". "The rapid deeping of the sedimentary basin is observed in the next Khunamuh Formation and is indicated by a development of limestone turbidite and pelagic shale"

Cowie, J. W. (1986) stressed in the "Guidelines and Statutes of the International Commission on Stratigraphy (ICS)":

"(e) Favourable facies for development of widespread reliable and time-significant correlation horizons: this requires that the GSSP should not be in or close to conglomerates, breccias, olistostromes, turbidites or remainie deposits." "......Even the "model" decision on the Silurian-Devonian Boundary has had, retrospectively at least, its weakness, the GSSP was placed within a turbidite on the basis of the "first occurrence of a species."

This has been made clear about the weakness of the Silurian-Devonian Boundary Stratotype, and it is also clear that the sediments near the boundary of the Guryul Ravine section are turbidites. This being the case, we can not imagine to vote for the GSSP of Permian-Triassic, showing a turbidity facies. It obviously violates the Guidelines and Statutes of the ICS.

2. During my stay in U.S.A. in the fall of 1985, I examined one specimen of *Hindeodus minutus* (from the *Otocears* Zone, Unit 52) from the Guryul Ravine section which is deposited in the Department of Geology and Geophysics, Wisconsin-Madison University. The color of the specimen is black. According to Clark, D. L. and myself, the Color Alternation Index (CAI) of this conodont specimen could be deter-

mined as 5. Dr. Matsuda sent me 10 additional conodont specimens from the Guryul Ravine section [5 specimens from Bed 59 (E_2) and 5 others from Bed 61(E_3)]. The value of CAI of all these specimens is also 5. This indicates that the Khunamuh Formation (Lower Triassic) in Kashmir underwent a process of thermal metamorphism of more than 300° C.

One of the pioneers of the study of conodont colors, Annita G. Harris, told me that if the value of CAI is above 3, the section should not be used as a boundary stratotype, because in this case all specimens of spores, acritarchs, chitinozoans and dinoflagellates become black and ambiguous, i. e. they can not be identified clearly.

As far as I know, when the Working Group on the Cambrian-Ordovician Boundary discussed and selected the candidate boundary stratotype, they considered the CAI as a key point for candidates and they excluded the sections, where their values are equal or above 3. Such metamorphic section can not provide any information about multispecies fossil zones, because all organic microfossils have been destroyed, while

"Multispecies fossil zones (e.g. faunal assemblages) are preferable biostratigraphic signatures for GSSP guidance" and "recognition of the boundary level should preferably be used on all available fauna and flora data." (Cowie et al., 1986).

If we take into account the presence of the Panjal volcanic rocks near the Guryul Ravine section and the activities of this volcano, it is not difficult to understand why the value of CAI is so high.

According to the guidelines and statutes for boundary stratotype of the ICS, a boundary stratotype should show "freedom from structural complication, metamorphism or other alternation". (Cowie et al., 1986). All the members of PTWC should take into consideration that the Guryul Ravine section suffered thermal metamorphism.

3. We have not got any data of magnetostratigraphy and geochronometry from the Guryul Ravine section. It is doubtful that one can get satisfactory results from the Guryul Ravine section which shows thermal metamorphism. Gowie has pointed out that results from magnetostratigraphy and geochronometry are important as well:

"(h) Amenability to magnetostratigraphy and geochronometry. Although these factors are mentioned last, they are probably the most important for future work and some would argue that no GSSP should be accepted without one or both." (Cowie et al., 1986).

"6.... An ideal GSSP would have the maximum possible correlation by magnetostratigraphic and geochronometric methods: this is increasing importance for future work." (Cowie et al., 1986).

4. The Upper Permian and Lower Triassic strata of Guryul Ravine are subdivided into two formation, The Zewan and the overlying Khunamuh Formation. "The Khunamuh Formation overlies conformably the Zewan Formation with a rather abrupt change in lithology." (Kapoor et Nakazawa, 1981, p. 10). The Khunamuh Formation is divisible into four members E to H at the Guryul Ravine. Member E of the Khunamuh Formation is composed of alternating shale and subordinate limestone, and it is divided into three units, Viz., E_1, E_2 and E_3 from the bottom to the top. The age of E_1 is disputed by Chinese and Japanese investigators. Unit E_1 belongs, according to the Japanese, to the uppermost part of the Permian, whereas the Chinese prefer to place it at the base of the Triassic. This is still an open question. But one fact should be noted: there is a rather abrupt change in lithology between the Zewan and Khunamuh Formation. Concerning the conformity, Cowie et al. (1986) have pointed out:

"(g) Freedon from unconformities: an obvious boundary should be suspect. Either it is obvious because there is a marked change in lithology or because there is a marked change in fauna or flora. In either instance the change may imply a time break, and consequently an unsuitable horizon at which to fix any time definition; no disconformities, unconformities, cryptic paraconformities or time-breaks in sedimentation any longer than a brief diastem can be tolerated close to a GSSP."

5. There is still a weak point for the Guryul Ravine section, as there are no ammonoids in E_1. nor even in the highest beds of Zewan, If we would place the boundary at the base of the *Otoceras* fauna, then we must have the forerunners of *Otoceras* in the underlying beds. How we could be sure that *Otoceras* begins only in E_2 and not earlier in E_1 or at the base of E_1. There is no section in the world, where *Otoceras* occurs in the ammonoid-bearing beds that contain the forerunner of *Otoceras* or even Otoceratidae. Therefore, nobody can determine the base of the *Otoceras*- bearing beds to be a synchroneous event with ammonoid as evidences.

On the basis of five facts mentioned above, it is obvious that the Guryul Ravine section shows some very weak points. Some would argue that there is no section that can satisfy all desired criteria. Nevertheless, we should not selected a section as a boundary stratotype which shows so many serious problems. This sould strongly violate the guidelines and statutes for boundary stratotype of the ICS.

I think it premature to decide on the Permian-Triassic boundary stratotype at the present day, for much work is still to be done before such a selection can be made.

PS: The munuscript of this paper was finished in 1986. The candidate sections as GSSP of the Permian-Triassic add more in recent years, at least, the following three sections in China have been recommended, namely,

1. Changjianggou section, Shangsi, Guangyuan, Sichuan Province.,
2. Huayun section, Lingshui, Sichuan Province.
3. Selong Xishan section in Tibet.

REFERENCES

Cowie, J. W. and others, 1986. Guidelines and Statutes of the International Commission on Stratigraphy (ICS). Cour. Forsch. -Inst. Senckenberg.

Epstein, A. G., Epstein, J. B. and Harris, L. D., 1977. Conodont color alternation-an index to organic metamorphism. U. S. Geol. Surv. Prof. Paper, 995. 1–27.

Nakazawa, K. and Kapoor, H. M., 1981. The Upper Permian and Lower Triassic faunas of Kashmir. Palaeont. Indica, N. S., 46.

Nakazawa, K., Kapoor, H. M., Ishii, K., Bando, Y., Okimura, Y. and Tokuoka, T., 1975. The Upper Permian and the Lower Triassic in Kashmir, India. Mem. Fac. Sci. Kyoto Univ., Ser. Geol. & Miner., 42 (1), 1–102.

Kapoor, H., Ishii, K. I., Bando, Y., Maegoya, T., Shimizu, D., Nogami, Y., Tokuka, T. and Nohda, S., 1970. Preliminary report on the Permo-Trias of Kashmir. Ibid., 37(2), 163–172.

Sheng Jinzhang, Chen Chuzhen, Wang Yigang, Rui Lin, Liao Zhuoting, Bando, Y., Ishii, K. I., Nakazawa, K. and Nakamura, K., 1984. Permian-Triassic boundary in Middle and Eastern Tethys. Jour. Fac. Sci., Hokkaido Univ., Ser. 4, 21(1), 133–181.

Yin Hongfu, 1985. On the transitional bed and the Permian-Triassic boundary in South China. Newsl. Stratigr., 15(1), 13–27.

PALAEONTOLOGIA CATHAYANA Volume 5, 1990, p.267–268

THE RISE, DEVELOPMENT AND EXTINCTION OF STROMATOPOROIDS

Dong Deyuan

(Nanjing Institute of Geology and Palaeontology, Academia Sinica)

When did the oldest stromatoporoid begin? Yavorsky, a Soviet geologist, described in 1932 the Middle Cambrian stromatoporoids: *Clathrodictyon* and *Anostylostroma* from western Siberia. Later on, V. Khalfina (1960) acquired some new materials, renamed *Anostylostroma vologdini* as *Praeactinostroma vologdini* and designated some species of *Clathrodictyon* as *korovinella*. Are they really stromatoporoids? Are the fossil-bearing beds correctly defined? Flügel (1957), after discussing them, pointed out that they are probably Archaeocyathida but not Stromatoporoidea. Galloway (1957) asserted that they are Archaeocyathida. Moreover, he pointed out that stromatoporoids probably evolved from Archaeocyathida, exactly from *Exocyathus* or from the related genera. Not only that, he thought that the oldest stromatoporoid began to appear in the Middle Ordovician and that in the Ordovician the simplest stromatoporoid, i. e. *Cystostroma vermontense* showed its coenosteum consisting of arched cyst plates but lacking in pillars. However, the first discovery of stromatoporoids in China is known from the lower member of the Majiagou Formation (late Early Ordovician) in Suxian and Xiaoxian, northern Anhui. Among them are *Labechia, Rosenella, Aulacera, Cystistroma* and *Cryptophragmus*, but *Cystostroma* is absent. In consideration of this, it is uncertain whether *Cystostroma* is the oldest stromatoporoid. Of the above forms, *Cryptophragmus* is simple in structure, and its coenosteum consists of large upward-arched cyst plates which are superposed one by one to take a columnar shape. The question whether it is the oldest stromatoporoid remains to be settled. Nevertheless, the above-mentioned fact indicates that stromatoporoid rose from the late Early Ordovician and developed from then on. Thus we should cast away the view that stromatoporoid first appeared in the Middle Ordovician.

The late Early and Middle Ordovician stromatoporoids were predominated by Labechiida. In the Late Ordovician, the forms of Labechiida were still abundant but a few genera and species of Cliefdenellidae and Clathrodictyida came forth. In the Silurian, stromatoporoids gradually diversified, but Clathrodictyida took the predominant position instead of Labechiida, which was strikingly diminishing in number. Up to the Late Silurian, stromatoporoids were further flourishing. Some genera and specied of Stromatoporida, Actinostromatida and Idiostromatida began to appear. In the Early Devonian, affected by the Caledonian Movement, a wide-spread regression took place, resulting in the ill-development of bioherm throughout the world, which more or less controlled the development of the stromatoporoids in this epoch. Taken as a whole, the stromatoporoids are fewer in taxa as compared with those in the Silurian.

In the evolutionary history of stromatoporoids, the late Early Devonian to the Middle Devonian is the important stage. Stromatoporoids were rapidly increasing both in number and in taxon. They are widely distributed in the world and especially often seen in the carbonate sediments. In many places abundant stromatoporoids were found to be associated with algae and corals and formed bioherm or biostrata. Labechiida, which was predominant in the Ordovician, became fewer in this period. Also Clathrodictyida, which held an important position in the Silurian, was gradually decreasing; while Actinostromatida, Stromatoporida and Idiostromatida were more and more increasing, or even developing to their peak. In the Late Devonian, the change of sedimentary environment had brought about the change of petrological facies, leading to the decrease of stromatoporoids. That is to say, stromatoporoids are much fewer both in number and in taxon in the Late Devonian than in the Middle Devonian.

It is worthy of notice that from the available materials, stromatoporoids were not many in the Late Devonian Famennian, but in many places of China many fossil stromatoporoids were found from the bed equivalent to the "Etroeungt" Bed, which was thought to belong to Early Carboniferous in age. Now it seems that these stromatoporoid-bearing beds are probably of Late Famennian stage. The stromatoporoid fauna has inherited some characters of the Devonian ones. but the commonest Devonian elements are never seen. Instead of them there occur a number of genera of Labechiida, especially the longitudinal skeleton further became complicated and many new genera and species rose, such as *pennastroma, parastylostroma, platiferostroma, Sichuanostroma, Spinostroma* and *Eopennastroma*. Similar phenomena can be seen in Novaya Zemlya, Donets Bassin and Omolon area of USSR. But it has not reported elsewhere in the world. What has been mentioned above indicates that stromatoporoids indeed occurred in the Late Devonian Famennian. However, whether they can extend to Carboniferous is still in dispute. In Yecheng of Xinjiang a species of stromatoporoid, *Pseudolabechia kunlunensis* Dong et Dong, was found to be associated with fusulinids and corals, which indicate that the fossil-bearing sediments are of Carboniferous Visean age. This stromatoporoid is the youngest Palaeozoic one so far discovered in the world.

From the foregoing statement, we can see that following the Devonian time stromatoporoids became extinct rapidly. Some one asserted that the Palaeozoic stromatoporoids did not exist in the post-Devonian. The assertion seems not to be correct as the individual species actually extends to the Carboniferous Tournaisian or even to the Visean. From the Middle Carboniferous to Permian or even to Early and Middle Triassic, none of stromatoporoids have been reported as yet. But in the Late Triassic stromatoporoids reappeared, though they were not many both in number and in taxon. Strange to say, in the Early and Middle Jurassic, rare stromatoporoids have been seen. This indicates that stromatoporoids were declining evolutionarily in this period. From the Late Jurassic on, stromatoporoids further developed not only in number but also in taxon. They are mainly distributed in the Tethyan Realm and in northern Xizang (Tibet) and southern Xinjiang. In the Early Cretaceous, some stromatoporoids still existed, but they were entirely extinct after that time.

The above statement is to briefly describe the rise and development of stromatoporoids as well as their main characters and the interrelationship in different geological time. Why they suddenly disappeared from the Early Carboniferous to the Middle Triassic? Why they reappeared in the Late Triassic? These questions are still unclear. Thus their evolution and interrelationship need to be further studied in the coming days.

REFERENCES

Bogoyavlenskaya, O. V., 1969a. Catalogus of Stromatoporoidea. Paleont. Zhur., 1, 147–148 (in Russian).

__________, 1969b. Classification of Stromatoporoidea. Ibid., 4, 12–26 (in Russian).

Dong Deyuan, 1964. Stromatoporoids from the Early Carboniferous of Kwangsi and kueichow. Acta palaeont. Sinica, 12 (2), 280–291 (in Chinese).

__________, 1981. Upper Jurassic Stromatoporoids from northern Xizang. In: Palaeontology of Xizang (3), 115–126. Science Press, Beijing (in Chinese).

__________, 1982. Lower Ordovician stromatoporoids of northern Anhui. Acta Palaeont. Sinica, 21(5), 577–583 (in Chinese).

__________, 1985. Silurian and Lower Devonian stromatoporoids from Darhan Mumingan Joint Banner, Inner Mongolia. In: Silurian and Devonian Rocks and Fauna of the Bateaobao Area in Darhan Mumingan Joint Banner, Inner Mongolia. Inner Mongolia People's Press, 57–77 (in Chinese).

__________ and Wang Baoyu, 1984. Paleozoic stromatoporoids from Xinjiang and their stratigraphical significance. Bull. Nanjing Inst. Geol. Palaeont., 7, 237–286. (in Chinese).

__________ and __________, 1985. Cnidarian fauna from the Mesozoic of South Xinjiang. Acta Palaeont. Sinica, 24(4), 449–452 (in Chinese).

__________ and Wang Chengyuan, 1982. Devonian stromatoporoids of eastern Yunnan. Bull. Nanjing Inst. Geol. Palaeont., 4, 1–40 (in Chinese).

Galloway, J. J., 1957. Structure and classification of the Stromatoporoidea. Bull. Amer. Paleont., 37(164), 345–470.

Gorsky, I., 1938. Some stromatoporoids from palaeozoic beds of Novay Zemlya. Tran. Arctic Inst. Leningrad, 101, 7–45 (in Russian).

Khalfina, V. K. and Yavorsky, V. I, 1974. On the evolution of stromatoporoids. Cnidaria (1). Acad. Sci. USSR. Siberia. Tran. Inst. Geol. Geoph., 201, 38–45 (in Russian).

Lecompte, M., 1951–1952. Les Stromatoporoidea du Devonien moyen et superieur du Bassin de Dinant. Inst. Roy. Sci. Nat. Belgique Mem., 116–117.

__________, 1956. Stromatoporoides, In: Treatise of Invertbrate Paleontology, part F. Coelenterata, 107–144. Geol. Soc. America and Univ. Kansas Press.

Mori, K., 1968. Stromatoporoids from the Silurian of Goltland. pt. 1. Stockolm contrib. Geol., 19, 1–100.

Nestor, H. A., 1974. On the Phylogeny of Palaeozoic stormatoporoids. Cnidaria (1). Acad. Sci. USSR. Siberia. Tran. Inst. Geol. Geoph., 201, 27–38 (in Russian).

Stearn, C. W., 1972. The relationship of the stromatoporoida to the Sclerosponges. Lethaia, 5, 369–388.

__________, 1980. Classification of the paleozoic stromatoporoids. Jour. Paleont., 54 (5). 881–902.

Turnšek, D., 1966. Upper Jurassic hydrozoan fauna from southern Slovenia. Razprave Slov. Acad. Znan. Umetr., 4 (9), 1–94.

Yang Jingzhi and Dong Deyuan, 1979. Devonian stromatoporoids from central and eastern parts of Guangxi, China. Palaeont. Sinica. (B). 14(in Chinese).

PALAEONTOLOGIA CATHAYANA Volume 5, 1990, p.269–276

TRACE ELEMENTS OF OSTRACOD SHELL USED AS AN INDICATOR FOR PALAEOENVIRONMENTARY RECONSTRUCTION

Zhao Yuhong

(Nanjing Institute of Geology and Palaeontology, Academia Sinica)

Abstract

Trace elements of forty-four ostracod shells have been accurately measured by using electron probe in conjunction with SEM. The specimens comprise four groups: living marine, living non-marine, fossil marine and fossil non-marine ostracods. The result reveals that these ostracod shells contain the following chemical elements: Calcium (Ca), Silica (Si), Aluminium (Al), Iron (Fe), Magnesium (Mg), Potassium (K), Chlorine (Cl), Sulphur (S), Sodium (Na) and Barium (Ba) (in their order of abundance). Ca is most abundant and accounts for 31–36%, and the total amount of other trace elements is less than 10%. There is obvious difference in trace elements between the marine ostracod shell and the freshwater one. This paper shows how they can be used to deduce the palaeoenvironment in which fossil ostracods once lived.

These analytical results of the chemical content of ostracod shells help to throw light on the cause of different colours in fossilised shells and the process of their formation.

INTRODUCTION

Hitherto, little research work has been done on the trace elements of ostracod shell, especially on their quantitative analysis. Müller (1894) confirmed that the ostracod shell consisted of calcium carbonate. Clark and Wheeler (1917) also worked on the composition of marine invertebrate skeletons, including some ostracod shells. Kesling (1951) analysed the freshwater ostracod *Cypridopsis vidua* (O. F. Müller) and found that the shell was composed of calcite. Chave (1954) analysed the composition of marine ostracod shells and pointed out that they contained about 1–10% magnesium carbonate. Sohn (1958) reported that the living ostracod shell was made up of 80–90% of calcium carbonate, 2–15% of chitinous substance, proteinaceous material and trace elements, namely potassium, magnesium, sodium, silica, alumnium, barium and strontium. In their work on some living ostracod shells Cadot, Kaesler and Schmus (1975). thought that the amount of magnesium was a variable constituent. Later Cadot and Kaesler (1977) put forward more details and felt that the magnesium content differed in the shells of four subfamilies as a result of the water temperature under which the ostracods once lived. Rosenfeld (1979) explained three curves of ostracod shell composition. He believed that among all the constituents of ostracod shell, calcium was most abundant, whilst the trace elements in order of abundance were phosphorus, silica, aluminum, potassium, magnesium and sulphur. Bodergat and Andreani (1981) analysed the shell composition of *Cyprideis torosa* (Jones). They held that although the shell composition had nothing to do with water salinity, the trace elements of the shell were affected by the chemical composition of the environment. Chivas, De Deckker and Shelley (1983) studied the shell composition of *Mytilocypris henricae* under certain conditions (water temperature, salinity, ph and water chemistry), which were scientifically controlled. The results revealed that the amounts of magnesium and strontium in the ostracod shell were certainly influenced respectively by the Mg/Ca ratio and the Sr/Ca ratio in the waters. There is no direct relationship between the amount of barium in the ostracod shell and Ba/Ca ratio in the waters. Bodergat (1983) has also published his article on ostracod shell composition.

In short, inorganic analysis (especially trace elements) of ostracod shell has only recently received the attention of scholars. The work done so far is still elementary and the results are not consistent. Some of the results have to await further confirmation. These studies dealt mainly with the living ostracod shell, and little work has been done on fossil shells. Previously, the emphasis has been mainly placed on the analysis of magnesium in the shell together with preliminary research work on the quantitative analysis of all the trace elements. No study and comparison of the shell composition of four groups, i. e. living marine ostracod (L. M. O.), fossil marine ostracod (F. M. O.), living non-marine ostracod (L. N. O.) and fossil non-marine ostracod (F. N. O.), have hitherto been made in literature.

MATERIAL AND METHOD

The specimens under study consist of *Leguminocythereis* sp., *Neomonoceratina* sp., *Abrotocythere quinquicornis, A. ovata, A. plegadis, Cyprinotus* sp., *Ilyocypris* sp., *I. caohaiensis, I. weiningensis, L. bradyi, I. subbiplicata, I. Leucocythere plena, Limnocythere xinanensis, Hungarocypris* sp., *Notodromas sinensis*, and *Cypris* sp., The living ostracods were collected from South China Sea, Bohai Sea, East Lake in Wuhan of Hubei, Caohai Lake in Guizhou and a pond near Daqing of Heilongjiang; while the fossil ostracods were from the Tertiary and Quaternary strata in Jiangsu and Guizhou. Details of various taxa, including the environment, geological age, date of collection and trace elements are also given in Table 1.

All the specimens were analysed by electron microprobe in conjunction with scanning electron microscope. The area examined was limited to the ostracod endocuticle. The endocuticle area was analysed repeatedly in some specimens in order to test the reliability of the scientific data obtained. All the data are listed in Table 1.

ANALYSED RESULT

Inorganic elements and their level of incidence in the ostracod shell

According to this initial research, Ca is the main inorganic constituent, but other elements, such as Mg, Si, Al, Fe, S, K, Cl and Ba are also present.

(1) Ca is the main constituent and makes up about 32–34% of the total composition. The percentage of Ca seems not to be related to the thickness of endocuticle in the ostracod carapace, the arrangement of calcite and the ostracod living environment. Instead, the Ca content seems inversely proportional to the amount of all the trace elements in the ostracod shell.

(2) Mg is a very important trace element and can be detected in the shells of most ostracods which lived in the marine environment. On the other hand, few specimens from fresh water showed the presence of Mg in their endocuticle. The highest amount (1.8%) of Mg is found in *Neomonoceratina* sp. The average content of Mg in marine ostracod shells is 0.988%, but it is only 0.24% in those from the non-marine environment. The mean content in fossil groups is lower than that found in living groups, probably due to the water filtration during the deposition.

(3) Si is a common trace element and occurs in almost all the examined specimens. The average amount is 2.69% in marine, and 0.829% in non-marine shells. It is very difficult to formulate a general rule about the appearance of Si in the ostracod shell because of the great variation.

(4) Al was found in all the examined specimens in which it varied generally between 1.10–1.30% of the composition. Only a few specimens showed higher values. There is little difference in the average content of marine and non-marine shells, which have 1.317% and 1.29% respectively.

(5) Fe and S. Fe is commonly present but varies from 1.65%–10.30%. Generally the amount of Fe is greater in fossil than in living shells in the case of both marine and non-marine ostracods. S always appears along with Fe, but in a small amount. These elements probably came from the ground water in the sedimentary process. It is vary obvious that in the fossil ostracods the amounts of Fe and S are closely related to the shell colour as shown in Table 2. Thus, the amounts of Fe and S probably reflect some conditions during and after the sedimentation.

(6) Cl is present in less than half of the specimens examined and only in a smaller amount. The mean error is great so that it is pointless to discuss it here.

(7) K is very rare in ostracod shells as in the case of Cl. It is only found in less than one third of the specimens analysed. But living specimens show a much higher potassium content than fossil specimens.

Table 1. The analysed data of elements in ostracod shells.

NUMBER	SPECIES NAME	ANALYSED TIME	ENVIRONMENT	GEOLOGICAL AGE	PERCENTAGE OF ELEMENTS IN OSTRACOD SHELL									
					Ca	Fe	S	Si	Al	Cl	Na	Mg	K	Ba
1	*Leguminocythereis sp.*	1984.7	South China Sea	Living Ostracoda	37.53 37.70			0.10 0.10	0.15 0.14		0.67 0.65	0.66 0.65		
2	*Leguminocythereis sp.*	1984.4	South China Sea	Living Ostracoda	40.54	0.29		0.92	0.73		0.83	0.83		
3	*Leguminocythereis sp.*	1984.4	Bohai Sea	Living Ostracoda	33.22 33.61			0.20 0.20	0.22 0.21	0.11 0.12	0.93 0.75	0.93 0.85		
4	*Leguminocythereis sp.*	1984.7	Bohai Sea	Living Ostracoda	26.2	11.23	0.20	0.49	1.91	0.09	0.93	0.93	0.62	
5	*Neomonoceratina sp.*	1984.4	South China Sea	Living Ostracoda	20.02	1.84	0.20	11.54	5.49	0.11	0.92	1.84	1.15	
6	*Neomonoceratina sp.*	1984.4	South China Sea	Living Ostracoda	37.37 37.74	0.24	0.13 0.13	0.47 0.47	0.45 0.42		1.59	0.69 0.48		
7	*Neomonoceratina sp.*	1984.7	Marine	Q	35.59 36.24	0.31 0.33	0.37 0.39	0.45 0.35	0.40 0.39	0.10 0.11		0.67 0.50	0.08	
8	*Neomonoceratina sp.*	1984.4	Marine	Q	31.68	0.15	0.27	0.68	0.49	0.14	0.52	0.56	0.06	
9	*Neomonoceratina sp.*	1984.4	Bohai Sea	Living Ostracoda	38.77 38.90	0.22 0.22	0.16	0.19 0.20	0.18 0.13	0.14 0.15	0.09	0.56 0.53		
10	*Neomonoceratina sp.*	1984.4	Bohai Sea	Living Ostracoda	26.03	2.04	0.19	5.88	3.14	0.07	0.86	1.49	0.68	
11	*Abrotocythere quinquiensis*	1984.7	Nonmarine	Miocene	41.22 41.15	0.16 0.18	0.10	0.40 0.40	0.34 0.33	0.15 0.12		0.19 0.15		
12	*Abrotocythere ovata*	1984.7	Nonmarine	Miocene	37.11 36.58		0.11	0.18 0.18		0.22 0.22	0.47	0.20 0.13		
13	*Abrotocythere ovata*	1984.7	Nonmarine	Miocene	35.71 35.72	1.08 1.10	0.47 0.77	3.49 3.58	2.60	0.90 0.74	1.67	0.49 0.27	0.77 0.83	
14	*Abrotocythere pligadis*	1984.4	Nonmarine	Miocene	35.11 34.58		0.11 0.11	0.35 0.34	0.24 0.22	0.34 0.34	0.81	0.53 0.40	0.14 0.15	
15	*Cyprinotus sp.*	1984.4	East Lake	Living Ostracoda	37.37	0.15		0.13		0.12	0.12	0.55	0.39	
16	*Cypria sp.*	1983.7	East Lake	Living Ostracoda	0.53	0.38	1.89	2.71	1.13	0.35	0.53	0.47	0.42	
17	*Ilyocypris sp.*	1983.7	Caohai Lake	Living Ostracoda	34.43	0.32	0.29	1.34	0.39	0.20	0.48	0.52	0.31	
18	*Cypria sp.*	1983.7	Caohai Lake	Living Ostracoda	34.49	0.20	0.07	0.21	0.17	0.17	1.04	0.39		
19	*Cyprinotus sp.*	1983.7	Caohai Lake	Living Ostracoda	36.31			0.12		0.10				
20	*Ilyocypris sp.*	1983.7	Rice Field	Living Ostracoda	30.78	0.11		0.98	0.73	0.14	0.99	0.99	0.54	
21	*Cyprinotus sp.*	1984.4	Rice Field	Living Ostracoda	33.29	0.17		0.06				0.22		
22	*Notodromas sinensis*	1984.8	Pool	Living Ostracoda	43.07			0.12						
23	*Ilyocypris gibba*	1983.7	Nonmarine	Oligocene	33.66	0.30		0.91	0.58	0.06	0.51	0.29		
24	*Cypria yanchengensis*	1983.7	Nonmarine	Q	32.99	1.73	2.28	0.07						
25	*Cypridopsis caohaiensis*	1983.7	Nonmarine	Q	29.07 29.99	0.59 0.61	0.33 0.34	1.67 1.62	0.90 0.92	0.19 0.20		0.38 0.39		
26	*Ilyocypris yanglinensis*	1983.7	Nonmarine	Q	36.32	0.63	0.19	1.39	1.10	0.09			0.31	
27	*Ilyocypris caohaiensis*	1983.7	Nonmarine	Q	34.79	0.39	0.31	0.78	0.77	0.19			0.07	
28	*Ilyocypris caohaiensis*	1983.7	Nonmarine	Q	32.79	4.04	0.26	1.11	0.90	0.24				
29	*Ilyocypris weiningensis*	1983.7	Nonmarine	Q	31.28	1.03	1.25	0.32	0.90	0.27	1.37	0.39		
30	*Leucocythere plena*	1983.7	Nonmarine	Q	30.17	0.50	0.14	1.40	2.14	0.25	0.43	0.23	0.11	
31	*Leucocythere plena*	1983.7	Nonmarine	Q	28.95	0.51	0.31	1.21	1.89	0.23	0.45	0.19		
32	*Leucocythere plena*	1983.7	Nonmarine	Q	35.67	0.24	0.23	0.26	0.25	0.27		0.11		
33	*Limnocythere xinanensis*	1983.7	Nonmarine	Q	22.90	10.31	1.05	1.97	1.92	0.64	1.52	0.43		
34	*Ilyocypris gibba*	1983.7	Nonmarine	Q	60.51	0.36		0.67	7.30	0.61	1.20			
35	*Ilyocypris subbiplicata*	1983.7	Nonmarine	Q	36.51	0.30	0.36	0.18	0.44	0.11	0.34	0.29		
36	*Ilyocypris subbiplicata*	1983.7	Nonmarine	Q	32.76	3.15	0.20	0.25	1.83	0.64	0.54	0.16		
37	*Ilyocypris subbiplicata*	1983.7	Nonmarine	Q	38.07	0.22		0.31	1.00	0.13				
38	*Ilyocypris subbiplicata*	1983.7	Nonmarine	Q	33.83	0.18	0.14	0.45	0.74	0.42	0.43	0.25		
39	*Ilyocypris subbiplicata*	1983.7	Nonmarine	Q	35.41	0.22		0.17	0.55	0.12				
40	*Ilyocypris bradyi*	1983.7	Nonmarine	Q	29.36	1.25	0.26	0.26	2.85	0.53	1.12	0.22		
41	*Ilyocypris bradyi*	1983.7	Nonmarine	Q	32.94	3.99		0.99	1.67	0.33	1.47			
42	*Ilyocypris bradyi*	1983.7	Nonmarine	Q	31.68	2.68		3.15	3.35	0.70	0.92			
43	*Hungarocypris sp.*	1984.8	Nonmarine	Oligocene	41.83		0.65	2.36	1.42			0.69		Rare
44	*Hungarocypris sp.*	1984.8	Nonmarine	Oligocene	38.28	1.46	0.29	1.27	0.35	1.41		0.24		Rare

Table 2. The relationship between the amount of Ca, Fe and S in fossil ostracod shells and their colours of shell.

ELEMENT	Ca			Fe			S		
AMOUNT / FOSSIL SHELL COLOUR	LOW	MED.	HIGH	LOW	MED.	HIGH	LOW	MED.	HIGH
WHITE			●	●			●		
BROWN	●					●		●	
BLACK		●			●				●

This is probably due to other factors in the fossilization process which reduced the amount of potassium in the fossil ostracod shell.

To sum up, calcium is the main chemical element in ostracod shell and appears in the form of calcite.The amount of magnesium reflects the environment in which they once lived. The amount of iron and sulphur may show some sedimentary factors. Chlorine is very rare and irregular in ostracod shell, so it is not discussed here.

Trace elements of ostracod shell used as an indicator for palaeoenvironmental reconstruction

(1) Frequency

If the number of specimens which contain element A is divided by the total number of specimens examined, and then multiplied by 100%, we get the frequency FA.

$$FA = \frac{\text{The number of specimens which contain element A}}{\text{Total number of specimens analysed}} \times 100\%$$

In this way, we can get FCa, FAl, FSi, FS, FFe, FK, FCl and FMg, they show frequencies of calcium, aluminium, silica, sulphur, iron, potassium, chlorine and magnesium respectively. The element frequency curves of the four groups of ostracods are illustrated in Fig. 1. This diagram shows that the element frequencies of the ostracod shells can be used as being indicative of the environment in which the ostracods once lived.

In different environments. the frequency of the various elements found in the ostracod shell differs. According to the analysis, in marine ostracod shells, Ca, Mg, Si and Al appear in every specimen (F= 100%). On the contrary, in non-marine ostracod shell, the frequency of Ca is 100%, but FMg is less than 50%, FSi and FAl are both lower than those in marine ostracod shells. Although FFe, FS and FK are similar in the two kinds of environment, in general, the frequency of appearence in marine ostracod shells is higher than in non-marine ostracod shells. Chlorine is the only exception.

(2) The average amount of elements in the ostracod shell is shown in Fig. 2.

From them it is very easy for us to determine the kind of environment by using the average amount of Ca, Mg, K, and Cl. Generally, the average amounts of Mg and K in marine ostracod shells are six times higher than in non-marine shells, whereas Ca and Cl show the opposite trend. The average amounts of Al and Si are different in the two environments, but they are rare, probably because they are not sensitive to environmental changes. The average amounts of Fe and S are probably less significant as indicators for determining the environment because they are easily influenced by taphonomy and sedimentation.

(3) Ratio of Ca: Mg

Fig.3 shows the ratio of Ca: Mg in all the analysed specimens. It is obvious that the curve can be divided into two parts.

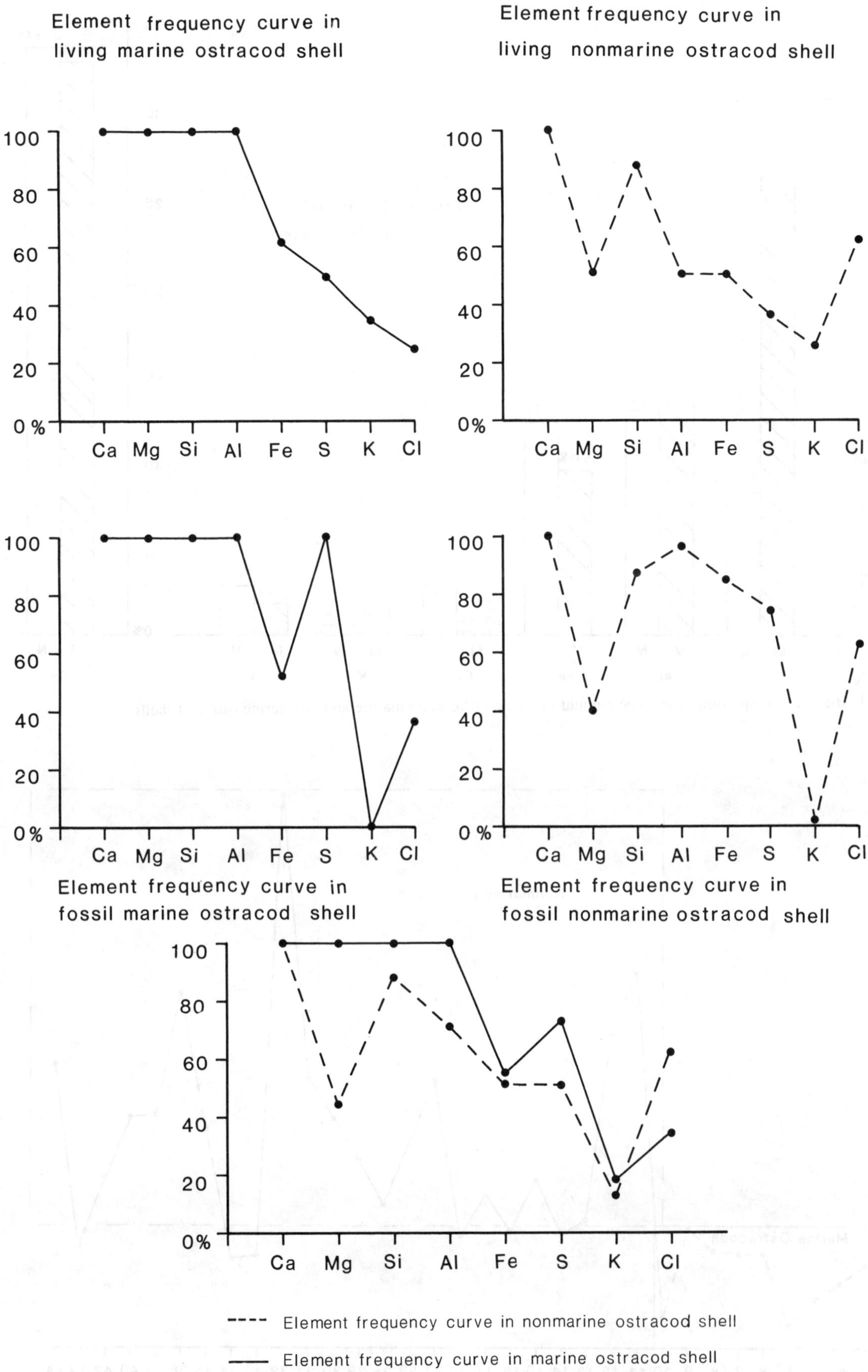

Text-fig. 1. Element frequency comparison between marine and nonmarine ostracod shells.

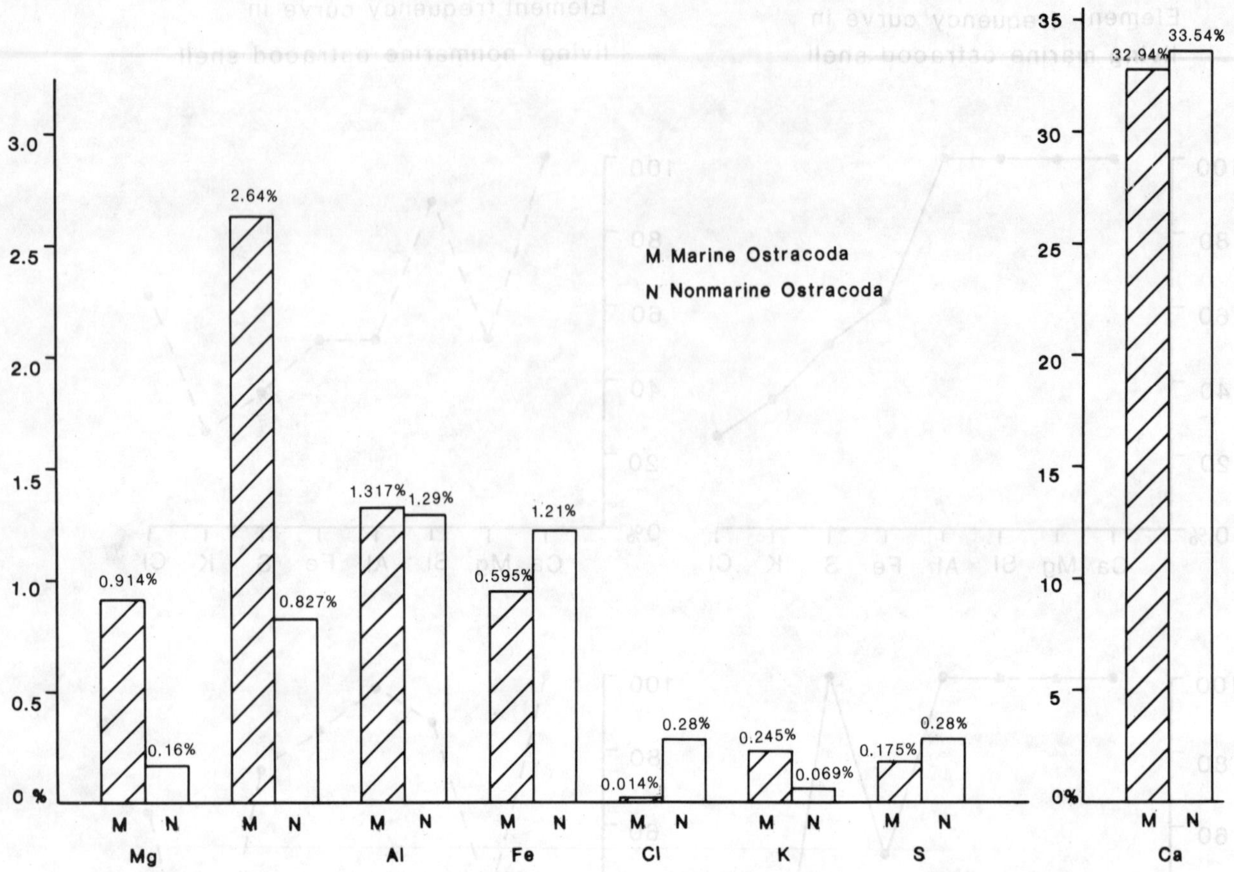

Text-fig. 2. Comparison of average amount of elements between marine and nonmarine ostracod shells.

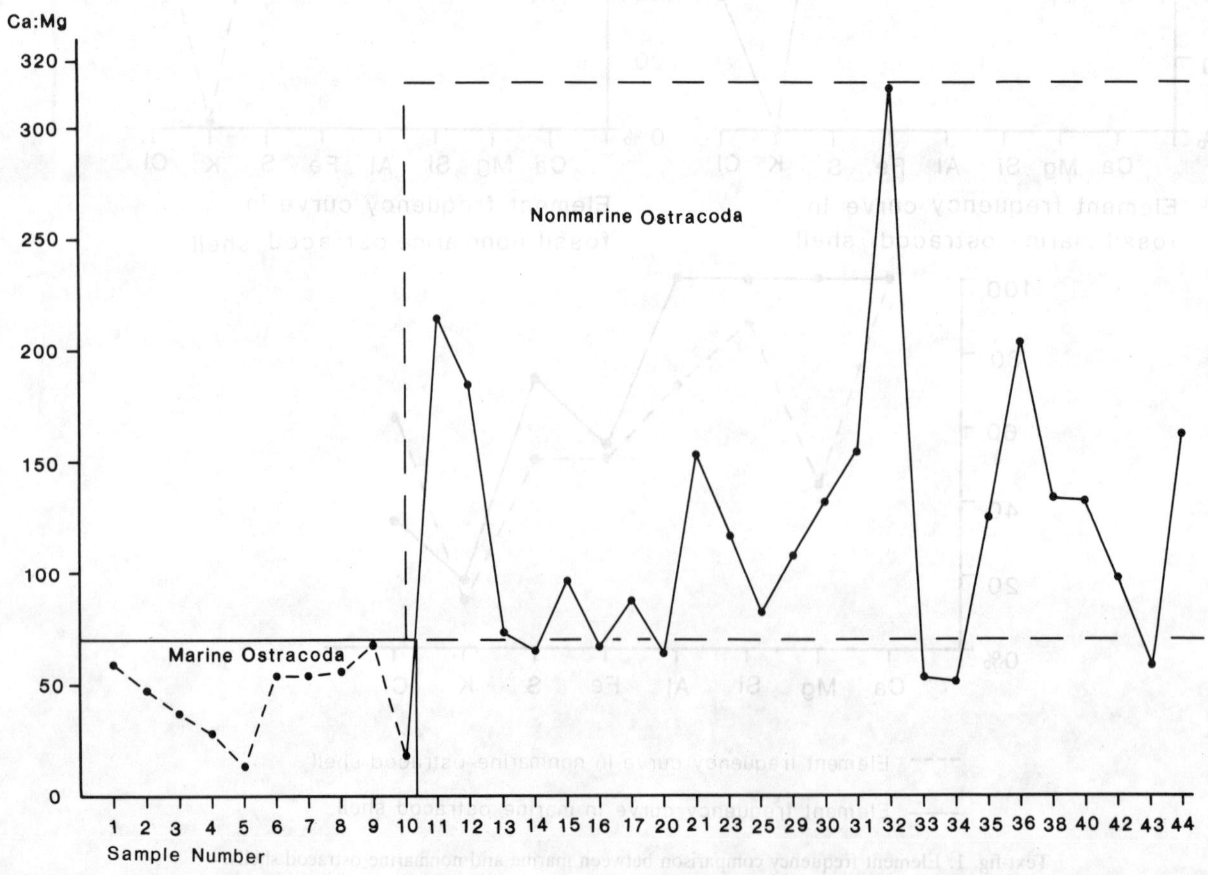

Text-fig. 3. Ca:Mg ratio curve in ostracod shell.

The lower ratio area is framed by a solid line and the high ratio area by a dotted line. It is very significant that this kind of division coincides with the pattern on the environment we know. For example, the-specimens in the lower ratio area were all collected in marine facies, whereas the specimens with the high Ca:Mg ratio were all collected from a freshwater habitat except numbers 11–14, which are species of Abrotocythere. The latter species probably lived in the oligohaline waters because *Youganospira costata* Youluo (Gastropoda) occurred in the same bed. It is certain that *Y. costata* occupied an oligohaline niche. In consequence, the ratio of Ca:Mg in Abrotocythere lies between 50–90 and occupies an intermediate position between the low and high areas.

These results show the importance of the Ca:Mg ratio, which not only differentiates marine from non-marine ostracods, but also helps to distinguish the ostracods in oligohaline from those in fresh water.

CONCLUSION

Trace elements in the ostracod shell (endocuticle) are very useful, which enable palaeontologists to deduce the palaeoenvironment in which the ostracods once lived. The most important indicators are the average amount, frequency of every element and the Ca:Mg ratio. Particularly, the Ca:Mg ratio is useful for judging whether species belongs to marine facies, freshwater facies or the intermediate oligohaline facies.

ACKNOWLEDGEMENTS

I wish to express my thanks to the late Prof. Mu Enzhi and Prof. Hou Youtang for their valuable guidance in my study and to Mr. Jiang Xiaojun of Shengli Oil Field for analysing the ostracod shell elements. Also I am grateful to Prof. J. W. Neale of Hull University, for his reading the manuscript critically and to Mr. P. McSherry for helping me to draw the figures and tables.

REFERENCES

Bodergat, A. M., 1983. Les ostracodes, temoins de leur environnement: Approche chimique et écologie en milieu lagunaire et oceanique. Docum. Lab. Geol., Lyon, 88, 1–246.

Bodergat, A. M. and Andreani, A. M., 1981. Mise en évidence de la réponse adaptative d'une espèce euryhaline *Cyprideis torosa* (Jones, 1950) à des conditions écologiques difficiles par l'analyse multi-elementaire en spectrometrie de masse à étincelle. In: Martinell, J. (ed.), Intern. Symp. On "Concept and method in palaeontology". Dept. Palaeontologia, Univ. Barcelona, 135–1139.

Cadot, H. M. and Kaesler, R. L., 1977. Magnesium content of calcite in carapaces of benthic marine Ostracoda. Palaeont. Contr. Univ. Kansas, Lawrence, 87, 1–23.

Cadot, H. M., Kaesler, R. L. and Schmus, W. R. Van., 1975. Application of the electron mocroprobe analyzer to the study of ostracode carapace. Bull. Amer. Paleont., Ithaca, 65(282), 577–585.

Chave, K. E., 1954. Aspects of the biogeochemistry of magnesium. 1. Calcareous marine organisms. J. Geol., Chicago, 62(3), 266–283.

Chivas, A. R., De Deckker, P. and Shelley, J. M. G., 1983. Magnesium, strontium and barium partitioning in nonmarine ostracode shells and their use in palaeoenvironmental reconstructions – A preliminary study. In Maddocks. R. F. (ed.), Applications of Ostracoda. Univ. Houston Geosc., Houston, 238–249.

Clarke, F. W. and Wheeler, W. C., 1917. The inorganic constituents of marine invertebrates. U. S. Geol. Surv. Prof. Paper, 124.

Kesling. R. V. 1951. The morphology of ostracode molt stages. Illinois Biological Monographs, Illinois, 21, 1–324.

Müller, G. W. 1894. Die Ostracoden des Golfes von Neapel und der angrenzenden Neeres Abschnitte: Fauna und Flora Golfes von Neapel, Berlin, 21.

Rosenfeld, A. 1979. Structure and secretion of the carapace in some living ostracodes. Lethaia, 12, 353–379.

Sohn, I. G. 1958. Chemical constituents of ostracodes; Some applications to paleontology and paleoecology, J. Paleont., 32(4), 730–736.

PALAEONTOLOGIA CATHAYANA Volume 5, 1990, p.277–294

COCCOID MICROFOSSILS FROM THE NIYAN FORMATION (UPPER PROTEROZOIC) AT QINGTONGSHAN OF SUXIAN, ANHUI AND THEIR SIGNIFICANCE

Yin Leiming

(Nanjing Institute of Geology and Palaeontology, Academia Sinica)

Abstract

Many tetrads of coccoid cells were well preserved in the cherts of stromatolites from the Late Proterozoic Niyan Formation at Qingtongshan of Suxian, Anhui. Some of them exhibit tetrahedral tetrads, a feature similar to the tetrads of eukaryotic organisms, but they are actually of prokaryotic organisms. Their formation is the final outcome of cell shifting affected by diagenetic degradation and compaction.

INTRODUCTION

Over the past two decades, many microfossils have been found in the cherts from the precambrian strata. In 1971, Schopf and Blacic reported the discovery of some specimens of tetrads of coccoid cells from the Late Proterozoic Bitter Springs Formation. In 1976, Oehler and Muir described some others from the Middle Proterozoic Amelia Dolomite. Later, Knoll & Golubic (1979) reexamined the Bitter Springs tetrads which Schopf and Blacic (1971) identified as eukaryotic tetrahedral tetrads, and reinterpreted them as *Gloeodiniopsis* (Schopf) em. Knoll & Golubic (1979), a probable chroococcoid cyanobacterium. In recent years, microfossils including hollow-tube-like filaments and coccoid cells have been obtained from siliceous stratiform stromatolites in the Upper Proterozoic Niyan Formation at Qingtongshan of Suxian, northern Anhui. Among them, many tetrads of coccoid cells are well-preserved. Some specimens are very similar to those of *Eotetrahedron princeps* Schoph & Blacic (1971). The study of them may help to resolve the problem whether they were eukaryotic or prokaryotic organisms.

GEOLOGICAL SETTING

Palaeogeographically, northern Anhui lies in the southeastern marginal belt of the Sino-Korean Platform between North China old land and Huaiyang old land. The Upper Proterozoic strata outcropping at Qingtongshan, about 65 km northwest of Suxian (Text-fig. 1), may be subdivided in ascending order into four units: the Jiayan Formation, the Zhaowei Formation, the Niyan Formation and the Jiudingshan Formation (Yang Qinghe et al., 1980). the Niyan Formation, about 370 m thick, consists of dolostone and limestone; its lower part is mainly thin- to medium-bedded calcareous dolostone containing chert lenses and argillaceous limestone; its upper part is mainly medium- to thick-bedded argillaceous limestone containing chert lenses. It contains the stromatolite *Stratifera*.

The underlying Zhaowei Formation is composed of limestone, with intercalations of dolostone, calcareous shale and mudstone. It contains abundant stromatolites, including *Inzeria, Baicalia, Tungussia, Gymnosolen, Jurusania, Anabaria, Conophyton* and *Archaeozon* (Yang Qinghe et al., 1980).

The Proterozoic rocks in the studied area have been determined by using radiometric method. According to Yang Qinghe and others (1980), the Jiayan Formation is dated at 738 my. The Niyan Formation is younger than the former formation, so the age of the Niyan Formation should be less than 740 my.

DEPOSITIONAL ENVIRONMENT

The Late Proterozoic strata at Qingtongshan of Suxian, northern Anhui were measured by Yang Qinghe et al. (1980). The upper part of the Zhaowei Formation is well stratified with many desiccation cracks. The depositional features show a shallow water environment either marine or lacustrine with periodic subaerial exposure. However, in consideration of the discovery of many organic-walled microfossils from the Zhaowei Formation and the Niyan Formation (Yang Qinghe et al., 1980), these deposits seem to be a marine rather than a lacustrine environment.

The boundary between the Zhaowei Formation and the Niyan Formation, can be hardly perceived. The Niyan Formation contains argillaceous limestone with intraclast conglomerate in the lower part and seems to have been deposited in the same environment as that of the Zhaowei Formation. Then, the depositional basin continually subsided, argillaceous and sandy matters decreased and marine water underwent salinization (MgO 18% and Ca 26–27%) as measured by Yang Qinghe et al. (1980), so the upper part of the Niyan Formation is characterized by medium or thick-bedded argillaceous or calcareous dolostone. In the relatively suitable environment, microbial communities were well developed and abundant stratiform stromatolites occurred. Meanwhile, many microfossils could be preserved in the chert stromatolites (Pl. 1, fig. 1). Based on the above facts, it seems probable that in an upper littoral or a supralittoral environment the Upper Niyan Formation was deposited.

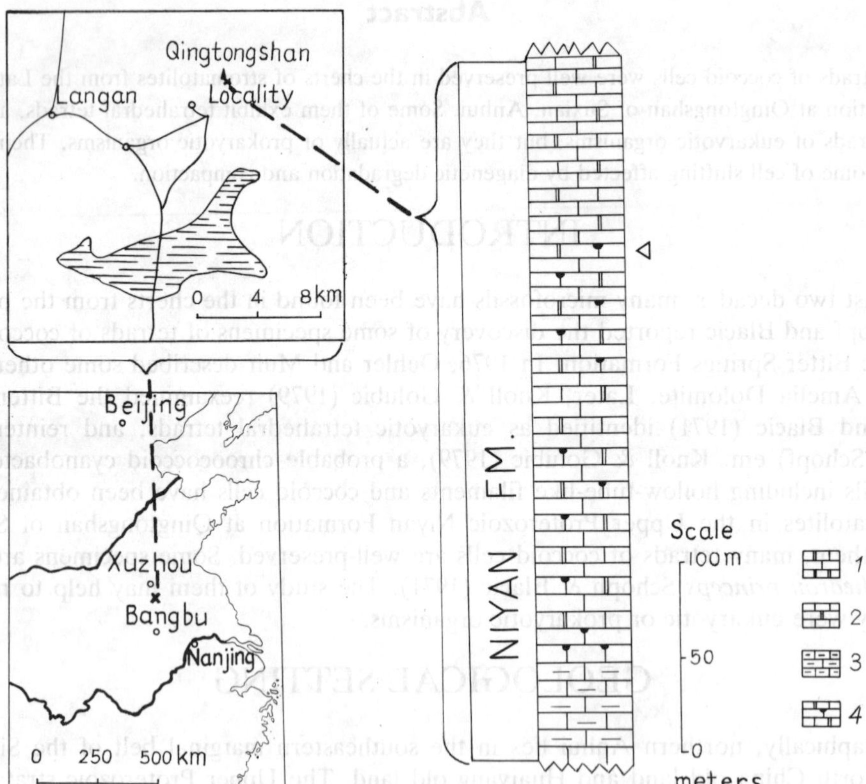

Text-fig. 1. Map showing East China, with an enlargement of the Suxian area of Anhui Province (Left), columnar section of Niyan Formation (Right). 1. calcareous dolomite, 2. dolomitic limestone, 3. argillaceous limestone, 4. cherty dolomite.

STATE OF PRESERVATION

The fossiliferous stratiform stromatolites from the Niyan Formation show distinct and fine laminations and appear dense, black, fine-grained and "waxy" (Pl. 1, fig. 1). The stromatolite laminae are alternately organic-rich and organic-poor (Pl. 1, fig. 2). Each lamina is about 0.3–0.5mm thick. The siliceous stromatolites from the Niyan Formation contain a low diversity of filament-dominated mocrobiota with sparse coccoid microfossils (Pl. 1, figs. 7–8). The microfossils studied here display different state of preservation owing to the diagenetic degradation and compaction.

The filamentous forms are dominantly hollow-tube-like and are often tangled together to form distinct "mats". Coccoid microfossils are disseminated through the "mats". But no distinct "mats" built by coccoid organisms are found in the present material. It is evident that the stratiform stromatolites of the Niyan Formation were built by filamentous organisms. In petrographic thin sections, it can be seen that tangled filaments are oriented parallel to bedding (shown in Pl. 1, figs. 2–3). These microbial mats are generally about 180–450 μm apart. Locally, the mats are broken and strongly curved, perhaps as a result of storms, currents or waves during the accretion of the stromatolites. Knoll and Golubic (1979) pointed out that the Proterozoic microfossils might be biased toward those communities in an inhabited environment where interstitial waters below the surface of the mat were poorly oxygenated and hypersaline or acidic. This may have also been true of the mat association in the Niyan Formation.

Fossilization was rapid, as proven in some specimens that apparently were "frozen" during cell division (Pl. 2, fig. 6; Pl. 4, fig. 5a, b). Thus, silicification might be either primary or very early diagenetic. Both the filamentous and coccoid microfossils are preserved in microcrystalline quartz. The quartz granules are about 5–40 μm in width. Besides, there is a remnant spherulitic fabric, like that in some specimens from the Gaoyuzhuang Formation in Quyang, Hebei Province (Yin Leiming, 1981), moreover, wherever the secondary diagenetic chert viens occur there are no microfossils.

SIGNIFICANCE OF THE NIYAN MICROBIOTA

In the past decades, many chinese publications on the Proterozoic microfossils discovered from siliceous stromatolites and from non-stromatolitic cherts came out (Yin Leiming & Li Zaiping, 1978; Yin Leiming, 1981; Zhang Yun, 1981; Liu Zhili, 1982; Zhang Zhongying, 1984; Wang Fuxing et al., 1984; Zhang Pangyuan & Yan Xiaoli, 1984; Luo Qiling & Wang Fuxing, 1984). It should be mentioned that most Proterozoic coccoid microfossils are prokaryotic cyanobacteria. They seem to have evolved very slowly, so that there are only minimal morphological differences between them. For example, according to Mendelson and Schopf (1982), the Sukhays Tunguska microbiota of the Siberian Platform of the USSR contains the members of cyanobacterial families (Oscillatoriaceae, Chroococcaceae and Entophysallidaceae), which are very similar to those known from the Bitter Springs Formation (Schoph, 1968; Schopf & Blacic, 1971), even to those known from the older sediments–the Belcher Supergroup (Hofmann, 1976). Evidently, the current data are insufficient for the biostratigraphic use.

The Niyan Formation is characterized by the occurrence of cyanobacterial (blue-green algae) mat-forming filaments and several types of colonial unicells. Among them, numerous mat-forming filaments (Pl. 1, figs. 3–4, 9) are very similar to those of *Eomycetopsis* Schoph (1968) and *Siphonophycus* Schopf (1968) which are known from the Middle and Late Pre-cambrian sediments, such as the Bitter Springs Formation (Schopf, 1968), Belcher Supergroup (Hofmann, 1976), Yudoma Formation (Lo, 1980), Draken Conglomerate (Knoll, 1979), Sukhaya Tunguska Formation (Mdndelson & Schopf, 1982), Wumishan Formation (Lin Chili, 1982), Jiudingshan Formation (Liu Xuexian, Liu Zhili et al., 1984). It is obvious that these tubular structures (*Eomycetopsis*-type, *Siphonophycus*-type) are not particularly useful for Proterozoic biostratigraphic correlation. But some colonial unicells may be helpful for discussing the age of the Niyan Formation. These coccoid fossils were previously known from the Bitter Springs Formation (Schopf, 1968; Schopf & Blacic, 1971), Draken Conglomerate (Knoll, 1979), Sukhaya Tunguska Formation (Mendelson & Schopf, 1982) and Jiudingshan Formation (Liu Xuexian, Liu Zhili et al., 1984). Moreover, in the Niyan assemblage, degradational variants and colonial unicells inferred to represent the life cycle of some living microorganisms are replete, as in the Bitter Springs assemblage (Text-fig. 2). However, time-restricted microfossil taxa used for correlating Proterozoic strata have not yet been identified. Based on some microfossil assemblages from the Proterozoic sediments elsewhere in the world, the stromatolites and the radiometric age determination of the underlying Jiayan Formation, the Niyan microbiota could be considered as of Late Proterozoic.

Furthermore, There are many well-preserved coccoid microfossils in the Niyan microbiota. According to their morphological features, they belong respectively to *Gloeodiniopsis* (Schopf, 1968), *Caryosphaeroides* (Schopf, 1968), *Eozygion* (Schopf & Blacic, 1971), *Bigeminococcus* (Schopf & Blacic, 1971), *Eotetrahedrion* (Schopf & Blacic, 1971), *Protosphaeridium* (Timofeev, 1966) and others. Many tetrads of coccoid cells have also been observed in one thin section of Niyan silicified stromatolites. Some assume "tetrahedral tetrads" (Pl. 2, figs. 5a-c; Pl. 3, figs. 6a, b; Pl. 4, figs. la, b). The morphologic features of tetrads in the Niyan assemblage are as follows:

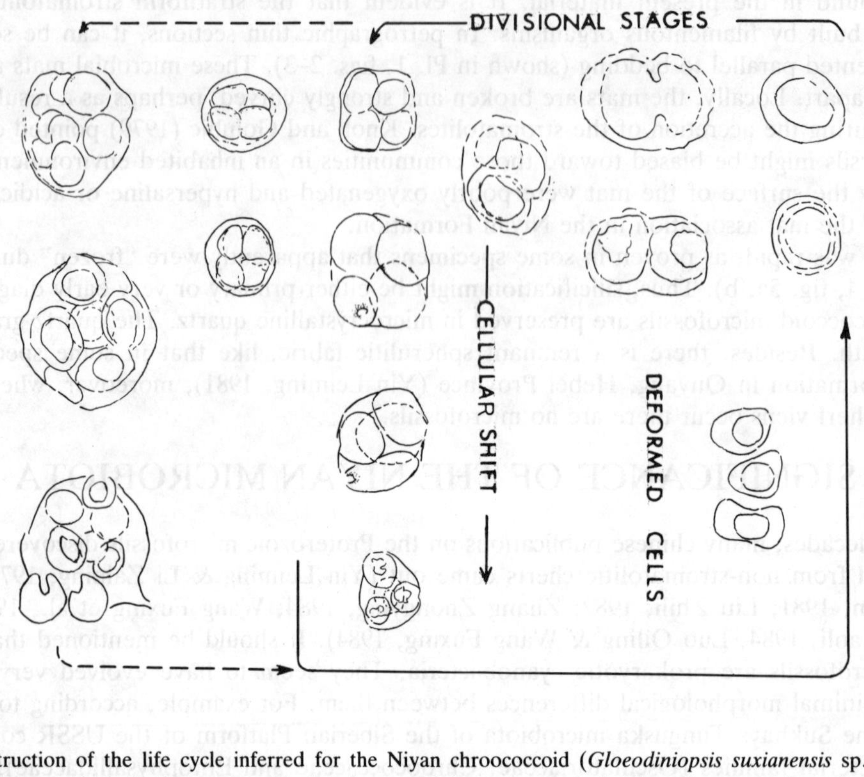

Text-fig. 2. Reconstruction of the life cycle inferred for the Niyan chroococcoid (*Gloeodiniopsis suxianensis* sp. nov.). Cell division proceeds successively in three perpendicular planes with intermittent production of extracellular gelatinous envelopes as in the modern cyanobacterium genus *Chroococcus*. The original configuration of the tetrad is planar one; in the process of cell separation within the confinement of the parental envelopes, cells shift to form vertically two pairs of cross tetrad and a nearly equidistant tetrahedral tetrad with complete separation.

1. Four spheroids are packed together in a common sheath or wall-like envelope. In some specimens, three cells rank on one plane and the fourth cell lies above them;

2. The spheroidal cells in the tetrads are variable in shape, showing spherical, elliptical and irregular morphology. They range in size from 12–33 μ m, mostly about 13–18 μ m. Some cells contain organic residues (shown in Pl. 1, fig. 5; Pl. 2, figs. 5a-c) and some unicells are enveloped with sheath-like material (Pl. 2, fig. 11; Pl 3, figs. 4–6).

3. All tetrads of coccoid cells are preserved in situ together with another several thousand individuals of coccoid cells.

4. No triradiate marks have been found on any of these cells.

These observations give rise to two problems:

1. The tetrads of coccoid cells are originally of tetrahedral ones.

According to the criteria proposed by Oehler et al. (1976), for distinguishing both cross tetrads and tetrahedral tetrads, some specimens of contiguous cross tetrads shown in Pl. 3, figs. 2a, b; 3a, b and 4a, b can be readily distinguished, because the two pairs of cross tetrads are obvious in vertical plane. Moreover, in three-dimensional mapping of spheroid positions, by measuring the actual distance between constituent spheroids in the same tetrad (Knoll & Golubic, 1979), the special arrangement of some tetrads has been reconstructed as shown in Text-fig. 3. Two specimens (Pl. 2, figs. 5a, b; Pl. 3, figs. 6a, b) show the typical form of tetrahedral tetrads, quite similar to *Eotetrahedrion princeps* from central Australia (Schopf & Blacic, 1971).

As summarized by Oehler et al. (1976), it is difficult to determine whether the tetrahedral geometry is original or has resulted from cellular shifting. Knoll and Golubic (1979), basing their interpretation on the three-dimensional mapping of spheroid positions and detailed examination of the taphonomic history of microcommunities, thought that the Bitter Springs tetrahedral tetrad represents one stage in a typically cyanobacterial life cycle, modified by diagenetic degradation and compaction.

Of significance is that the above-mentioned tetrad cells are parts of an assemblage of several thousand spheroids, many of which show "tetrahedral tetrads" in morphology. Various stages of cellular division (Pl. 2, fig. 6, 13; Pl. 4, figs. 5a, b) in individuals, pairs and planar tetrads are very well-preserved, indicating that most microorganisms of Niyan microbiota are reproduced simply through cell division, which can be comparable with that of extant coccoid cyanobacteria. On this basis, it can be said that the tetrahedral tetrads were reproduced through cell shifting, as reconstructed in Text-fig. 2. The tetrads of the Niyan assemblage are strikingly similar in configuration to those in the Bitter Springs assemblage, and their geometric configuration varies as a result of the diagenetic degradation and compaction. Also Knoll & Golubic (1979) thought that a tetrahedral configuration is the final outcome of packing under confinement in order to occupy the most stable cell arrangement within the thallus.

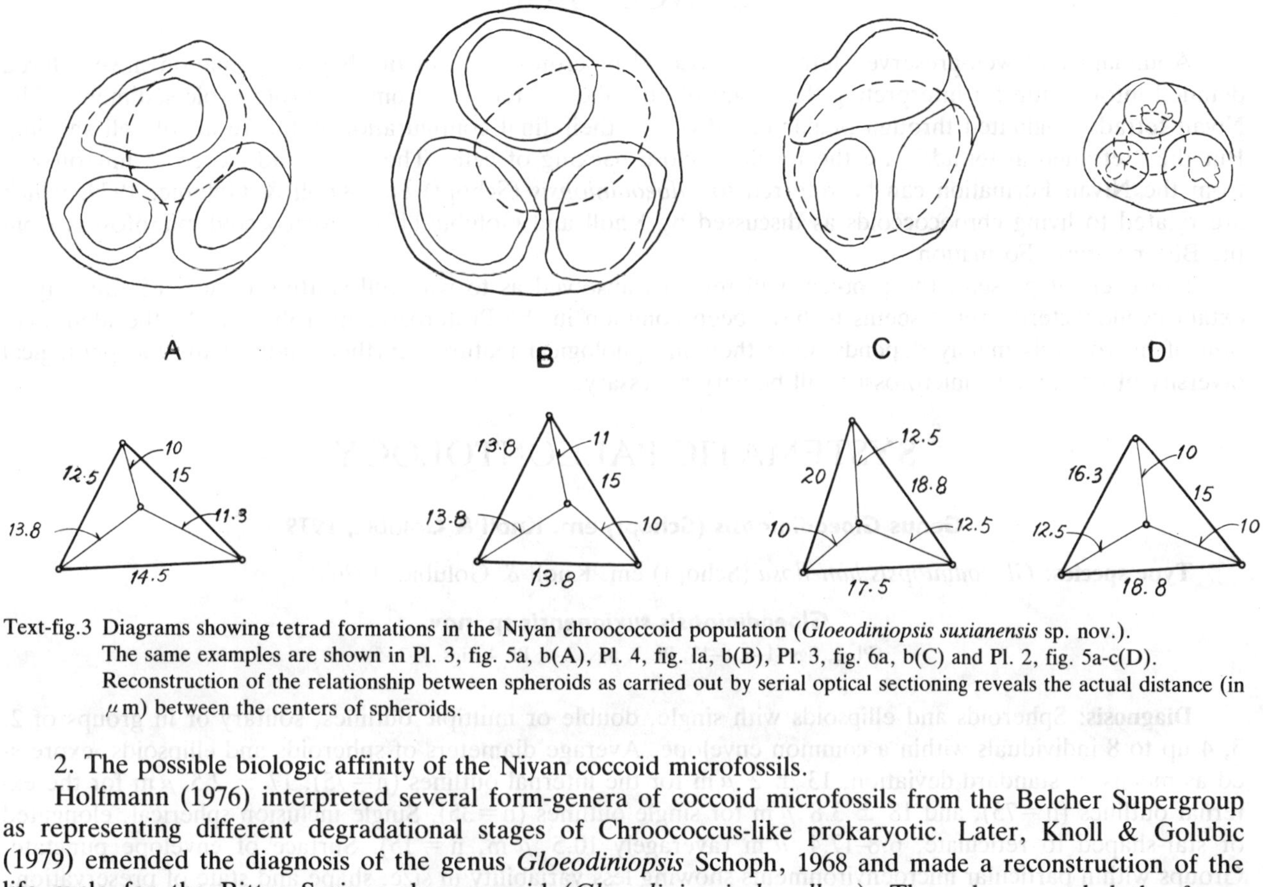

Text-fig.3 Diagrams showing tetrad formations in the Niyan chroococcoid population (*Gloeodiniopsis suxianensis* sp. nov.). The same examples are shown in Pl. 3, fig. 5a, b(A), Pl. 4, fig. la, b(B), Pl. 3, fig. 6a, b(C) and Pl. 2, fig. 5a-c(D). Reconstruction of the relationship between spheroids as carried out by serial optical sectioning reveals the actual distance (in μm) between the centers of spheroids.

2. The possible biologic affinity of the Niyan coccoid microfossils.

Holfmann (1976) interpreted several form-genera of coccoid microfossils from the Belcher Supergroup as representing different degradational stages of Chroococcus-like prokaryotic. Later, Knoll & Golubic (1979) emended the diagnosis of the genus *Gloeodiniopsis* Schoph, 1968 and made a reconstruction of the life cycle for the Bitter Springs chroococcoid (*Gloeodiniopsis lamellosa*). They also regarded the form-genera–*Bigeminococcus* Schopf & Blacic, *Eozygion* Schopf & Blacic, *Eotetraherdrion* Schopf & Blacic and *Caryosphaeroides* (partim) as synonyms of the genus *Gloeodiniopsis* (Schopf) em. Knoll & Golubic. For the same reason, a reconstruction of the life cycle of Niyan coccoid microfossils is shown in Text-fig. 2. This is strikingly similar to that proposed by Knoll & Golubic (1979) for *Gloeodiniopsis lamellosa*.

The coccoid microfossils of the Niyan assemblage are similar mostly to living cyanobacteria of the family Chroococcaceae. Nevertheless, there are still some doubts concerning the coccoid cells of the Niyan assemblage.

1. Cellular shifting has been observed in certain cyanobacteria (Oehler et al., 1976, Liu Chili, 1982), but no reports have been made on the cell shifting which resulted in the formation of enveloped cyanobacterial tetrahedral tetrads. There is still a problem as to why cell shifting so commonly happened within chroococcoid cells during the Proterozoic period as seen in the present Niyan assemblage, the Bitter Springs assemblage (Schopf & Blacic, 1971) and the Amelia Dolomite assemblage (Oehler et al., 1976). Knoll and Golubic (1979) proposed their views that cells were often secondarily shifted because of the pressure of the common envelope, in order to occupy the most stable packing arrangement. In the coccoid cells of the Niyan assemblage, singly deformed cells (Pl. 2, fig. 10) and deformed clusters of cells (Pl. 3, figs. 6a-b) have been commonly observed, which indicates that coccoid cells were strongly affected by diagenetic degradation and compaction.

2. Amongst the Niyan coccoids, a few unenveloped ones exhibit a spheroidal or elliptical shape and a smooth surface. Similar specimens from the Proterozoic Jiudingshan Formation of Suining County, Jiangsu Province were described as *Protosphaeridium densum* Tim. and their affinity to living forms of Chlorococcaceae was discussed by Liu Xue-xian et al. (1984). According to the criteria given by Liu Xue-xian et al. (1984) for distinguishing both the sporangium of eukaryotic alga and sheath-like envelopes of cyanobacteria, the wall of the sporangium is characterized by a distinct outline, showing finely reticulate ornaments the same as those on its autospore and a straight split excystment structure. Their viewpoint has to be confirmed with more microfossil evidence and more detailed study.

CONCLUSION

Abundant and well-preserved tetrads of coccoid mocrofossils from the Niyan Formation have offered detailed information for interpreting the origin of "tetrahedral tetrads" from the Proterozoic sediments. The Niyan tetrads originated through planar cell division; their final configuration is the result of cell shifting. But the "tetrahedral tetrads" are the result of close packing of cells. Therefore, most coccoid microfossils from the Niyan Formation can be referred to *Gloeodiniopsis* (Schopf) em. Knoll & Golubic, 1979, which are related to living chroococcoids as discussed by Knoll and Golubic (1979) for coccoid microfossils from the Bitter Springs Formation.

However, at present, the problem still remains unsolved as to why cell shifting occurs infrequently in extant cyanobacteria, yet it seems to have been common in the Proterozoic microbiotas. As the identification of microfossils mainly depends upon their morphological features, further study of the morphological diversity of Proterozoic microfossils will be very necessary.

SYSTEMATIC PALEONTOLOGY

Genus *Gloeodiniopsis* (Schopf) em. Knoll & Golubic, 1979

Type species: *Gloeodiniopsis lamellosa* (Schopf) em. Knoll & Golubic. 1976

Gloeodiniopsis suxianensis sp. nov.

(Pl. 2, figs. 1–6, 8–13; Pl. 3, figs. 2–6; Pl. 4, figs. 1–3, 5)

Diagnosis: Spheroids and ellipsoids with single, double or multiple outlines, solitary or in groups of 2, 3, 4 up to 8 individuals within a common envelope. Average diameters of spheroids and ellipsoids, expressed as means ± standard deviation, 13 ± 2 μm for the internal outlines (n=75), 17 ± 2.5 μm for the external outlines (n=75), and 18 ± 3.8 μm for single outlines (n=35). Single inclusion spherical, elongated or star-shaped to reticulate, 6.8–17.4 μ m (averagely 10.5 μ m, n = 15). Surface of envelope punctate. Groups within particular microenvironments showing less variability in size, shape and state of preservation.

Remarks: This species is similar to *G. lamellosa* (Schopf) em. Knoll & Golubic, but the latter is generally smaller and has fewer tetrads.

Etymology: With reference to the microfossil-bearing locality.

Occurrence: The Niyan Formation, at Qingtongshan of Suxian, Anhui Province.

Genus *Myxococcoides* Schopf, 1968

Type species: *Myxococcoides minor* Schopf, 1968

Myxococcoides cf. *cantabrigiensis* Knoll, 1982

(Pl. 4, figs. 4, 6, 7)

Diagnosis: Vesicle spherical, 12–18 μm in diameter, sometimes elongated by compression, with a thick, well-defined vesicle wall approximately 1 μm thick; wall smooth to granular. Extracellular mucilage indistinct, but mucilage matter commonly occurring in populations of several to dozens of cells in close proximity. In the case of populations, individuals closely packed, but adjacent cells generally several microns apart. Cell division unclear.

Remarks: The species is distinguished by the size frequency distribution, the geometry of cell aggrega-

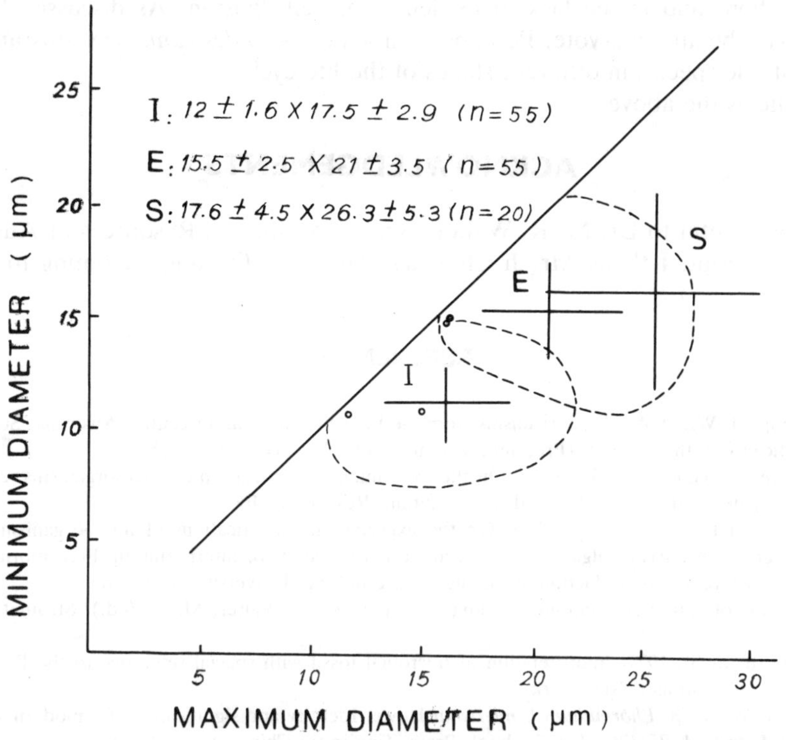

Text-fig. 4. Size distribution (Minimum vs. maximum diameter) of chroococcoid unicells population (*Gloeodiniopsis suxianensis* sp. nov.). Diagonal line plot away from this line. Individual measurements fall within the field outlined with dash lines, the means and standard deviations of external envelope (E), internal envelope (I) and single envelope (S) are represented as dots and crosses. The four cells (shown in Pl. 2, fig. 5a, b) of tetrahedral tetrads are marked by circle and fall close to the inner envelope plot for all chroococcoids measured.

tions and the presence or absence of extracellular mucilage (Knoll, 1982). In size, M. cf. *cantabrigiensis* bears a strong resemblance to *M. reticulata* Schopf (1968) and *M. inornata* Schopf (1968), but *M.* cf. *cantabrigiensis* differs from these species in its thickness of wall and in the spacing of individuals within populations. It is closely similar to *M. cantabrigiensis* Knoll (1982) in the thick vesicle wall and in the spacing of individuals within populations. The differences between *M.* cf. *cantabrigiensis* and *M. cantabrigiensis* lie in that the latter has a smaller size and no defined mucilage matter in the group of cells.

Occurrence: Same as the above.

Genus *Protosphaeridium* Timofeev, 1966

Type species: *Protosphaeridium densum* Timofeev, 1966

Protosphaeridium? densum Timofeev, 1966

(Pl. 2, Fig. 7; Pl. 3, Figs. la, b)

Diagnosis: Single spheroidal vesicles, 23 μm in average diameter (7 cells measured), 18–29 μm in maximum diameter, with a thick, well-defined vesicle wall approximately 1 μm thick. Wall smooth. In the case of populations, individuals closely packed within a sheath-like envelope, but adjacent cells several microns apart.

Remarks: Many specimens of *Protosphaeridium* have been collected from shale by maceration. They occur as solitary individuals or groups of vesicles, having size frequency distribution of 20–50 μm (Yin Leiming & Li Zaiping, 1978). The present specimens are closely comparable to *Protosphaeridium densum* Timofeev, 1966 excepting that there is mucilage around groups of vesicles. In respect of the biologic affinity of *Protosphaeridium*, palaeontologists and botanists have suggested that it is a planktonic eukaryote (Vidal & Knoll, 1982), having an affinity with living green algae, some members of Chlorococcaceae (Liu Zhili, 1982; Liu Xuexian & Liu Zhili, 1984). Moreover, the present specimens of *Protosphaeridium? densum* are closely comparable to *M. cantabrigiensis* Knoll (1982) in the possession of a thick wall, the spacing of individuals within populations and in the lack of evidence for cell division. As discussed by Knoll (1982). *M. cantabrigiensis* may also be an eukaryote. Perhaps both *Myxococcoides cantabrigiensis* and *Protosphaeridium densum* are simply of one species in different stages of the life cycle.

Occurrence: Same as the above.

ACKNOWLEDGEMENTS

I am particularly grateful to Dr. M. R. Walter, Bureau of Mineral Resources of Australia, for his critical reading of the manuscript. I thank Mr. Ji Chengdao and Mrs. Ouyang Qiaoming for their technical and drafting assistance.

REFERENCE

Barghorrn, E. S. and Schopf, J. W., 1965. Microorganisms from the Late Procambrian of central Australia. Science, 150, 337–339.

Cloud, P. E., 1965. Significance of the Gunflint (Precambrian) microflora. Science, 148, 27–35.

Francis, S., Margulis, L. and Barghoorn, E. S., 1978a. On the experimental silicification of microorganisms. 2. On the time of appearance of eukaryotic organisms in the fossil record. Precambrian. Res., 6, 65–100.

__________, __________ and __________, 1978a. On the experimental silicifications of microorganisms. 3. Implications of the preservation of the green prokaryotic algal *Prochloron* and other coccoids for interpretation. Precambrian Res., 7, 377–383.

Fritsch, F. E., 1956. The structure and reproduction of the algae. 2. Cambrian University Press. England.

Golubic, S., 1976. Taxonomy of extant stromatolite-building cyanophytes. In: Walter, M. R. (ed.), Stromatolites. 127–140. Elsevier, Amsterdam.

__________ and Barghoorn, E. S., 1977. Interpretation of microbial fossil with special reference to the Precambrian. In: Flugel, E. (ed.), Fossil Algae. 1–14. Springer, New York.

__________ and Focke, J. W., 1978. *Phormidium hendersonii* Howe: Identity and significance of a modern stromatolite-building microorganism. J. Sed. Petrol., 4, 751–764. Sci. Technol. Press, Tianjin (in Chinese).

__________ and Hofmann, H. J., 1976. Comparison of Holocene and Mid-Precambrian Entophysalidaceae (Cyanophyta) in stromatolitic alga mats: cell division and degradation. J. Paleontol., 50, 1074–1082.

Hofmann, H. J., 1976. Precambrian microflora, Belcher Islands, Canada: significance and systematics. J. Paleontol., 50, 1040–1073.

Horodyski, R. J., Bloeser, B. and Vonder Harr, S., 1977. Laminated algal mats from a coastal lagoon, Laguna Mormona, Baja California, Mexico. J. Sed. Petrol., 47, 680–696.

Horodyski, R. J. and Donaldson, J. A., 1980. Microfossils from the Middle Proterozoic Dismal Lakes Group, Arctic Canada. Precambrian Res., 11, 125–159.

Knoll, A. H., 1982. Microfossils from the late Precambrian Draken Conglomerate, Ny Friesland, Spitsbergen. J. Palaeontol., 56, 755–790.

__________ and Barghoorn, E. S., 1975. Precambrian eukaryotic organisms: a reassessment of the evidence. Science, 190, 52–54.

__________ and Golubic, S., 1979. Anatomy and taphonomy of a Precambrian algal stromatolite. Precambrian Res., 10, 115–151.

Liu Chili, 1982. Microfossil algal communities from the Wumishan Formation in Jixian, China, and their geological significance. Acta Nanjing Univ. (Monograph of algae), 121–166 (in Chinese).

Liu Xuexian, Liu Zhili, Zhang Lun and Xu Xue-si, 1984. A Study of Late Precambrian microfossil algal community from Suining County, Jiangsu Province. Acta Micropalaeont. Sin., 1(2), 171–182.

Mendelson, C. V. and Schopf, J. W., 1982. Proterozoic microfossils from the Sukhays Tunguska, Shorikha, and Yudoma Formations of the Siberian Platform, USSR. J. Paleontol., 56, 42–83.

Oehler, J. H., 1976. Experimental studies in Precambrian paleontology: structural and chemical change in blue-green algae during simulated fossilization in synthetic chert. Geol. Soc. Am. Bull., 87, 117–129.

__________, Oehler, D. Z. and Muir, M. D., 1976. On the significance of tetrahedral tetrads of Precambrian algal cells. Origins of Life, 7, 259–267.

Schopf, J. W., 1986. Microflora of the Bitter Springs Formation, Late Precambrian, central Australia. J. Paleontol., 42, 651–688.

__________, 1977. Biostratigraphic usefulness of stromatolitic Precambrian microbiotas: a preliminary analysis. Precambrian Res., 5, 143–173.

__________, and Blacic, J. M., 1971. New microorganisms from the Bitter Springs Formation (Late Precambrian) of the north-central Amadeus Basin, Australia. J. Paleontol., 45, 925–960.

Timofeev, B. V., 1966. Mikropaleofitogicheskoye issledovaniye drevnikh svit. Nauka, Leningrad.

Yang Qinghe, Zhang Youli, Zheng Wenwu, Xu Xuesi, 1980. Subdivision and correlation of Sinian Suberathem in northern Jiangsu and Anhui. PP. 231–265. In: Sinian Suberathem in China (Research on Precambrian Geology). Sci. Technol. Press, Tianjin (in Chinese).

Yin Leiming, 1981. Precambrian microfossils and Pseudofossils from the Gaoyuzhuang Formation in Quyang of Hebei. Acta Palaeont. Sin., 20(3), 188–198.

__________ and Li Zaiping, 1978. Precambrian microfloras of Southwest China, with reference to their stratigraphical significance. Nanjing Inst. Geol. Palaeont. Mem., 10, 41–108 (in Chinese).

Zhang Yun, 1981. Proterozoic stromatolite microfloras of the Gaoyuzhuang Formation (Early Sinian: Riphean), Hebei, China. J. Paleont., 55, 485–506.

EXPLANATION OF PLATES

The figured specimens photographed under transmitted light microscope with co-ordinates by using the England Finder Graticule, are deposited in the Nanjing Institute of Geology and Palaeontology, Academia Sinica.

Plate I

1–2. Silicified stromatolitic microbial mat.
 1. × 2, 2. × 55; thin section, Chm-010-3.
3. *Eomycetopsis* sp.
 Empty sheaths of a filamentous oscillatoriacean cyanobacterium. × 140; thin section, Chm-010-4:S/38.
4. Remains of a filamentous oscillatoriacean cyanobacterium. × 700; thin section, Chm-010-3:0/48-3.
5–6. Degradational variants of coccoid microfossils.
 × 250; thin sections, Chm-010-1:P/28-1; Chm-010-4:M/27-3.
7–8. *Gloeodiniopsis suxianensis* sp. nov.
7. Transmitted-light photo.
8. interpretative drawing, showing cherty pocket in dolomite. × 60; thin section, Chm-010-1:N/28.
9. Degraded filamentous oscillatoriacean cyanobacterium. × 200; thin section, Chm-010-3:P/32-3.

Plate 1

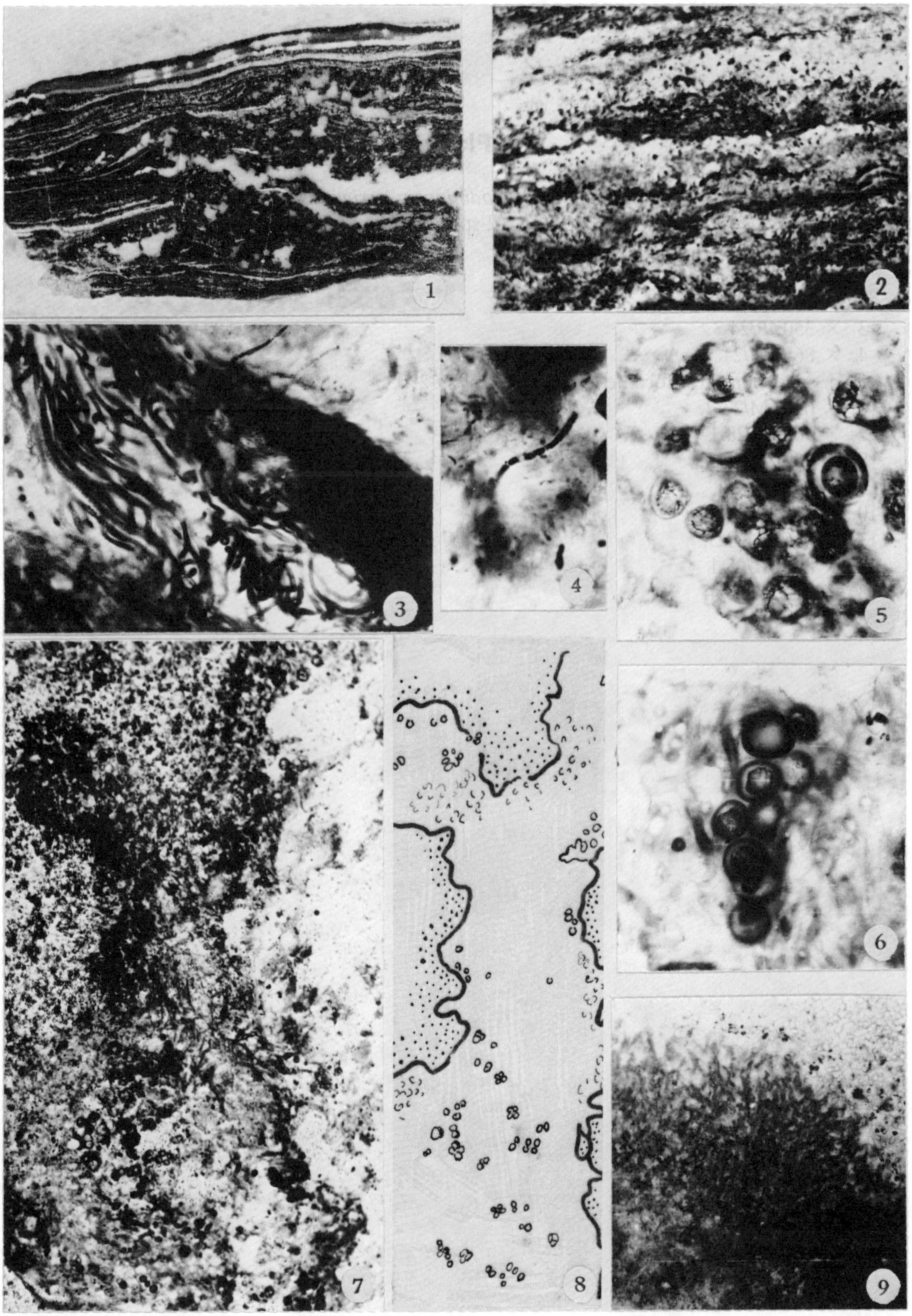

Plate 2

All the figures in this plate belong to *Gloeodiniopsis suxianensis* sp. nov. with the exception of fig. 7(*Protosphaeridium? densum* Timofeev, 1966). All specimens are of thin sections (Chm-0l0-1).

1. × 865; P/23-1.
2. × 865; K/26-2.
3. × 350; P/24-3.
4. × 865; M/24.

5a-c. specimens at 3 focal levels. × 545; N/29-2.

6. × 620; M/27-2.
7. × 100; N/28-1.
8. × 850; M/29-3.
9. × 850; M/27.
10. × 865; N/28-2.
11. × 865; R/23-3.
12. × 850; R/25-1
13. × 850; K/26-4.

Plate 2

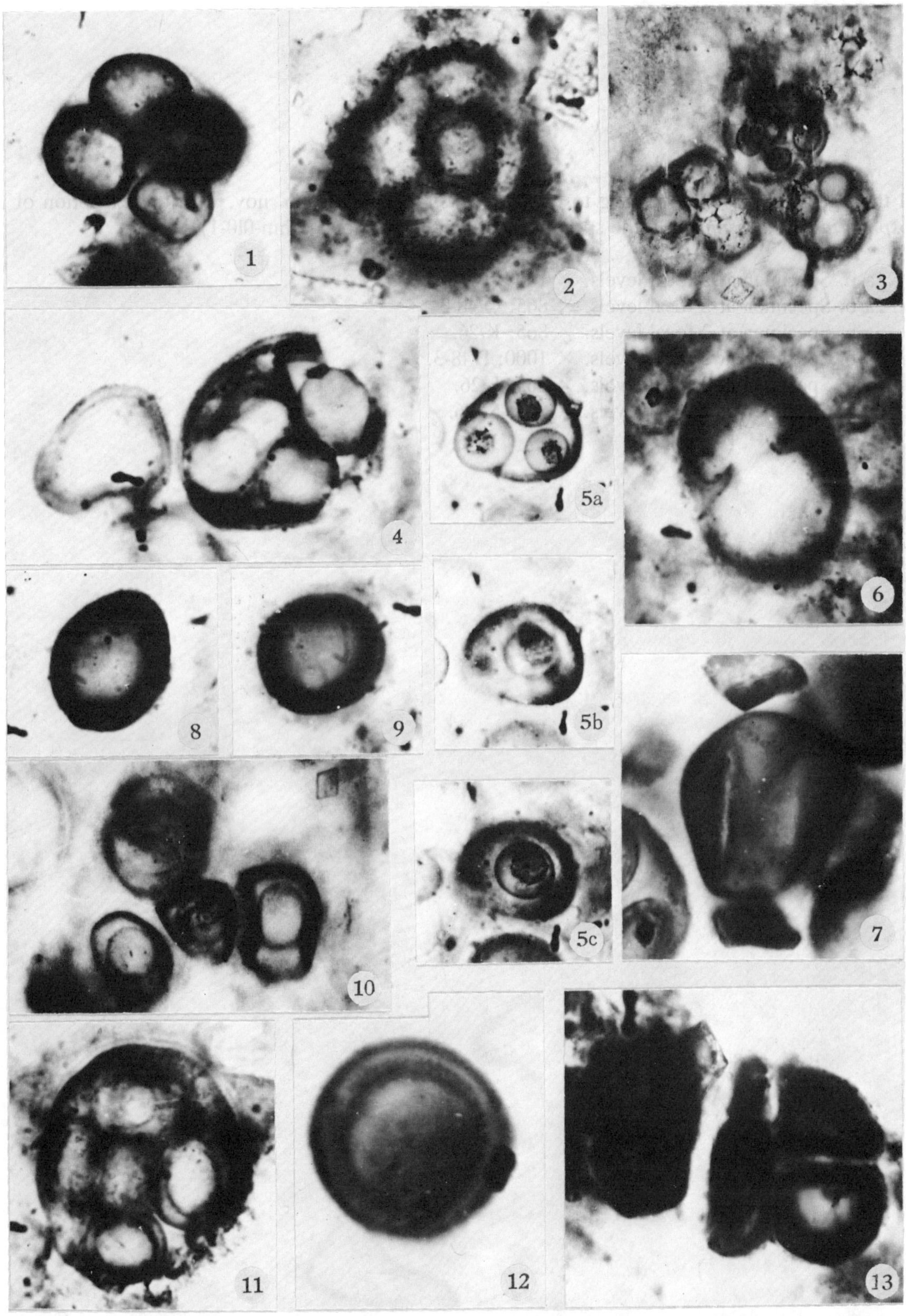

Plate 3

All the figures in this plate belong to *Gloeodiniopsis suxianensis* sp. nov. with the exception of fig. la-b(*Protosphaeridium? densum*). All the specimens are of thin sections (Chm-0l0-1).

1a-b. Specimen at 2 focal levels. × 1000; N/28-1.
2a-b. Specimen at 2 focal levels. × 865; O/48-3
3a-b. Specimen at 2 focal levels. × 865; K/26.
4a-b. Specimen at 2 focal levels. × 1000; P/48-3.
5a-b. Specimen at 2 focal levels. × 865; L/26.
6a-b. Specimen at 2 focal levels. × 865; M/26.

Plate 3

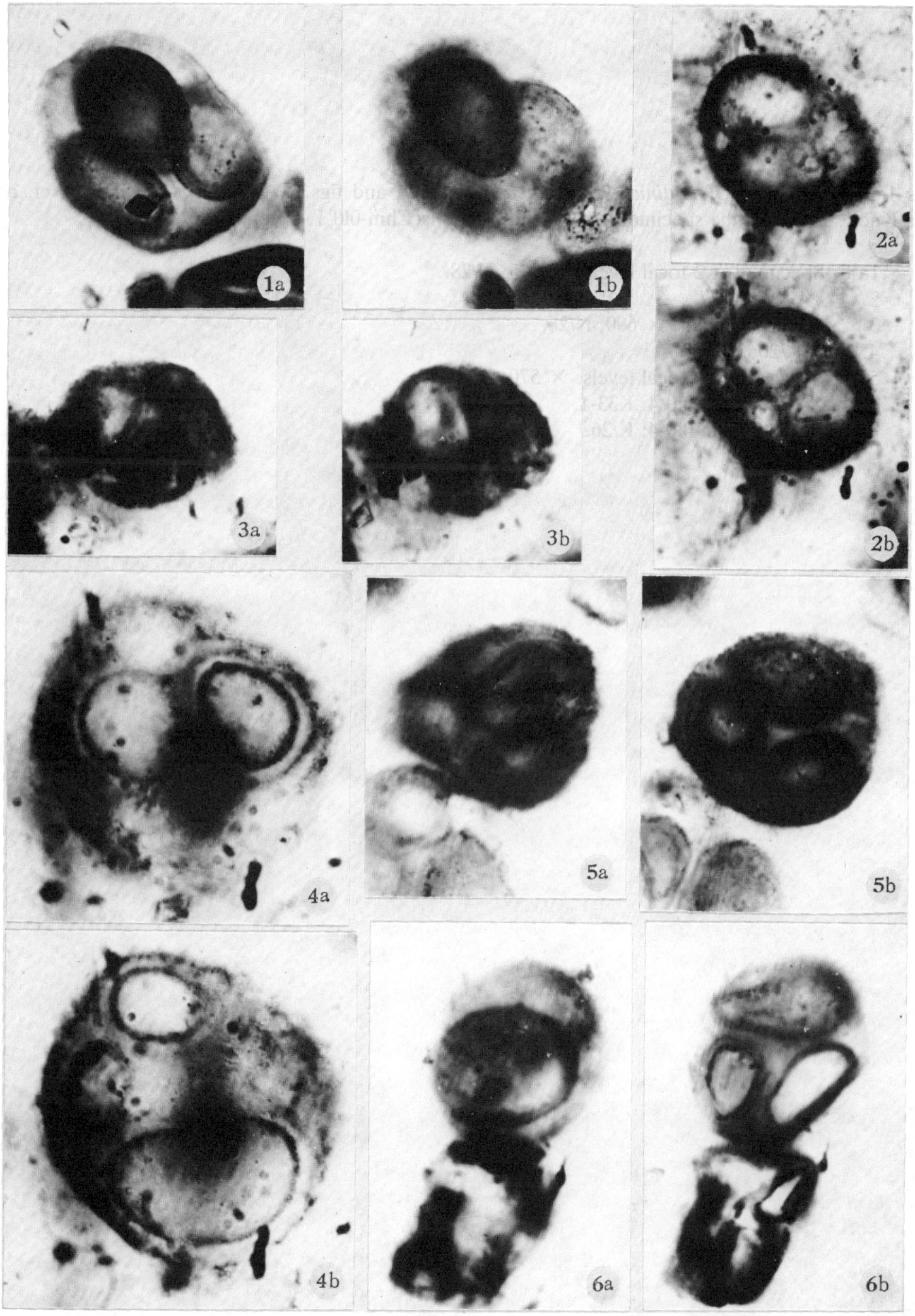

Plate 4

Figs 1–3, 5 belong to *Gloeodiniopsis suxianensis* sp. nov. and figs. 4, 6–7 to *Myxococcoides* cf. *cantabrigiensis* Knoll, 1982. All the specimens are of thin sections(Chm-0l0-1, except figs. 6–7).

1a-b. Specimen at 2 focal levels. × 865; N/28.
2. × 865; M/27.
3. Type specimen. × 600; N/28.
4. × 720; N/29-4.
5a-b. Specimen at 2 focal levels. × 570; M/28-4.
6. × 720; Chm-0l0-4: K33-1.
7. × 720; Chm-0l0-4: K/26.

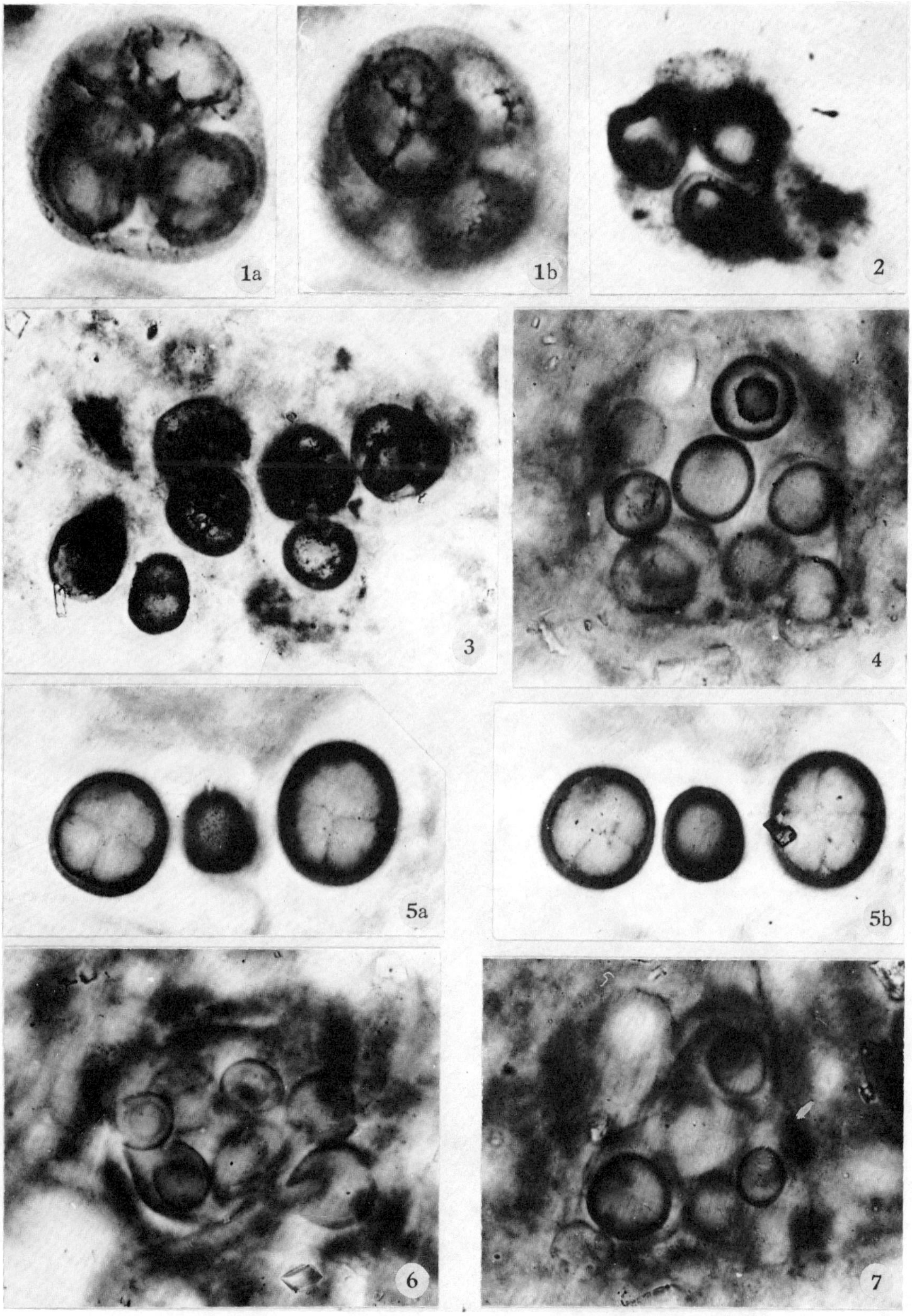
1a
1b
2
3
4
5a
5b
6
7

PALAEONTOLOGIA CATHAYANA Volume 5, 1990, p.295–310

UPPERMOST CRETACEOUS AND LOWEST TERTIARY STRATA IN THE NANXIONG BASIN OF GUANGDONG WITH REFERENCE TO THEIR PETROGRAPHIC CHARACTERS AND SEDIMENTARY ENVIRONMENT

Liu Yun and Wang Zongzhe

(Nanjing Institute of Geology and Paleontology, Academia Sinica)

Abstract

The purpose of this paper is to study the mineralogical composition, texture and geochemistry of the red beds in the Yangmeikeng-Nilongkeng and Fengmenao sections in the Nanxiong Basin by means of grain-size parameter, X-ray, infrared spectrum, spectrometry, petrographic microscope and electron microscope.

Based on the mineralogical characteristics and geochemical analyses, the following conclusions have been obtained: 1. Uppermost Cretaceous and Lowest Tertiary strata may be divided into three parts: The Shanghu Formation, the middle and upper members of the Nanxiong Formation. 2. During the Latest Cretaceous and Earliest Tertiary, there were deposited a series of red clastic sediments of piedmont plant-lake facies, the climate was dry and hot, the water medium was salinized with rich magesium. 3. There exist depositional breaks between the middle and upper parts of Nanxiong Formation, and between the Nanxiong Formation and the Shanghu Formation.

While making a field survey in 1983–1984, the authors systematically collected rock samples from the Cretaceous-Tertiary boundary strata in the Yangmeikeng-Nilongkeng and the Fengmenao sections, the former being located about 25km northeast of Nanxiong county town, Guangdong and the latter about 2km southwest of the town. After studying the rock samples, their lithological characters, geochemistry and sedimentary environment are treated below.

LITHOLOGICAL CHARACTERS

Detrital Rocks

According to data of grain-size parameter and the observation under petrographic microscope, it is shown (fig. 1) that in this area the sediments are characterized by sandy argillaceous siltstone and sandy silty mudstone (fig. 1-A). and the detrital rock consists mainly of calc-argillaceous sandstone or conglomerate and calc-sandy. mudstone (fig. 1-B), with clay accounting for $< 50\%$ and carbonate $< 25\%$. The sandstone is complex in composition and poor in sorting, and cemented by calcareous pelitic materials (matrix $> 15\%$), thus being of apogritic type. Its detritus is composed chiefly of feldspathic lithic-apogrite, and subordinately of lithic-apogrite and lithic feldspathic apogrite (fig. 1-C). In the apogrite detritus amounts to 10–60%, feldspar to 7–15%, and quartz to 30–70%. The feldspar is fresh on surface (pl. 1, fig. 5). The detritus is composed of quartzite, metamorphosed quartzite, haplite, silicolits, carbonatite, clay stone, etc.

In this area the grain-size of sediments show great change of mean-size, a high standard deviation, poor-sorting, positive skewness and low kurtosis, a feature indicating fluvial facies. The laminated bedding, mud cracks and hybrid condensated structure can be seen in the middle part of Nanxiong formation. The laminated bedding (pl. 1, fig. 3) is the deposits resulting from the suspended turbid matters during the

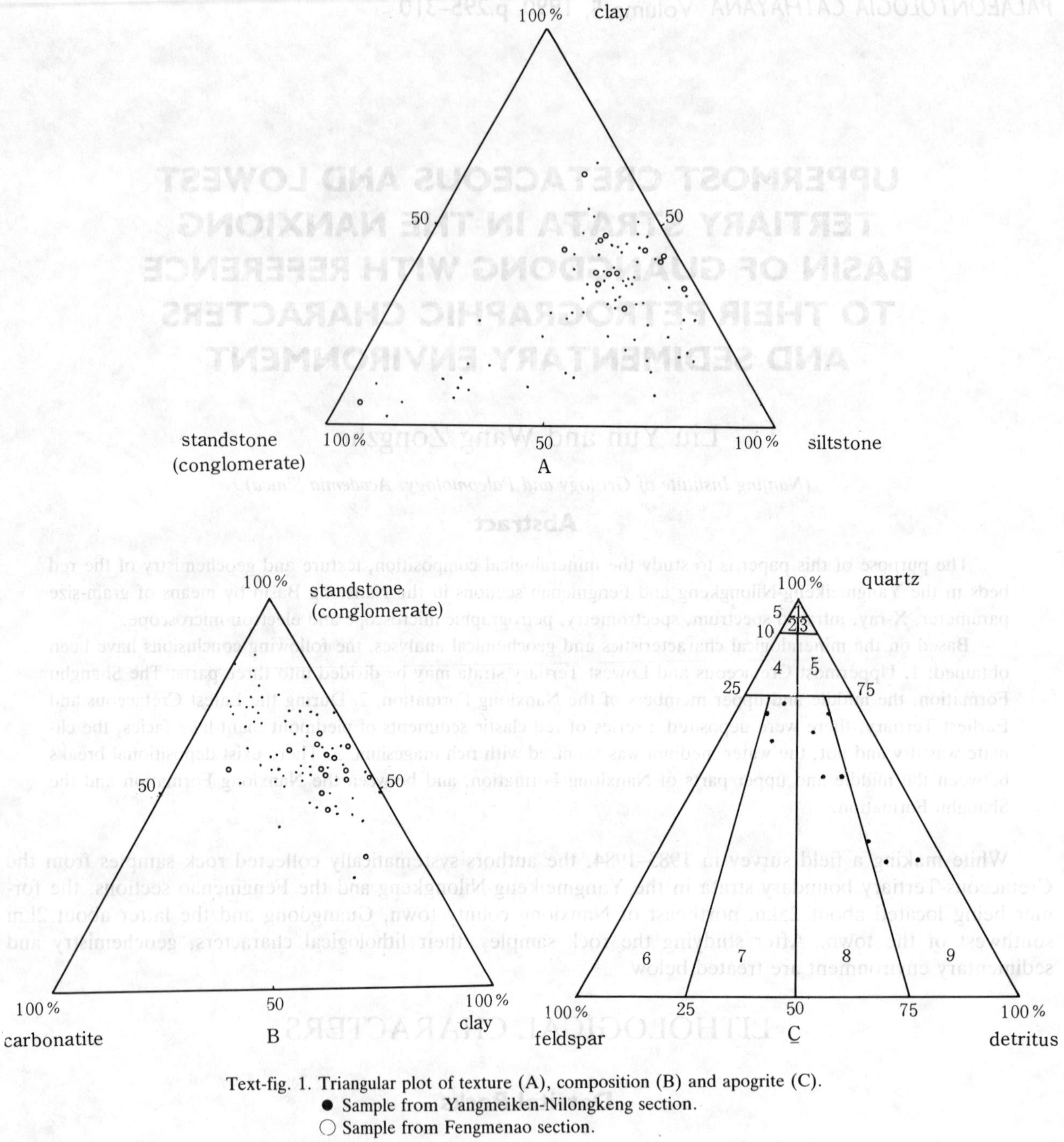

Text-fig. 1. Triangular plot of texture (A), composition (B) and apogrite (C).
● Sample from Yangmeiken-Nilongkeng section.
○ Sample from Fengmenao section.

transportation in the shallow water, which is of the sediments of shallow lake facies. The hybrid condensated structure is frequently seen in the delta and shallow lake areas, known as a kind of non-homogeneous sand-mud mixtum. When sand, mud and colloidal clay were carried from the river into the lake, the concentration of cation mediun is higher in it. Thus colloidal clay materials are deposited, and so does the sand. Under the arid climatic condition, mud cracks (pl. 1, fig. 4) were developed frequently in the valley flat, the littoral plain and the perched temporary ground-water area. The mud flow inclusion (pl. 1, fig. 7) and the clay gravel (pl. 1, fig. 6) can be observed in the sediments of the upper part of Nanxiong Formation. Under polarization microscope, the mud flow inclusion is encircled by a calc-argillaceous light girdle. The clay gravel may be an indicator of the sediments elevated above the surface. In the Shanghu Formation, besides clay gravel and mud flow inclusion, there occurs cross-stratification (pl. 2, fig. 8) which resulted from increasing water current energy.

Quantitative measurements of quartz, feldspar and calcite by means of infrared spectrum are used to interprete the maturity of detrital rocks, which probably reflect the change of landform and climate at that time. In figs. 2–3, it can be seen that optical-density ratio of AC/AQ and AM/AQ is lower in the upper part of Nanxiong Formation and the lowest values lie between Beds 6–7 and 10–12 in the Yangmeikeng-Nilongkeng section. In the Fengmenao section, the AC/AQ value is zero and the AM/AQ value is lower in the upper Nanxiong Formation and between Beds 6–7, indicating a higher maturity of rocks and strong weathering. As indicated from the lithological characters, the then climate was hot and arid, with physical weathering as predominant. The intensification of weathering is primarily related to the change of landform and the elevation-subsidence of the basin, indicating that the Nanxiong Basin might have elevated at that time and the sediments were exposed above the water level.

Clay minerals

Among the 47 clay samples from the two sections, those, less than 0.002mm in grain size, were used and analysed by means of X-ray, infrared spectrum and electronic microscope. The clay minerals in this area are composed of illite, chlorite, mixed-layer minerals, attapulgite, etc. The mineralogical characters are shown in Table 1 and their distribution in the sections is as follows:

Shanghu Formation: I-Ch-I/M-At-Cg
Upper Nanxiong Formation: I-Ch-(rich), I-Ch-I/M–At(less)
Middle Nanxiong Formation: I-Ch, I-Ch-I/M-At
(I-illite, I/M-irregular mixed layer of illite-smectite, Ch-chlorite, At-attapulgite, Cg-36 Å swelling chlorite).

Most illite and chlorite are inherited minerals and derived from source rocks. Under the dry-hot climatic condition, due to the slow water current, less amount of R_2O, RO, Si_2O, etc. were carried away, and water medium became salinized. As a result, the illite was montmorillonnized with cloudy margin (pl. 2, fig. 2), and was thus transformed into I/M mixed laryer minerals. Besides, under the condition of salinization, water medium with rich magnesium, illite and smectite may change into magnesian clay minerals, such as fibrous attapulgite (pl. 2, figs 3, 4) and 36 Å swelling chlorite (Thorez, J., 1976). a feature reflecting a salinized lake basin under the hot and dry climatic condition.

Carbonate concretions

The Shanghu Formation contains a large number of Carbonate concretions, which are massive- or gin-

Table 1. Characters of clay minerals.

Mineral	X-ray diffraction analysis	Infrared spectrum analysis	TEM analysis
Illite	natural sample d(001) = 10 Å, no change after glycerolsaturated treatment and after 500°C heat treatment	AL-OH 3624 cm^{-1}, Si-O-Si 1023cm^{-1} AL-OH 912 cm^{-1}, 830 cm^{-1} Si-O-Al 750 cm^{-1}, 530 cm^{-1} Si-O 475 cm^{-1}	flake-shaped, outline clear
Chlorite	natural sample d(001) = 14 Å, no change after glycerolsaturated treatment d(001). After 500°C heat treatment, the reflection peak is intensified and its position is shifted to lower value	Fe-OH 3420 cm^{-1} (Fe-Mg) OH 3550 cm^{-1} Si-O-Si 987 cm^{-1} Si-O-Mg 660 cm^{-1} Si-O 458 cm^{-1} Si-O-Mg 440 cm^{-1}	thick flakeshaped or granular, with angular edges
I/M mixed layer minerals	natural sample d(001) = 10 Å, asymmetrical peak-shape, raised at low angle. After glycerol-saturated treatment, the reflection peak is widened and separated into two peaks; 10 Å, 11 Å. d(001) is contracted to 10 Å after 500°C heat tratment.	(AL-Mg)OH 830, 845 cm^{-1}	illite transformed into smectile with cloudy edges
swelling chlorite with regular mixed layer	d(001) = 36 Å, after glycerolsaturated treatment		
Attapulgite			fibrous or bundle-shaped

ger-shaped and about 1-5cm in diameter, (pl. 2, figs. 5–6). These concretions are distributed horizontally but indistinctly around the surrounding rocks with crack fillers of secondary calcite but without concentric structure. Under the microscope they are composed mainly of carbonate. clay and angular detritus or sometimes with shelly fragments in them (pl. 2, fig. 7). Chemical analysis reveals that the elemental component is homogeneous between concretions and surrounding rocks, except for CaO, the content of which is higher in concretions (Table 2). At present, there are different views on the genesis of carbonate concretions. It is generally believed that the concretion was formed in the early diagenetic stage, which can be demonstrated by the occurrence of shelly fragments in it. Under the non-homogeneous physio-chemical conditions, the void space of unconsolidated sediments was filled with soluble carbonate, and through local precipitation the concretion took shape. It is generally known that the concretion was formed at places, where CO_2 and NH_3 were decomposed from organic matters. CO_2 accelerated the solution of carbonate to form $Ca(HCO_3)_2$, while the formation of ammonia and ammonia salt made the PH value of water medium increase to 7. Thus, the carbonate will precipitate like this: $NH_3 + Ca^{++} + HCO_3^- \longrightarrow CaCO_3 + NH_4^+$. Since the ammonia content is low in the organic matter, it is insufficient to form concretions by relying simply on its decomposition. In shallow water area under arid and hot climatic condition, strong evaporation resulted in that water and CO_2 were diminishing and $Ca(HCO_3)_2$ change into $CaCO_3$. Hence, large numbers of carbonate concretions reflect a lake shore sedimentary environment with alkaline water medium under dry climatic condition.

Table 2. Chemical composition of concretions and surrounding rocks.

element% / composition	SiO_2	CaO	MgO	Fe_2O_3	Al_2O_3	P_2O_5	MnO	TiO_2	K_2O	Na_2O	FeO	H_2O^+	CO_2	Total
concretion	29.08	33.16	1.35	1.04	4.94	0.130	0.463	0.21	0.77	0.037	0.27	2.60	25.69	99.74
concretion	19.54	37.55	1.12	2.24	5.44	0.058	0.043	0.22	1.02	0.014	0.19	3.50	29.39	100.32
surrounding rock	29.52	28.63	1.91	2.79	7.60	0.268	0.180	0.40	1.57	0.04	0.30	3.66	22.66	99.528
surrounding rock	36.32	25.31	1.91	1.94	6.84	0.065	0.218	0.41	1.21	0.065	0.31	2.86	21.36	99.318

GEOCHEMICAL CHARACTERS

The ICP was used for the analysis of elements CaO, Al_2O_3, P_2O_5, SiO_2, Fe_2O_3, MnO, MgO, TiO_2, K_2O, Na_2O and also for the trace elements of Zn Ni, Co, Ca, Cr, V. A comparison was made of the average chemical composition with crustal clarke value and sandstone average value (F. J. Pettijohn, 1963) (see Table 3). The elemental changes in the section are listed in Figs. 2–3.

Table 3 shows that the average chemical composition is less than the crustal clarke value except for CaO and K_2O. These elements (CaO, Al_2O_3, P_2O_5, Fe_2O_3, MgO, K_2O, MnO) are rich in sandstone and so are SiO_2 in siltstone and TiO_2 in mudstone. In general, the distribution of different elements is dependent chiefly on the "carrier". For instance, those carried in the form of detritus are seen in sandstone and siltstone, and those carried in the form of colloid-absorbing are contained in mudstone. The enrichment of elements in sandstone demonstrates that the deposits in this area suffered in the large from physical weathering and mechanical slacking, as the chemical decomposition was weak. Befor being decomposed incompletely, the mineral elements were accumulated quickly in the form of detritus. To compare the average chemical composition of sandstone with that of F. J. Pettijohn (1963), the content of SiO_2 is low and those of Al_2O_3, Fe_2O_3, K_2O, Na_2O, MgO are high. According to B. B. Polinov (1960), the "active series" of elements in as follows: CaO$>Na_2O > K_2O >$ MgO $> Fe_2O_3 > Al_2O_3 > SiO_2$. SiO_2 was quite stable is sediments, while those unstable elements were easily soluble during deposition. Hence, the increase of Al_2O_3, K_2O, Na_2O, MgO contents and the decrease of SiO_2 content indicate the low maturity of sandstone near the eroded sourceland in this area, Fig.2 shows the decrease of CaO, CaO/MgO, P_2O_5, MnO, MgO and the increase of K_2O/Na_2O, SiO_2, TiO_2 between Beds 6–7 and 10–12 in the Yangmeikeng-Nilongkeng section; Fig. 3 shows the dicrease of CaO, CaO/MgO and the increase of K_2O/Na_2O, Al_2O_3, TiO_2, Fe_2O_3 between Beds 6–7 in the Fengmenao section. In the dry region, the decrease of easily soluble elements in the deposits and the increase of inactive elements might be somewhat related to the depositional break, exposure of surface and weathered erosion.

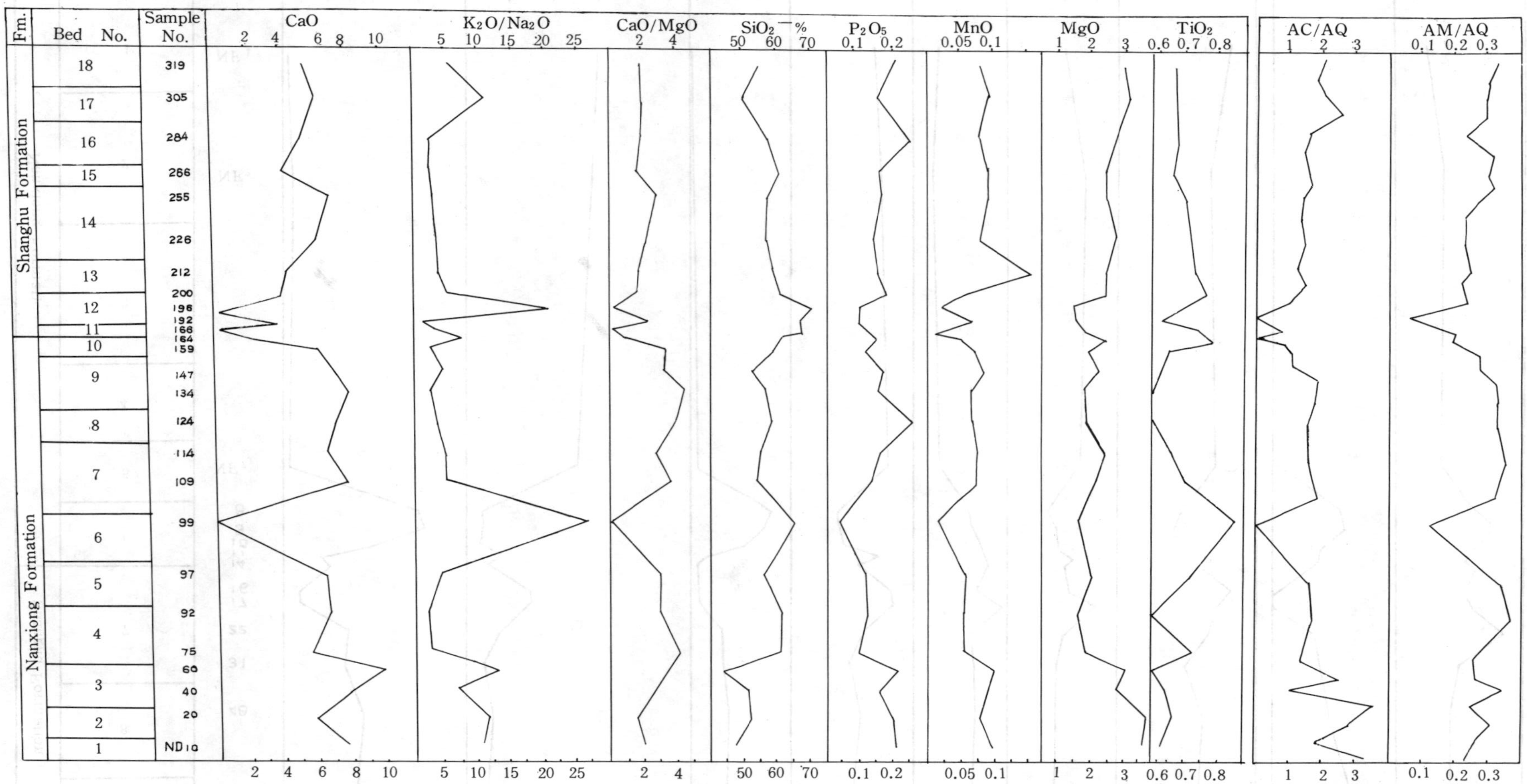

Text-fig. 2. Distribution of chemical constituents of Cretaceous-Tertiary boundary strata in the Yangmeikeng-Nilongkeng section of Nanxiong basin.
AC/AQ and AM/AQ indicate the optical density ratio of quartz, feldspar and calcite.
AC: optical density value of calcite; AM: optical density value of feldspar; AQ: optical density value of quartz.

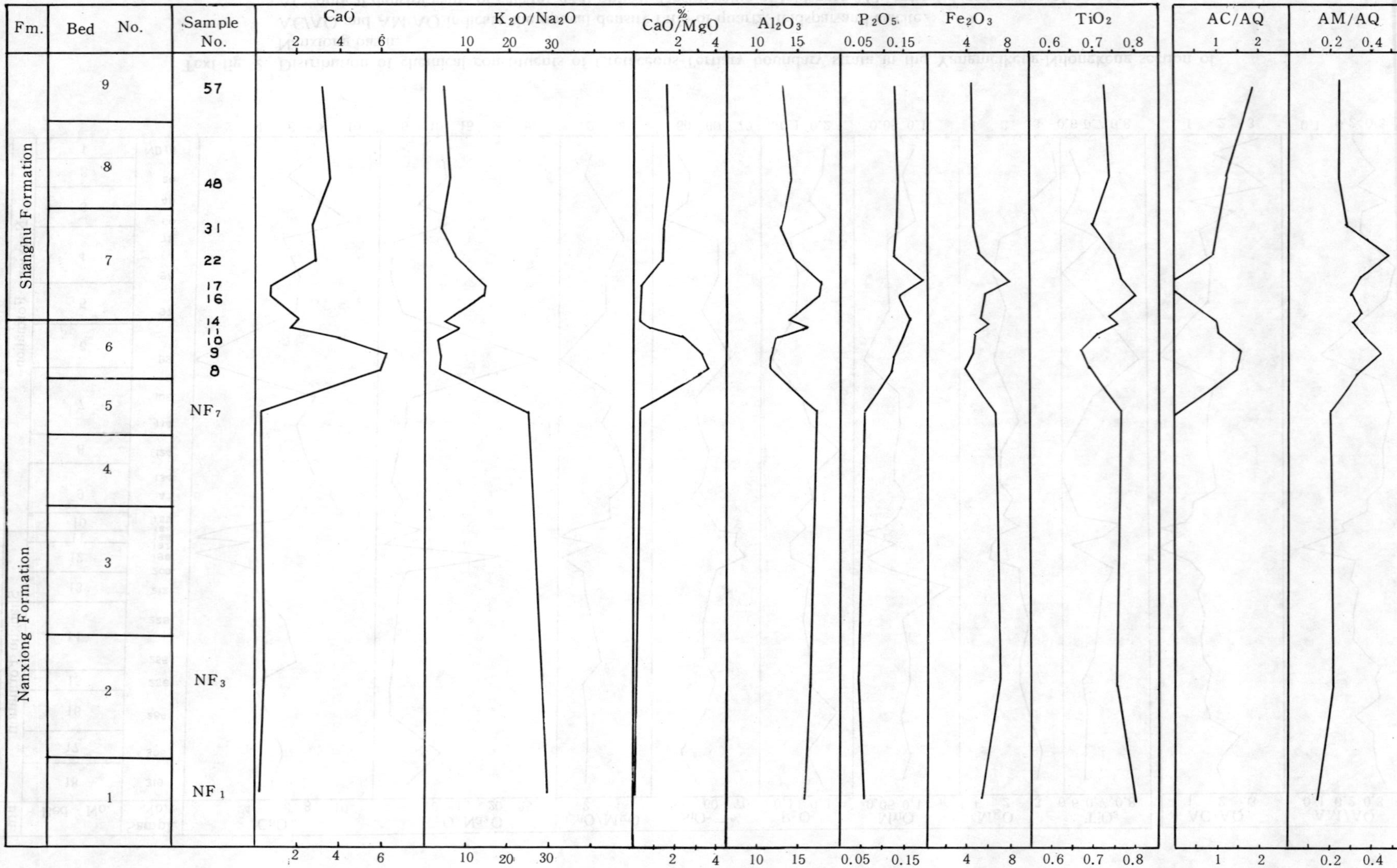

Text-fig. 3. Distribution of chemical constituents of Cretaceous-Tertiary boundary strata in the Fengmenao section of Nanxiong basin.

Table 3. Average chemical composition.

rock type (number) \ component %	CaO	Al_2O_3	P_2O_5	SiO_2	Fe_2O_3	MnO	MgO	TiO_2	K_2O	Na_2O	Source
sandstone (3)	5.86	16.9	0.196	52.3	6.06	0.085	3.30	0.687	4.22	0.44	Nanxiong Basin under study
siltstone (12)	4.169	13.31	0.136	62.17	4.71	0.07	2.05	0.695	2.73	0.69	Nanxiong Basin under study
mudstone (3)	3.412	15.26	0.140	56.57	5.20	0.065	2.49	0.734	3.32	0.606	Nanxiong Basin under study
average chemical composition	5.074	14.17	0.152	58.23	4.996	0.076	2.31	0.685	3.148	0.638	Nanxiong Basin under study
crustal clarke value	4.14	15.21	0.18	62.14	6.64	0.13	3.12	0.75	3.01	3.37	A.P. Vinoglatov. 1962
average value of sandstone	5.47	4.73	0.04	78.75	1.40	0.01	1.16	0.25	1.29	0.44	F. J. Pettijohn, 1963

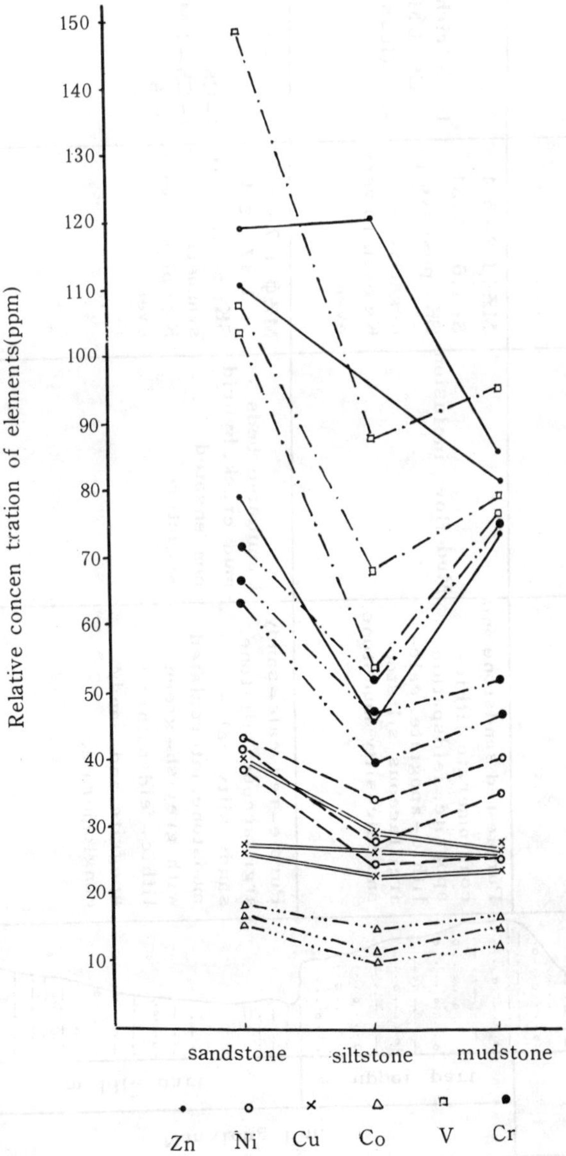

Text-fig. 4. Distributive curve of trace elements.

Fm.		section	lithological character	texture & structure	grain size parameter	clay mineral association	fossil	water medium condition	palaeo-climate	sedimentary environment
Shanghu Fm.		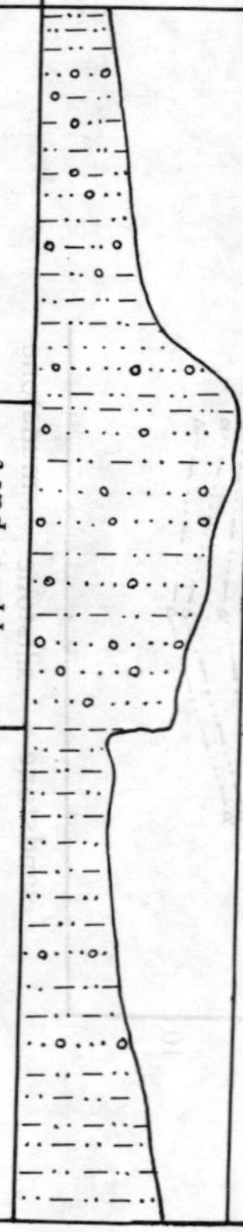	Brown calc-argillaceous siltstone, calc-silty mudstone, intercalated with sandy conglomerate, containing many mudstone detritus.	Carbonated concretion, clay gravels, mud flow inclusion, cross-stratification.	MZϕ1.3–6.1 δ_1 1.12–3.24 SK_1 positive to negative Kg very spire to very even.	I–CH–I/M –At–Cg	Ostracoda Gastropoda Charophyta Pelecypoda bone fossil Conchostraca	water medium satinization of rich magnesium	dry and hot	lake shore, diluvial plain facies
Nanxiong Fm.	upper part		Purple-red sandstone and conglomerate lithic-apogrite, feldspathic lithic-apogrite, calc-argillaceous siltstone, and calc-silty mudstone.	Mud flow inclusion	MZϕ1.2–5.4 δ_1 1.67–2.31 SK_1 positive to negative Kg even to very even	I–Ch (rich) I–Ch–I/M –At (less)	Dinosaurian eggs			dry delta facies
Nanxiong Fm.	middle part		Purple-red calc-sandy argillaceous siltstone, sandy silty calc-mudstone, intercalated with greyish-green lithic-feldspathic apogrite, and sandy conglomerate.	Laminated beds, mud crack, hybrid condensated structure.	MZϕ 1.7–5.6 $_1$ 1.37–2.4 SK_1 positive to symmetry Kg spire to very even	I–Ch I–Ch–I/M – At	Dinosaurian ichnite Ostracoda Charophyta			lake basin bog facies

Text-fig. 5. Comprehensive sedimentary characters of the Cretaceous-Tertiary boundary strata in the Nanxiong basin. MZ–Mean size; δ–Standard deviation; SK_1–Skewness; Kg-Kurtosis.

Fig. 4 shows the change of trace elements in various rock types. The contents of Zn, Ni, Co, Cu, Cr, V. are high in sandstone, but low in siltstone, which is inconsistent with the generality that the trace element content is more in mudstone than in sandstone. In general, the better the sorting of sediments is, the less the trace element contents are in the sandstone but the higher in the mudstone. The above characters also indicate that the sediments in this area are of low maturity and poor-sorting.

SEDIMENTARY FACIES AND CHARACTERS

According to the sedimentary composition, texture and organic assemblages, the Uppermost Cretaceous and the Lowest Tertiary may be divided into the following sedimentary facies with their lithological characters shown in fig. 5.

The middle part of Nanxiong Formation is composed of purple-red calc-sandy argillaceous siltstone and sandy silty calc-mudstone, intercalated with grayish-green sandstone and conglomerate, containing fossil ostracods, charophytes (pl. 1, figs. 1–2) and dinosaurian ichnits. In the sandstone quartz accounts for 65% on the average, feldspar 20% and detritus 15%. Detritus are those of quartzite, silicolite and haplite, etc. and feldspar shows fresh surface (pl. 1, fig. 5). The high AC/AQ, AM/AQ ratio indicates a low maturity of rock. Geochemistrically, the soluble elements are rich, while the insoluble elements are relatively poor, which suggest that the rocks are poor-sorting. Clay minerals are mainly composed of illite, chlorite, mixed-layer smectite-illite and attapulgite. Attapulgite occurs commonly in salinized water medium, rich in magnesium. The laminated bedding and mud cracks indicate that the middle part of Nanxiong Formation belongs depositionally to lake and bog facies.

The upper part of Nanxiong Formation consists of purple-red sandstone, conglomerate, calc-argillaceous siltstone and calc-silty mudstone, containing no fossils other than dinosaurian eggs. In the sandstone, quartz accounts for 56% on the average, feldspar 11%, detritus 31%. Detritus are poor in sorting and coarse in size. Mud flow inclusion can be seen. Clay minerals are composed of illite, chlorite, mixed layer smectite-illite and attapulgite. The upper part of Nanxiong Formation seems to be of delta facies.

The Shanghu Formation is composed of dark brown calc-argillaceous siltstone and calc-silty mudstone, intercalated with sandsotne and conglomerate. Besides, there is a great amount of elliptic mudstone detritus, suggesting a proluvial plain facies. The carbonate concretions are also very abundant. Furthermore, there occur mud flow inclusion and cross-stratification, which indicate that the water current has increased, fossils of charophytes, ostracods, mammals, pelecypods, conchostracans are contained. Between the Nanxiong and Shanghu Formations there exists a bed of clay gravel. The appearance of the 36 Å swelling chlorite of clay minerals shows that the climate was hotter and the water medium was more salinized at that time. All these indicate the deposits of lake shore-proluvial plain facies.

DEPOSITIONAL BREAK

It is shown from the petrographic characters and element analytical data, there is much difference between the middle and upper parts of Nanxiong Formation and between the Nanxiong and Shanghu Formations. there might have been depositional breaks between them as viewed from the following reasons.

The depositional break between the middle and upper parts of Nanxiong Formation:

1. Lithologically, the middle part of Nanxiong Formation is mainly made of calc-sandy argillaceous siltstone and calc-sandy silty mudstone, while the upper part of the formation is composed of sandstone conglomerate, revealing a discontinuous sedimentary rhythm.

2. From sedimentary environmental viewpoint, the middle part is of lake and bog facies, while the upper part is of delta facies.

3. From the maturity of rock and curve of elemental change between Beds 6–7, calcite, feldspar and easily soluble elements CaO, CaO/MgO, P_2O_5, MnO, MgO decreased rapidly, while SiO_2, TiO_2 K_2O/Na_2O increased greatly.

The depositional break between the Nanxiong and Shanghu Formations:

1. A bed of yellowish gray clay gravelstone between the Nanxiong and Shanghu Formations indicates that the sediments have been exposed above the earth surface.

2. Carbonate concretions, clay detritus and 36 Å swelling chlority are seen in the Shanghu Formation, but not in the Nanxiong Formation.

3. From maturity of rock and the elemental contents, the change of peak value is clearly shown be-

tween Beds 10–12 in the Yangmeikeng-Nilongkeng section and between Beds 6–7 in the Fengmenao section.

All of the above-mentioned evidences show that these changes are related to the dry and hot climatic condition and to the depositional breaks or to the tectogenesis.

REFERENCES

Gard, J. A., 1971. Electron-optical investigation of clays. Mineralogical Society, London.

Lu Yanchou, Liu Gaokui and Peng Wenshi, 1982. The ratios of microcline to quartz and calcite to quartz for loess and their palaeoenvironmental implications. Geochemica, 2, 198–204 (in Chinese).

Pettijohn, F. J., 1963. Chemical composition of sandstones-excluding carbonate and volcanic sands. U. S. Geol. Surv. Prof. Paper, 440-S.

Polinov, B. B., 1960. Crust of weathering and geochemistry. Akad. Nauk U.S.S.R. (In Russian).

Thorez, J., 1976. Practical identification of clay minerals. Dison, Belgique.

Vinoglatov, A. P. 1962. The average content of chemical element of essential volcanic rock in crust. Geochemistry, 7, 555–571 (in Russian).

EXPLANATION OF PLATES

Plate I

1. Fossil charophyte in silty calcareous mudstone, thin section, × 20, middle part of Nanxiong Formation, ND 10.
2. Fragments of charophyte and ostracod in silty calcareous mudstone, thin section, × 20, middle part of Nanxiong Formation, ND 10.
3. Laminated bedding, thin section, × 20, middle part of Nanxiong Formation, ND 41.
4. Mud crack structure, field photograph, middle part of Nanxiong Formation.
5. Feldspathic lithic-apogrite, feldspar fresh on surface, polarized light, × 40, middle part of Nanxiong Formation, ND 93.
6. Clay gravel, polarized light, × 40, Shanghu Formation, ND 166.
7. Mud flow inclusion, detritus encircled with calc-argillaceous light girdle, polarized light, × 40, Shanghu Formation, ND 166.
8. Feldspathic lithic-apogrite, containing a number of mudstone detritus, thin section, × 40, Shanghu Formation, ND 174.

Plate 1

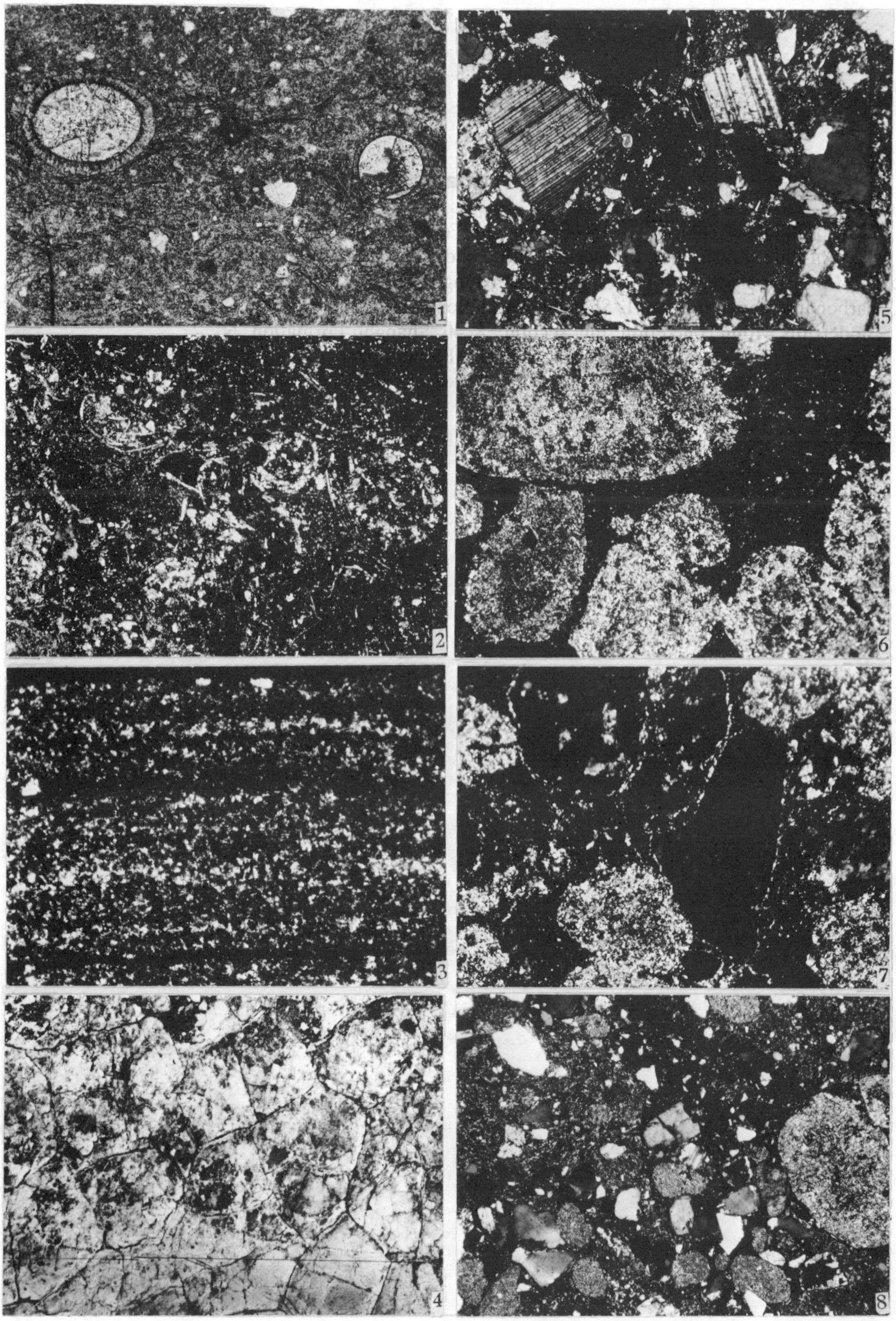

Plate 2

1. TEM of illite and chlorite, × 15000, Shanghu Formation, NF 22.
2. TEM of illite transformed to smectite with cloudy margin, × 15000, Shanghu Formation, NF 22.
3–4. TEM of attapulgite, fibrous, × 30000, Nanxiong Formation, NF 11.
5–6. Carbonatite concretion, × 1, Shanghu Formation.
7. Biodetritus within concretion, thin section, × 40, Shanghu Formation, NF 27.
8. Cross-stratification, field photograph, Shanghu Formation.

Plate 2

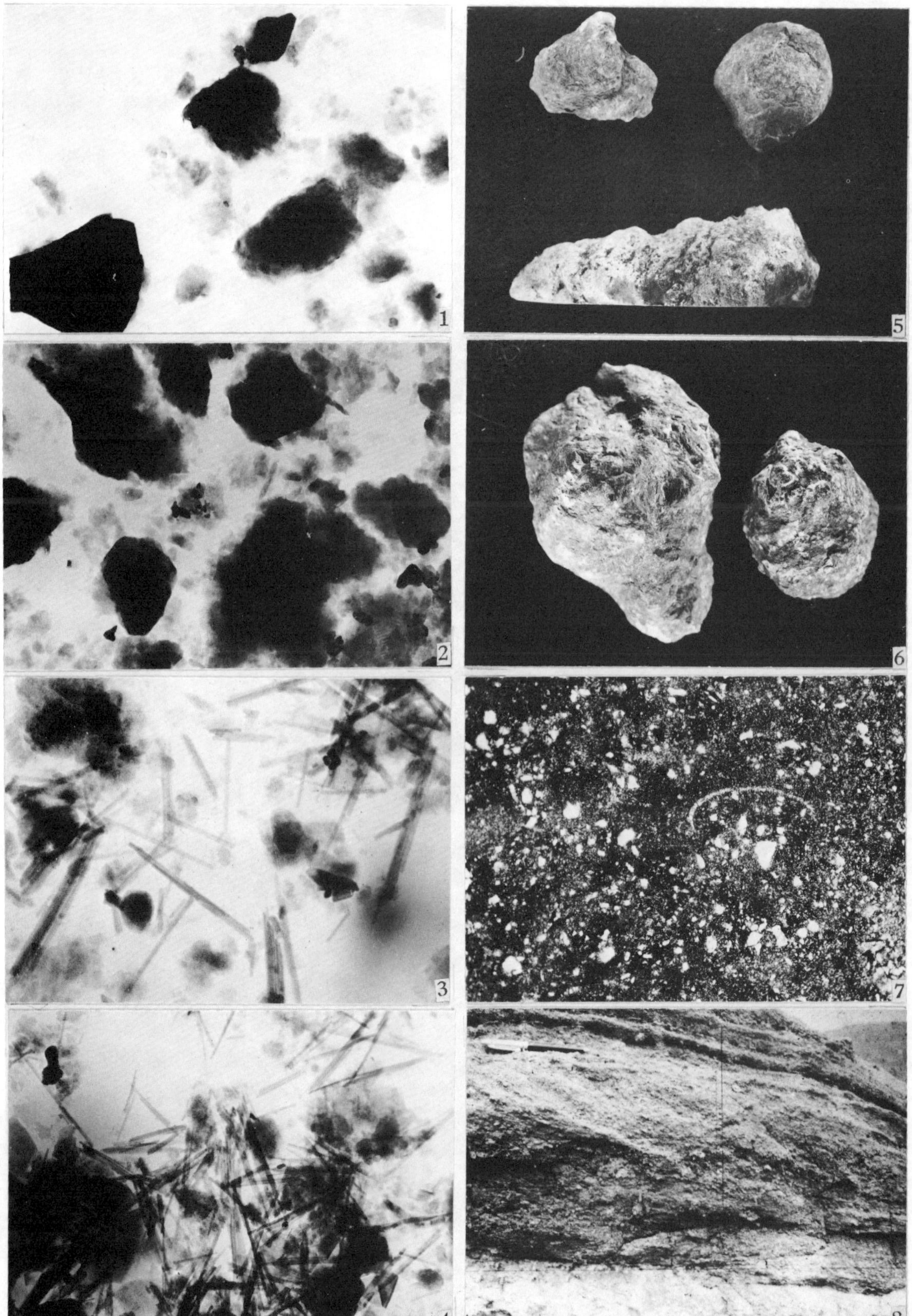

PALAEONTOLOGIA CATHAYANA Volume 5, 1990, p.311–352

DISCOVERY OF JURASSIC CONCHOSTRACANS WITH WELL-PRESERVED SOFT PARTS AND NOTES ON ITS BIOLOGICAL SIGNIFICANCE*

Zhang Wentang and Shen Yanbin
(Nanjing Institute of Geology and Palaeontology, Academia Sinica)

Niu Shaowu
(Tianjin Institute of Geology and Mineral Resources)

Abstract

Conchostracans with well-preserved soft parts were found from the Middle Jurassic Jiulongshan formation at Zhouyingzi of Luanping, northern Hebei. Four types of preserved patterns in the collections show the carapace valve, carapace valve with soft parts, soft parts alone and post-larvae respectively. The cephalon, mandibula, antennae, pinnate setae, claspers, eggs, appendages, digestive tube, caudal segment (telson) and caudal furca are clearly visible. All the specimens are considered to represent a new species, *Euestheria luanpingensis* sp. nov., which occurs in association with fossil insects and plant remains.

As compared with living concostracans, the new species should be referred to the suborder Spinicaudata. In consideration of the similarities in the cephalon and telson between the living and fossil materials, it is possible that the new species represents an ancestral form of living family Cyzicidae.

Following the study of the higher classification for the living and fossil conchostracans, a unified scheme proposed here comprises 3 orders: Laeviscaudata Linder, 1945; Spinicaudata Linder, 1945; and Leaiina Kobayashi, 1972.

Based on the sedimentary characters, the occurrence of minor kaolinite and the associated insects, the waters in which the conchostracans lived seemed to be a small, shallow and quiet pond in the Jurassic time. It might have existed for at least one month or more, and the new species could reproduce from generation to generation. Besides, in the waters there were no carnivorous and saprophagous animals.

The discovery of fossil animals with well-preserved soft parts is of great help to us in understanding the morphological structure, physiological function, ecology, origin and classification of animals during the geological time. These valuable materials, undoubtedly, make it possible to study ancient organisms in more detail and systematically. In the past, few fossil conchostracans with soft parts were made known or, if any, they were almost ill-preserved or fragmentary.

Recently, very abundant fossil conchostracans with well-preserved soft parts were found from the Middle Jurassic Jiulongshan formation at Zhouyingzi of Luanping, northern Hebei. It is interesting to note that the soft parts were isolated from carapace valves and preserved separately as fossils. For instance, such morphological structures as cephalon, mandibula, antennae, pinnate setae, claspers, caudal segment (telson) and caudal furca can be clearly seen. What is more, the soft parts of larval stage are also well-preserved. It is out of the question that these materials are of great importance to the study on the life history, classification, comparative anatomy, palaeoecology and taphonomy of conchostracans.

---

* This paper is translated by Qi Baoji.

FOSSIL-BEARING HORIZON AND LOCALITY

The Zhouyingzi section in Luanping County is located in about 40 km north of Chengde (fig. 1), where the Middle Jurassic volcanic sedimentary series is well developed and may be divided ascendingly into three formations: The Jiulongshan formation (mainly variegated clastic deposits), the Tiaojishan formation (volcanic rock intercalated with a little clastic rock) and the Houcheng formation (red clastic rock intercalated with a little volcanic rock). On the southern side of the highway, northwest of Zhouyingzi village there is a small-sized quarry, where the Middle Jurassic strata, about 16 m thick, are composed of deep grey, yellow-green shale and sandstone containing, in addition to insects and plants, abundant conchostracans with well-preserved soft parts. According to lithological characters, the series of strata may be subdivided descendingly into 4 beds (fig. 2) (measured by Niu Shaowu):

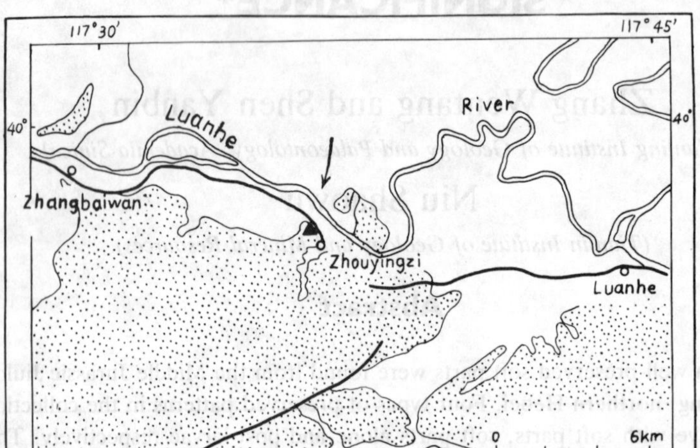

Text-fig. 1. Map showing the location of collecting sites.

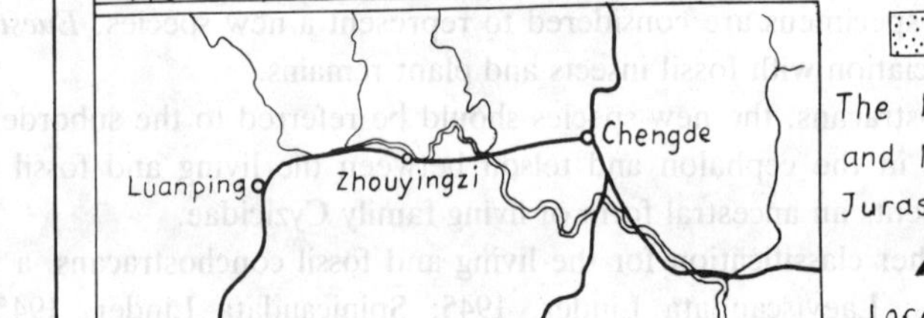

Text-fig. 2. Profile of the Middle Jurassic Jiulongshan Formation at Zhouyingzi village of Luanping, Hebei.

Middle Jurassic Jiulongshan Formation

4. Grey-yellow, grey-green medium- or fine-grained sandstone, covered by mantle rock >0.5m
3. Grey-green, yellow-green argillaceous fine sandstone, calcareous fine sandstone and argillaceous siltstone, containing conchostracan valves without soft parts 3.6m
2. Yellow-green plank-like calcareous siltstone and mudstone, alternating with deep grey leafy silty mudstone and shale, containing a large number of conchostracan valves and soft parts together with plants and insects 6.7m
1. Yellow-green medium-or thick-bedded sandstone, alternating with deep grey thin-bedded argillaceous siltstone, containing conchostracans, base unseen 5.5m

In the Jiulongshan formation occur many fossil insects including *Anapetophlebia chinensis, Mesobattina densivenosa, Mesonemura* sp. *Karataviella chinensis* and *Haplotrichocera invalida*, and fossil plants *Coniopteris hymenophylloides, Cladophlebis* cf. *denticulatus*, etc.. But in association with them is only one species of conchostracans called *Euestheria luanpingensis* sp. nov..

PRESERVED PATTERN OF SOFT PARTS

Based on the attachment or detachment of soft parts to the carapace valves, four types of preserved patterns may be distinguished:

1. Most soft parts are enclosed in the carapace valves (fig. 3). Generally the flagella of second antennae protrude out of the anterior part of the valves. The male claspers are clearly seen in the valves or extend slightly beyond the margin of the valves. The telson is located within the valves or slightly outside the valves. All these features show the close attachment of soft parts to the carapace valves, and the adductor muscles are linked up with the two valves.

2. Most soft parts protrude out of the carapace valves (fig. 4). For instance, the cephalon, second antennae, male clasper, appendages and telson are all shown outside the carapace valves. There is a close relation between soft parts and carapace valves, and the adductor muscles exert an effect. This preserved pattern is of considerable interest to the overall understanding of the attachment of soft parts to the carapace valves.

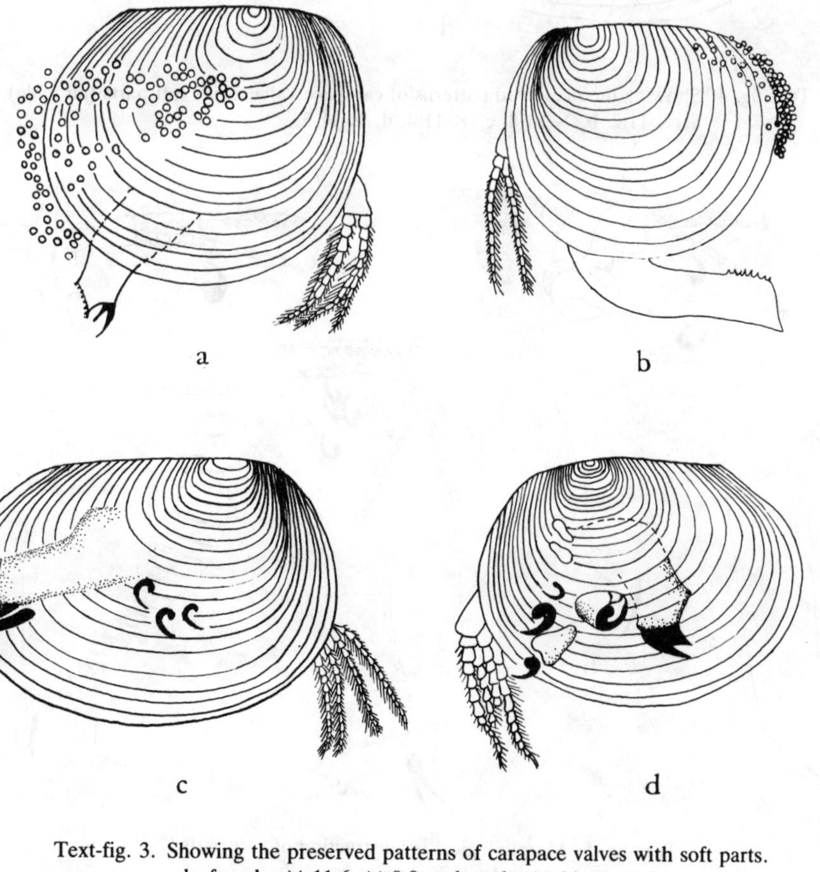

Text-fig. 3. Showing the preserved patterns of carapace valves with soft parts.
a-b. female, × 11.6, × 8.9; c-d. male, × 14, × 11.8.

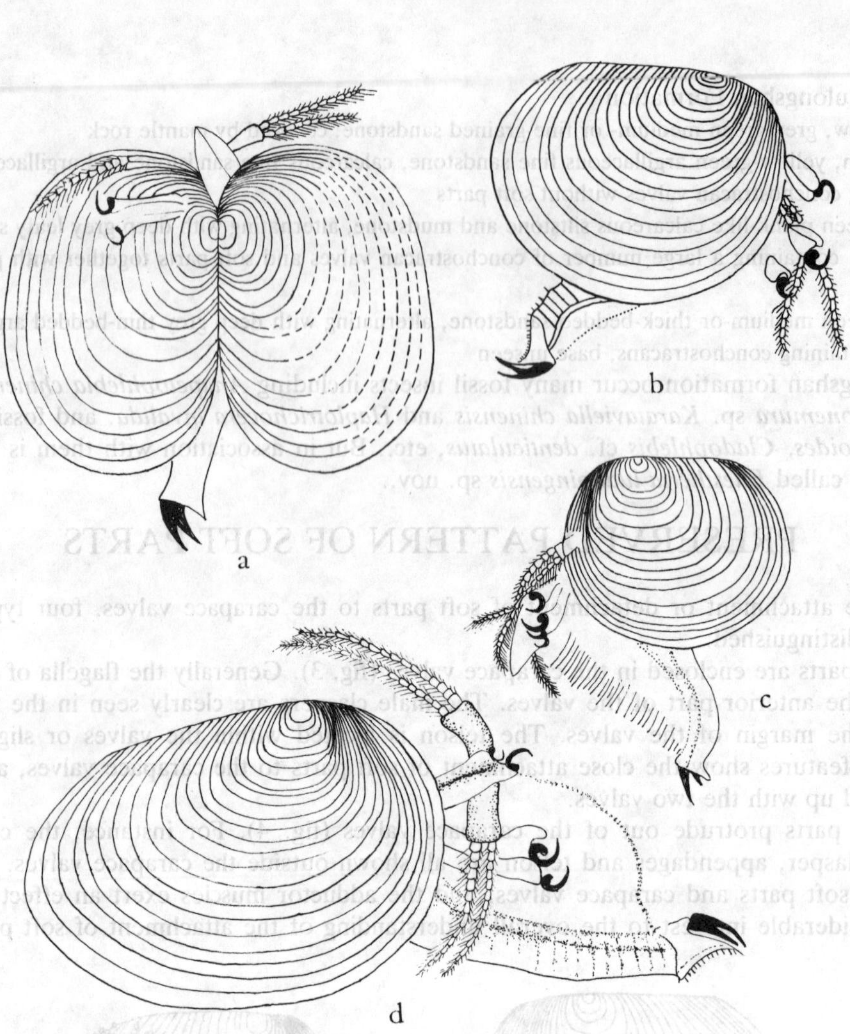

Text-fig. 4. Showing the preserved patterns of carapace valves with soft parts (all male). a. × 11.2; b. × 10.8; c. × 11.4; d. × 11.5.

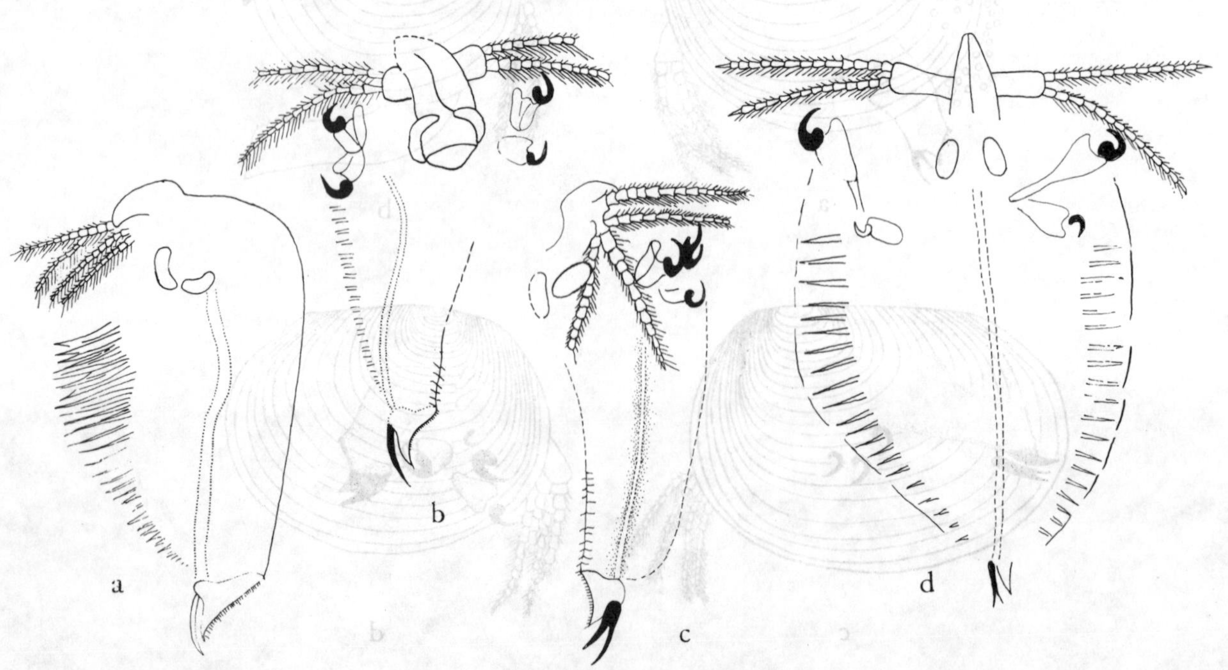

Text-fig. 5. Morphological reconstruction of soft parts. a. female; b-d. male.

3. Soft parts are isolated from the carapace valves and preserved separately (fig. 5). The counterparts of soft parts can be indiscernible in the carapace valves. Of these soft parts, some are rather complete, some are parts of cephalon. This pattern is of considerable importance to the understanding of morphological structure of soft parts as well as to the discussion on the classification of conchostracans. In the collections there are a large number of isolated soft parts which account for a large percentage or sometimes overnumber the carapace valves.

4. Specimens of larval stage (fig. 6). These fossils were preserved in the larval stage after the eggs were hatched. None of these fossils have as yet been reported in the world. And surely these valuable fossils are of great importance to the study of life history of conchostracans.

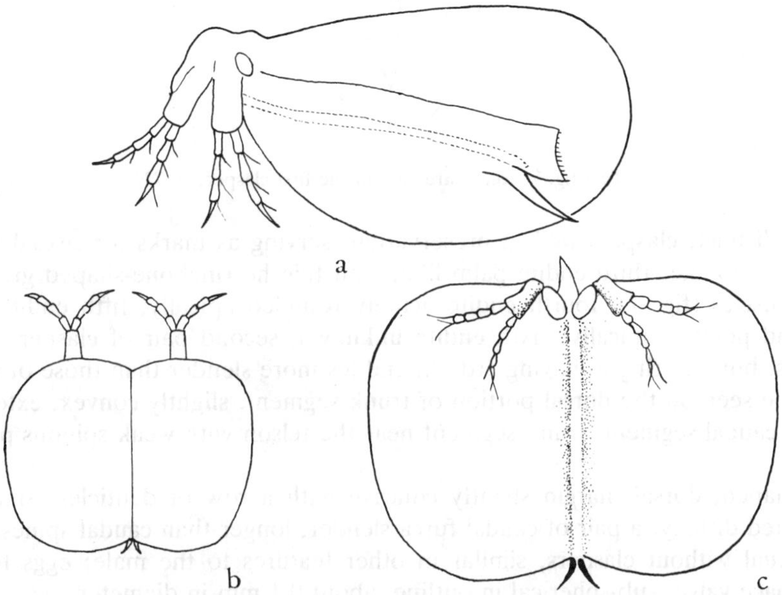

Text-fig. 6. Shape of larval stage. a. lateral view, × 18.9; b. dorsal view, × 16; c. internal view, × 22.8.

DESCRIPTION OF SPECIES

Euestheriidae Defretin 1965

Genus *Euestheria* Deperet et Mazeran, 1912

Euestheria luanpingensis Shen et Niu (sp. nov.)

(Pl. 1–12; Text-figs. 3–8, 10)

Material: There are nearly a hundred specimens of larval stage. Both carapace valves and soft-part specimens amount to 1,000 or more. Male soft-part specimens are double the number of female ones. There are in 1 m^2 2,500–8,000 individuals (including larvae, soft parts and carapace valves).

Carapace valve: Usually small, 3.5–5 mm in length, but less than 1 mm in larval stage, elliptical or ovate in outline; dorsal margin straight and long; umbo situated forward; anterior margin broadly arched; ventral margin slightly curving downward; posterior margin of male individual acutely bending backward; while that of the female broadly arcuated; anterior height nearly equal to posterior height; posterior dorsal angle 150°; growth lines on the carapace valve of adult individual ranging from 20 to 30; growth band medium in width, ornamented with fine reticulated sculptures; fine radial striae on the peripheral bands.

Soft-parts: Male individual 4–6mm in length; rostrum triangular in lateral view, slightly bending downward, with a pointed apex, but gradually widening in top view, with carinae in the middle; occipital notch broad, shallow but indistinct; mandibula situated at the back of rostrum, ovate in shape, rounded on both ends, slightly concave anteriorly; a triangular swelling seen at the back of mandibula, possibly linking cephalon and trunk.

First antennae rarely seen or, if present, with only 2–3 serrate sensory papillae; second antennae well-preserved, biramous; scape stout and cylindrical; flagellum with 11–14 joints; joint robust, trumpet-shaped in the lower part, with pinnate setae, which are short and sparse on outer side but long and dense on inner side.

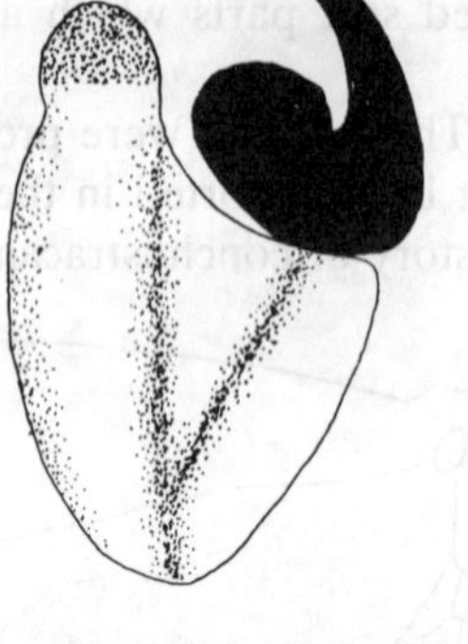

Text-fig. 7. Reconstruction of the first clasper, × 72.

Appendages indistinct; claspers well in preservation, serving as marks for sexual organ; exopodite unseen; second endite club-like; third endite palm-like, with fine herringbone-shaped grooves which are possibly the scars of muscles (fig. 7); fourth endite acutely rounded apically; fifth endite clear, hook-shaped, rounded distally and pointed apically; sixth endite unknown; second pair of claspers nearly similar to the first pair of claspers, but differing in having 3rd-5th endites more slender than those of the first pair.

A digestive tube seen on the dorsal portion of trunk segment, slightly convex, extending from the postcephalic portion to caudal segment; trunk segment near the telson with weak spinous projection on the dorsal portion.

Telson apex-shaped; dorsal margin slightly concave with a row of denticles (small spines) which are long and slightly tilted distally; a pair of caudal furca slender, longer than caudal spines.

Female individual without claspers, similar in other features to the male; eggs found in the posterodorsal part of carapace valve, sub-spherical in outline, about 0.1 mm in diameter.

Larva: Strenth and Sissom (1975) studied the development of *Eulimnadia texana* Packard and proposed 11 stages: 1). first nauplius stage, 2). second nauplius stage, 3). first metanauplius stage, 4). second metanauplius stage, 5). third metanauplius stage, 6). fourth metanauplius stage, 7). first postlarval stage, 8). second postlarval stage, 9). third postlarval stage, 10). fourth postlarval stage and 11). adult stage. In the fourth metanauplius stage carapace valves are formed. In the postlarval stage they increase in size but bear no growth lines, and the internal organs gradually develop perfectly. Sars (1896) thought that three stages may be separated for the ontogenesis of *Estheria packardi* Brady, 1) larval stage, 2) post-larval stage and 3) adualt stage. The post-larval stage is characterized by the formation of carapace valves. The present specimens are probably the post-larval individuals, in which the carapace valves are subelliptical in outline, small in size, 1.5-2 mm in length but bear no growth lines. In the soft parts, there are triangular rostrum, slightly concave occipital notch, small and ovate mandibula. Besides, the scapes of second antennae are stout and branch into two flagella. Earch flagellum has 4 joints with sparse pinnate setae on them. A digestive tube extends from the postcephalic portion to the telson. The appendages of trunk segment are indistinct. The telson is developed incompletely with slender caudal furca (fig. 6).

Ontogenesis: The life cycle of conchostracans, beginning from egg-hatching, lasts for three weeks or more. In the developmental process the carapace valves and soft parts of the same individual underwent a lot of changes. But in the adult stage the outer shell is not cast off. So no specimens of different developmental stages of the same individual have really been found in the strata. Here we suggest a morphological link between different developmental stages of the same species and sketch in part the ontogenesis of the species under study, for these specimens were collected from the same bed (fig. 8).

Through the post-larval stage the internal organs of canchostracans grow completely. According to the number of growth lines and the size of individual, 3 developmental stages may be recognized:

1. Juvenile stage: Carapace valves about 0.5–2 mm long; growth lines numbering 1–20; protoconch rather large. In comparison with those of the post-larval stage, the valve, being composed of chitinous substance, is obviously much thicker and ornamented with growth lines on the surface.

2. Adult stage: Valves continue to increase in size, generally 3.5–5 mm long; growth lines ranging from

20 to 30; protoconch diminishing in size.

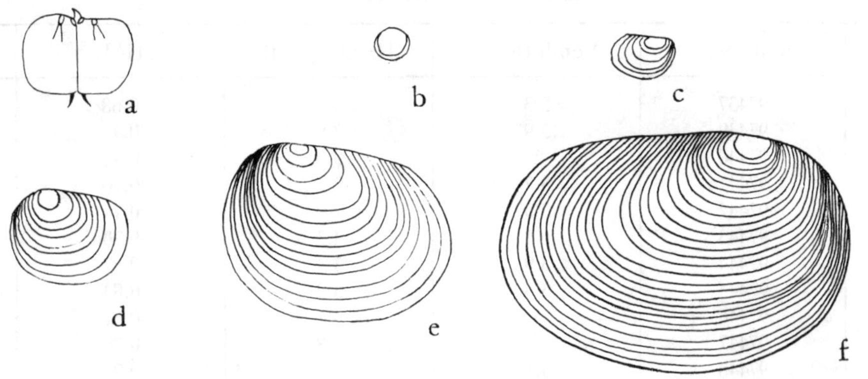

Text-fig. 8. Showing the ontogenetic development of *Euestheria luanpingensis* sp. nov.
a. post-larval stage, × 6.4; b-d. juvenile stage, × 8.6; e. adult stage, × 8.6; f. post-adult stage, × 8.6.

3. Post-adult stage: Valves not increasing in size; physiological function declined; growth bands narrow and dense on the margin, more than 30 in number.

Dimorphism: Of the extant conchostracans, the male and female valves are different in shape. For example, in *Eulimnadia inflecta* and *E. diversa* (Mattox, 1939), the male valves are narrow and long with dorsal margin rather straight; while the female valves are broader with dorsal margin strongly arched. Similar features can also be seen in the genera, such as *Cyzicus, Leptestheria, Caenestheria, Limnadia* and *Limnadiopsis* (Daday de Dees, 1925; Barnard, 1929; Tasch, 1969; Ueno, 1967). Male soft parts can be easily distinguished from female ones. The first and second pairs of appendages of male individual become claspers. In the mature female, eggs are enclosed on the dorsal portion of the valves, and the dorsal lobes of exopodites of 11th and 12th appendages are modified to form brood pouch for carrying eggs.

Attention to the dimorphism of fossil conchostracans has been aroused among paleontologists, but the difficulty in determination of dimorphism is due to rarity of fossil conchostracans with soft parts. The change of shells arising from dimorphism is often ascribed to different species. In the specimens from the Zhouyingzi section quite a few carapace valves were found with male claspers or eggs. The sexuality can be easily discernible. The H/L proportion of 25 valves is shown in fig. 9, in which the female valve is > 0.8 and the male one is < 0.8. As a result, eight individuals with claspers fall on the right of line, 3 individuals with eggs on the left of line. Of other 13 valves (with no soft parts attached) the sexuality can be determined on which side of the line they fall. Moreover, as is shown in pl. 5, fig. 7, the H/L proportion of valve is 0.81, the valve can be justified to be female individual, although it lacks soft parts. Around it are also found female soft parts and a bunch of eggs. Furthermore, in the external features, the valve seems to carry eggs.

The male valve is elliptical in outline and rather large, with the H/L proportion being 0.67–0.78. The dorsal margin is straight and long, and the posterior margin is acutely rounded with a conspicuous postero-dorsal angle. The female valve is nearly ovate in outline, but shorter than the male one. The H/L propor-

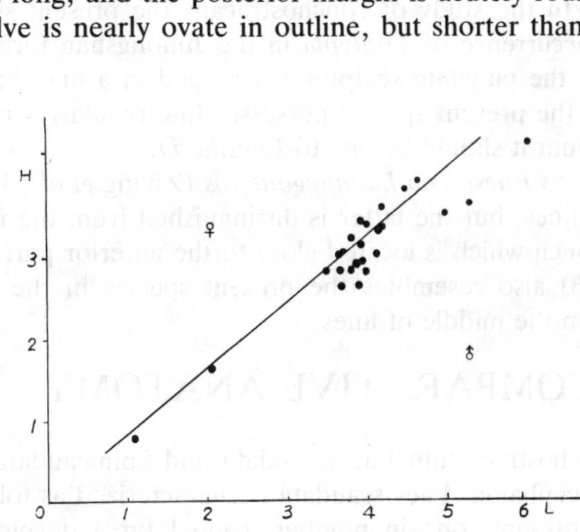

Text-fig. 9. Height-length proportion of sexual carapace valves.

Table 1. Measurement of Carapaces.

	Cat. No.	Lenth (mm)	Height (mm)	H/ L	Sex
1	97437	5.3	3.6	0.68	male
2	97439	3.9	2.9	0.74	male
3	97440	5	3.5	0.70	male
4	97441	3.7	2.6	0.70	male
5	97442	3.9	2.9	0.74	male
6	97443	6.1	4.3	0.70	male
7	97444	3.9	3.4	0.87	female
8	97445	3.8	3.2	0.84	female
9	97446	5	4.1	0.82	female
10	97447	4	2.8	0.70	male
11	97448	4.1	3	0.73	male
12	97449	3.8	2.8	0.73	male
13	97450	4	3.4	0.85	female
14	97452	3.6	2.9	0.81	female
15	97454	4.5	3.9	0.87	female
16	97455	4.8	3.9	0.82	female
17	97456	3.9	3.1	0.73	male
18	97457	3.9	2.6	0.67	male
19	97460	1.1	0.71	0.65	male
20	97461	2.15	1.6	0.78	male
21	97462	3.7	3.0	0.81	female
22	97464	3.7	3.0	0.81	female
23	97465	4.2	3.3	0.79	female
24	97466	4.5	3.8	0.84	female
25	97467	3.7	2.8	0.76	male

tion is over 0.8. The ventral margin is strongly arched downward, and the posterior margin is broadly arcuated (fig. 3). The sexuality of soft parts can be determined on whether they bear claspers.

Bisexual proportion: In the same waters the extant conchostracan populations vary in the bisexual proportion. Sometimes the male are prevailing, sometimes the female are overwhelming, or occasionally in a population only a single sex can be seen. Conchostracans belong to primitive Crustacea. In the adult stage the male and female are heteromorphic. But under the unfavorable environmental condition they can change sexually or become parthenogenetical (Webb, 1979). Of the fossil-bearing beds in the Zhouyingzi section, the second bed is composed of yellow-green silty mudstone in which larvae, soft parts and carapace valves are well-preserved, and nearly all the adults are male. The third bed consists of deep grey silty mudstone, in which individuals of different developmental stages are preserved and mostly belong to female. In other beds male individuals usually hold the predominant part. The reason why in the second and third beds occurred a single sexual distinction may be due to the change of living environments.

Discussion and Comparison: In the study of conchostracans the present species is described in detail. Wang Sién (1984) reported the occurrence of *Triglypta* in the Jiulongshan formation. Under SEM, Wang's genus shows on the growth band the punctate sculptures arranged in a line, each of which is composed of two rows of pores. Under SEM, the present species possesses fine regularly-arranged reticulated sculptures on the growth band. On this account it should belong to *Euestheria*.

The present species is similar to *Euestheria haifanggouensis* (Zhang *et al.*, 1976) in the outline of female valve and the number of growth lines, but the latter is distinguished from the former by the more arcuated dorsal margin and by the protoconch which is located close to the anterior part of dorsal margin. *Euestheria shandanensis* (Zhang *et al.*, 1976) also resembles the present species in the valve, but differs in having coarse growth lines with a notch in the middle of lines.

COMPARATIVE ANATOMY

Linder (1945) classified Conchostraca into Laeviscaudata and Spinicaudata, based mainly on the morphological features of telson and cephalon. Laeviscaudata is characterized as follows: Valve spherical in outline; growth line lacking or, if present, one in number; caudal furca developed incompletely; cephalon spoon-like; trunk segment without dorsal spines. Spinicaudata is characterized as follows: Valve with growth lines; caudal furca large; trunk segment with dorsal spines. The present species is characterized as follows: Cephalon nearly triangular in lateral view; caudal furca well-developed on the telson; trunk segment with dorsal spines in the posterior part. On these characters it should be ascribed to the Spinicaudata (fig. 10).

According to the morphological features of cephalon and telson, Spinicaudata may be subdivided into four families: Cyclestheriidae Sars, 1888; Limnadiidae Sars, 1896; Cyzicidae Stebbing 1910 and Leptestheriidae Daday de Dees, 1923. Of them Cyclestheriidae is easily distinguished from other families by having a row of stout spines on the dorsal margin of telson (Zhang *et al.*, 1976, p. 20).

The morphological comparison of the present species with the living species of Cyzicidae, Leptestheriidae and Limnadiidae is shown in fig. 11. In *Eulimnadia inflecta* the rostrum is squarish in outline, bearing a trumpet-shaped frontal organ but without any apical spines; while in *Leptestheria nanjingensis*, it is rather broad with fine apical spines. Whether the rostrum bears the frontal organ and apical spines is considered as one of the important taxonomical criteria for the familiar level. In *Eocyzicus mongolianus*, the male rostrum is rather broad, whereas the female rostrum is narrow and nearly triangular in outline. In dorsal view the male rostrum is broad distally and gradually widening upwards; while the female one is narrow distally and rapidly broadening upwards. In the present species the rostrum, nearly triangular in dorsal and lateral views, has neither frontal organ nor apical spines, a feature similar to that of *Eocyzicus mongolianus*. Especially its male individual is more related to the female of the latter species in the cephalon as the rostrum is narrow distally but gradually widening upwards.

The mandibulae of *Eocyzicus mongolianus* and *Leptestheria nanjingensis* are large in size, long-ovate in outline, pointed apically and rounded distally as shown in fig. 11. *Eulimnadia inflecta* was described by Mattox, but no figure of mandibula was attached with it. When Berry (1926) described *Eulimnadia stoningtoensis*, he showed a long-ovate mandibula in it. So did Sayce (1903) when he recorded *E. rivolensis* Brady. In the present species the mandibula is small in size, elliptical in shape, rounded apically and distally, a feature quite similar to those of *Eocyzicus mongolianus* and *Leptestheris nanjingensis*.

The second antennae of conchostracans serve as an organ for locomotion. In the present species the scape and two flagella are rather stout. Each flagellum is composed of 11–14 short and stout joints with long pinnate setae on the inner side. So it is also similar to *Eocyzicus mongolianus* and *Leptestheria nanjingensis* in the scape and the number of joints. But *Eulimnadia inflecta* can be easily distinguished from the present species in having 7–9 slender joints (Strenth, 1977; Nayar and Nair, 1968; Berry 1926). In Spinicaudata, it is difficult to define whether the first and second pairs of thoracopods all become claspers. For instance, Cyclestheriidae only bears a pair of claspers. The other three families possess two pairs of claspers. Morphologically, the first pair of claspers are longer and thicker than the second pair. In some species of *Eulimnadia* the sixth endite is slender and stripe-like and consists of two segments; while, in *Eocyzicus mangolianus* and *Leptestheria nanjingensis*, the sixth endite is stout and short, being composed of only one segment. Moreover, the fourth endite of the clasper of the extant form bears a small lateral lobe. But in the present species the third endite is ovate in shape, rounded apically and distally; the fourth endite is rounded apically and bordered by the third endite with a concavity, without a small lateral lobe distally; the fifth endite is hook-shaped, rather stout distally but pointed apically; the sixth endite is unseen. The second and the first claspers are alike in morphological features, though the former is more slender, about two-thirds the size of the latter. So the present species resembles *Eocyzicus mongolianus* and *Leptestheria nanjingensis* in the general outline.

The claspers serve as an organ for grasping in the copulation of male and female individuals. The opening and closing or grasping are controlled by the muscles. In the present species herringbone-shaped grooves can be seen on the third endite, linking with the fourth and the fifth endites anteriorly. Ueno (1935, pl. 4, fig. 5, G. H.) showed the orientation of muscles on the third endite of the first and the second claspers. The present species bears herringbone-shaped grooves on the third endite which must be the scars of muscles for extension and contraction. If so, it bears an analogy to *Eocyzicus mongolianus* in the extension of muscles which are connected with the fourth endite and are nearly parallel to the ventral margin, but differs from the latter in the muscles which are connected with the fifth endite.

From the physiological activity of conchostracans, it is inferred that the clasper performs the function with the fifth endite as main organ for opening and closing, and the sixth endite serves as auxiliary one. The fourth endite can not move freely. It seems that the clasper can't exert the function without the fifth endite. Hence the strength and orientation of muscles of the fifth endite have a close relation to physiological function of conchostracans. Muscles are powerful in *Eocyzicus mongolianus,* being in a flabelliform orientation from the fifth endite to the dorsal margin and the posterior portion of the third endite. As is shown in fig. 11, even if half of muscles are cut off, they can perform the function as before.

Cyzicidae, Leptestheriidae and Limnadiidae all belong to Spinicaudata, but they are different in the caudal segment. In Cyzicidae, the spines on the dorsal margin of the telson are large and stout, especially those on its terminal tilt strongly and upwards and the caudal furca is slightly longer than telson. In Leptestheriidae, the spines on the dorsal margin of the telson are less developed, those on its terminal tilt inconspicuously, and the candal furca is longer than the telson. In Limnadiidae, the spines on the dorsal margin of the telson are also undeveloped. But the caudal furca is much slender, obviously extending beyond the

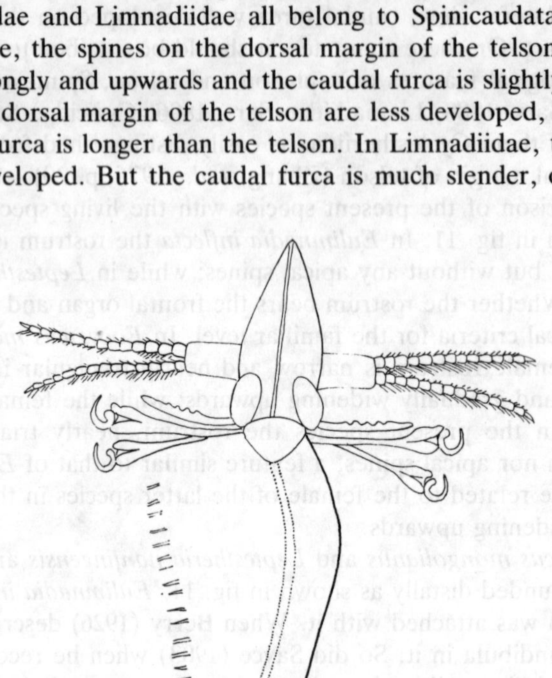

Text-fig. 10. Reconstruction of male soft parts of *Euestheria luanpingensis* sp. nov.

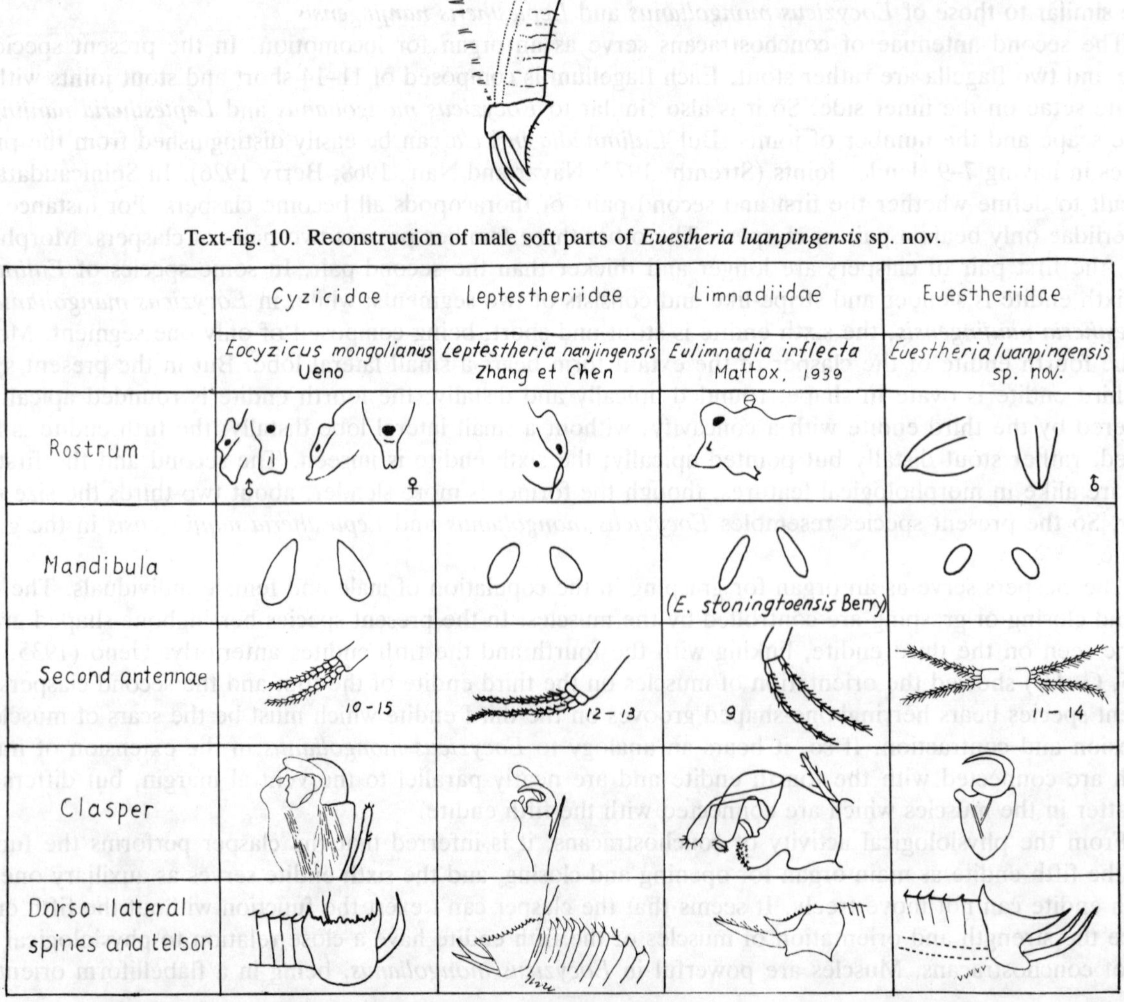

Text-fig. 11. Morphological comparison of *Euestheria luanpingensis* sp. nov. with some living species.

telson. In the present species the telson is rather short, bearing spines on the dorsal margin. But those on its terminal are stout and slightly tilt upwards. The dorsal margin is concave and long. The ventral margin is narrow, and the caudal furca is slightly longer than the telson. It seems that the present species bears a close similarity to the species of Cyzicidae in the telson. From what has been stated above, we can see that the present species closely resembles *Eocyzicus mongolianus* in the cephalon and telson, both of which are considered to be the important diagnoses for the determination of high taxa. But it differs from the latter in that the rostrum is nearly triangular in dorsal and lateral views and pointed apically, the mandibula is rather small and subovate in outline, and the telson is short and the spines on the terminal of dorsal margin are stout but tilt ambiguously. On this account, the present species might be the ancestral form of the extant family Cyzicidae.

In the life cycle the extant conchostracans underwent molting for several times. Usually the outer test is not shed, but following the molting, the individual increases in size. The shed skin is a fabric transparent membrance. However the membrance involves the membranous lining on the inner edge of valve, cephalon, appendages, trunk and telson. Thus arose a conjecture whether those soft parts preserved alone are the specimens of shed skin in the life duration. But from these fossils we can see that the impressions on the rock bedding of the scape and joints of second antennae, mandibula, clasper and telson are in such a salient relief that they can not be formed with a fabric membrance. In the diagenetic process, many chitinous test were compressed to some extent. The skin is light and soft. In the waters the oscillation can, in any way, make the skin to be crumpled into a ball. It is hardly imaginable to be preserved in a flattened state. Moreover, digestive tubes can be seen in the fossil soft parts. Also it proves that they are not the specimens of shed skin.

TAXONOMICAL SIGNIFICANCE

Raymond (1946) first proposed the scheme for the classification of conchostracans. In the past forty years, many scholars expressed different views on the classification of conchostracans. Therefore, there have been diverse opinions even on the suborder level (Kobayashi, 1954, 1972; Novojilov, 1960; Tasch, 1956; Defretin-Lefrance, 1965; Zhang Wentang *et al.*, 1976; Chen Piji & Shen Yanbin, 1985) (Table 2). Why the controversy arose is because the problem on classification between the extant and fossil conchostracans has

Table 2. Taxonomical comparison of conchostracan suborder.

Kobayashi (1972)	Novojilov (1960)	Tasch (1969)	Zhang Wentang *et al.* (1976) Chen Piji et Sheng Yanbin (1985)	This paper
Suborder Praestheritina Kobayashi (1972)				
	Superfamily Lynceioidea Stebbing,1902	Suborder Laeviscaudata Linder, 1945	Suborder Laeviscaudata Linder, 1945	Suborder Laeviscaudata Linder, 1945
Suborder Estheritina Kobayashi, 1972	Superfamily Cyzicoidea Stebbing, 1902	Suborder Spinicaudata Linder, 1945	Suborder Estheritina Kobayashi, 1972	Suborder Spinicaudata Linder, 1945
	Superfamily Limnadiopseoidea Novojilov, 1958 Superfamily Leaioidea Raymond, 1946			
Suborder Leaiina Kobayashi, 1972			Suborder Leaiina Kobayashi, 1972	Suborder Leaiina Kobayashi, 1972

not been solved. Some scholars made an attempt to assign fossil conchostracans in accordance with the classification of extant ones (Raymond, 1946; Tasch, 1956, 1969; Novojilov, 1960); other thought that fossil and extant conchostracans should be classified separately (Kobayashi, 1972; Defretin-Lefrance, 1965). Zhang Wentang *et al*. (1976) held that "Fossil conchostracans, especially those with well-preserved soft parts, are now insufficient. Before a unified classification for the fossil and extant conchostracans is adopted, we think it better to classify them separately. What we proposed is one of the important principles for the classification of conchostracans. Why we asserted different classification for fossil and extant conchostracans or at least tentatively, is to scheme out a unified and perfect classification for both of them in the future."

Following the constant and detailed study, more and more reports on the discovery of conchostracans with soft parts came out. It is sttractive that Martens (1983) described the Permian conchostracans form the western part of G. D. R., altogether totalling 3 genera and 4 species. Among them *Lioestheria monticula* was erected on soft parts, which were preserved alone as fossils, bearing a similarity to those from the

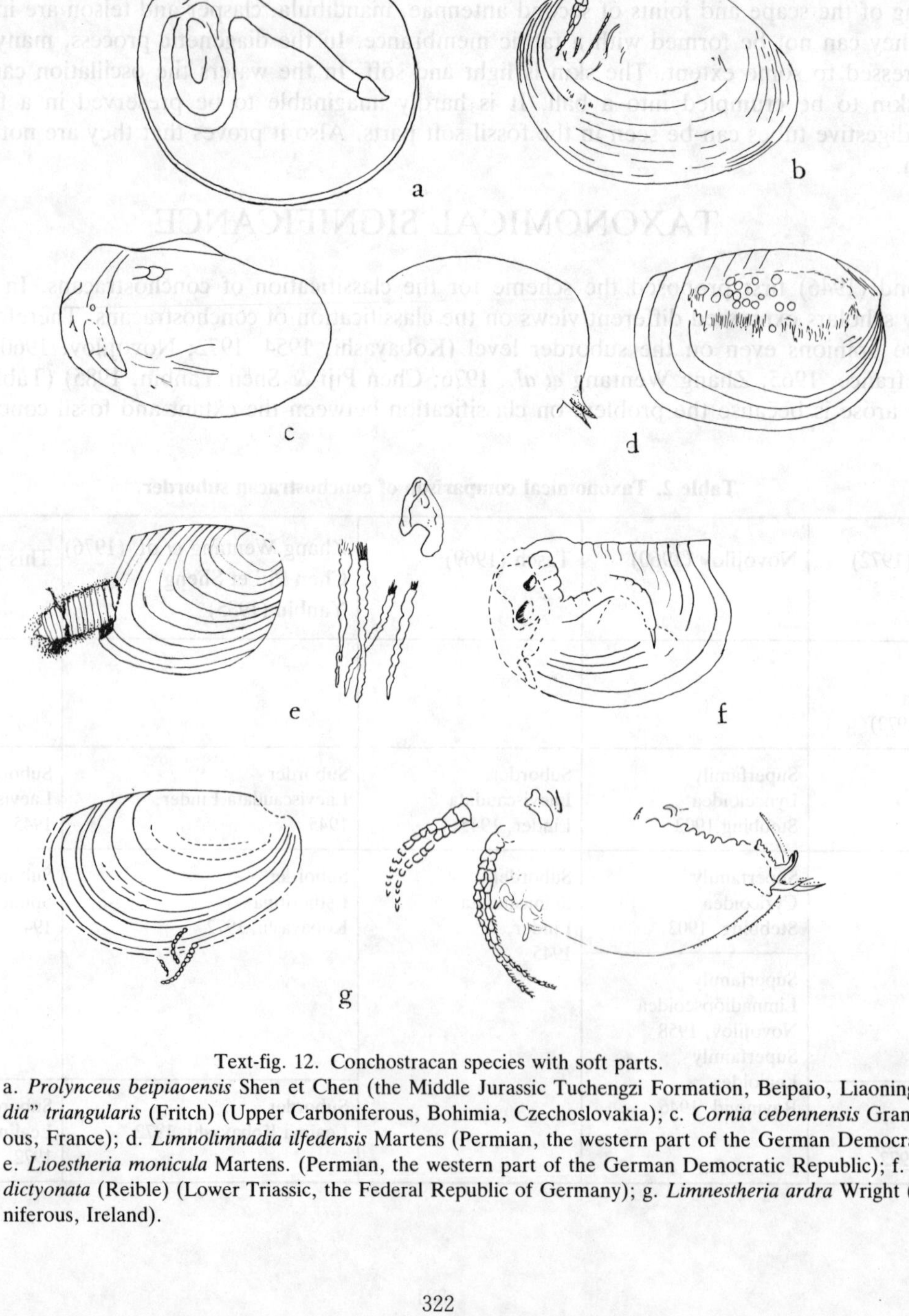

Text-fig. 12. Conchostracan species with soft parts.

a. *Prolynceus beipiaoensis* Shen et Chen (the Middle Jurassic Tuchengzi Formation, Beipaio, Liaoning); b. "*Limnadia*" *triangularis* (Fritch) (Upper Carboniferous, Bohimia, Czechoslovakia); c. *Cornia cebennensis* Grand (Carboniferous, France); d. *Limnolimnadia ilfedensis* Martens (Permian, the western part of the German Democratic Republic); e. *Lioestheria monicula* Martens. (Permian, the western part of the German Democratic Republic); f. *Leptolimnadia dictyonata* (Reible) (Lower Triassic, the Federal Republic of Germany); g. *Limnestheria ardra* Wright (Upper Carboniferous, Ireland).

Zhouyingzi section. These valuable fossils, of course, provide important evidences for the discussion on the classification for fossil and extant conchostracans.

About 15 species of conchostracans with soft parts have hitherto been recorded over the world (Wright, 1920; Novojilov, 1965, 1970; Martens, 1983; Zhang Wentang *et al.*, 1976; Shen Yanbin, 1982, 1986). Of them, two species: *Paleolynceus stchukinii* Chernyshev, 1940, *Prolynceus beipiaoensis* Shen et Chen, 1986, belong to Laeviscaudata, eight species exhibiting their cephalons or telsons, are of significance taxonomically (fig. 12); others are eggs, digestive tube or indistinct specimens and thus are of little interest to the classification of conchostracans.

Prolynceus beipiaoensis occurred in the Middle Jurassic Tuchengzi formation in Beipiao of Liaoning. Fossil soft parts include the postero-dorsal part of cephalon, shell gland, trunk segment and telson. Since the valve bears only one growth line and the caudal furca is absent, the species seems to be identical with the extant representative of Laeviscaudata. According to Zhang Wentang *et al.* (1976), the remaining seven species fall into 4 families or 2 superfamilies, namely, 6 species belong to Vertexioidea, the present species to Eosestherioidae (Chen Piji *et al.*, 1985). Though they were found from the sediments of different geological ages at different places, they all show a common character, belonging to Spinicaudata. This provides us with an evidence for a unified classification of the fossil and extant conchostracans on higher level. On this basis, we think it practicable to erect 3 suborders for the fossil and extant conchostracans (Table 2).

Laeviscaudata, as a peculiar population, was first found from the late Middle Jurassic (Shen Yanbin *et al.*, 1986). In the geological history or the recent days, this suborder, not forming a predominant population, is few both in species and in number. In view of the nature of carapace valve and fossil soft parts, the familiar taxa of the suborder can be considered in light of the classification of extant conchostracans. On the contrary, Spinicaudata, as a large population, is an indispensible branch for tracing the orgin, development, evolution, succession, prosperity and extinction of conchostracans. It ranges from Early Devonian to the modern time. Especially in the geological time it reproduced so much that it might overnumber the extant 4 families. Therefore for the classification of the suborder there is much to be desired until more fossil soft parts have been found.

Leaiid, as an extinct type, bears stout radial carinae on the carapace valve. No soft parts of it have been found. But from its peculiar carapace valve, it may be regarded as an independent suborder. Only in this way can it conform with the developmental process of conchostracans. Meanwhile, it indicates a fact that the classification of extant conchostracans can't hold true to all the fossil ones in the long geological time.

A unified classification for the fossil and extant conchostracans, particularly on the suborder level, is now in the preparatory stage. We firmly believe that it will be schemed out, accepted and adopted unanimously among the geologists. The finding of fossil conchostracans with soft parts from the Zhouyingzi section has brought us an important enlightment on this point.

HABITAT AND TAPHONOMY

Habitat: Around the Zhouyingzi village the Jiulongshan formation outcrops widely, in which conchostracans are the commonest fossils. But those with soft parts were found only at the quarry where the fossil-bearing sediments extend for about 30 m and are composed of yellow-green, deep grey silty mudstone with the development of fine bedding and strong rhythem, reflecting that the waters was small and rather tranquil, but the water level changed periodically.

Through the analysis of infrared spectrum (made by Liu Yun) the rock sampled from the soft-part-bearing bed consists mainly of illite with minor kaolinite. The occurrence of kaolinite indicates that at that time the sediments were laid down under a warm and moist climatic condition. The materials seem not to undergo a long-distance transportation but to derive from the vicinity. The landform is gentle or slightly undulatory.

At the Zhouyingzi quarry the conchostracan soft parts were found in association with fossil insects and plant remains, which were carried from slow current. Fossil insects are more in taxa. According to Lin Qibin, *Mesoblattina densivenosa* is adaptable to warm, moist or rainy subtropical and tropical environments. It often lived on the edge of the waters, but the larva and adult liked to inhabit on the land. *Karatavjella chinensis* is an aquatic insect, the larva lived in the waters and the adult liked to row on the waters, feeding on algae and waterweed. The waters is shallow and still with sufficient organic matters. In our collection the larva and adult specimens were found together. Among the insects is *Haplotrichocera invalida*, the adult of which might fly freely in the air under the hot and moist climatic condition. But the larva lived in the shallow and still waters. The larva and adult were all preserved as fossils. Moreover, the larvae of *Anapelophelia chinensis* and *Mesonemura* sp. like to live in the waters but the adult ones to inhabit on the land. The insect fauna indicates that at that time they lived under the warm and moist climatic condition in the waters, such as quiet swamp and lake.

The extant conchostracans are mostly living in the small and shallow ponds, ditches or bogs, or in temporary rainwater pools. Generally, eggs may be hatched above 4°C. The ideal hatching temperature is 13°–24°C (Bishop, 1967), beginning in the spring and ending in the summer or early autumn. The waters with a PH range of 6.9–9.5 is generally alkaline. If the water level changes greatly, the saline solusion changes rapidly. They have no adjustment capacity for the penetration of abrupt variation. From the ecological condition of conchostracans, it suggests a small and quiet bog environment under the warm and moist climatic condition.

In view of the fossil groups, the aquatic organisms were not abundant in taxa at that time. In the waters there were no carnivorous and saprophagous animals, but the filter and algal feeders like arthropods, especially conchostracans which held a predominant position. Conchostracans and insects feed on the pieces of plants and algae and, in turn, their dead bodies are organic nutritions for the growth of plants.

At Yaogou, west of Zhouyingzi the Jiulongshan formation is well exposed, containing various groups of fossils. Besides insects and conchostracans (without soft parts) there also occurred fossil plants: *Coniopteris hymenophylloides, C. burejensis, C. tatuensis, Cladophlebis* cf. *denticulatus, C.* cf. *asiatica, Gingoites sibiricus, Baira* cf. *gracilis, B. furcata, Sphenobaira longifolia, S.* cf. *crasinevis*; Lamellibranchs: *Ferganoconcha subcentralis* and *F. sibiria*; pisces: Ptycholepoidei and Palaeonisidae. In this area the aquatic organisms are more in taxa, the lake area is rather wide and the waters is deeper. Plants are very luxuriant, surrounding the lake, especially filices and ginkgophytes grow on the shore. Accordingly, the Zhouyingzi quarry was located close to a small lake or bog.

Taphonomic condition: When extant conchostracans were caught from the waters and placed into alcohol, we can see they struggle violently. The second antennae, appendages and telson shack back and forth. Most individuals close up two valves with soft parts huddled up within the carapace, some others with the second antennae extending to the anterior margin of the valve or the telson slightly outside the posteroventral margin as stated in the first type of preserved patterns (fig. 3). A living individual was caught and put in the open air for over 10 hours or 1–2 days. After it died out, it was then placed into alcohol, if slightly shacking the alcohol solution, we can see that most soft parts of the individual were outside the carapace valves or the two valves opened up, and that the flesh began to decay but the soft parts were still attached to the valve by adductor muscles, not isolated completely from the valves, a manner like the second type of preserved patterns (fig. 4).

The opening or closing of conchostracan valves and the connection of carapace to soft parts are supported by a pair of adductor muscles, which are located at the back of cephalon and are attached on the inner surface of two valves. The extant forms of *Leptestheria* and *Cyzicus* bear a pair of strong adductor muscles. In general, they can't be separated from the valves. What is the prerequisity for the isolation of soft parts from the valves is that the adductor muscles must be so disabled as to make them to be separated from each other. From the morphological structure of extant forms, it may be inferred that the conchostracans with well-preserved soft parts from the Zhouyingzi section had experienced the following taphonomic processes:

1. The waters in which the conchostracans lived was gradually dimishing in scope or temporarily dried up. Under these circumstances, conchostracans and insects died away, being short of water. On the brink of dying, some individuals struggled violently that the second antennae or telson extended outside the carapace valves. The muddy sediments formed beneath the tranquil waters were elevated above the water surface for only a short time, so the mud cracks are rather fine and shallow.

2. Following the increase of rain-fall, fine sand and mud flowed from the neighbouring areas into the lake, the gentle landform changed into slow current. Conchostracan individuals were buried *in situ*. Under the activity of water current, the adductor muscles of individual were disabled and the soft parts were separated wholly or at least mostly from the carapace valves. As is shown in pl. 1, fig. 1, the cephalon and telson are both directed anteriorly. Obviously, the water current flows forwards and this makes the soft parts in such an orientation. Soft parts were isolated from the carapace valves. Some sorts of muscles for connecting various structures could not exert their effect. Thereby they were preserved in a stretching state.

The interval between the two processes is very important for the preservation of soft parts. No sooner the individual was buried than it died away, so the soft parts could not be isolated from the carapace valves. If the individual closed its life for over 10 days, the soft parts became rotten and could not be preserved at all. When collecting the extant conchostracans in the dry-up pool, we have found carapace valves and eggs, but the soft parts were either rotten or hardly discriminating. On this account, it seems that a span of 2–3 days is favorable for the preservation of soft parts. In the tempest or heavy rain, the hydrodynamic pressure is strong, the soft parts could not preserved, even the carapace valves were washed away by the current.

3. The specimens of different developmental stages of the same species are found together, which seem to be preserved in the stable waters. The preservation of soft parts in multi-layers has a close relation with a long-term stable geographical environment. The conchostracan eggs may be divided into two kinds: summer eggs and winter eggs. The former, if under a favorable condition can be hatched in the same year; while the latter were even exposed to the sun in the field or frozon or eroded for years and, if under a suitable condition, could still be hatched (Belk, 1970). From hatching to growth the same individual underwent constant changes, belonging to monomorphic development. Therefore in the same bedding plane we found individuals of post-larval stage together with those of adult stage, a feature indicating that they represent at least 4 different generations. Among them there might have been individuals which were hatched from summer eggs. The reproduction from generation to generation requires stable waters and suitable condition. Conchostracans reproduced so rapidly (or abruptly) and in great number (with many eggs scattered) in the ecological realm. In some temporary pools or ponds the eggs can be hatched and grow rapidly. But it is impossible to reproduce from generation to generation in such waters. The life span of *Eulimnadia stoningtoensis* is about 23 days, but this species passed from hatching to post-larval stage for a duration of 6–7 days. It proves a fact that the waters remained in a stable state at least for a month so that the conchostracans could reproduce once again. In the deep and broad lake, conchostracans could develop on the margin of the waters, however, under the influence of strong hydrodynamics the soft parts could hardly be preserved. But the shallow and small waters, which are easily affected by rainfall and seasons, are favorable to the preservation of soft parts. After the soft parts were buried, the deposits were accumulated continuously and rapidly, which make them to be free from air under the reducing environment. Thus, the soft parts were well preserved in the diagenetic process.

4. So far as the fossil conchostracans with soft parts are concerned, a considerable number of specimens of *Lioestheria monicula* are known to be found from the Permian in the western part of G. D. R. (Mattens, 1983). Why the soft parts can be separately preserved is because of internal factors of organic body in addition to external condition mentioned above. Yao Yimin measured the surrounding rocks and the carapace valves by using the electric probe and obtained a conclusion that they are all composed of Na, Ca, Mg, Si, Fe, Ti, but the contents of Na and Mg are slightly higher on the valves than on the surrounding rocks. Whether they bear any importance to the physiological function of conchostracans is still indistinct. But from the functional viewpoint, the adductor muscles of extinct forms are less strong than those of the extant ones. Therefore, after the individual died off, the muscles could not be linked with the carapace valves, and the soft parts were isolated from the carapace valves under the external effect.

CONCLUSION

1. The fossil conchostracans with soft parts from the Zhouyingzi section are so far known to be many in number, diverse in type and well in preservation. These valuable materials enable us to understand the morphology, ontogenesis, bisexuality and population of *Euestheria luanpingensis*. They also afford important evidences for further studying fossil conchostracans.

2. Through the study of comparative anatomy. We came to know that *Euestheria luanpingensis* sp. nov. shows a resemblance to the representative of Cyzicidae in the cephalon, telson and appendages and, in addition, to presume that the species may be the ancestor of the family.

3. Fossil conchostracans with soft parts have been more and more accumulated, especially the materials reported in the present paper have laid a sound foundation for a unified classification of the fossil and extant conchostracans at a high level and thus create a good condition for solving the long-term dispute. Here we proposed to divide the conchostracans into 3 suborders – Laeviscaudata, Spinicaudata and Leaiina, and to scheme out a unified classification for fossil and extant conchostracans at the suborder level.

4. The better preservation of conchostracan soft parts has a close relation with the living environment on which they depend closely. Through the analyses of sedimentary characters, associated insects and the living condition of conchostracans, the Zhouyingzi quarry was located under the tropical and subtropical

climatic condition close to a small lake in which the waters was shallow and tranquil without carnivorous and saprophagous animals, and around which the vegetation was luxuriant, the landform was gentle and the climate was warm and moist.

5. Considering the occurrence in the same bed of the specimens of different developmental stages of the same species it may be assumed that the waters had been stable at least for a month or more and thus individuals of different generations could live in the same waters. Since the soft parts were isolated from the carapace valves and preserved separately as fossils, the lake might be short of water temporarily. When the individuals died off, the soft parts were detached from the carapace valves under the effect of weak current and later they were buried as timely as possible. Only in this case can they formed diverse preserved patterns as we see today.

ACKNOWLEDGEMENTS

The authors are thankful to Hu Shangqing and Xu Jizhou for their technical assistance. They are also grateful to Liu Yun for her affording data by analysis of infrared spectrum of rocks, to Yao Yimin for his providing elemental measurements by electric probe and to Lin Qibin for identification of fossil insects.

REFERENCES

Barnard, K. H., 1929. A revision of the South African Branchiopoda (Phyllopoda). Ann. S. Afr. Mus., 29(1), 181–272.

Belk, G. D., 1970. Functions of the conchostracan egg shell. Crustaceana, 19, 105–106.

Berry, E. W., 1926. Description and notes on the life history of a new species of *Eulimnadia*. Am. Jour. Sci., 2, 429–433.

Bishop, J. A., 1967. Seasonal occurrence of a branchiopod crustacean *Limnadia stanleyana* King (Conchostraca) in Eastern Australia. Jour. Anim. Ecol., 36, 77–95.

Chen Piji and Shen Yanbin, 1985. Conchostraca. Science Press, Beijing (in Chinese).

Chernyshev, B. I., 1940. Mesozoic Branchiopoda from Turkestan and Transbaikal. Jour. Inst. Geol. Akad. Nauk Ukrain. SSR, 7(3), 5–27 (in Russian).

Daday de Dees, E. (Jeno), 1925. Monographie systematique des phyllopodes conchostraces. Ann. Sci. Nat. Zool., 10(8), 143–184.

Dechascaus, C., 1951. Contribution a la connaissance des Estheries fossils. Ann. Palaeont., 37, 125–132.

Defretin-Lefrace, S., 1965. Etude et revision de phyllopodes conchostraces en provenance d'U. R. S. S. Ann. Geol. Soc. Nord, 85, 15–48.

Dong Yumao et al., 1982. Crustacea. In: Animal atlas of China (2nd ed.). Science Press, Beijing(in Chinese).

Hu Weixing, 1985. A new species of the genus *Eocyzicus* Daday from Shandong. Acta Zootaxonomia Sinica, 10(4), 357–362.

Kobayashi, T., 1954. Fossil estherians and allied fossils, Jour. Fac. Sci. Univ. Tokyo., Sec. 2, 9(1).

_______, 1972. On the two discontinuities in the history of Order Conchostraca. Proc. Japan Acad., 48(10), 725–729.

_______, 1973. On the classification of the fossil Conchostraca and the discovery of Estheriids in the Cretaceous of Borneo. Geol. Palaeont. SE Asia, 13, 47–72.

Linder, F., 1945. Affinities within the Branchiopoda, with notes on some dubious fossils. Arkiv Zool. Stockholm, 37a(4), 1–28.

Martens, T., 1983. Zur taxonomie und biostratigraphie der Conchostraca (Phyllopoda, Crustacea) des Jungpalaozoikums der DDR. Teil. 1, Freiberger Forschunshefte.

Mattox, N. T., 1939. Description of two new species of the genus *Eulimnadia* and notes on the other Phyllopoda of Illinois. Am. Midl. Nat., 22(3).

_______, 1950. Notes on the life history and description of a new species of conchostracan Phyllopoda, *Caenestheriella gynecia*. Trans. Amer. Micros. Soc., 69(1), 50–53.

Nayar, C. K. G. and Nair, K. K., 1968. On a collection of Conchostraca (Crustacea: Branchiopoda) from Southern India, with a description of two new species. Hydrobiologia, 32, 219–224.

Novojilov, N., 1954. Crustaces Phyllopodes du Jurassque superieur et du Cretace de la Mongolie. Inst. Paleont. Acad. Sci. URSS, Travaux, 48, 7–124 (in Russian).

_______, 1960. Conchostraca. In: Orlov U. A. (ed.), Osnoby Paleontologii, 8, 220-253 (in Russian).

Pai, P. G., 1959. On post-embryonic stage of Phyllopod Crustaceans, *Triops* (*Apus*), *Stretocephalus* and *Estheria*. Indiana Acad. Sci. Proc., B, 48, 229–250.

Raymond, P. E., 1946. The genera of fossil conchostracan Order of bivalved Crustacea. Bull. Mus. Comp. Zool. Harvard Coll., 96 (3), 217–307.

Reible, P., 1962. Die Conchostraken (Branchiopoda, Crustacea) der Germanischen Trias. N. Jb. Geol. Pal. Abh., 114(2). 169–244.

Sars, G. O., 1896. Development of *Estheria packardi* Braby, as shown by a artificial hatching from dried mud. Arch. Math. Naturv, 18.

Sayce, O. A., 1903. The Phyllopoda of Australia, including descriptions of some new genera and species. Proc. Roy. Sco. Vict., 15, 244–261.

Shen Yanbin, Wang Sien and Chen Piji, 1982. Conchostraca, In: Palaeontological atlas of Northwest China (Shaanxi-Kansu-Ningxia). 3, 52–69. Geol. Publ. House, Beijing (in Chinese).

_______, 1985. Jurassic conchostracans at Hanshan County, Anhui Province. Acta Pal. Sinica, 24(3), 293–299.

_______ and Chen Piji, 1986. Late Middle Jurassic conchostracans from the Tucheng Formation of W. Liaoning. Bull. Nanjing Inst. Geol. Palaeont., 9(in Chinese).

Strenth, N. E. and Sissom, S. L., 1975. A morphological study of the postembryonic stage of *Eulimnadia texana* Packard (Conchostraca, Crustacea). Tex. Jour. Sci., 26, 137–154.

_______, 1977. Successional variation of sex ratios in *Eulimnadia texana* Packard (Crustacea, Conchostraca). Southwestern Naturalist, 22(2), 205–212.

Tasch, P., 1956. Three general principles for a system classification of fossil conchostracans. Jour. Pal., 30(5), 1248–1257.

_______, 1969. Order Conchostraca (in Branchiopoda). In: Treatise on Invertebrate Paleontology (R). Arthropoda, 141–191. Geol. Soc. Amer and Univ. Kansas.

Ueno, M., 1935. Phyllopoda, Crustacea of Jehol. Rept. lst Sci. Exp. Ser. 5, Div. 1, Pt, 2, 1–16.

_______, 1967. Two new species of Conchostraca (Branchiopoda) from Nepal and Iran. Crustaceana, 13, 249–256.

Wang Sien, 1984. New Jurassic-Cretaceous conchostracans from northern Hebei and Nei Mongol. Acta. Pal. Sinica, 23(6), 726–736.

Webb, J. A., 1979. A reappraisal of the palaeoecology of conchostracans(Crustacea: Branchiopoda). N. J. Geol. Pal. Abh., 158(2), 259–275.

Wright, M. C., 1920. *Limnestheria*, a new conchostracan genus from the Kilkenny coal measures. Roy. Irish. Acad., 35, sec. B, 187–203.

Zhang Wentang, Chen Piji and Shen Yanbin, 1976. Fossil conchostracans from China. Science Press, Beijing (in Chinese).

EXPLANATION OF PLATES

All the specimens described in this paper are housed in Nanjing Institute of Geology and Palaeontology, Academia Sinica. They all belong to *Euestheria luanpingensis* Shen et Niu (sp. nov.) collected from the Middle Jurassic Jiulongshan Formation of Zhouyingzi village in Luanping County of Hebei.

Plate I

1. External mold of male left valve with soft parts. × 16, paratype. Coll. No. 46; Cat. No. 97437a.
2. Internal mold of left valve of the same specimen as in fig. 1. × 12. Coll. No. 37; Cat. No. 97437b.
3. Two valves opening along dorsal margin, with second antennae in the anterior part. × 10. Coll. No. 33; Cat. No. 97438.

Plate 1

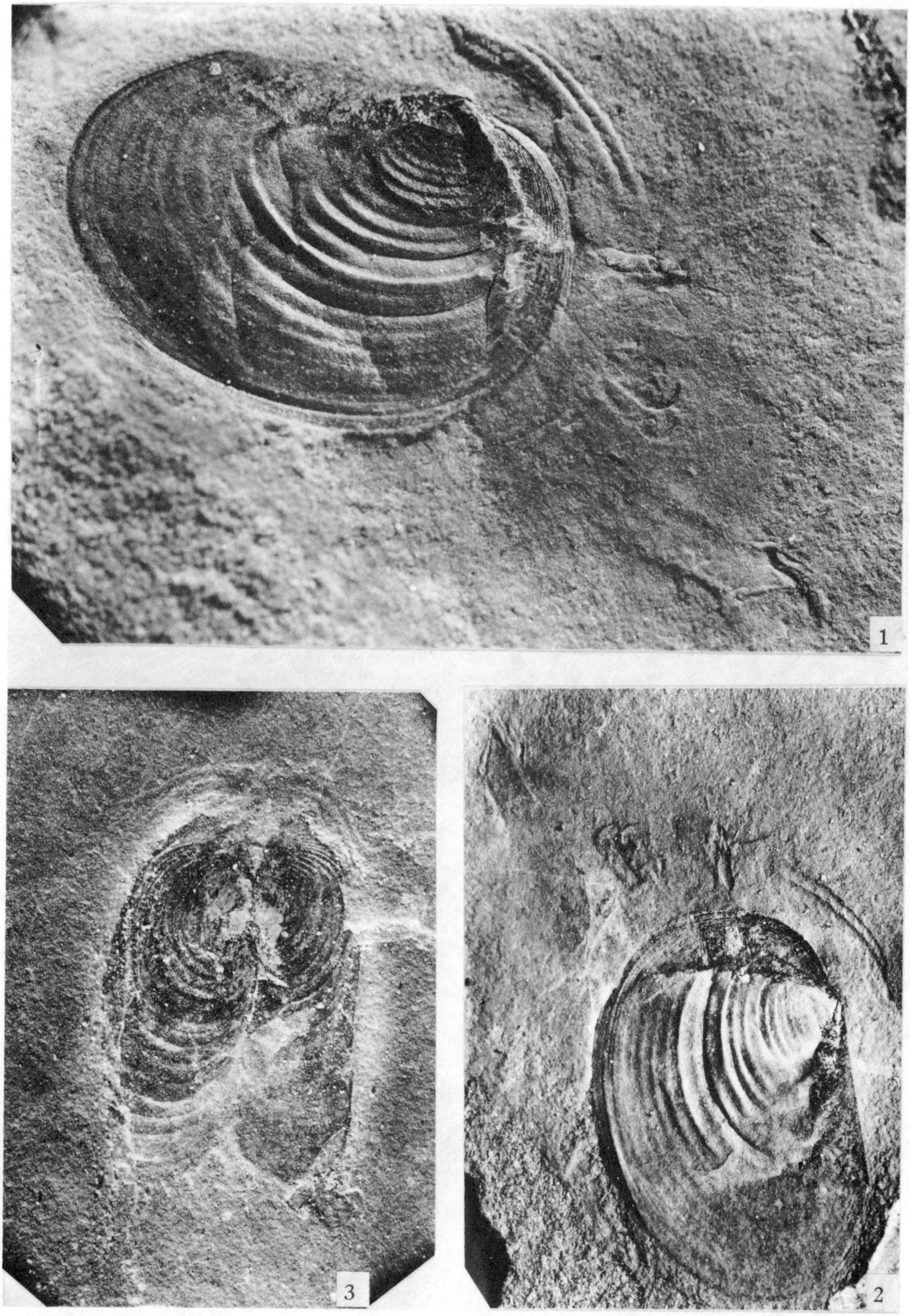

Plate 2

1. Male left valve with soft parts.
   a. × 16.2.
   b. × 14.2(SEM photo).
   c. × 20.
   Holotype. Coll. No. Y1413; Cat. No. 97439.
2. Male left valve with soft parts. Internal view. × 20. Coll. No. Y1452; Cat. No. 97440.
3. Male left valve with soft parts. × 10. Coll. No. Y1451; Cat. No. 97441.
4. Male valve, showing second antennae, clasper and telson.
   × 10. Coll. No. Y1406; Cat. No. 97442.

Plate 2

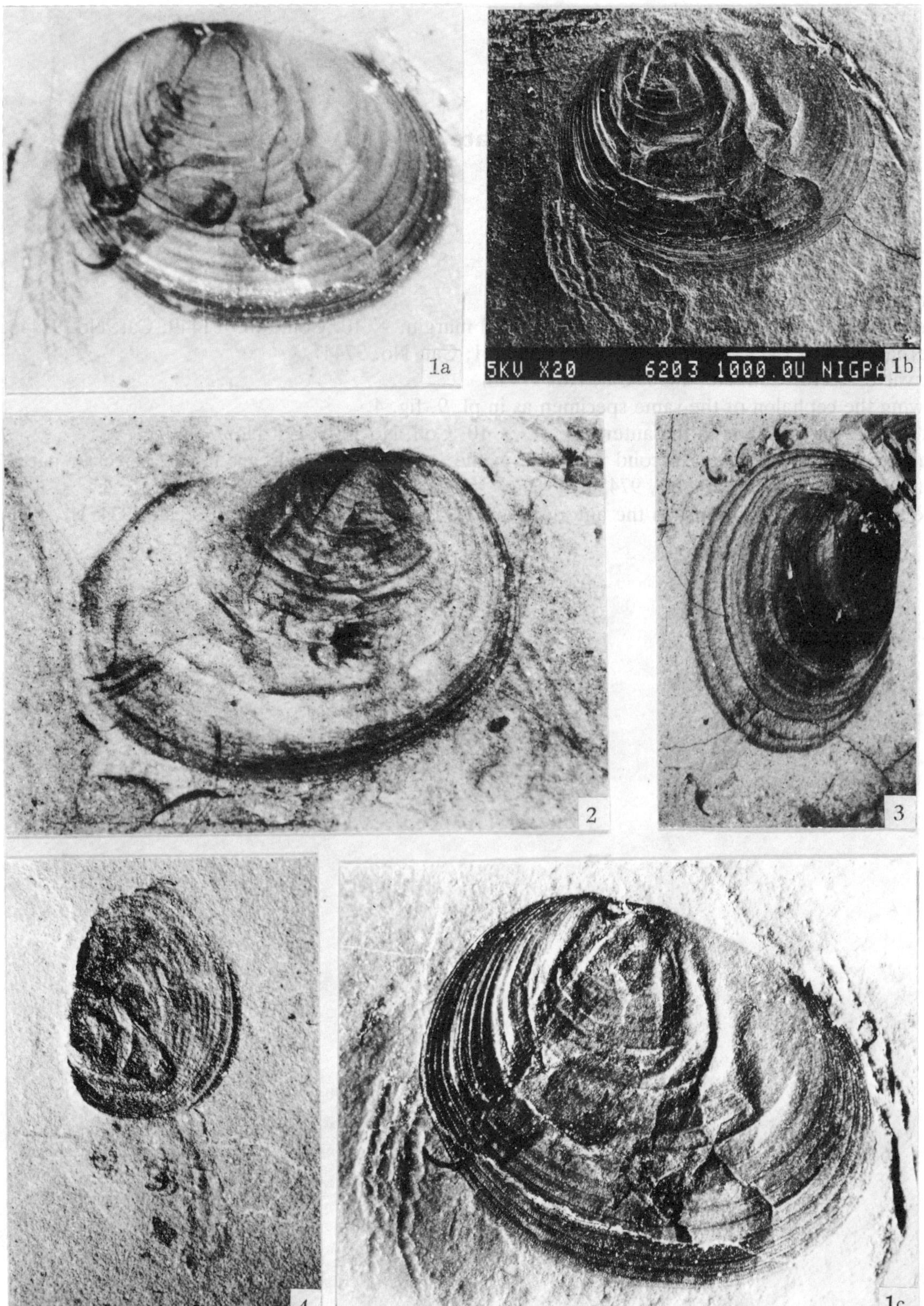

Plate 3

1. External mold of left valve (post-adult stage).
   × 10. Coll. No. Y1438; Cat. No. 97443.
2. External mold of right valve.
   × 10. Coll. No. 11; Cat. No. 97444.
3. Right valve. × 10. Coll. No. 11; Cat. No. 97445.
4. External mold of two valves, opening along dorsal margin. × 10. Coll. No. Y1449; Cat. No. 97446.
5. External mold of right valve. × 10. Coll. No. 1431; Cat. No. 97447.
6. Left valve. × 10. Coll. No. 1430; Cat. No. 97448.
7. Showing the cephalon of the same specimen as in pl. 9, fig. 4.
8. Left valve with soft parts in the anterior part. × 10. Coll. No 1401; Cat No. 97449.
9. Female right valve, showing second antennae in the anterior part and eggs in the posterior part. × 16, paratype. Coll. No. 22; Cat. No. 97450.
10. Male left valve with soft parts in the anterior part. × 16. Coll. No. 1403(1); Cat. No. 97451.

Plate 3

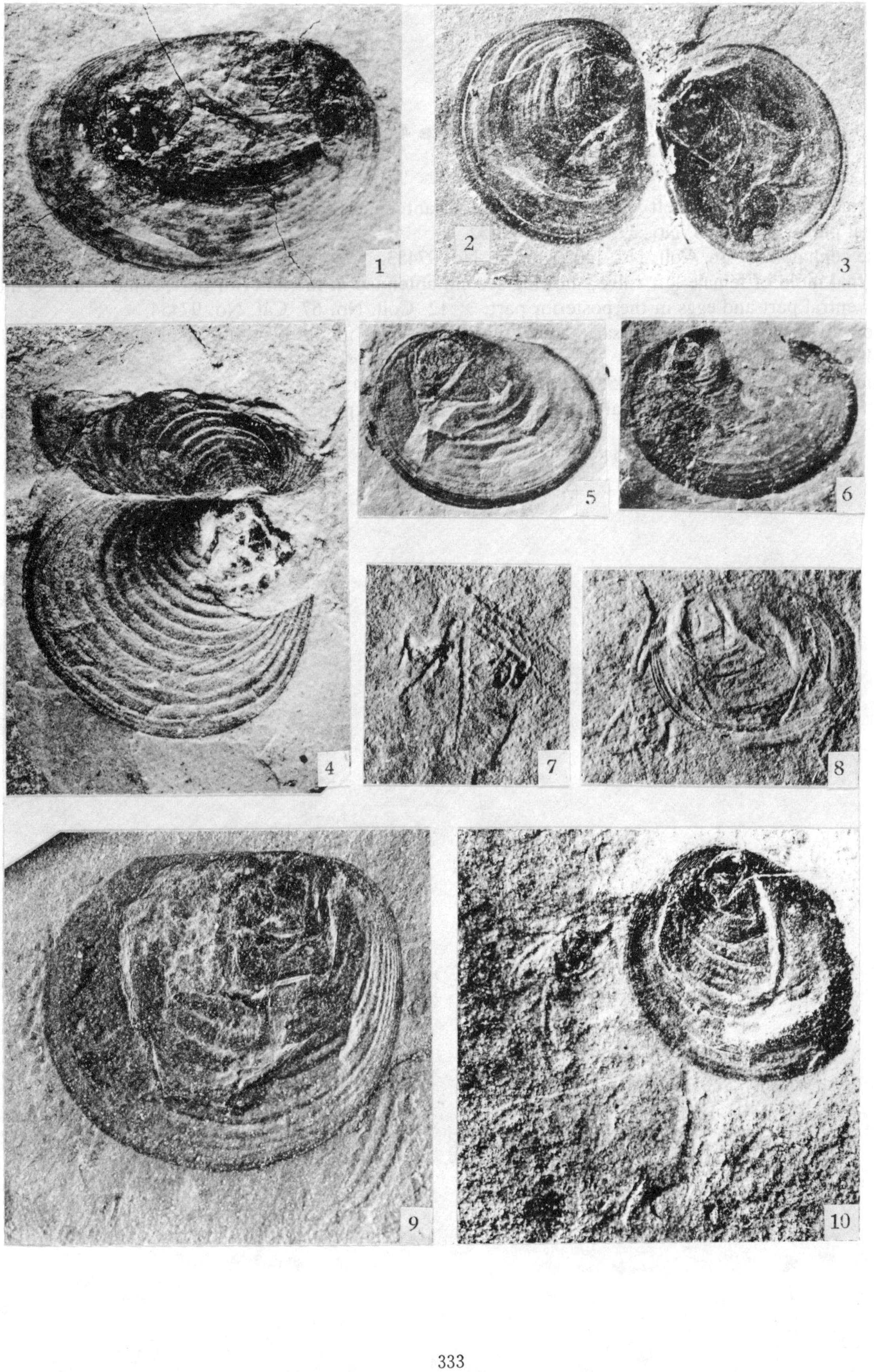

Plate 4

1. Internal mold of female left valve, showing second antennae in the anterior part and eggs in the postero-dorsal part. × 12. Coll. No. 39; Cat. No. 97452.
2. Male cephalon. × 16. Coll. No. 1405(1); Cat. No. 97453.
3. Internal mold of female left valve, showing second antennae in the anterior part, some trunk segments in the ventral part and eggs in the posterior part. × 12. Coll. No. 67. Cat. No. 97454.
4. Two valves opening along dorsal margin, with many eggs in the postero-dorsal part. × 12. Coll. No. Y1409; Cat. No. 97455.
5. Male valve with soft parts. × 15. Coll. No. 81; Cat. No. 97456.
6. Two valves opening along dorsal margin, with telson in the posterior part. × 16. Coll. No. 24; Cat. No. 97457.

Plate 4

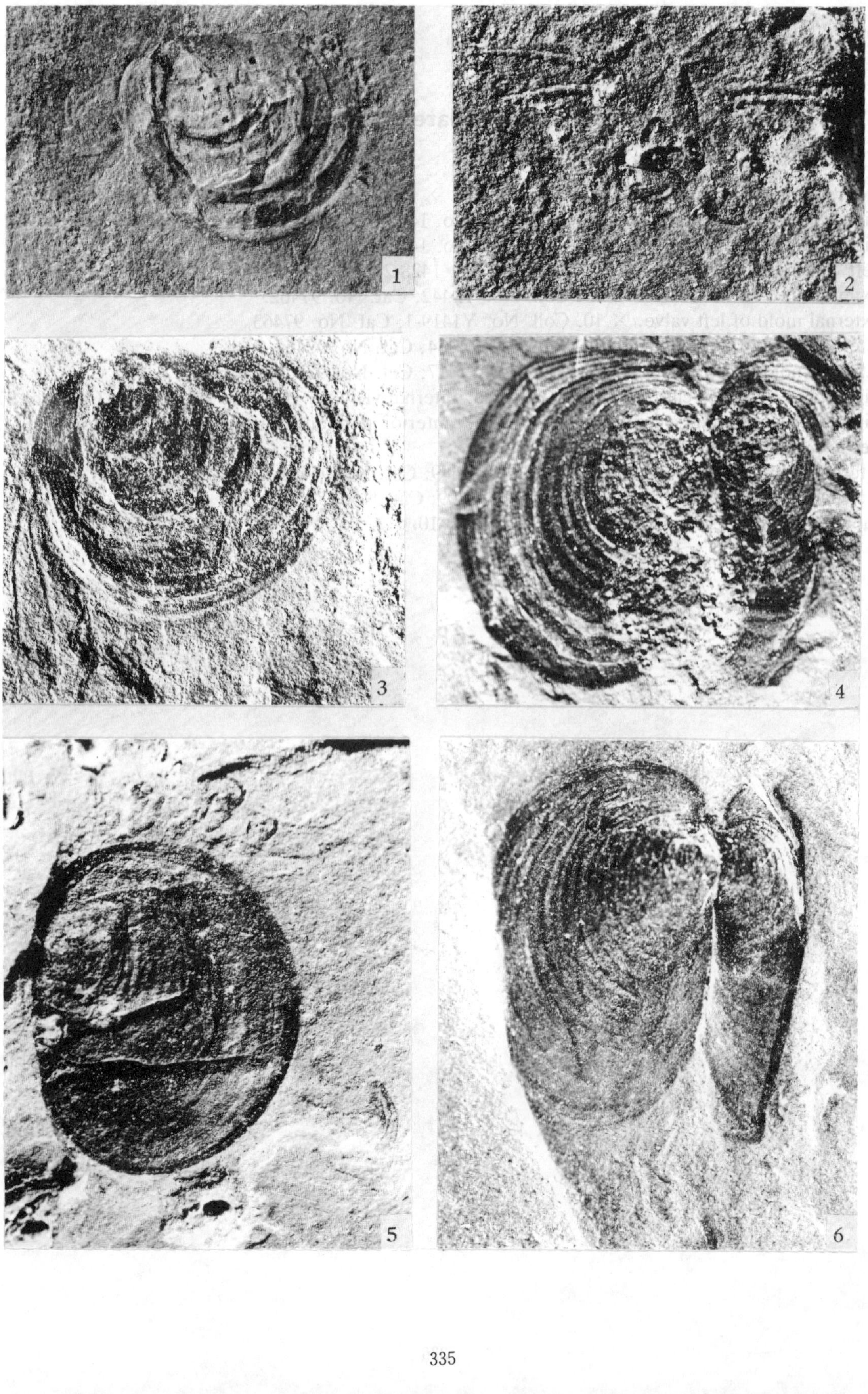

Plate 5

1. External mold of left valve (juvenile individual). × 10. Coll. No. 1428-3; Cat No. 97458.
2. Right valve (juvenile individual). × 10. Coll. No. 1428-1; Cat. No. 97459.
3. Right valve (juvenile individual). × 10. Coll. No. 1419-2; Cat. No. 97460.
4. Left valve (juvenile individual). × 10. Coll. No. 1428-2; Cat. No. 97461.
5. External mold of left valve. × 10. Coll. No. Y1442; Cat. No. 97462.
6. External mold of left valve. × 10. Coll. No. Y1419-1; Cat. No. 97463.
7. Internal mold of left valve. × 10. Coll. No. Y1434; Cat. No. 97464.
8. Internal mold of left valve. × 10. Coll. No. Y1417; Cat. No. 97465.
9. Right valve, showing clasper and caudal furca. Internal view. × 10. Coll. No. Y1443; Cat. No. 97466.
10. Right valve, showing second antennae in the anterior part and telson in the posterior part. Internal view. × 10. Coll. No. 1409; Cat. No. 97467.
11. Internal mold of right valve, × 10. Coll. No. 1409; Cat. No. 97468.
12. Internal mold of righ valve, × 10. Coll. No. 1409; Cat. No. 97469.
13. Showing the same specimen as in pl. 2, fig. 3, × 10.

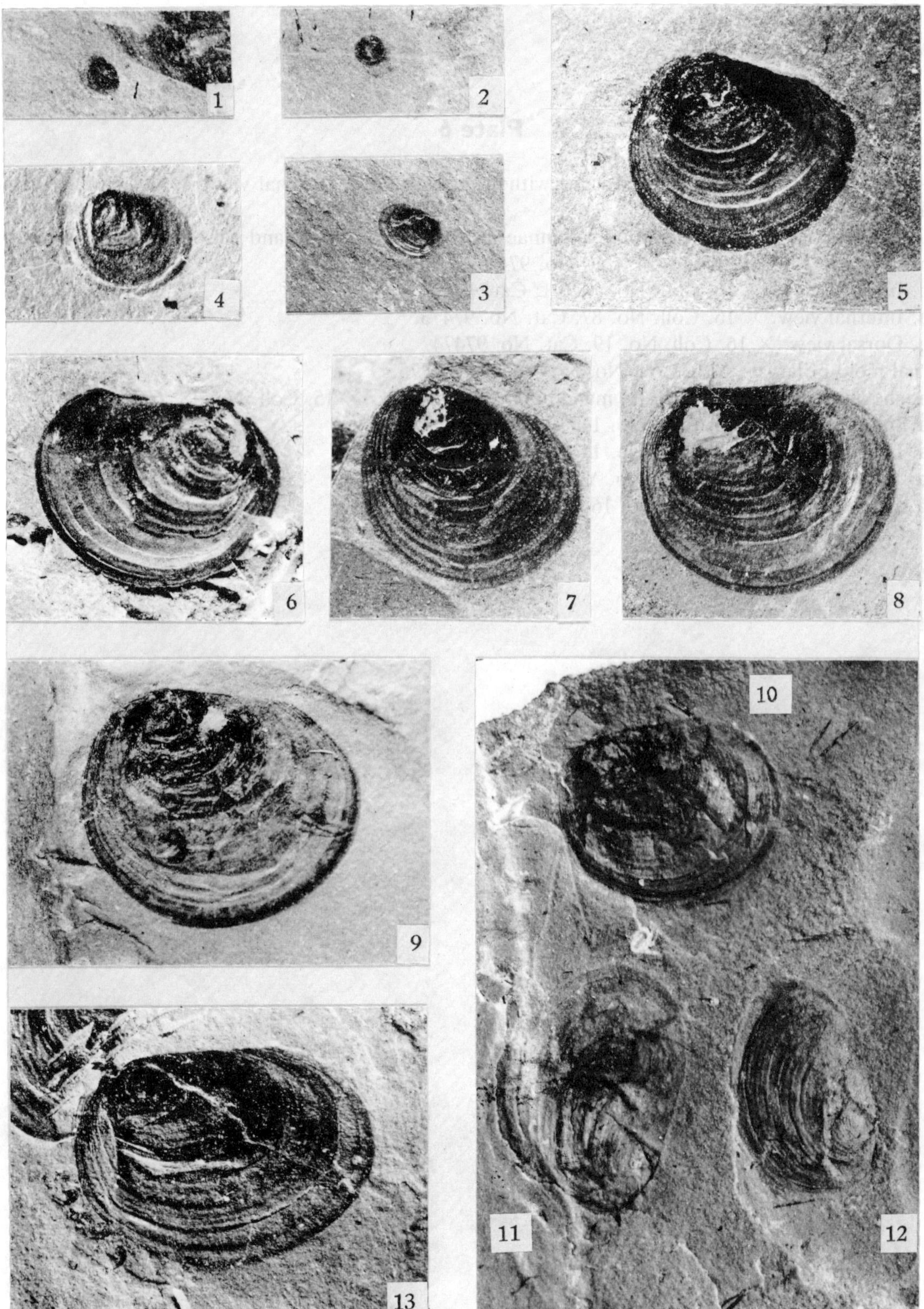

Plate 6

1. Larva, showing second antennae and valve without growth lines. Internal view. × 20. Coll. No. 1403-3; Cat. No. 97470.
2. Larva, showing cephalon and second antennae in the anterior part and telson in the posterior part. Lateral view. × 15. Coll. No. 1414; Cat. No. 97471.
3. Larva. Internal view. × 16. Coll. No. 1426-1; Cat No. 97472.
4. Larva. Internal view. × 16. Coll. No. 87; Cat. No. 97473.
5. Larva. Dorsal view. × 16. Coll. No. 19; Cat. No. 97474.
6. External mold of larva. × 16. Coll. No. 35; Cat. No. 97475.
7. Larva, showing second antennae and mandibula. Lateral view. × 15. Coll. No. 18; Cat. No. 97476.
8. Larva. Internal view. × 15. Coll. No. 1409-1; Cat, No. 97477.
9. Larva. Internal view. × 20. Coll. No. 1426-2; Cat. No. 97478.
10. Larva. Internal view. × 15. Coll. No. Y1410; Cat. No. 97479.
11. Larva. Dorsal view. × 15. Coll. No. 16; Cat. No. 97480.

Plate 6

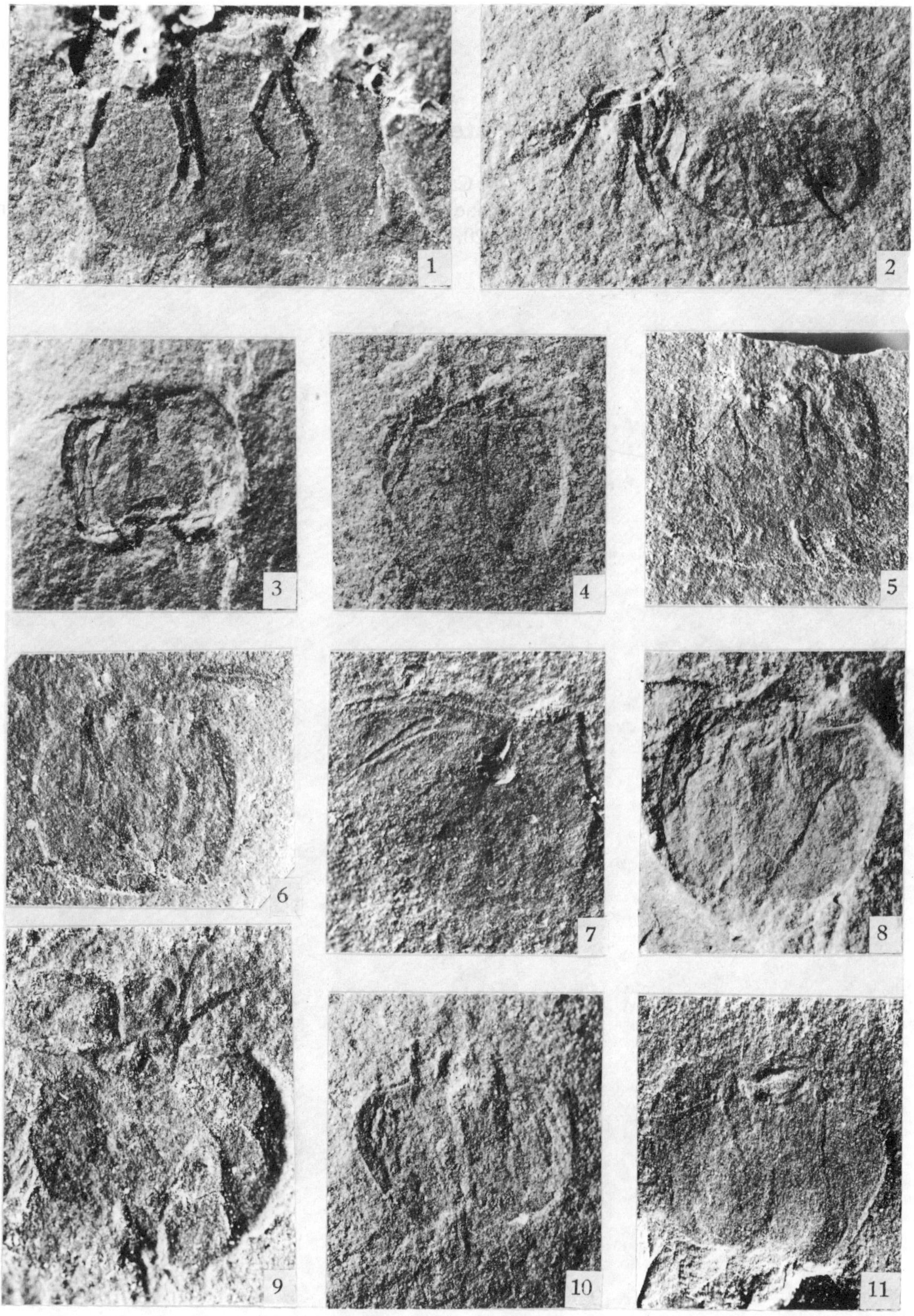

Plate 7

1. Male soft parts. Lateral view. × 20. Coll. No. 76; Cat. No. 97481.
2. Male soft parts, showing rostrum, second antennae, mandibula, clasper, appendage of trunk segment, digestive tube and caudal furca. Ventral view. × 20. Coll. No. Y1456; Cat. No. 97482.

1

2

Plate 8

1. Female individual, showing cephalon, second antennae mandibula, appendage of trunk segment and telson. Lateral view, × 16. Coll. No. 45; Cat. No. 97483.
2. Male soft parts, showing mandibula clearly. Dorsal view. × 16. Coll No. 1403-2; Cat. No. 97484.
3. Male soft parts, showing cephalon. × 20. Coll. No. Y1408; Cat. No. 97485.
4. Male soft parts. Lateral view. × 16. Coll. No. Y1439; Cat. No. 97486.

Plate 8

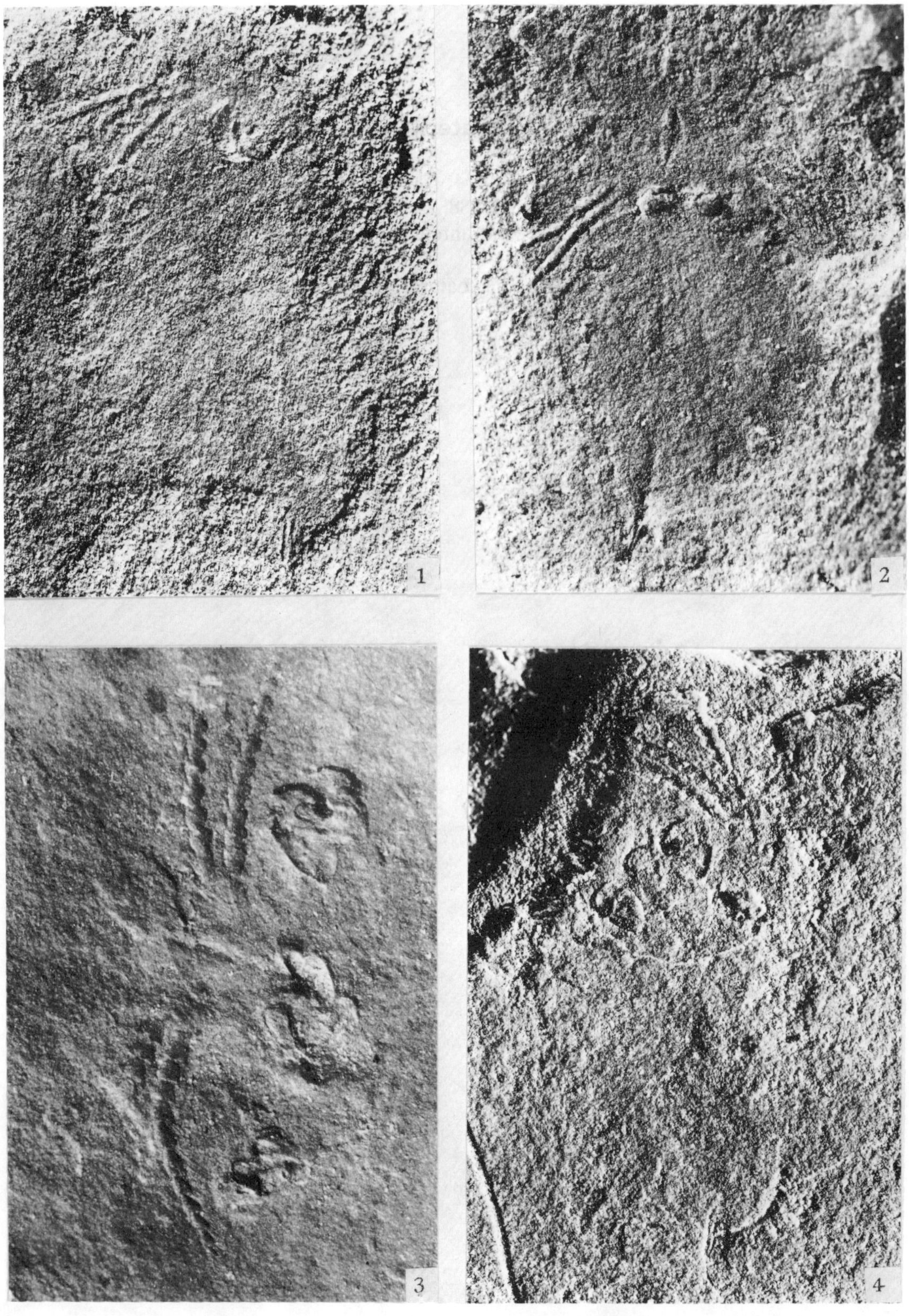

Plate 9

1. Female soft parts. Lateral view. × 15. Coll. No. 34; Cat. No. 97487.
2. Male soft parts. × 15. Coll. No. 86; Cat. No. 97488.
3. Male soft parts, showing four flagella of second antennae with pinnate setae. Dorsal view. × 15. Coll. No. 1426; Cat. No. 97489.
4. Male soft parts, showing telson and caudal furca. Lateral view. × 15. Coll. No. Y1419; Cat. No. 97490.

Plate 9

Plate 10

1. Male soft parts, showing the dorso-lateral spines on the trunk segment clearly. × 15. Coll. No. Y1437; Cat. No. 97491.
2. Showing the cephalon of the same specimen as in fig. 1. × 10.
3. Male soft parts, showing rostrum clearly. Dorsal view. × 10. Coll. No. Y1447; Cat. No. 97492.
4. Male soft parts. × 20. Coll. No. 71; Cat. No. 97493.

Plate 10

Plate 11

1. SEM photo, showing male cephalon. Dorsal view. × 20. Coll. No. Y1445; Cat. No. 97494.
2. Female soft parts, showing the dorso-lateral spines of trunk segment clearly. × 10. Coll. No. Y1446; Cat. No. 97495.
3. Male cephalon. × 16. Coll. No. 54; Cat. No. 97496.
4. Male cephalon. × 15. Coll. No. 86; Cat. No. 97497.
5. Eggs. × 10. Coll. No. Y1436; Cat. No. 97498.
6. Male soft parts. × 20. Coll. No. 78; Cat. No. 97499.

Plate 11

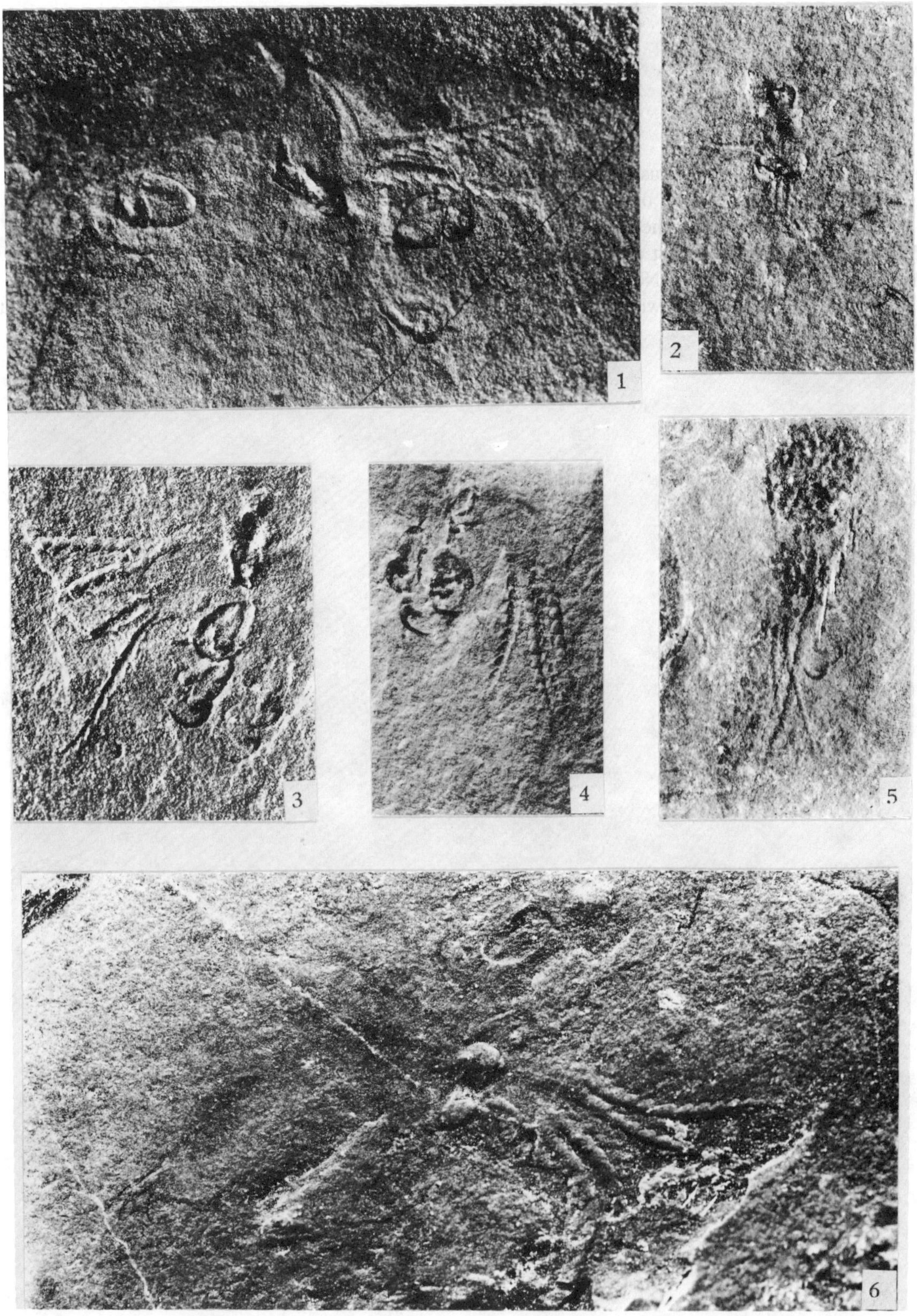

Plate 12

1–3. Showing the same specimen as in pl. 2, fig. 1.
   1. Showing the ornaments on the posterior part of the valve. × 40.
   2-3. SEM photos, showing the ornaments on the posteroventral part. × 65, × 195.

4. SEM photo, showing the first clasper of the same specimen as in pl. 10, fig. 1. × 100.

5. Male cephalon. × 16. Coll. No. Y1403; Cat. No. 97500.

6. External mold of valve, showing the ornaments on the growth band. × 40. Coll. No. 1413; Cat. No. 97501.

Plate 12

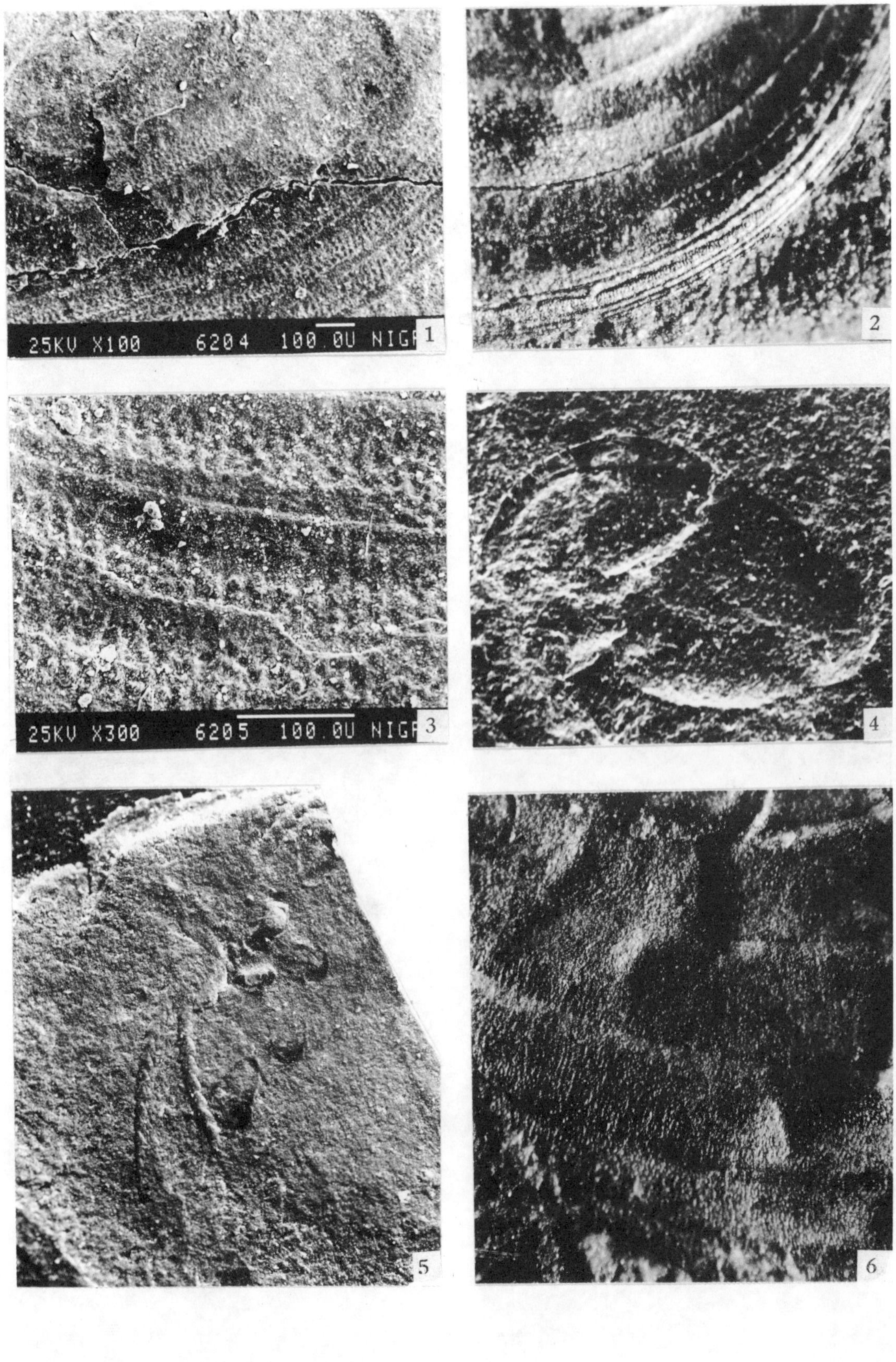

ORDER FORM

PALAEONTOLOGIA CATHAYANA No. 1-4

are distributed by Science Press, orders can be sent directly to :
International Sales Marketing Department
Science Press
16 Donghuangchenggen Beijie Street
Beijing 100707, China
Telex: 210247
Telephone: 443389
Fax: 4012180

No.1	US$ 120.00
No.2	US$ 80.00
No.3	US$ 80.00
No.4	US$ 80.00

☐Payment enclosed ☐Bill me

Name: (Please print)____________________

Address:____________________

____

Date/Signature:____________________

No.1, 1983, 452 pp. US $ 120.00

Lu Yanhao and Qian Yiyuan: Cambro-Ordovician Trilobites from Eastern Guizhou
Mu Enzhi: On the Boundary between Ordovician and Silurian in China
Lu Yanhao and Wu Hongji: Ontogeny of the Trilobite *Dalmanitina (Dalmanitina) nanchengensis* Lu
Mu Enzhi and Ni Yunan: Uppermost Ordovician and Lowermost Silurian Graptolites from the Xainza Area of Xizang (Tibet) with Discussion on the Ordovician-Silurian Boundary
Sheng Jinzhang et al.: Permian-Triassic Boundary in South China
Zhang Wentang: On the Subfamily Coronocephalinae (Trilobita)
Yang Jingzhi and Lu Linhuang: The Bryozoans from the Upper Carboniferous and Lower Permian in Kalpin, Western Xinjiang
Li Xingxue and Yao Zhaoqi: Current Studies of Gigantopterids
Zhang Lujin: On the Characters of the Permian Microflora in the Junggar Basin of Xinjiang
Wu Wangshi and Kong Lei: Rugose Corals from the Carboniferous-Permian Boundary Beds in Yunnan, Guangxi and Guizhou Provinces
Wang Yujing and Mu Xinan: Upper Carboniferous and Lower Permian Strata in the Gondwana-Tethys Province in Tibet
Cao Ruiji et al.: Precambrian Classification and Correlation in China

No.2, 1985, 278 pp. US $ 80.00

Mu Enzhi and Ni Yunan: Researches on the Graptolites of Xizang (Tibet)

Jin Yugan: Permian Brachiopoda and Paleogeography of the Qinghai - Xizang (Tibet) Plateau

Zhang Wentang: Current Biostratigraphic Scheme of Chinese Cambrian

Zhang Shanzhen and He Yuanliang: Late Palaeozoic Palaeophytogeographic Provinces in China and their Relationships with Plate Tectonics

Chen Xu and Stanley C. Finney: *Isograptus* and Its Descendant Graptolites

Hou Youtang: Morphology and Biogeography of Nonmarine Cristid-ostracods

Li Jijin, Ge Meiyụ and Chen Xu: Xinchangian (Early Ordovician) Anisograptid Fauna from Taishan, Guangdong

Qian Yiyuan: Late Late Cambrian Trilobites from the Tangcun Formation of Jingxian, Southern Anhui

Guo Shuangxing: Preliminary Interpretation of Tertiary Climate by Using Megafossil Floras in China

Zhang Wentang: Synonyms of Ordovician and Cambrian Trilobites

Zhou Zuren: Several Problems on the Early Permian Ammonoids from South China

Sun Dongli: Discovery of Early Cretaceous *Peregrinella* (Bachiopoda) in Xizang (Tibet) and Its Significance

Yin Leiming: Microfossils of the doushantuo Formation in the Yangtze Gorge District, Western Hubei

Mu Enzhi, Li Jijin, Ge Meiyu, Chen Xu, Ni Yunan and Lin Yaokun: Late Ordovician Palaeogeographical Maps of Central China Region with Their Explanatory Notes

Huang Baoyu and Guo Shuyuan: Discussion on Stratigraphic Division, Geological Period and Palaeogeography of the Nihewan Deposits (Cenozoic) According to the Mollusca

No. 3, 1986, 368 pp. US $ 80.00

Li Xingxue, Ye Meina and Zhou Zhiyan: Late Early Cretaceous Flora from Shansong, Jiaohe, Jilin Province, Northeast China

Chen Junyuan and Curt Teichert: The Ordovician Cephalopod Suborder Cyrtocerinina (Order Ellesmerocerida)

Zhang Wentang: Silurian Encrinurid Trilobites from Southwest China

Zhang Wentang: Correlation of the Cambrian of China

Yue Senxun and Liao Weihua: Some Mesozoic Scleractinia Corals from the Mount Xixabangma Region, Southern Xizang (Tibet)

Zhang Shanzhen, Yeh Dehshyan and Bian lizeng: On Seed and Polyphylesis of Seed Plants

Zhang Lujin: The Application of Palynostratigraphy to the Prospecting of Energy Resources in Sichuan

No. 4, 1988, 474 pp. US $ 80.00

Sun Weiguo: Subdivisions and Correlations of the Upper Precambrian in China and Australia

Derek V. Ager and Sun Dongli: Distribution on Mesozoic Brachiopods on the Northern and Southern Shores of Tethys

Sun Xiaowen: Cambrian Agnostids from the North China Platform

Zhou Zhiyan: Late Triassic Plants from Shaqiao, Hengyang, Hunan Province

Zhu Zhaoling and Heinz W. Wittke: Upper Cambrian Trilobites from Tangshan, Hebei Province, North China

Wang Zhihao and A.C. Higgins: Conodont Zonation of the Namurian – the Lower Permian Strata in South Guizhou, China

Lu Linhuang: Bryozoans from Chouniugou Formation of Late Early Carboniferous on the Northern Slope of Mt. Qilian

Zhang Lujin: On the Classification and Nomenclature in Mesozoic Palynology

Ouyang Shu and Chen Yongxiang: Palynology of Devonian-Carboniferous Transition Sequences of Jiangsu, E.China